Overnight
Loan

The Biology of Horticulture

AN INTRODUCTORY TEXTBOOK

SECOND EDITION

The Biology of Horticulture

AN INTRODUCTORY TEXTBOOK

SECOND EDITION

JOHN E. PREECE
Southern Illinois University

PAUL E. READ
University of Nebraska

John Wiley & Sons, Inc.

Acquisitions Editor	Rebecca Hope
Marketing Manager	Clay Stone
Associate Production Manager	Kelly Tavares
Designer	Kevin Murphy
Chapter opener cover art	Hadel Studio
Cover Design	Nancy Field
Illustrations	Gene Aiello

This book was set in 10/12 pt. Sabon by John Wiley & Sons Australia, Ltd and printed and bound by R. R. Donnelley & Sons. The cover was printed by Phoenix Color Corp.

To order books or for customer service please call 1-800-CALL WILEY (225-5945).

Recognising the importance of preserving what has been written, it is a policy of John Wiley & Sons, Inc. to have books of enduring value published in the United States printed on acid-free paper, and we exert our best efforts to that end.

Library of Congress Cataloguing in Publication Data:

Preece, John E.
    The biology of horticulture : an introductory textbook / John E. Preece, Paul E. Read.
        p.   cm
    Includes bibliography references and index.
    ISBN 0-471-46579-8 (cloth)
    1. Horticulture.   2. Botany.   I. Read, P. E. (Paul E.)
II. Title.
SB318.P74   1993
635—dc20   92-37678
                                               CIP

Printed in the United States of America

10 9 8 7 6 5 4 3 2 1

To Linda, Christine, Ellen, Molly, Jessica, Emma, and Peter,
without whose patience, love, and understanding this book
could not have been accomplished.

This book is by design, purposefully organized in a different fashion than is the case for other introductory horticulture textbooks. We feel that there is a scientific basis for growing and caring for plants that is common to all horticulture commodities. Such reasoning is rooted in the fact that in horticulture production, growers generally attempt to produce the very best plants within their capabilities. Initially, this involves selecting the best possible crop, cultivar, or clone (plant genotype), and then providing the proper growing conditions (environment). Plant characteristics of size, shape, color, flavor, and texture are governed by the interactions between the genotype and the environment. It is through an understanding of these interactions and their manipulations that we can produce and maintain high quality horticultural crops. An understanding of this principle is fundamental to all areas of horticulture, whether for the professional horticulturist, or the home gardener.

It is our belief that the basic understanding of horticultural principles covered in this book will prepare students for specialized courses in commodity areas of horticulture, including vegetable science, fruit science, nursery management, floriculture, turfgrass science, landscaping, and others. A sound foundation in the biological principles behind horticultural practices will help develop knowledge that can be used to understand horticultural practices and solve problems. It is by acquisition of fundamental knowledge about plants and how this knowledge is applied to horticulture that a horticulturist can become successful.

During the course of this book, we have remarked to each other that if students at any level, from undergraduate to graduate, could learn general horticultural principles and make these a part of themselves, they would be well-prepared for careers in this interesting field. In this textbook, we have attempted to include the most important general principles with some of their applications.

Horticulture is much more than simply tossing a seed on the ground and watching it grow. It combines an understanding of plant biology, chemistry, physics, mathematics, zoology, economics, and other disciplines in order to grow or maintain plants successfully. Although horticulture is a branch of agriculture, it incorporates many aspects of fundamental science to produce plants that are essential for maintaining our high standard of living and extraordinary quality of life.

We did not design this text to serve as a "how to" book. Rather, we attempted to incorporate the principles behind the methods described in such "how to" books. In the second edition we have updated principles and practices, and added illustrations to reflect advances made since publishing the first edition. By learning the principles many of the methods incorporated in horticulture should become logical and thus easy to remember. Therefore, we believe that this text will be valuable to beginning students as well as more experienced people who wish to learn more about the joy of horticulture.

ACKNOWLEDGMENTS

Our experiences and success largely have been attributed to our teachers as well as our students. In this book we have incorporated many of the principles that we learned from our teachers when we were students, because these have served as the basis of our knowledge of horticulture—we are grateful to them for this. As we have approached the second edition, this concept has been further reinforced by the contributions of an additional large number of students with whom we have worked since publishing the first edition of *The Biology of Horticulture*. It has been said that you never really know a subject until you teach it. Not only do you need to understand a subject sufficiently well to explain to students, but we have been blessed with many undergraduate and graduate students who have had such high levels of interest that their questions and observations caused us to stretch and increase our understanding. To us, teaching and learning is a two-way street, so students: thank you. The interactions that we have had with our teachers and students have made major contributions to this book.

Many people have contributed to this book in special ways. We have listed in our figure legends those who have provided us with photographs. We thank all of you for your generosity. Various administrators at the University of Minnesota, Southern Illinois University at Carbondale, and the University of Nebraska-Lincoln deserve thanks for their encouragement and support in writing this book. People deserving special mention are: Donna Michel and Amanda Myers for their dedicated assistance with preparation of the second edition; Jay Fitzgerald, University of Nebraska; Orville Lindstrom, Jr., University of Georgia; Oval Myers, Jr., Bradley Taylor and Alan Walters, Southern Illinois University for helpful suggestions; Helen Kraus, North Carolina State University; Christopher Walsh, University of Maryland; Sven Verlinden, West Virginia University; Judy Burgholzer, College of DuPage; Angus Murphy, Purdue University; and David M. Reed, Texas A&M University for their thoughtful critique of the first edition; Shirley Munson and Daniel and Jean Preece for assistance in staging and facilitating photographs; and Linda Preece and Christine Read for patience and understanding.

The illustrations in this textbook are by Ellen Starr. We thank her for working with us.

Photographs were taken at many locations. We thank the following for allowing us to use their facilities to take several of the photographs that were used throughout the text: Bailey's Nursery, Newport, MN; Len Busch Greenhouses, Plymouth, MN; Melvin Gerardo, Union County State Tree Nursery, IL; Monrovia Nursery, Azuza, CA; Plant Sciences, Inc., Watsonville, CA; and NASA Space Life Sciences and the Bionetics Corporation, Kennedy Space Center, FL.

CONTENTS

HORTICULTURAL
BIOLOGY

INTRODUCTION

Horticulture is a field of study in which career opportunities abound; in fact, recent indications from several sources suggest that there are many career openings available for graduates of baccalaureate programs in horticulture. Entrepreneurial opportunities include owning and managing horticulture businesses such as orchards and vegetable farms, nurseries, floral shops, and landscape businesses. A strong need exists for professional horticulturists to fill positions in public horticulture, such as golf course superintendents and managers of parks, public gardens and arboreta. Consider also the myriad opportunities in communicating horticultural information to the public—writing for the various newspapers, journals, and television and radio programs—and in the pursuit of careers in teaching, extension, and horticultural sales.

Challenging career opportunities in *research* are clamoring for attention. The first practical applications of biotechnology were developed and commercialized with horticultural plants by horticultural scientists, and rapid growth in this exciting arena of opportunity will occur as the 21st century unfolds. Robots and computers, genetic engineering, and basic science—all will come together as horticultural scientists lead the way toward a better future for this planet's burgeoning population.

Just what is this field called **Horticulture?** It's the science and art of producing nutritious food for the body—fruit, nut, and vegetable crops—and beautiful food for the soul—flowers and ornamental plants, landscapes, and lawns. In short, horticulture impacts us all every day, day in and day out (Figure 1-1). Indeed, it would be nearly impossible to go through a single day without horticulture influencing our lives. Try to picture a day in the life of an average citizen without horticulture—what would it mean?

❖ No orange juice for breakfast, no strawberries on your cereal, no blueberry muffins.
❖ No flowers on the table, no Monet or van Gogh prints framed on the wall.
❖ No shrubbery or lawn to enhance the aesthetics and value of your home (did you know that landscaping enhances the value of a home by as much as 20%?).
❖ No fruits or vegetables to enrich your diet (diseases such as scurvy and nightblindness would be rampant because of a lack of vitamins C and A).

❖ No alternatives to cereals as sources of the basic needs for calories and protein, since potatoes, cassava, beans, and peas would be unavailable.
❖ No golf courses for your leisure time, no sports-turfs for football, soccer, croquet, and lawn tennis.
❖ No coffee breaks, no chocolate bars, no afternoon tea!

HORTICULTURE DEFINED

The word horticulture was first used in the 1600s. It is derived from two latin words: *hortus*, which means garden, and *cultura*, which means cultivation. Therefore, in its strictest sense, horticulture means "cultivated garden," or more commonly, "culture of garden plants."

Horticulture is a branch of agriculture that is different from agronomy and forestry for the following reasons: (1) In general, horticulture requires more intensive management and higher labor inputs than the other branches. In horticulture the individual plant is important. For example, the street or landscape tree has a much greater value than

FIGURE I-I. The impact of horticulture is evident in the beauty of this inviting landscape.

a single tree in the middle of the forest, and that street or landscape tree receives much more intensive care, such as pruning and fertilization. Comparison of a horticultural crop such as strawberries and an agronomic crop such as wheat is also appropriate. Each strawberry has value, unlike the individual plant in a field of wheat. It is not uncommon to see a farmer back his or her truck onto the field of wheat so that a few hundred plants are crushed. That would be inconceivable on a field of strawberries because each strawberry plant is too valuable. (2) Horticulture offers a higher gross return per unit area per unit time. A good greenhouse grower can obtain total sales of over $20.00 per square foot ($215/square meter) of bench space per year. This profit level can be accomplished through intensity of production and careful scheduling of crops to take advantage of the entire 365-day year. If that grower has about an acre (0.4 hectare) of production, this translates to between $800,000 and $1,000,000 worth of sales a year. Certainly that is *intensive* production and a very *high value* crop!

The purpose for growing the crop often determines into what commodity area or field of study it is placed. For example, Kentucky bluegrass grown as a forage or pasture crop is considered an agronomic crop, whereas Kentucky bluegrass in a lawn is considered horticultural. A maple grown for its wood is considered under forestry, whereas a maple grown as a shade tree is considered a horticultural plant. Custom and tradition of a geographical area or a specific state will often determine the commodity area into which a particular crop is placed. However, in this textbook we will consider most commodities that require intensive management under the broad umbrella of horticulture.

Within horticulture there are several branches, divisions, or topic areas. These include:

OLERICULTURE The growing and study of vegetables.

POMOLOGY The growing and study of fruits and nuts (from *Pomona* the Roman goddess of fruit trees).

VITICULTURE The growing and study of grapes or vines. *Vitis* is Latin for vine, hence viticulture is vine culture. Viticulture may be included under pomology.

FLORICULTURE The growing and study of flowers (from *Flora* the Roman goddess of flowers). Floral design and production of indoor foliage plants are usually included under floriculture.

GREENHOUSE MANAGEMENT The growing and study of plants in greenhouses. The principles of greenhouse management are also employed in other controlled-environment growing systems.

TURFGRASS MANAGEMENT The growing and study of turfgrasses. This includes home, municipal and commercial lawns; sports turf maintenance; highway rights-of-way; and seed and sod production.

NURSERY MANAGEMENT The growing and study of trees and shrubs that are produced primarily for landscape purposes.

ARBORICULTURE The growing and study of trees (*arbor* means tree in Latin, so "arboriculture" means tree culture; it is termed **silviculture** in forestry). Arboriculture is essentially synonymous with **urban forestry**.

LANDSCAPE HORTICULTURE The application of design and horticultural principles to the placement and care of plants in the landscape. This term implies the close tie between horticulture and landscape architecture.

INTERIORSCAPING The application of design and horticultural principles to placement and care of plants in indoor environments.

HORTICULTURAL THERAPY The use of horticultural plants and methods as therapeutic tools with disabled and disadvantaged people.

There often is a distinction between pure, or basic, science and applied science. In the past, horticulture has sometimes been called "applied botany." Horticulture applies principles of many other pure and applied sciences. In order to understand how to successfully grow a horticultural crop, for example, we apply the science of botany, including the study of plant structure—morphology and anatomy. We also use plant physiology, another division of botany, that explains how plants function. A horticulturist also needs a knowledge of chemistry, because chemical reactions are important for understanding why we use particular cultural practices, such as fertilizer application and specific pest control practices. Biochemistry helps the horticulturist to explain metabolic reactions within and among cells to understand how plants will respond to external stimuli. We use mathematics in calculating spray rates and for a host of other computations; we use physics to understand light and plant structure; we use plant pathology to understand and cure or prevent plant diseases; we use a knowledge of soil science because plants are commonly grown in the soil; and we use a knowledge of genetics because the genes in a plant interact with the environment to control the makeup of the organism.

Is horticulture a science or an art? This is an age-old question and horticulture is undoubtedly both; one may employ the science but there is an art involved in successfully cultivating plants or placing them in a landscape design. This is where practical experience will be helpful. This concept may be better understood with an analogy between the art of horticulture and piano playing. One can know technically how a piano functions and attend concerts for years, but that will probably not prepare one to perform like Beethoven. Grafting is an example of combining the art and science of horticulture. One can understand the mechanics of grafting as well as the biology (anatomy of cells at the graft union), but it is only through practice that the art is mastered and a high degree of success is achieved. Seldom do all horticulturists agree on everything relating to plants. Surprisingly, however, two scientists can both be correct when they differ in opinion about a cultural practice or its application. This certainly suggests that our applied science of horticulture is also an art form.

HISTORY OF HORTICULTURE

Prehistoric people were primarily hunters and gatherers; that is, they took what their environment provided for them but did not attempt to modify what nature provided. These early gatherers simply collected fruits, seeds, and nuts that they found to be edible and readily available in their immediate surroundings. Subsequently, primitive people began to study plants in an early attempt to control their environment and began to adapt the environment to their advantage. First studies of plants considered practical questions.

❖ Is it edible? Poisonous?
❖ Does eating it modify well-being?
❖ Does it taste good?
❖ Can it be used to keep me warm? As fuel? As clothing?
❖ Is it useful to combat pain? Disease?

People first began cultivating edible plants in the Neolithic Age, about 7000 to 10,000 years ago. By 3000 B.C., which is approximately 5000 years ago, land preparation, irrigation, and pruning were all practiced in Egypt. People east of Egypt, in Mesopotamia, Babylonia, and Assyria, made irrigation canals lined with burnt brick with sealed asphalt joints. This extensive system helped keep 10,000 square miles under cultivation, which in 1800 B.C. fed over 15,000,000 people. These societies cultivated many ornamental, medicinal, and orchard species, including roses, figs, dates, grapes, and olives.

As early people's curiosity increased, they began to seek answers to questions about plants. How do they grow? How do they reproduce? How are they constructed? How are they nourished? How are they related to one another? How are various traits passed from one generation to the next?

Much of the early information about plants was related to agricultural and medicinal uses. Such information was often passed on by word of mouth, but various ancient records exist. Early manuscripts, hieroglyphics, and pictures painted on caves and tombs or carved in stone give clues to the early knowledge about plants. Most commonly, the plants depicted or described were horticultural plants, such as the lotus and other flowers seen in paintings from ancient Egyptian tombs.

Most information was of a practical nature, pertaining to various cultural practices, selection of the best clones, irrigation styles, and harvest methodologies. However, some discoveries of a scientific nature were emerging in this period, including an understanding of sexuality in date palm and recognition of the need for pollination for fruit development (noted in Mediterranean cultures). The Chinese had developed highly advanced cultural practices for tea, oranges, and ephedra, a source of a medication for relief of nasal congestion.

Concomitantly, early civilizations in the Americas were developing crops destined to hold great significance in the world of horticulture. The Pre-Incas of Peru are thought to be the first to cultivate maize (corn). Records indicate that they were growing corn more than 5000 years ago. Potatoes, sweet potatoes, squashes, several types of beans, tomatoes, peppers, avocadoes, cocoa (cacao, the source of chocolate), and many other important species were cultivated by the various aboriginal American peoples. Objectives again were primarily to produce food, clothing, shelter, beverages, fuel, and medicines.

The use of plant products as drugs or for medicinal purposes eventually gave rise to specialists who were drug sellers, the forerunners of both physicians and plant scientists. The early Greek drug sellers, as they strove to learn more about medicinal values of plants, also made discoveries related to plant structure and function and found apparent relationships among plants.

One could argue that Theophrastus was one of the first scientific horticulturists. A student of Plato and Aristotle, Theophrastus wrote books entitled *History of Plants* and *The Causes of Plants*. In his *History of Plants*, he described morphology of roots, flowers, leaves, and other structures and also gave details of anatomical features such as bark, pith, fibers, and vessels. The relationship of weather to soils and agricultural practices, the importance of seeds, the value of grafting, the tastes and fragrances of plants, and the death of plants were subjects he treated in *The Causes of Plants*. Early in the Christian era, Dioscorides, following the lead of Theophrastus, wrote a valuable treatise on the medicinal uses of plants. In this book Dioscorides also proposed some advanced ideas about the relationships of plants, primarily the composite, mint, and legume families.

Relatively little advancement of plant science (or any other science, for that matter) occurred in the Middle Ages. However, Arabian cultures established botanical gardens, primarily for the study of medicinal plants, in the period between A.D. 800 and 1300. Also during this period, much of the scientific advances of antiquity and Greek and Roman cultures were preserved in the monasteries and convents of medieval Europe. As the Middle Ages came to a close, the Renaissance Period signalled a re-birth of energetic attention to scientific discovery. Taxonomy, morphology, and anatomy began to expand as branches of botany, with many of the studies involving horticultural species. English anatomists Robert Hooke, Nathaniel Grew and the Italian scientist Marcello Malpighi made significant contributions during this period.

As civilization and agriculture continued to advance, more and more plants were recognized as being useful to people, and a system of plant classification became necessary. During his lifetime, from 1707 to 1778, the Swedish botanist, Linnaeus developed a binomial classification scheme for plants based on their sexual or flowering systems, which is the basis of all modern plant classification systems today. Linnaeus's system built upon the information presented in the works of the Greek drug purveyors and writers (especially Dioscorides) and other emerging plant scientists. For more information on this subject see Chapter 2.

As the Renaissance period evolved and developed, the budding flower of horticulture burst forth. Formal gardens of various sorts were established in all parts of Europe; those at Versailles and Belvedere in Vienna exemplify some of the best of this art. Meanwhile, systematic improvements were being made in fruit, nut, and vegetable production throughout most of Europe. Flowers, vegetables, fruits, and other plants from "the colonies"—the Americas, Africa, and Australia—became subjects of much attention and in some cases, (e.g., potato, tomato), became mainstays of the diet in many European countries.

Horticulture in America

When European colonists arrived in the Americas, they naturally brought seeds, cuttings, and plants of familiar horticultural and agricultural species with them. Orchards and other horticultural plantings were established, employing both imported and native species. Many early Americans, including George Washington and Thomas Jefferson, experimented with the cultivation of a wide range of species. Indeed, Jefferson, a noted wine enthusiast, made exhaustive attempts to establish vineyards in Virginia for the production of wine grapes.

Horticulture in the United States received new stimulus following the creation of land grant universities by the Morrill Act of 1862. These institutions encouraged growth of all agricultural knowledge, with horticulture emerging as an early beneficiary of educational opportunities.

In the latter part of the nineteenth century, Liberty Hyde Bailey began a career destined to earn him the accolade "Father of American Horticulture" (Figure 1-2). Born in 1858 in South Haven, Michigan, Bailey graduated in 1882 from the Michigan Agricultural College (now Michigan State University), and then studied under Asa Gray at Harvard University from 1883 to 1885. He became Professor of Horticulture and Landscape Gardening and Superintendent of the Horticulture Department at Michigan Agricultural College in 1885, where he remained until Cornell University, Ithaca, NY, lured him away in 1888 to become Professor of General and Experimental Horticulture. He remained at Cornell in various capacities until his retirement in 1913.

FIGURE I-2. Liberty Hyde Bailey, a prolific author, renowned taxonomist, respected educator and administrator, deserves the oft-suggested title, "Father of American Horticulture."
Photograph courtesy L.H. Bailey Hortorium.

Bailey was a prodigious writer. He authored a great number of books containing much horticultural information, including *Hortus*, a taxonomic index of horticultural plants. He also wrote his famous *Cyclopedia of Horticulture*, which contains cultural as well as taxonomic information. He helped found the first horticulture department distinct from a botany program in the United States. Bailey's philosophy was to live on the "25-year plan": devoting the first 25 years of his life to his education, the second 25 years to gainful employment and public services, and the last 25 years of his life to retirement, doing as he pleased. Fortunately for horticulture, the last 25 years extended to more than 40, and were a period of intense activity—of writing, editing, plant exploration and description, and the establishment of the Bailey Hortorium, a tremendous contribution to the field of horticulture. *Hortorium* was a word coined by Bailey to be a repository for "things of the garden," including his vast collection of plant specimens dating back to his days at Michigan Agricultural College. Who knows how extensive his contributions might have become had he not fallen and broken his leg in a New York bank in December, 1949, for in his pocket he had airline tickets to tropical Africa to collect more horticultural plants. Although his injury confined him to his home, he continued writing and editing, completing a book on bellflowers in the year preceding his death, in December, 1954.

Liberty Hyde Bailey, along with several other leading horticulturists of the day, was instrumental in founding the American Society for Horticultural Science (ASHS) in 1903. Many states had established State Horticultural Societies in the early 1800s, but these societies focused primarily on the utilitarian matters of culture, orchard establishment, fertilizer practices, and so on. The founding of ASHS finally gave the field of horticulture a national organization and a solid scientific base. Since that time, many important developments have revolutionized horticulture. The following paragraphs briefly describe some of the most important achievements for the horticultural industry.

NEW CULTIVARS. Breeders have developed new cultivars with improved quality, greater yield potential, improved growth characteristics, increased pest resistance, and greater tolerance to environmental extremes (see Figure 1-3). Genetic engineering and other modern technologies employed for cultivar development and improvement are covered in Chapter 4.

FIGURE I-3. 'Sugar Snap' pea, an example of an eminently successful new cultivar.
Photograph courtesy All America Selections.

PLANT-WATER RELATIONSHIPS. We have a better understanding of irrigation (Figure 1-4) and plant-water relationships, and of new application methods such as trickle or drip irrigation. Intermittent mist and fog systems have aided plant propagation. Improvements also have been made in hydroponic plant production, which still remains a minor, but important, horticultural industry.

TEMPERATURE. Our knowledge of plant responses to temperature has expanded phenomenally. Bottom heat is now used with many crops to increase their growth. We now more fully comprehend the phenomenon of *vernalization*, which is the direct effect of low temperature on flower initiation. Understanding the relationship between temperature and dormancy has enabled commercial producers to schedule their crops more effectively. New developments in temperature and energy management have helped horticulturists to use fuel reserves more wisely (Figure 1-5). Recent advances in the adjustment of night and day temperatures, for example, enable the grower to increase the production efficiency of greenhouse crops. We also now have a much better understanding of stresses caused by extremes of temperature.

LIGHT. We now have a better understanding of how light influences the all-important process, photosynthesis, and how light can trigger or delay flowering. Many

FIGURE 1-5. Use of waste heat: hot water from an electric power generating plant heating a greenhouse.

crops can therefore be programmed to flower in a timely fashion or they can be managed to avoid flowering when no flowers are desired (Figure 1-6). Supplemental lighting can be utilized to increase yields of some species during periods of low light. Because of our knowledge of light, we now have successful indoor gardening and interiorscaping to beautify and enhance large public areas such as shopping malls.

PLANT NUTRITION. Many new and different types of fertilizers have been developed in recent years. Formulations are now available for varying needs, including different rates of availability to plants. However, it wasn't until about 1920 that trace element fertilization began to be better understood. Great strides also have been

FIGURE 1-4. Supplying water by sprinkler irrigation.

FIGURE 1-6. Chrysanthemums may be induced to bloom in any season of the year by manipulating photoperiod.

made in our comprehension of the fixation of atmospheric nitrogen within plant roots by microorganisms (symbiosis).

THE RHIZOSPHERE (ROOT ZONE). Properties of soils, the value of organic matter, and interaction among soil components, soil microflora, and microfauna have led to advances in commercial and home horticulture. Increased container plant production has been facilitated by the development of improved pots, flats, multipacks, and starter blocks. Root aeration has been improved by the development and use of peat-lite media, bark mixes, and other soil-less mixes (Figure 1-7). Our knowledge of the importance of organic matter in field soils has improved along with erosion control practices such as reduced tillage. Mulches have been developed to control soil moisture, temperature, erosion, and weeds.

INTEGRATED PEST MANAGEMENT (IPM). The management and control of insects, diseases, and weeds involves a program utilizing genetic pest resistance within plants, a knowledge of pests' life cycles and epidemiology, crop scouting reports, growing plants under appropriate environmental conditions, and the proper use of pesticides. This integrated approach can emphasize the control or elimination of plant pests and minimize negative impact on the environment. The necessity for pest control has led to the growth of a large agrichemical industry that conducts research and markets a wide variety of products that aid in effective pest control. The strategy of integrated pest management has led to substantially increased yields and has given the public a much higher quality of agricultural products (Figure 1-8).

PLANT GROWTH REGULATORS. Horticulturists and plant physiologists have gained a better understanding of the nature of growth regulation in plants. This knowledge has led to the development of chemicals that enable growers to control plant growth responses more efficiently and thus to improve crops or adapt them for new uses. Plant growth substances have revolutionized the propagation of horticultural plants by increasing rooting of cuttings and facilitating effective application of tissue culture techniques. Growth regulators are also used to control plant height, plant branching, flowering, fruit set, fruit size, and fruit and leaf drop, along with many other responses (Figure 1-9).

FIGURE I-7. Soilless mixes such as peat-vermiculite combinations have become popular for growing transplants and pot plants.

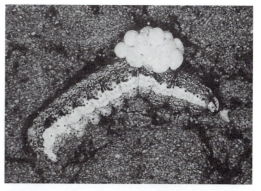

FIGURE I-8. Encouraging natural predators can be an important part of an integrated pest management (IPM) program, as shown by this parasitized cutworm larva.

FIGURE I-9. Growth retardant chemicals are often used to produce more aesthetically pleasing plants, such as these poinsettias. The plant at right was treated with paclobutrazol, a chemical growth retardant.

Photo courtesy Kenneth Sanderson, Auburn University, Auburn Alabama.

HORTICULTURAL ENGINEERING AND MECH-ANIZATION. Engineering advances in the field of horticulture have been tremendous. Controlled-environment growing facilities including greenhouses, have been greatly improved in recent years. Labor saving machinery is now available for soil prepara-tion, cultivation, spraying, fertilizer applica-tion, mechanized seeding, transplanting, and harvesting. Aerial applications of pesti-cides and even remote sensing of stresses and other factors by satellites (global infor-mation systems, GIS) have become practices of importance in many areas of horticulture (Figure 1-10).

POSTHARVEST FACTORS. Expanded know-ledge of the postharvest physiology of crops has led to longer shelf life of our horticultural products, such as fruits, vegetables, plants and cut flowers. Controlled-atmosphere cold storage, for example, enables good-quality apples to be available many months after harvest. Longevity of cut flowers has been enhanced so that a bouquet may provide beauty for two weeks or more, instead of only for a few days. Transportation has been revolutionized. Refrigerated trucks, air-planes, and railroad cars speed improved fresh fruits, vegetables, flowers, and plants to the consumer for year-round enjoyment (Figure 1-11).

FIGURE I-11. Modern refrigerated transport speeds fresh produce to market at the peak of quality.

FIGURE I-10. Computers have become a valuable tool for horticultural enterprises. The personal computer (*above*) may be employed in a variety of tasks; other equipment (*right*) may be used to monitor and control environmental factors in a greenhouse.

PLANT CLASSIFICATION

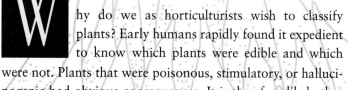

W hy do we as horticulturists wish to classify plants? Early humans rapidly found it expedient to know which plants were edible and which were not. Plants that were poisonous, stimulatory, or hallucinogenic had obvious consequences. It is therefore likely that even in the days before recorded history, people had applied a rudimentary naming system to plants in order to identify these differences.

EARLY CLASSIFICATION SYSTEMS

Theophrastus (370–285 B.C.), a student of Aristotle and later his assistant, is widely credited with the first significant attempt to provide an orderly classification of plants. Following extensive study of the plants found in the gardens established by Aristotle, Theophrastus published *Historia Plantarum* or *Enquiry into Plants*. In this extensive work, he covered characteristics of seeds, germination, plant distribution, and environmental adaptations, but he is best remembered for his attempt to place plants into categories. He divided plants into four categories: trees, shrubs, half shrubs, and herbs. He also categorized all plants as either annuals, biennials, or perennials and listed several plant "families" among the flowering plants. In fact his description of the parsley family (*Apiaceae/Umbelliferae*) categorized for the shape of its flower cluster, is still essentially the same family in today's modern classifications.

Although many subsequent descriptive lists, "herbals," and other attempts at classification of plants appeared over the years, it was not until the middle of the 18th century that a workable and broadly applicable system was developed. A binomial (or two-word) system for scientifically classifying plants according to their sexual systems or flower structures was developed in the 18th century by the Swedish naturalist and physician, Carl von Linné (1707–1778). Better known as Carolus Linnaeus, the Latinized version of his name, he first published *Genera Plantarum* in 1737, a work in which he suggested the concept of classification of plants based on the number of sexual parts. Subsequently, in 1753, he published *Species Plantarium*, which established what is generally referred to as the "sexual" system. He described over 1300 plants (and a like number of animals as well), placing these plants in 23 "classes" of flowering plants and lumping nonflowering plants (mosses, algae, ferns, and fungi) into a 24th class. He placed the flowering plants in their respective classes based on number of stamens, stamen characteristics, and their relationship to other floral parts. This system serves as the basis of plant scientific nomenclature in use today and utilizes botanical names in the Latin form.

Plants have also been given horticultural names (cultivars) and nonscientific common names. Although many people are most comfortable with the use of common names for plants, they offer several disadvantages when used alone because they are not governed by any accepted rules, are variable from region to region, and may not be understood by persons who speak other languages. For example, in Minnesota, the state tree, *Pinus resinosa*, is know as the Norway Pine, whereas in other areas of the United States it is known as red pine. In another example, we see that Bailey lists six different species as "bluebell" or "blue-bells," but only two are even in the same family! Different "common" names for the same plant or several distinctly different species called by the same common name are just two examples of how use of common names may create problems in communication because of the obvious potential for misunderstandings.

BOTANICAL CLASSIFICATION/TAXONOMY

A *taxonomist* is a plant scientist who studies plant classification, or more specifically, plant categories or *taxa* (singular, *taxon*—hence the name taxonomy). Understandably, Linnaeus is generally accorded the title, "Father of Taxonomy."

Today's system of plant classification is largely based on his precepts, with modifications of this "artificial" system based on "natural" principles of classification. A natural system of classification relies on overall similarities in morphology. The theory of evolution, espoused in Charles Darwin's *Origin of Species* published in 1859, was an important step towards combining a phylogenetic approach with natural classification and the artificial system developed by Linnaeus. Darwin's theory of evolution was based on the concepts of gradual development and evolution by both plants and animals. Thus a phylogenetic classification system relies on relationship by descent; that is, plants within a genus or family are related by descent from a common ancestral plant and would have many morphological similarities. It is therefore reasoned that a high number of morphological similarities indicates close relationship. A 20-volume natural plant classification system based on the work of August Eichler, was published by Adolph Engler, Karl Prantl, and others beginning in the late 19th century. Their system is based on combining Darwin's evolutionary theories with genetic concepts and uses as many morphological characteristics as possible. This exhaustive work provided the backbone of the most commonly accepted system of plant classification today. It divides the plant kingdom into divisions, which are then divided into classes, orders, families, genera (singular *genus*), and species. Further division into several minor taxa may be followed for some plants. (A discussion of plant nomenclature follows.)

As in any science, all taxonomists do not agree entirely on what constitutes the most accurate and best system. For this reason, plant taxonomy is a dynamic and continually changing field. Furthermore, as new methods and technologies emerge, taxonomists employ more sophisticated approaches to gain precision in plant identification. Physiological, biochemical, cytological, ecological, genetic, and more recently biotechnological approaches provide information that enables taxonomists to more accurately identify plants than ever before. Chromatographic and electrophoretic techniques are physicochemical methods that can help identify similar and dissimilar protein constituents of plants, thus enabling the taxonomist to "fingerprint" such plants as an aid to identification of taxonomic relationships.

Plant nomenclature

Botanical names, in addition to having been derived from one language (Latin), are governed by specific rules that are internationally accepted and are published in the *International Code of Botanical Nomenclature*. Students need to become familiar with this hierarchical naming system, because it is universal, and because it will enable them to

Major taxa used for horticultural plants

TAXON	NAME(S) AND COMMENTS
Kingdom	Plant Kingdom (Embryonta, or "Land Plants")
Division	Tracheophyta, higher plants with vascular or treacheary system. (Although edible mushrooms are considered a horticultural commodity, they are members of the Division *Thallophyta*.) The division is equivalent to the Phylum of animal classification systems.
Class	Filicinae (Ferns), Gymnospermae (ginkgoes, cycads, taxads, conifers), and Angiospermae (the flowering plants).
Subclass	Subclasses in the Angiospermae (see also Table 2-1, Figure 2-1): ❖ Dicotyledoneae (or dicots): Two cotyledons in their seeds; net leaf venation; flower parts in fours, fives, or multiples thereof. ❖ Monocotyledoneae (or monocots): One cotyledon in their seeds; parallel leaf venation; flower parts in threes or multiples thereof.
Order	Some orders in subclass Diotyledoneae: ❖ Polemoniales, which has the families Solanaceae (nightshade family) and Convolvulaceae (morning glory family); Sirophulariales has the family Oleaceae (olive family). Some orders in subclass Monocotyledonae: ❖ Liliales (lilies), Palmales (palms), Graminales (grasses).
Family	*Solanaceae.* Family names generally end in *aceae* and often are named for a "type genus," (e.g., *Solanum* for Solanaceae). The *International Code* directs that all family names should be based on a type genus, but the older descriptive family names continue to be commonly used. Examples: Poaceae is the proper family name for Gramineae, the grass family; Asteraceae for Compositae, the composite or sunflower family; and Apiaceae for Umbelliferae, the parsley family.

Minor taxa

TAXON	NAME(S) AND COMMENTS
Genus	*Capsicum* (always capitalized)
Specific epithet	*annuum* (normally not capitalized, but may be if it is a proper name, such as *Lonicera Albertii*). The genus and specific epithet together are referred to as the *species* of a plant, or the binomial or scientific name; *Capsicum annuum*, for example, is the species of the garden pepper. When properly written, an additional name (often abbreviated) follows the specific epithet such as *Capsicum annuum* L., *Cypripedium reginae* Walt., and *Prunus persica* (L.) Batsch. This name is referred to as the **authority**, the taxonomist responsible for naming this plant by the specific binomial; in these examples L. stands for Linnaeus who named the garden pepper; Walt. is Thomas Walter, who named the showy ladyslipper orchid, and Batsch renamed peach by its present name after Linnaeus had given it an incorrect name. The genus and specific epithet are properly italicized in printing (or underlined in writing).
Forma	A sub-division of a species not differing sufficiently to be called a variety, often only by a slight genetic variation giving rise to a different appearance or "form."
Variety	Subclassification of the traditional species (e.g., *Scilla sibirica* var. *alba*).
Cultivar	A group of plants within a species with one or more common characteristics. The initial letter is normally capitalized. The word is correctly enclosed in single quotes, as in 'Tropic' tomato; or preceded by cv., as in tomato, cv. Tropic; never both.

Related Terms

Hybrid plants	Names of hybrid plants should be preceded by the capital letter X. A hybrid genus example is X *Fatshedera*, a hybrid between the genera *Fatsia* and *Hedera*, and hybrid species examples are *Pelargonium* X *hortorum* and *Petunia* X *hybrida*. The capital X should not be pronounced as "ex," but rather stated as "the hybrid genus" or "the hybrid species."
Cultigen	This taxon is used to describe a plant or group of plants that is presumed to have originated in domestication with no clearly evident native form. Examples are Brussels sprouts, cabbage, and corn (maize).
Line	A sexually reproduced cultivar. 'New Yorker' tomato is an example of a *pure* or *inbred* line that is maintained by allowing it to self-pollinate naturally and avoiding cross-pollination. 'Celebrity' tomato is a *hybrid* cultivar produced by plant breeders systematically crossing a select pollen parent with a specific seed parent. The parents of hybrid cultivars such as 'Celebrity' tomato are usually inbred lines.

more effectively work with, sell, purchase, or communicate about horticultural plants.

The term *cultivar* is a contraction of the two words *cultivated variety*. It is a group of plants within a species that has a particular character or group of characteristics and can refer either to plants propagated sexually or asexually (clones[1]). The term

---

[1]A **clone** is a population of plants derived asexually from one original individual (see Chapter 14).

variety, however, is a taxonomic term and refers to the botanical variety. Historically, what we now call "cultivars" were called "*varieties*"; in fact, some people still use this terminology. The term cultivar was coined by L. H. Bailey to avoid confusion between the botanical variety and cultivated variety (cultivar) by agricultural scientists and botanists. An example of the proper use of both terms is in the following scientific name: *Gleditsia triacanthos* var. *inermis* 'Skyline.' This is the 'Skyline' clone of honeylocust. 'Skyline' is the cultivar name, *inermis* is the botanical variety and means that this particular honeylocust is thornless. Therefore, in many cases the cultivar is actually a subdivision of the variety and the two terms are not synonymous. However, many species when classified are not subdivided further into botanical varieties. They are just named with the genus and specific epithet and then are followed by the cultivar name, e.g., *Vitis vinifera* 'Chardonnay,' a common winegrape and *Cucurbita pepo* cv. Papaya Pear, an All American Selection of summer squash.

TABLE 2-1. A comparison of dicotyledonous plants (dicots) and monocotyledonous plants (monocots)

Dicots	Monocots
Cotyledons 2 (seldom 1, 3, or 4)	Cotyledon 1 (or the embryo sometimes undifferentiated, as in Orchidaceae)
Leaves mostly net veined	Leaves mostly parallel veined
Vascular bundles in stem usually borne in a ring that encloses a pith (herbaceous and young dicots)	Vascular bundles in stem usually scattered, or in two or more rings
Vascular cambium usually present	Vascular cambium lacking; usually no cambium of any sort
Floral parts, when of definite number, typically borne in sets of five, less often four, or multiples of five or four (carpels often fewer)	Floral parts, when of definite number, typically borne in sets of three or multiples, seldom four, never five (carpels often fewer)
Mature root system primary, adventitious, or both	Mature root system wholly adventitious

Plant keys

An analytical key is a written diagnostic device designed to enable the student to identify a particular plant by elimination of characters not present in the plant and ultimately to arrive at the name of the only plant that can be correct. The most common type is called a **dichotomous** key. In such keys, opposite conditions are placed in paired couplets, with each contrasting statement termed a **lead**. By examining the first couplet, the student will reject one statement and accept the other, which will lead to the next couplet to be examined.

In the following example, the student would be able to identify a tulip flower as being a member of the Liliaceae. Upon examination of the flower, statement A would be rejected and AA accepted, which would **lead** one to consider leads C and CC. C would then be rejected and CC accepted. When reaching this conclusion, the student would then be directed to a more detailed key of the lily family (*Liliaceae*) to determine the genus and eventually to identify the plant in question.

> **A.** Flowers with no stalks and arranged along a central axis, individual flower parts not showy.
>> **B.** Plant herbaceous, flowers in one compact group per plant: Araceae
>> **BB.** Plant woody, flowers in one to several compact groups: Arecaceae (*Palmae*)
> **AA.** Flowers attached by short stalks, to a central axis, flower parts showy.
>> **C.** Ovary inferior, stamens three: Iridaceae
>> **CC.** Ovary superior, stamens six: Liliaceae

Many manuals and guides for plant identification are available and in most cases they contain one or more keys. It is preferred, where possible, to employ keys

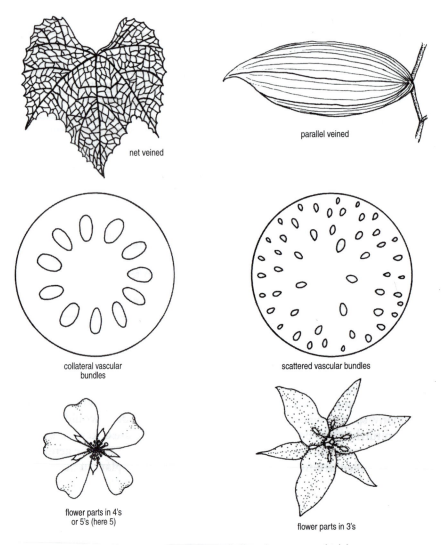

net veined

parallel veined

collateral vascular bundles

scattered vascular bundles

flower parts in 4's or 5's (here 5)

flower parts in 3's

FIGURE 2-1. Some characteristics of dicots (*left*) and monocots (*right*).

developed specifically for the local flora, such as *Keys to Spring Plants* developed by Muenscher and Petry for the spring flora of the Cayuga Lake basin in New York State, or for specific plant categories, such as Backberg's *Cactus Lexicon* or Gould's *Grass Systematics*. However, once a student identifies a plant by use of a dichotomous key, he or she should seek corroboration by comparing the plant in hand with a published description and/or by consulting one or more additional resources such as *Hortus Third*.

HORTICULTURAL CLASSIFICATION

Numerous other systems of classifying horticultural plants have been developed since ancient times, as noted earlier. Most of these approaches are less scientific and systematic than botanical classification, but they are often useful because they rely on practical criteria such as environmental adaptation, form, growth habit, and usage.

Morphological features (form) are often among the most practical because they are easily observed. Plants may be categorized on the basis of their physical hardness: Nonwoody plants are generally termed **succulents** or more often **herbaceous** and they are usually short-lived; **woody** plants have more dense, sturdy tissues and tend to be long-lived. The stature and need for supplemental support helps further delineate plant types. Herbaceous plants that climb or twine around a vertical support are termed **vines** and woody climbing plants are correctly called **lianas**. Self-supporting herbaceous plants are **herbs**, and woody plants that require no supplemental support are considered **trees** or **shrubs**. Shrubs generally are low in stature and have several essentially upright stems, whereas trees have a single upright central stem and often are very tall. Many woody plants are difficult to categorize as trees or shrubs, because some smaller statured trees may appear shrubby, especially if modified by pruning or other cultural practices.

Some plants lose all of their leaves for a portion of the year and are known as **deciduous**. Plants that retain a portion of their leaves throughout the year are called **evergreens**. Evergreens shed their leaves a few at a time; some species may completely replace their leaves over a period of a few years, but they do so in a gradual fashion so that they are never fully leafless. Some tropical and subtropical plants are deciduous, losing their leaves during a dry season; in temperate zones deciduous trees and shrubs are leafless during a cold period (winter). Furthermore, some evergreen species from warm climates may become deciduous if grown in temperate zones.

Environmental adaptation encompasses another group of factors involved in classification of horticultural plants. Tolerance to extremes of temperature, resistance to water deficits, and adaptation to varying light levels and soil characteristics are among the most common of these factors. Plants are often categorized as **cold-hardy** if they can

survive low winter temperatures and may be called **tender** if they are readily damaged by cold. In fact, cold hardiness zones for plants have been delineated for most of North America, based on plants' abilities to survive annual minimum temperatures typical of a given zone. In many cases roots are more easily damaged than the above-ground parts of plants. Cold hardiness is sometimes further divided into **flower-bud hardiness, wood hardiness,** and **root hardiness**.

Many herbaceous plants, especially annual flowers and vegetables, are grouped according to the temperatures most favorable for their survival and optimum growth. For example, sweet potato (*Ipomoea batatas*), muskmelon (*Cucumis melo* var *reticulata*), and lima bean (*Phaseolus lunatus)* are considered **warm season** vegetables and broccoli (*Brassica oleracea* var *Italica*), lettuce (*Lactuca sativa*), and peas (*Pisum sativum*) are placed in the **cool season** category.

Plants adapted for growth in water or in very wet soils, such as water lilies (*Nymphaea* spp.) and water cress (*Nasturtium officinale*), are termed **aquatic plants** or **hydrophytes,**. Plants adapted to seasonal or persistent drought, such as cacti (*Opuntia* spp.) and succulents such as certain members of the Crassulaceae, are termed **xerophytes**. Most horticulturally important plants are **mesophytes**, or intermediate in their water requirements and tolerances. Plants adapted to soils high in salinity (salts) are called **halophytes**, and include, for example, members of the goosefoot family (Chenopodiaceae) such as beet (*Beta vulgaris*), spinach (*Spinacia oleracea*), and saltbush (*Atriplex* spp.). Plants adapted to extremes of soil pH are referred to as **basophilic** ("base-loving") or **acidophilic** ("acid-loving"). An example of a basophilic plant is coltsfoot (*Tussilago farfara*) and examples of the acidophilic type abound in the Ericaceae, including such horticulturally important species as blueberry and cranberry (*Vaccinium* spp.) and rhododendron and azalea (*Rhododendron* spp.).

Classification by growth pattern

Three life cycle patterns exist among higher plants and plants exhibiting these patterns are categorized as **annuals, biennials,** and **perennials.** An annual plant is one that completes its life cycle in one growing season; that is, the seed germinates, vegetative and reproductive growth follow, and then the plant dies. Examples include corn (*Zea mays*) and spinach (*Spinacia oleracea*). In cold climates, some perennials, such as tomato, geranium, and petunia are treated as annuals even though they are true perennials, because they are killed by cold temperatures at the end of the growing season. Biennials require all or parts of two growing seasons to complete their life cycles. They grow vegetatively the first growing season and then flower and die the second growing season. This flowering usually occurs in response to the cold temperatures experienced by the plant during the winter period. In some biennials, such as beets (*Beta vulgaris*), long days (short nights) may also be required to facilitate flower development after the cold requirement has been satisfied. Examples of biennials include several important horticultural crops: carrot, beet, onion (*Allium cepa*), honesty (*Lunaria annua*), canterbury bells (*Campanula medium*), and foxglove (*Digitalis purpurea*) (Figure 2-2).

Some biennials are referred to as **winter annuals.** Their seeds commonly germinate in the fall and the plants grow vegetatively and overwinter in a nongrowing state, then flower in the spring or early summer. Examples include winter grains and several common temperate zone weeds, such as yellow rocket (*Barbarea vulgaris*) and shepherd's purse (*Capsella bursapastoris*).

Perennial plants live for more than two growing seasons (often for many years) and may flower as early as the first growing season. Many species fall into this category, including all trees and shrubs. In bramble fruit species (*Rubus* spp.), although the

FIGURE 2-2. *Top*: Rosette (vegetative) stage, *left*, and flowering (reproductive) stage, *right*, of wild carrot (*Daucus carota*), a typical biennial. *Bottom*: fruiting inflorescence of beet (*Beta vulgaris*), another biennial.

plant is perennial, the above-ground portions are functionally biennial. The first year's stems, or **primocanes,** are usually unbranched and vegetative. In the second year, these stems (called **floracanes**) branch, flower, fruit, and then die. Some cultivars produce fruit on the tips of the primocanes in late summer, in addition to the next season's floracane crop, and are therefore often called "everbearing" or fall-bearing types.

Classification by usage

One of the oldest approaches to plant classification is to categorize the plant or plant part by the way it is used. Horticultural plants readily fall into two broad groups, those that are **edible** and important for their food value and those that are important for their **amenity** uses, (i.e., plants that are grown for their ornamental and aesthetic values).

Edible horticultural plants

Edible plants are usually divided into **vegetables** and **fruits**. Vegetables are plants or plant parts consumed as part of the main portion of the meal, but fruits are most often eaten as a dessert. Vegetables are usually annual or biennial herbs. Exceptions include *Asparagus officinalis* and globe artichoke, *Cynara scolymus;* these plants are perennials, but their above ground portions are still primarily herbaceous. Most vegetables are eaten raw or cooked simply and the parts eaten may be vegetative, such as leaves (spinach, lettuce), stems (asparagus), and roots (carrot), or reproductive tissues, such as flower parts (broccoli), fruits (tomatoes), and seeds (peas).

The true botanical **fruit** or closely related structure is the more or less succulent part taken from a fruit plant and eaten raw or cooked as a dessert or a snack. **Tree fruits** are those harvested from trees and include apples, pears, mangoes, and most citrus fruits. Fruits born on shrubby or herbaceous plants are termed **small fruits** and include strawberries, currants (*Ribes* spp.), grapes (*Vitis* spp.), blueberries, and the brambles (*Rubus* spp., raspberry, blackberry, cloudberry). Pecans (*Carya illinoensis*), walnuts (*Juglans* spp.), almonds (*Prunus dulcis*), and pistachio (*Pistacia vera*) are examples of a special category of fruits called **nuts**. They typically have a hard fruit wall or shell enclosing the edible portion or kernel. Specifics of fruit structures and categories are covered in more detail in Chapter 3.

Amenity plants

This category includes all plants grown to provide aesthetic enhancement, especially visual enhancement, to our lives. These include such categories as cut flowers, pot crops, bedding plants, nursery crops, and turfgrasses. All may be grown as part of a landscape planting, visually increasing the beauty of one's surroundings. **Cut flowers** may be made into corsages or boutonnieres and worn for ornament, fabricated into special displays, or simply placed in a vase on the dinner table. **Pot crops** include both flowering types such as chrysanthemums and cyclamen and **foliage plants**, those plants grown for the special features the foliage provides for the indoors ("houseplants"). **Bedding plants** are flowering plants grown to be transplanted while young into flower "beds," primarily to augment and add color to a landscape design.

Nursery crops include those trees and shrubs produced in special locations, or nurseries, prior to placing them in a landscape setting. Both deciduous and evergreen trees and shrubs are grown by commercial nurseries; they may also grow **ground covers**, other **herbaceous perennials**, and bedding plants.

Turfgrasses or lawn grasses, or just turf, are essential components in most temperate zone landscapes, and useful in many warmer areas. Turfgrasses are also critical components of most sports turfs, including golf courses, football pitches, and baseball fields.

Plants often have multiple uses, especially in the landscape. Plants such as blueberries and crabapples may be prized in the spring for the beauty and fragrance of their flowers, valued later in the year for their fruits harvested for food, and again enjoyed for their autumnal colors in the fall and winter. Likewise a grape arbor may be both ornamental and yield delicious fruits for direct consumption or production of beverages such as wine.

Many other plants are grown for specialty products such as **beverages** (coffee, *Coffea* spp.; tea, *Camellia sinensis*), **medicinals, herbs,** and **spices.** Medicinals are herbaceous or woody plants that yield substances of pharmaceutical, or medicinal, value. Examples include: Madagascar periwinkle (*Catharanthus roseus*), a plant often used as a bedding plant, but possessing medicinal properties for humans; however it is poisonous to cattle; foxglove, a herbaceous perennial (*Digitalis purpurea* and *D. lanata* are sources of an important cardiac drug); and quinine. The valuable antimalarial alkaloid known as quinine is extracted from the bark of several species of the genus *Cinchona*. These and other potential uses of plants for medical purposes underscore the importance of learning more about plant diversity; it is imperative that this diversity be preserved, for the next cure for cancer or other life-threatening disease may be resident in a plant that has not yet been discovered!

Herbs for flavoring are usually made from fresh or dried vegetative parts of herbaceous plants (e.g., sage, *Salvia officinalis*; thyme, *Thymus* spp.) whereas spices are generally dried fruits or parts of tree species (often tropical in origin, such as pepper, *Piper nigrum*, and cinnamon, *Cinnamomum zelanicum*).

The horticulturist is constantly finding new uses for plants, so horticultural classification systems are unlikely to remain static. Thus it behooves the student to continue to follow these changes and integrate them into their knowledge of both botanical and horticultural systems of classification.

REFERENCES

Backberg, C. 1977. *Cactus Lexicon* (English ed.). Blandford Press, London.

Bailey, L. H., E.Z. Bailey and the Staff of the L.H. Bailey Hortorium. 1976. *Hortus Third: A Concise Dictionary of Plants Cultivated in the United States and Canada.* MacMillan, New York.

Bailey, L.H. 1949. *Manual of Cultivated Plants Most Commonly Grown in the Continental United States and Canada,* (rev. ed.) MacMillan, New York.

Cronquist, A. 1988. *The Evolution and Classification of Flowering Plants* (2nd ed.). The New York Botanical Garden, Bronx, New York.

Flora of North America Editorial Committee. 1993. *Flora of North America North of Mexico, Volume 1: Introduction.* Oxford University Press, New York.

Flora of North America Editorial Committee. 1997. *Flora of North America North of Mexico, Volume 3: Magnoliophyta: Magnoliidae and Hamamelidae.* Oxford University Press, New York.

Gould, F.W. 1968. *Grass Systematics.* McGraw-Hill, New York.

Hickey, M. and C. King. 2000. *Cambridge Illustrated Glossary of Botanical Terms.* Cambridge University Press, UK.

Lawrence, G. H. M. 1951. *Taxonomy of Vascular Plants.* MacMillan, New York.

Muenscher, W. C. and L. C. Petry. 1949. *Keys to Spring Plants.* Cornell Univ. Press (Comstock), Ithaca, N.Y.

Stearn, W. T. 1995. *Botanical Latin: History, Grammar, Syntax, Terminology and Vocabulary.* David & Charles, London.

PLANT STRUCTURE

A good knowledge of plant structure is essential for understanding how horticultural plants grow. An effective grasp of the vocabulary of plant structure will arm one with the tools requisite to fully appreciate and apply the information presented in the chapters that follow. Furthermore, development of an understanding of plant structure and gaining a strong background in such botanical vocabulary will facilitate effective communication on a wide variety of horticultural subjects. Details of plant structure and related terminology are therefore presented in this chapter.

THE CELL

The basic unit of a plant is the **cell** (Figure 3-1). Living cells contain various organelles and structures. Plant cells differ from animal cells in that they are surrounded by a rigid cell wall, which may have one or two layers. The basic wall material is called **cellulose**. Cellulose is a polysaccharide composed of unbranched chains of beta-glucose molecules that organize into crystalline subrods called **micelles,** which in turn organize into slender **microfibrils**. These microfibrils comprise the majority of the cell wall material and impart strength and rigidity to the cell.

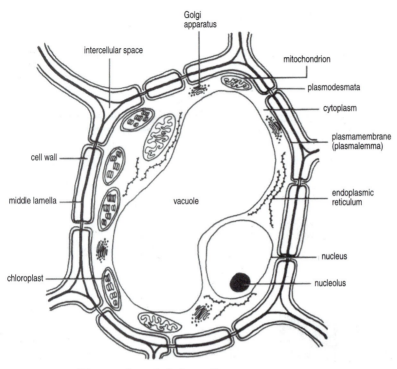

FIGURE 3-1. Diagram of a typical plant cell

On the inner side of the cell wall is a **plasma membrane** that surrounds the cytoplasm and is called the **plasmalemma**. It is differentially permeable, and therefore governs what substances may enter or leave the cell. The **protoplast** is comprised of the contents of the cell (**cytoplasm**) surrounded by and including the plasmalemma. The protoplasts of adjacent living cells are commonly continuous because they are connected by fine strands of cytoplasm (the contents of the cell excluding the nucleus) that penetrate the cell wall. These strands are termed **plasmodesmata** and function as a transport system through which substances can move from cell to cell. The continuous internal system within a plant that constitutes the plasmodesmata and membrane-bound cytoplasm is referred to as the **symplast**. The continuum outside of the symplast that includes nonliving cells, cell walls, and intercellar spaces is called the **apoplast**.

Many structures reside within the plasmalemma; the types vary with the degree of

cellular specialization. Commonly comprising 90% or more of the cellular volume of mature cells is a membrane-bound **central vacuole**, which contains the cell sap. The **vacuolar membrane** is known as the **tonoplast** and is differentially permeable; thus it governs which substances may enter or exit the vacuole. The primary component of the cell sap is water, although many solutes and colloids are also present. Among the constituents of the cell sap are pigments, which are responsible for the colors in many flowers and other plant parts. The central vacuole also stores other cellular substances, including waste materials.

Living plant cells contain a large membrane-bound organelle called the **nucleus**. The nucleus contains the plant's genetic material, which is composed of the nucleic acids **deoxyribonucleic acid (DNA)** and **ribonucleic acid (RNA)**. The DNA bears hereditary information and forms the plant's genes and chromosomes. The DNA of the genes governs the formation of RNA, which moves into the cytoplasm and controls protein bio-synthesis and cellular development. The **nucleolus** is a densely staining structure within the nucleus that consists of proteins that are closely associated with the regions of DNA that code for specific types of RNA. The importance of the genetic material is discussed more fully in Chapter 4.

Contiguous with the nuclear membrane is a loose network of membrane material known as the **endoplasmic reticulum**. Associated with the endoplasmic reticulum, or sometimes free in the cytoplasm, are organelles call **ribosomes**. Ribosomes contain RNA and participate in the manufacture of proteins.

Various organelles called **plastids** are also present in living cells. Among the most important plastids are **chloroplasts**, which contain the green pigment **chlorophyll**. Photosynthesis occurs in the chloroplasts and is described in detail in Chapter 5. Some other plastids are colorless and are called **leucoplasts**, whereas some contain pigments and are termed **chromoplasts** (because chloroplasts contain chlorophyll, they are a type of chromoplast). The pigment **carotene**, which is discussed in Chapter 5, is present in chromoplasts and contributes to the stability of chloroplasts.

Plant cells also contain **mitochondria** (Figure 3-2), in which the processes of respiration and energy transfer occur. Living plants and animals respire continuously, both during the daylight hours and night hours (discussed further in Chapter 6).

Within living plant cells are organelles know as **dictyosomes** or **Golgi bodies**. Dictyosomes have several important functions, one of the most significant being their involvement in the synthesis of cell walls.

Adjacent cells are cemented together in a manner similar to concrete blocks in the wall of a building (Figure 3-3). The area where the cell walls are in contact is called the **middle lamella**. The middle lamella is rich in pectins, which bind the cells together much as mortar holds concrete blocks together.

Plant cells undergo various degrees of specialization or differentiation, and thus fulfill diverse roles and functions. If their cells failed to differentiate and organize but continued to divide and enlarge, plants would resemble amorphous, tumorous masses.

The least differentiated plant cells are those that are considered "embryonic" because their primary function is cell division. Such **meristematic** cells are responsible for cell formation and subsequent growth in length and girth. The cells that form from meristems undergo various developmental changes. The organization and union of similar or varied cells leads to formation of

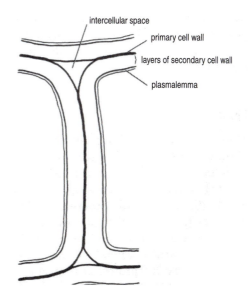

FIGURE 3-3. Cell wall sketch illustrating two cells with middle lamella and cell wall with the plasmamembrane.

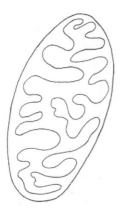

FIGURE 3-2. Sketch of mitrochondrion.

plant **tissues** and tissue systems organized into structurally and functionally cohesive units known as **organs**.

Tissue and Cell Types

Meristems

Apical meristems (Figure 3-4a) are located in main and lateral shoot and root tips and are responsible for elongation growth. Most plants grow in length from the apical region; this is why fences that are attached to tree trunks, remain at the same height and do not move higher as the tree grows.

Grasses, however, often have **intercalary meristems** (Figure 3-4b), meristematic tissue located between two non-meristematic regions near their leaf bases. These mono-cots therefore grow from below, so the mer-istem is not removed when lawns are mowed, the meristem is not removed as it would be in species in which elongation growth occurs at the apex.

Inside of the stem and roots there is often a **vascular cambium**, which is a **secondary** meristem that divides and is responsible for growth in girth, (see Figures 3-5 and 3-6). The vascular cambium gives rise to conduc-tive tissue, forming xylem to the inside and phloem to the outside. Growth in diameter of monocots is of a different type, due to cell expansion alone.

Epidermis

The epidermis initially covers all plant parts. On organs that increase in girth, however, the epidermis sloughs off and is replaced by bark. Located inside of the bark (periderm) of woody dicots and gymno-sperms is a **cork cambium** (phellogen) that divides and gives rise to the bark tissue (see Figure 3-5). The epidermis controls water loss and gaseous exchange, and offers protection from environmental stresses, such as heat, insects, diseases, and pollut-ants. Epidermal cells are packed closely together with few intercellular spaces.

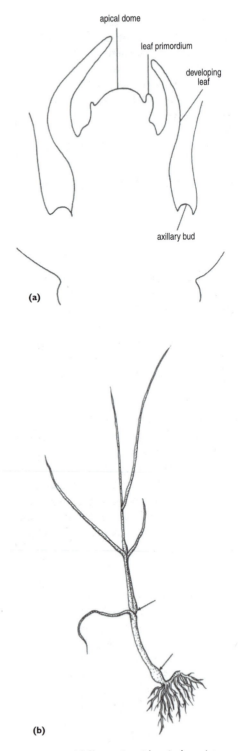

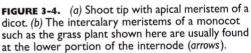

FIGURE 3-4. (a) Shoot tip with apical meristem of a dicot. (b) The intercalary meristems of a monocot such as the grass plant shown here are usually found at the lower portion of the internode (*arrows*).

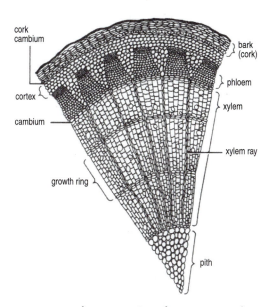

FIGURE 3-5. A cross section of a mature woody dicot stem, illustrating cork cambium.

Some epidermal cells are specialized, such as the **guard cells**, which have differentially thickened cell walls. They become turgid during daylight hours (during the dark for plants exhibiting crassulacean metabolism, see Chapter 5) and expand, thus opening tiny pores known as the **stomata** (shown later in Figure 3-37). Stomata tend to be primarily located on the undersides of leaves, but often are also present on upper leaf surfaces and stems. Stomata function in gaseous exchange (see Chapter 5) and control water loss (see Chapter 7). Epidermal cells can also have outgrowths known as **trichomes**, which commonly appear as hairs on leaves, stems, or roots. The epidermis is generally covered with a complex material containing cutin, waxes, and other substances, known collectively as **cuticle**. Amounts and types of cuticle will vary depending on species. A major function of the cuticle is the prevention of water loss from a plant.

Bark (periderm)

The bark, like the epidermis, serves as a protection against water loss and damage by external agents, (see Figures 3-7, 3-8, and 3-9). The bark consists primarily of cork cells; the corks for wine bottles and other uses are from the bark of the cork oak (*Quercus suber*). Cork cells lack protoplasts when they are mature and their walls contain a substance called **suberin**, which repels water.

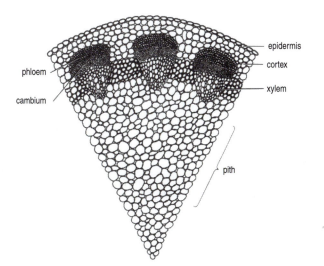

FIGURE 3-6. A cross section of a young woody dicot stem. Note the location of the vascular cambium.

(b)

(a)   (c)

FIGURE 3-7. Cork oak, *Quercus suber.* The thick bark is harvested for various uses. (*a*) The bark (cork) has been removed from the lower trunk of this tree. (*b*) Leaves, acorns (fruits), and a segment of bark. (*c*) The cork is cut into various shapes to be used as gaskets and stoppers for bottles.

Phloem

Phloem is a living vascular (conducting) tissue located throughout the plant and functions for support and the downward (basipetal) movement (or movement toward storage tissues) of the products of photosynthesis and other substances synthesized by the leaves. Phloem consists of sieve cells, sievetube members, associated parenchyma cells, companion cells, and fibers (Figure 3-10). The phloem fibers from flax, hemp, and jute have important commercial value and are used for such items as cloth and rope.

Xylem

Xylem tissue is continuous throughout the plant and, when mature, is nonliving. It conducts water and nutrient elements from the roots to the remainder of the plant. Xylem serves to provide support to a plant and also stores food. The woody portion of trees and shrubs consists of xylem tissue. Several types of xylem tissues are depicted in Figure 3-11.

Parenchyma

Parenchyma cells are the primary components of such tissues as the **cortex, pith,** and **leaf mesophyll**; they also occur in vascular tissues. These cells are generally living, but are more specialized than meristematic cells. Parenchyma cells serve various functions within the plant, including storage, wound healing, adventitious organogenesis,

FIGURE 3-8. Exfoliating (peeling) bark of river birch (*Betula nigra*).

Collenchyma

Collenchyma are living cells closely related to parenchyma cells, but are specialized in that they function to support young organs. The strings in celery petioles, for example, are strands of collenchyma cells that give strength and flexibility to this plant part. Collenchyma cells characteristically have unevenly thickened cell walls.

Sclerenchyma

Sclerenchyma cells have greatly thickened walls and can have diverse shapes and functions. They can be present in two general forms: **fibers** and **sclereids**. Long slender fibers add strength and support to many plants, especially woody plants. Sclereids are variable in shape (Figure 3-12). Many people have experienced such stone cells as the grit in pears or the pit of stone fruits such as cherries, peaches, and plums.

secretion, and excretion. Some contain chloroplasts and are therefore major sites of photosynthesis. Parenchyma cells are important horticulturally, because they form the greater part of leaves, flowers, and the edible portion of fruits and vegetables.

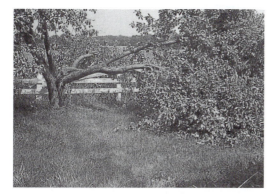

FIGURE 3-9. This apple tree (*right*) split because it had a narrow crotch angle, causing bark inclusions seen in close-up of damage (*above*) that made it vulnerable when laden with fruit.

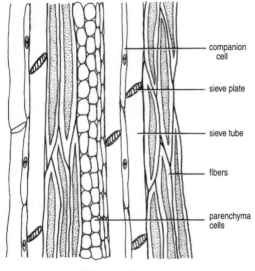

FIGURE 3-10. Phloem cells.

companion
cell

sieve plate

sieve tube

fibers

parenchyma
cells

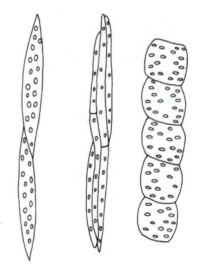

FIGURE 3-11. Types of xylem cells: tracheid, *left*; fibers, *center*; and vessel elements, *right*.

ORGANS

The complexity of horticultural plants is illustrated by the diversity of the various organs and their many modifications. Each organ has one or more functions and is composed of numerous tissues. Plants are divided into three primary organ types: roots, stems, and leaves. Flowers and fruits are believed to have evolved from stem and leaf tissues. Because these specialized reproductive structures have so much importance in horticulture, they will be discussed separately.

Roots

Roots are marvelous organs that are responsible for several characteristics essential to the growth and development of the plant. Uptake of water and nutrients from the soil and provision of anchorage and support are major functions of roots. Water and nutrients are slowly absorbed in the region of maximum cell elongation but more rapidly absorbed in the **root hair** zone, where the epidermis is specialized as an absorbing tissue and where often thousands of root hairs occur. Root hairs tremendously increase the root surface area, thus

enhancing rapid uptake of water and nutrients (Figures 3-13, 3-14, 3-15, 3-16, and 3-17). Root hairs are single-celled epidermal modifications that commonly function and survive for one or two days. It is important to recognize that very little absorption occurs in mature roots or root tips; maximum absorption takes place within the first 10 cm from the tip in rapidly elongating roots and decreases towards the base and root tip. However, because most plants have large numbers of young roots and millions of root hairs, a great amount of water and nutrients can be absorbed by an actively growing plant.

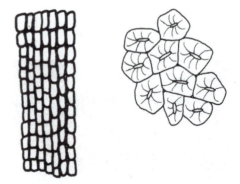

FIGURE 3-12. Sclerenchyma cells have highly thickened cell walls. The sclereids at right are typical of the stone cells found as grit in ripe pear fruits.

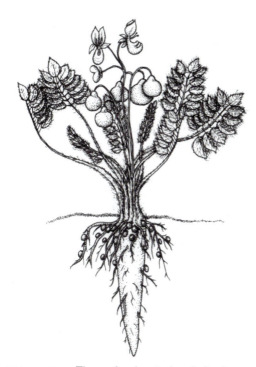

FIGURE 3-13. The perfect horticultural plant?

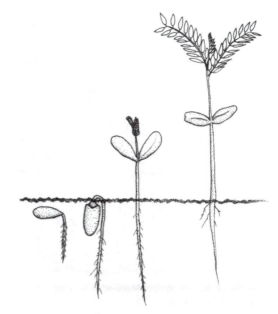

FIGURE 3-14. Germination of a honeylocust (*Gleditsia triacanthos*) seed.

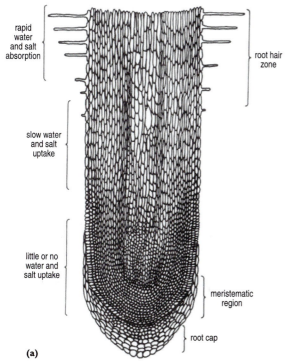

rapid
water
and salt
absorption

root hair
zone

slow water
and salt
uptake

little or no
water and
salt uptake

meristematic
region

root cap

(a)

(b)

FIGURE 3-15. (*a*) Diagram of a longitudinal section through a root tip. (*b*) Scanning electron micrograph of the root tip of *Hibiscus Rosa-sinensis*. Note root cap and numerous root hairs. Bar = 10 microns.

Photo courtesy of Humberto Puello, Southern Illinois University, Carbondale, IL.

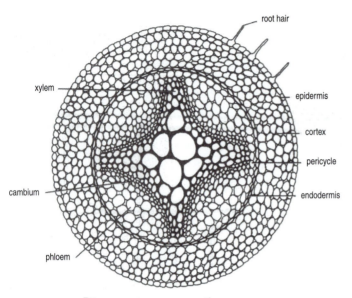

root hair

xylem

epidermis

cortex

pericycle

cambium

endodermis

phloem

FIGURE 3-16. Dicot root transverse section.

Plants vary significantly in their root system characteristics. Some plants have many fibrous multi-branched roots that may penetrate to only shallow depths, as in some grasses, lettuce, and petunia, while others such as corn, may explore great depths of soil. Indeed, corn roots have been excavated to depths of five meters (15 feet) and more. In contrast, other types of plants rely upon root systems that contain a dominant **tap root**, the primary root that develops from the original seedling root or radicle. Such roots penetrate to

FIGURE 3-17. Germinating radish seeds, showing root hairs (fuzzy) on emerging radicles.

varying depths: shallow, as in carrot and radish or deep, as in alfalfa, oak, and maple.

The depth and extension of a plant's root system depends largely on soil aeration. Oxygen supply to roots is governed by such factors as soil type, compaction, and the depth of the water table.

These diverse root morphologies govern anchorage characteristics of a given plant as well as its ability to absorb water and withstand drought. Details of water uptake, movement and utilization in the plant are discussed further in Chapter 7.

In addition to their primary functions of anchorage, support, and water and nutrient absorption, plant roots are often modified to perform other functions. Some roots are enlarged and serve as reservoirs for stored foods. Fleshy tap roots of carrot, parsnip, and beet store significant quantities of sugars and other carbohydrates. The enlarged fleshy or thickened root of the dahlia stores **inulin**, a polymer of fructose. The dahlia does not have a fleshy tap root, but because it is a swollen structure that developed from a fibrous root, it is referred to as a **tuberous root**. The sweet potato

FIGURE 3-18. Tuberous roots of the sweet potato, *Ipomoea batatas.*

(*Ipomoea batatas*)[1] also has similar fleshy tuberous roots which, depending on cultivar and type, store various amounts of sugars and starches (Figure 3-18).

It is interesting to note that, although usually referred to as "fleshy tap roots," such structures as the edible part of the radish, beet, turnip, and rutabaga (swede) are actually composed of both fleshy tap root and significant proportions of fleshy hypocotyl (seedling stem) that has been modified for storage of food (Figure 3-19). The radish sold in the United States, for example, is often more than 50% hypocotyl.

Another root type that has evolved in many plant species is termed **adventitious**. Adventitious roots arise from parts other than roots, such as stems and leaves, frequently as a result of injury. In fact, the majority of the root system of monocot plants is appropriately called adventitious, because the primary root from the seed is often short-lived and inconsequential. The

....
[1]In the United States, the sweet potato is often mistakenly called a yam. True yams are completely different plants that are found in the monocot family Dioscoreaceae.

"prop" or brace roots of corn, **contractile roots** of certain monocots such as *Lilium* and *Freesia*, roots forming at nodes on climbing vines (holdfasts) and the roots forming on stem and leaf cuttings used in propagation are all correctly referred to as adventitious roots (Figure 3-20). Roots formed from callus or plant parts in tissue culture, or *in vitro* micropropagation, are also adventitious.

Stems

A stem is the main axis of the plant body, typically bearing leaves and buds. Unlike roots, stems have **nodes**, places where leaves are (or have been) located, and **internodes**, the stem sections between pairs of adjacent nodes. **Buds** are undeveloped or compressed stems, flowers, or both, and are known as leaf buds (vegetative), flower buds, or **mixed** buds (containing both leaf and flower primordia). They often have protective coverings known as **bud scales** (modified leaves). A bud can be either **terminal**, positioned at the apex, or **lateral** (**axillary**), occurring in the **leaf axil**. The leaf axil is the angle formed by the junction of a leaf and stem at a node. Some plants grow tall and relatively unbranched, whereas others have bushy growth habits. The apical bud, to varying degrees, controls the growth of axillary buds by inhibiting their elongation. This is called **apical dominance** and explains why plants become more bushy when pinched back or pruned, because the inhibiting effect of the apical bud is removed by these procedures. Buds arising from leaves, roots or internodal stem locations are termed adventitious buds. Such adventitious bud formation may occur naturally and spontaneously or as a result of injury, and its frequency varies with species. During certain seasons, buds may be in an arrested state of development or **dormant**. This mechanism helps plants survive unfavorable growing conditions (for details on dormancy, see Chapter 6).

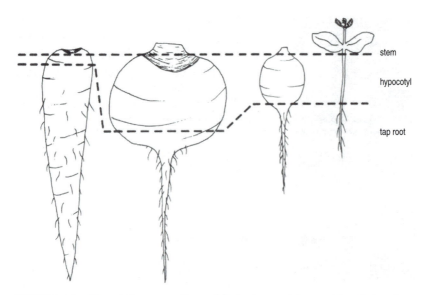

stem

hypocotyl

tap root

FIGURE 3-19. The relative proportions of fleshy tap root, hypocotyl, and stem in some common root vegetables. The figure at the left is mostly fleshy tap root, represented by carrot or parsnip; the next figure has a much greater proportion of fleshy hypocotyl, as in rutabaga and turnip; and the common radish is mostly hypocotyl tissue. A germinated seedling is included for comparison purposes.

Buds have various shapes, forms, and positions. Indeed, such morphological features may form the basis for identification of deciduous trees and shrubs during their dormant (winter) period. If only one bud is present at a node, the arrangement is said to be **alternate**, while two or more buds present at a node may be termed **opposite** (if two), or **whorled** (if three or more). Since these buds are found in leaf axils, the bud placement is essentially the same as the leaf arrangement. Such leaf arrangement (or bud arrangement) is referred to as **phyllotaxy** and is another useful tool in plant identification, because phyllotaxy tends to remain constant in a given taxon (Figure 3-21).

Stems of horticultural plants serve a wide array of purposes and functions. They bear the flowers and fruits and place the leaves in a position to intercept sunlight, which is crucial for the process of photosynthesis. Stems also provide a conduit, through the xylem, for necessary water and nutrients to reach the leaves, flowers, and fruits. Food and metabolites manufactured in leaves are translocated to flowers, fruits,

and roots through the phloem of the stems. The development of secondary xylem forms the **wood** of woody dicots and gymnosperms (see Figure 3-5). Stems also store food reserves, particularly carbohydrates, and indeed some stems are modified dramatically for such storage functions (Figure 3-22). An example of such a modified stem is the **tuber** of potato (*Solanum tuberosum*) or the Jerusalem artichoke (*Helianthus tuberosus*). A tuber is an enlarged or swollen underground stem, and the "eyes" of the potato are actually axillary buds occurring at nodes subtended by vestigial leaf scars (the "eyebrows") (Figure 3-23). The true tuber should not be confused with the tuberous root (e.g., sweet potato), because a tuber has true stem characteristics including nodes with axillary buds. The root does not have nodes or lateral buds. The tuber of the potato stores food as starch, a glucose polymer, and Jerusalem Artichoke stores inulin, a polymer of fructose. Inulin is considered to be more readily utilized than starch by people who have diabetes.

(a)

(b)

(c)

FIGURE 3-20. Examples of various adventitious root types. (*a*) Brace (prop) roots of corn. (*b*) Roots of poison ivy forming all along the stem, enabling it to cling to the bark of this tree. (*c*) Roots of Boston ivy modified to help climb a wall (hold-fasts).

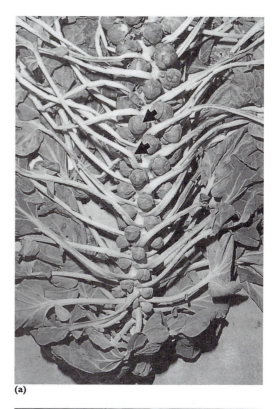

(a)

(b)

FIGURE 3-21. (*a*) Brussels sprouts plant showing spiral phyllotaxy of buds (sprouts). (*b*) Spiral leaf arrangement in *Yucca* sp.

A **corm** is a solid, compacted vertical stem formed at the base of plants such as gladiolus, freesia, and crocus (Figure 3-24). One can readily observe that the corm is true stem tissue, because it has easily discernible nodes and internodes (see Figure 3-22). **Cormels** are small cormlike structures that form laterally on corms.

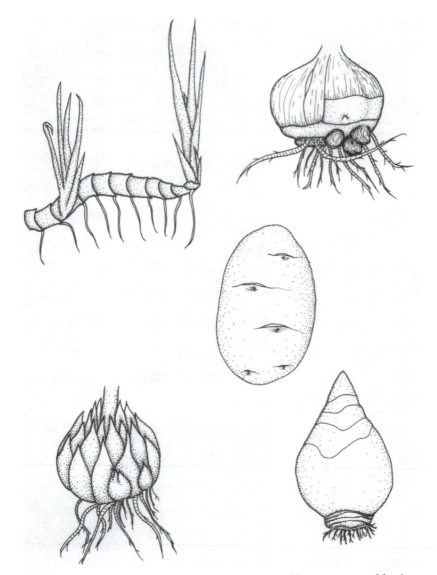

FIGURE 3-22. Stem modifications that enable storage of large amounts of food reserves. Clockwise from upper left, fleshy rhizome (e.g., *Iris* spp.); corm (e.g., *Gladiolus*, *freesia*); tuber (e.g., potato, *Solanum tuberosum*). tunicate bulb (e.g., *Narcissus* spp., onion); scaly bulb (e.g., *Lilium* spp.).

Corms and bulbs are only found in mono-cotyledonous plants.

A **bulb** is a structure resembling a large bud comprised of a short, thick stem (the basal plate) with basal roots and fleshy or membranous overlapping **leaf bases** (often called **scales**). The bulk of most bulbs is therefore primarily leaf tissue and the morphology and arrangement of the fleshy leaf bases help determine the bulb type. In **tunicate** bulbs such as onion or tulip, each fleshy leaf base completely encloses all parts of the bulb within it (Figure 3-25). Onion "rings" are therefore cross sections of fleshy leaf bases. Tunicate bulbs also have a papery covering called a tunic. The lily (*Lilium* spp.) is an example of a plant that produces a **scaly** bulb: that is, one composed of overlapping swollen leaf bases that do not completely encircle the interior part

FIGURE 3-23. Photo of tuberizing potato plant.

FIGURE 3-25. A comparison of the internal structure of flowering (hyacinth, *left*) and vegetative (onion, *right*) tunicate bulbs. Note that the bulk of the bulb structure is *leaf* tissue.

of the bulb and do not form complete circles in cross section. Scaly bulbs also are called nontunicate bulbs because they lack a papery tunic. Small bulbs that naturally form underground, or are induced to form, from intact bulbs or portions of bulbs are termed **bulblets**, whereas the small bulblike structures that form aerially in leaf axils (e.g., in certain *Lilium* species) are correctly referred to as **bulbils** (see Figure 3-26).

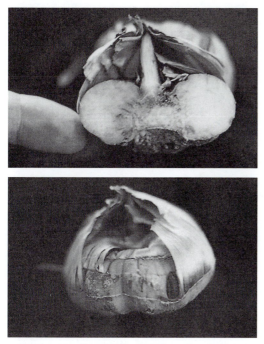

FIGURE 3-24. Longitudinal section of a corm (*above*) and a half-peeled corm (*below*).

FIGURE 3-26. (*Top*) scale of lily bulb with *bulblets*. (*Bottom*) lily stem with *bulbils* in the leaf axils. A few harvested bulbils are also shown.

(a)

(b)

(c)

(d)

FIGURE 3-27. Stems modified for spread and propagation. (*a*) Slender or wiry rhizome of quackgrass (*Agropyron repens*). (*b*) Stolon (runners) of strawberry, (*Fragaria* spp.) complete plantlet (*inset*) forms at a node. (*c*) *Chlorophytum* stolons. (*d*) Suckers (stems arising from root tissue) (*Rubus* spp.).

Another stem adapted for food storage is the fleshy **rhizome** of plants such as iris. A rhizome is a more or less subterranean and usually horizontal main stem that produces roots and shoots at the nodes. Some rhizome types are not thickened, as in quackgrass (*Agropyron repens*), but may grow through many feet of soil and reproduce prolifically by the plantlets that develop at the nodes. Indeed, when quackgrass rhizomes are cut into pieces and dispersed by cultivation equipment, the scattered pieces become new plants, thus making quackgrass a particularly difficult to control

(e)   (f)

FIGURE 3-27 (continued). (e) Watersprouts of apple arising near a severe pruning cut.
(f) Suckers coming from roots of understock near the base of an apple tree.

and troublesome weed (Figure 3-27a). The **stolon** is similar to the rhizome in that it is a horizontal stem that may produce roots and shoots at the nodes, but unlike a rhizome, a stolon arises from leaf axils and is above-ground. A **runner** is a specialized slender stolen. When a runner contacts the soil, generally at the tip, a new plant is formed that roots and produces its own runners from axillary buds. Strawberries (*Fragaria* spp.) and spider plants (*Chlorophytum* sp.) exhibit this creeping habit of spreading vegetatively by stolons (Figure 3-27b and 3-27c).

The **crown** has two distinct definitions: the horticulturists' definition and the foresters' definition. Horticulturists often use the term to mean the part of the stem near or at the soil surface. It may include the transition zone from root to shoot, or it may be composed of compacted stem tissue. Crowns can be divided during plant propagation, as in *Hosta*, rhubarb, and many bromeliads. When transplanting strawberries, the crown should be placed at the same depth as it was previously. If it is planted too deep, decay will ruin the planting, and if planted too shallow, the plants tend to dry out and die (Figure 3-28a). Foresters use the term crown to mean the top of a tree where the branches are located (Figure 3-28b).

A **spur** is a stem with very short internodes found on a mature woody plant and it is often modified for flower and fruit production, as is seen in apple, pear, and quince (Figure 3-29). Care must be taken when harvesting an apple crop because if the spurs are broken off, the next year's crop will be reduced. This is the reason that many fruit growers are hesitant to implement

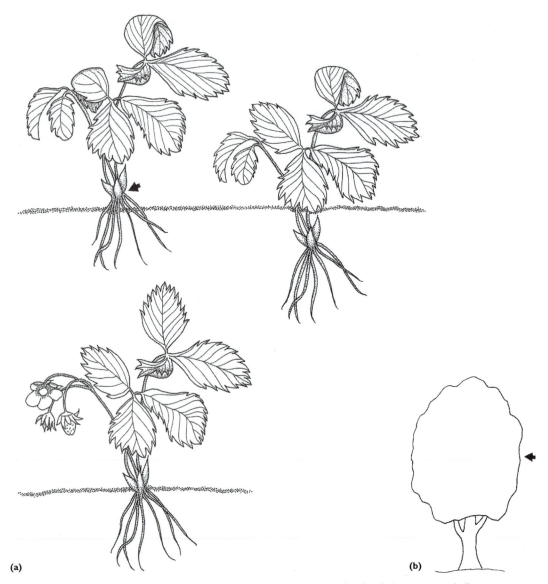

FIGURE 3-28. (*a*) "Crown" (*arrow*) of strawberry, illustrating proper depth of planting: too shallow, *top left*; too deep, *top right*; properly planted, *bottom*. (*b*) "Crown" of a tree (as used in forestry).

pick-your-own apple marketing operations. In spite of being modified in this way for fruit production, spurs can resume more vigorous vegetative growth, especially if dramatic environmental changes such as pruning take place.

An interesting stem type useful for propagation purposes (divisions) is the **sucker**, an adventitious shoot arising from *root* tissue. Suckers are particularly common, and valuable as a propagation device in red raspberry, blackberry, and sumac (*Rhus* spp.). A **watersprout**, however, is a shoot that arises from *stem* tissue, often from **latent buds** located deep within the tissues of branches or the trunk (not at nodes). Watersprouts generally grow in response to pruning—the greater the amount of stem material removed, the greater the number and growth of these new shoots. Watersprouts are so named because of their very succulent and rapid growth. Although watersprouts are sometimes incorrectly called suckers, there is a specific definition for each shoot type, depending on their point of origin. Watersprouts are frequently utilized as the source of wood for plant propagation via grafting and budding for crops such as apple.

Some stems possess adaptations that facilitate climbing walls, trees, or other supports. These may take the form of stems that merely twist or twine around the adjacent support, as in morning glories (*Ipomoea* spp.), or may take the form of stem or leaf modifications called **tendrils**. Grapes, cucumbers and peas are examples of plants that have tendrils. Such features greatly influence the methods utilized in culture of these and similar crops; grapes and morning glories are grown on trellises, for example, and peas are often trained on fences (Figure 3-29).

FIGURE 3-29. Other stem modifications. Clockwise from *top left*: a dormant spur of apple, a spur with developing fruits, tendrils of grape, spine of *Robinia*, prickle (*Rosa*), and twining stems of morning glory.

FIGURE 3-30. Thorns on juvenile *Citrus*. Note that it is a true stem modification, forming in the axil of a leaf.

Plants of some species are said to be armed; that is, they have stem modifications such as **thorns, spines,** or **prickles.** Prickles are usually small, more or less hardened, pointed outgrowths of the epidermis or bark, while spines are sharp, hard modified leaves. Thorns are modified stems. Spines and thorns may reach several centimeters in length, as they do in the spines of some cacti and thorns of some hawthornes (*Crataegus* sp.) (Figure 3-30).

Other stems are highly evolved for photosynthetic activity, having inconsequential or nonexistent leaves. Such stems are referred to as **cladophylls** or **cladodes,** and are typically found on plants such as asparagus and *Ruscus*.

Leaves

Leaves are flattened or expanded appendages of the stem of vascular plants where photosynthesis typically takes place. Leaves of many Angiosperms are composed of a **petiole** (leaf-stalk), which bears the flattened expanded portion or leaf **blade,** sometimes called the **lamina.** The petiole may have flared, flattened, or scalelike outgrowths known as **stipules** (Figure 3-31), which may appear more or less leaflike. In the leafless pea cultivars such as 'Novella,' much of the photosynthesis takes place in the stipules. If the leaf blade lacks a petiole, but is attached directly to the stem, it is said to be **sessile.** Leaves normally occur singly (**alternate** leaf arrangement), in pairs (**opposite** leaf arrangement), or in groups of three or more (**whorled** leaf arrangement), and they subtend buds (Figure 3-32). The position of this bud helps distinguish between a **simple** leaf and a **compound** leaf. Simple leaves have **entire** leaf margins—that is, the blade is not divided into leaflets—whereas compound leaves are composed of two or more leaflets, or blade portions, which often appear leaflike but do not have a bud at their bases (see Figure 3-33).

FIGURE 3-31. Stipules of sycamore (*Platanus* sp.), *left* and alder (*Alnus* sp.) *right*.

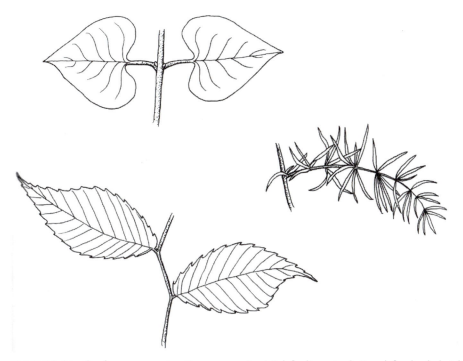

FIGURE 3-32. Leaf arrangement patterns: opposite, *top left*; alternate, *bottom left*; whorled, *right*.

Simple leaves (and leaflets) have one of three kinds of **venation,** or pattern of the veins, in the blade; **parallel** venation, as in monocot leaves (Figure 3-34); **palmate** venation, where the main veins arise from a common point of origin (as in the **palm** of a human hand); and **pinnate** venation (literally, "feather-form"), with the veins placed laterally on either side of a main vein, which usually appears to be simply a continuation of the petiole (Figure 3-33). Compound leaves may be either **palmately compound** or **pinnately compound,** with the leaflets arranged in a fashion similar to the main veins in a palmately veined, or pinnately veined, simple leaf, respectively. A compound leaf composed of three leaflets is said to be **trifoliolate** (Figure 3-35) **not trifoliate** (literally, trifoliate means three-leaved, as in *Trillium* spp.). The veins of a leaf are a continuation of the vascular system of the stem, connected through the petiole.

The petiole may develop an adaptation at its base called the **abscission zone,** which is where the leaf **abscises,** or breaks from the stem (Figure 3-36). Leaf-fall occurs as a result of a build-up of materials suberized (converted into corky tissue) in the abscission zone; in deciduous plants suberization is stimulated by the lengthening of the night period as autumn approaches. This suberized layer heals and protects what would otherwise be a wound open to water loss and invasion by pathogenic organisms. Abscission can also be brought about by various stresses, such as drought, waterlogging, root damage, excess heat, damage to the leaf by pests (see Chapter 16), and other agents. Application of ethylene-producing chemicals can also artificially induce abscission. **Evergreen** plants keep all or part of their leaves through one or more winters before abscission occurs. The length of time an evergreen keeps its leaves varies with species, with some types dropping all or part of their leaves at the beginning of the second growing season and others keeping them for four or more years.

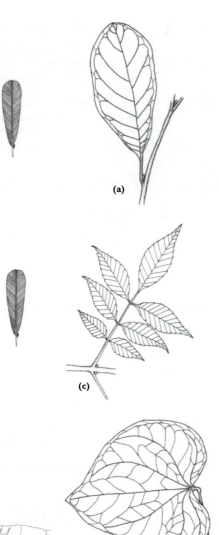

(a)

(b)

(c)

(d)

(e)

(f)

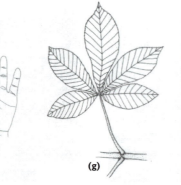

(g)

FIGURE 3-33. Some leaf shapes, margins and venations, showing progression from the simplest to most complex (or advanced) forms. (*a*) Simple, entire, pinnate venation. (*b*) Simple, lobed, pinnate venation. (*c*) Pinnately compound. (*d*) Twice pinnately compound. (*e*) Simple, entire, palmate venation. (*f*) Simple, lobed, palmate venation. (*g*) Palmately compound. (*Insets*: the term pinnate comes from the pinnae of a feather and palmate from the shape of a hand.)

FIGURE 3-34. Some examples of parallel venation found in different types of monocot species: blade and sheath of corn, *Zea Mays*, *left*; leaf of orchid, *middle*; and compound leaf of bamboo, *right*.

The anatomy of a leaf is illustrated in Figure 3-37. The **upper epidermis** and **lower epidermis** enclose the **palisade** cells and the **spongy mesophyll**, together with the veins (vascular system) that supply them with water, mineral nutrients, and metabolites. A protective noncellular layer of a waxy substance called **cutin** covers the epidermis of most leaves and this layer is termed **cuticle**. The palisade cells appear to be tightly packed, whereas the spongy mesophyll is loosely arranged, but both layers contain intercellular spaces, which are connected to the external atmosphere by pores in the epidermis called the **stomata** (the singular form is **stomate**). Each stomate is enclosed by two highly modified epidermal cells called **guard cells**. These guard cells pull apart when they are turgid and collapse toward one another when flaccid, thus opening and

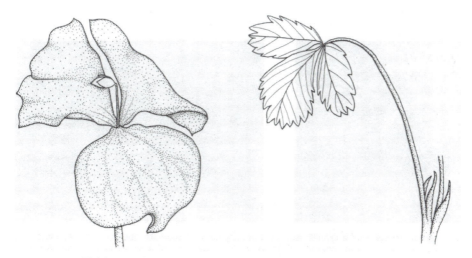

FIGURE 3-35. Trifoliate leaf arrangement (3 *leaves* arising at one point on the stem, as in *Trillium*), *left*; and a trifoliolate leaf (compound leaf with 3 *leaflets*, as in clovers and strawberry), *right*.

closing the stomate. Gaseous exchange takes place through such open stomata, thus providing carbon dioxide for photosynthesis and facilitating **transpiration**. Transpiration is the evaporation of water from internal leaf cell surfaces and the subsequent loss of water, primarily through the stomata. Water is to a lesser extent also lost directly through the leaf's surface ("cuticular transpiration") or through specialized structures composed of one or more permanently open pores called **hydathodes**. This latter process, called **guttation**, is thought to be a result of root pressure and usually occurs at night during periods of high humidity, when the stomata are closed (Figure 3-38).

A vast array of leaf shapes and modifications exist in addition to the flat expanded

FIGURE 3-36. The abscission zone commonly found in leaf petioles of deciduous plants.

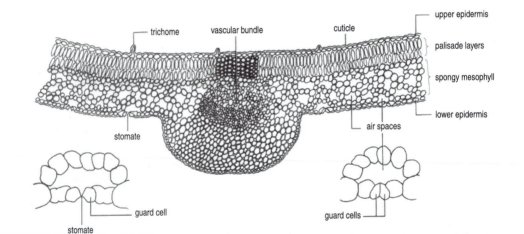

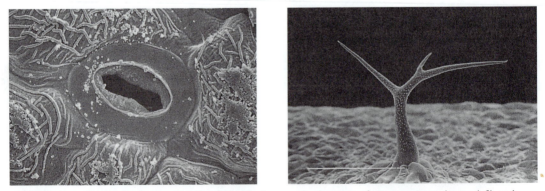

FIGURE 3-37. The internal structure of a typical leaf, top and schematic of stomate open (*center, left*) and closed (*center, right*). Scanning electronmicrograph (SEM) of an open stomate of *Arabidopsis thaliana; bottom left,* SEM of a trichome of *Arabidopsis thaliana, bottom right.* Trichome magnified approximately 200 ×, stomate guard cell exterior width magnification is approximately 10 μm and magnified about 6000 ×.

Photos courtesy of Peter Frederick, Southern Illinois University, Carbondale, IL.

FIGURE 3-38. Leaves of corn, illustrating *guttation*.

FIGURE 3-40. Longitudinal section of the head of cabbage. Note that the core is a greatly compressed stem.

type just described. Leaves of coniferous species are frequently awl-shaped or needle-shaped and are referred to as **needles** (Figure 3-39). A thick cuticle on such persistent (evergreen) leaves may prevent water loss in winter. Other leaves may be cylindrical, as in the green leaves of onion, or modified for food storage, as in cabbage and brussels sprouts (Figure 3-40), and the fleshy leaf bases already described for tunicate bulbs. The fleshy petioles of celery and rhubarb are popularly eaten and the **bracts**

that enclose the immature inflorescence (bud) of the globe artichoke (*Cynara scolymus*) are considered a delicacy by many vegetable connoisseurs (Figure 3-41). A bract is a leafy structure that usually subtends a flower, and in the case of many ornamentals such as poinsettia (*Euphorbia pulcherrima*) and dogwood (*Cornus* sp.), may be large and colorful.

Phase Change and Juvenility

Another morphological difference often observed in leaves is related to phase change—that is, the transition from **juvenility** to **adulthood**. Many experts define a juvenile plant as a plant in its vegetative stage that is incapable of responding to the flower induction stimuli that would normally lead to flowering in an adult plant. Thus apple trees may not bear fruit for several years following planting. Such juvenile plants often exhibit morphological characteristics that are distinctly different from those of the adult form of that species. The juvenile form of English ivy (*Hedera helix*) is the classic example. It has lobed leaves and prostrate stems that readily produce roots adventitiously. Its adult form, however, which has entire leaf margins and erect growth habit, may flower and produce

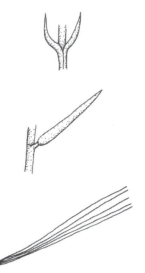

FIGURE 3-39. Examples of leaf types typical of coniferous species: awl-shaped (*Juniperus*) *top*; solitary needles, (*Picea, Abies*) *middle*; and needles in fascicles (*Pinus, Larix*), *bottom*.

fruits but has no propensity to produce adventitious roots (Figure 3-42). Other morphological features found in the juvenile forms of some species include thorns in *Citrus*, entire leaves in *Monstera deliciosa* (leaves are split in the adult form, hence the common name, "split-leaf philodendron"), and leaf retention in the fall by juvenile oak (*Quercus* spp.) and beech (*Fagus* spp.). The juvenile period has different durations in many woody species and can extend for many years after germination (Table 3-1).

FIGURE 3-41. Globe artichoke, *top left* and *right.* Note that the immature inflorescence, or head, *left,* is the edible stage. When in bloom, *right,* the head is past edible maturity, but may be used in dried flower arrangements. The poinsettia "flower," *bottom left* and *right,* is really colorful bracts, enclosing inconspicuous flowers called *cyathia* (close-up photo).

FIGURE 3-42. Juvenile English ivy, *left*, and the mature form, *right*. Note the lobed leaves and adventitious roots of the prostrate, juvenile form and the entire leaf margins and erect stem character of the mature form.

Fall Color

Autumnal coloration of foliage is brought about concommitently with chlorophyll loss and leaf aging, or **senescence**. It occurs prior to leaf fall and is characterized by the bright yellow **xanthophyll** and orange **carotene** pigments (which are generally present in the plastids but are no longer masked by chlorophyll) and the reds of some of the vacuolar pigments, including **anthocyanins**. The anthocyanins intensify in the fall, especially under conditions of cool nights and short bright days (see Chapter 5). It

should be noted that the process of leaf abscission—which leads to leaf aging and chlorophyll loss—is triggered by short days.

FLOWERS

Although varying vastly from species to species, a "typical" flower (Figure 3-43) normally is composed of **sepals, petals, stamens,** and **pistils**. These four floral organ types are arranged in whorls and are attached to a stem tip referred to as the **receptacle**. The

TABLE 3-1. Duration of juvenile period in forest trees

SPECIES	YEARS
Pinus sylvestris (Scots pine)	5–10
Larix decidua (European larch)	10–15
Pseudotsuga Menziesii (Douglas fir)	15–20
Picea abies (Norway spruce)	20–25
Abies alba (Silver fir)	25–30
Betula pubescens (Birch)	5–10
Fraxinus excelsior (European ash)	15–20
Quercus rober (English oak)	25–30
Fagus sylvatica (European beech)	30–40

After Wareing, P. F. and I. D. J. Phillips. 1981. *Growth and Differentiation in Plants*. 3rd ed. Pergamon Press, Oxford. Used with permission.

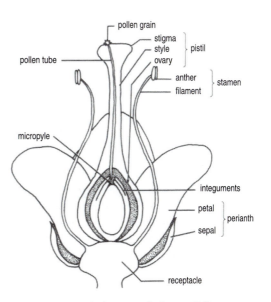

FIGURE 3-43. A diagram of a "typical" flower.

numbers of sepals, petals, stamens, and **car-pels** (leaflike structures that comprise the body of the ovaries) are relatively constant for a species. The number is typically five (or multiples of five) for dicots and three (or multiples of three) in monocots, although other numbers may occur (e.g., four sepals and four petals in the Cruciferae/Brassicaceae). Sepals tend to appear leaflike and enclose the bud, and they may be fused at their bases. Collectively, the sepals are called the **calyx**. The petals, collectively the **corolla**, are usually showy, white, or colored, and are attractive to insects and other animals that could be potential pollinating agents. **Nectaries** may also occur, usually at the base of the petals, and they normally secrete a sweet liquid known as nectar, which also tends to attract pollinators. Sometimes the petals are fused into a tubular, funnel-like or belllike shape, as in tobacco, petunia and *Campanula*, respectively. When sepals and petals are showy and colorful but indistinguishable from one another, such as in Liliaceae members tulip and lily, they are called **tepals**, rather than sepals or petals (Figure 3-44). The calyx and corolla are collectively termed the **perianth** and when fused together (often with the stamen and ovary bases also), they are called the **hypanthium**. An example of the ontogeny of floral organs is presented in Figure 3-45.

FIGURE 3-44. A flower of a hybrid lily, *Lilium* sp. Note that the three sepals and petals appear similar, they are therefore referred to as *tepals*.

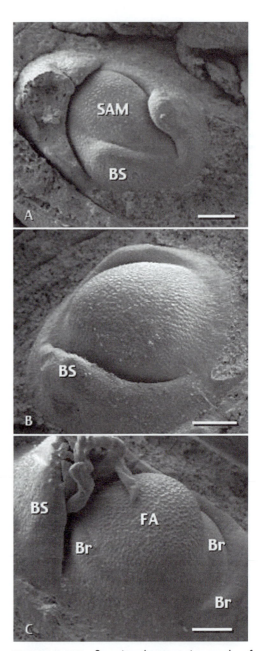

FIGURE 3-45/I. Scanning electron micrographs of almond bud apices showing stages of development through floral initiation. (*a*) Stage 0. The apex terminates in a shoot apical meristem that produces vegetative organs, the bud scales. (*b*) Stage I. This micrograph shows a shoot apical meristem in transition from the vegetative to reproductive condition. Note the increase in shoot apex diameter and elevation of the apex relative to the lateral organ primordia. (*c*) Stage 2. Transition to the reproductive state is complete. The floral apex and three bracts are shown. Br = bract; BS = bud scale; FA = floral apex; SAM = shoot apical meristem. Scale bars = 50 μm.

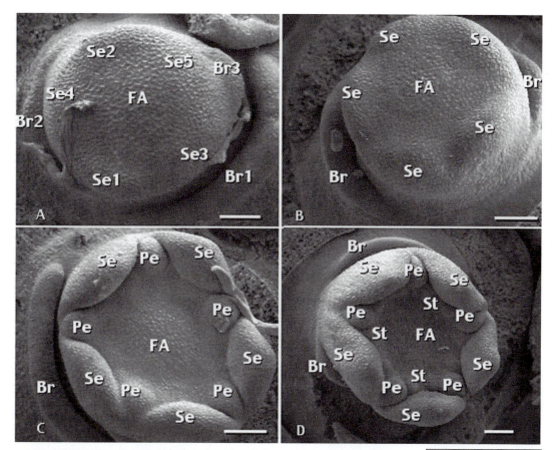

FIGURE 3-45/II. Scanning electron micrographs of almond bud apices showing initiation of floral organs through stamen initiation. (*a*) Stage 3. Bract and sepal initiation at the floral apex follows a spiral phyllotaxy. The three bract primordia are numbered from the oldest to youngest (Br1–Br3). The sepal primordia are labeled similarly. The site of initiation of the final sepal primordium is indicated (Se5) but the primordium had not yet formed when this bud was dissected. (*b*) Stage 3. The floral apex, subtended by bracts, with initiation of the five sepal primordia complete. (*c*) Stage 4. Petal primordia form alternate to the sepal primordia. (*d*) Stage 5. The stamen primordia form acropetal to the petal primordia. Note the fused bases of the sepal, petal, and stamen primordia that form the hypanthium. Br = bract; FA = floral apex; Pe = petal; Se = sepal; St = stamen. Scale bars = 50 µm.

Notes on figures:
All scale bars are 50um.
SAM: Shoot Apical Meristem
FA: Floral Apex
Br: Bract
Se: Sepal
Pe: Petal
St: Stamens
C: Carpel
Arrowheads: Carpel Margins

FIGURE 3-45 CONTINUED ON PAGE 52

The stamens are the male organs of the flower and consist of two parts, the **pollen**-bearing part called the **anther** and the usually slender stalk upon which the anther is born, called the **filament**. The pollen is shed by the splitting, or **dehiscence**, of the anther (**anthesis**). The function and development of the pollen are discussed later in Chapter 4. Stamens of many plants may appear as highly modified structures that are petal-like in appearance called **staminodes**. Flowers with numerous staminodes are the "double" flowers of many important horticultural species such as cultivated roses. Indeed, some botanists consider the normal petals of many flowers to have evolved from stamens and therefore to be staminodes.

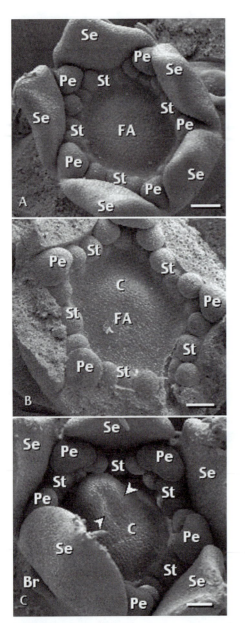

FIGURE 3-45/III. Scanning electron micrographs of almond bud apices showing initiation of the carpel. (*a*) Stage 6. Stamen primordia are present and the floral apex has not yet begun to differentiate a carpel primordium. (*b*) Stage 7. A terminal carpel primordium emerges at the periphery of the floral apex. Sepals have been removed in this sample. (*c*) Stage 7. As the terminal carpel primordium differentiates, it consumes the entire floral apex. Arrowheads indicate the carpel margins, which will fuse to enclose the carpellary locule as differentiation progresses beyond this stage. Br = bract; C = carpel; FA = floral apex; Pe = petal; Se = sepal; St = stamen. Scale bars = 50 μm. Images for Figures 3-45I, II & III courtesy of Vito Polito, Univ. Calif.-Davis; J. Amer. Soc. Hort. Sci. 126:689–696, B. M. Lamp, et al.

The pistil is the female part of the flower and is comprised of one or more carpels, which are modified leaves bearing **ovules** along each edge. If the pistil has a single carpel, it is called a **simple pistil** and the carpel is folded lengthwise so that the ovules are protected and enclosed. In ovaries with more than one carpel (a **compound pistil**), the carpels are fused along adjacent edges, thus forming a tubular structure, somewhat analogous to tubular forms of a corolla or calyx. Whether simple or compound, a pistil is normally differentiated into three parts: the basal portion, or **ovary**, contains the ovules; the **stigma** is the upper portion and receives the pollen; and the **style** is the usually slender part of the pistil that connects the stigma to the ovary. In compound pistils, the stigma may be split or lobed into as many parts as there are carpels—three for lily, for example (Figure 3-46). The ovary chambers are referred to as **locules**. The placement of the ovules in the locules, or on the interior of the carpels, is termed **placentation** and is a useful tool in plant identification. (Figure 3-47).

If all four kinds of floral organs are present, the flower is said to be **complete**; if one or more floral organ types is missing, the flower is **incomplete**. The stamens and pistil are considered the **essential parts** and if both are present, the flower is considered to be a **perfect** flower; if either is absent, the flower is termed **imperfect**. Perfect flowers are also called **hermaphroditic** (from the Greek mythological figure, Hermaphroditus, who was the son of Hermes and Aphrodite and was united when bathing into a single body with the nymph Salmacis). They are typical of flowers found in many angiosperm families.

If the stamens are absent, and only pistils present, the flower is said to be **pistillate**, and it is **staminate** if stamens are present and pistils absent. A plant with both staminate and pistillate flowers is referred to as **monoecious** (literally "one house" meaning both sexes in one house); examples are oak, corn (maize), begonia, and squash.

(a)

(b)

FIGURE 3-46. (*a*) Lily ovary cross-section, illustrating the three locules that compose the ovary. Note the axile placentation. (*b*) Three-lobed stigma of a lily pistil (*arrow*).

Conversely, a **dioecious** species is one in which staminate flowers and pistillate flowers occur only on separate plants; examples are asparagus, holly (*Ilex*), *Ginkgo*, and papaya (*Carica papaya*). If a plant has both perfect and staminate flowers, as in some Cucurbitaceae, it is said to be **andromonoecious** and if only pistillate flowers exist, the plant is **gynoecious**. This latter character (*gynoecy* or all-femaleness) has been actively sought in cucumber breeding programs, resulting in numerous commercial gynoecious pickling cucumber cultivars that set their fruit in a concentrated fashion, thus facilitating once-over

machine harvest of desirably sized fruits. It is necessary, however, to include seeds of some stamen-bearing (andromonoecious) cultivars when planting a field of gynoecious cucumbers to provide a pollen source for adequate fruit set.

Position of the ovary is another useful parameter in classification and identification of plants. The types of ovary position are illustrated in Figure 3-48, and include **hypogynous** (hypo = below) wherein the origin of the bases of the sepals, petals and stamens is below, and distinct from, the ovary base (**superior** ovary). A modification of this condition is also illustrated; the

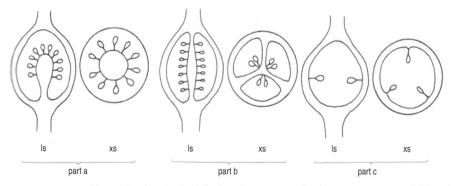

ls   xs   ls   xs   ls   xs

part a   part b   part c

FIGURE 3-47. Alternating longitudinal (l.s.) and transverse (x.s.) representations of (*a*) radial, (*b*) axile and (*c*) parietal types of placentation.

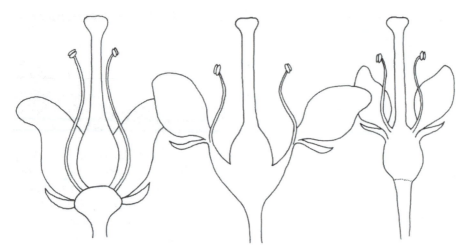

FIGURE 3-48. Types of ovary positions: hypogynous flower of lily, *left*; perigynous flower of cherry, *center*; and epigynous flower of muskmelon, *right*. Hypogynous and perigynous flowers are superior ovary types and the ovaries of epigynous flowers are called inferior.

perigynous (peri = around or encircling) flower is still an example of a superior ovary, but the fused bases of sepals, petals, and stamens (hypanthium) surround the ovary, without fusing with the ovary as they do in the **epigynous** flower (epi = above). In an epigynous flower, the hypanthium is fused to the ovary walls so that the sepals, petals, and stamens appear to arise from the top of the ovary, hence the term **inferior** ovary. Lily and raspberry flowers are hypognyous; cherry flowers are perigynous; and flowers of the members of the Cucurbitaceae are epigynous.

The placement or arrangement of flowers on a stem can also vary greatly from species to species. Flowers may be borne singly (**solitary** flowers) or may be grouped in various clusters called **inflorescences** (Figures 3-49, 3-50, 3-51, and 3-52). Solitary flowers and inflorescences are borne on stems called **peduncles**, the tip of which is the receptacle. The smaller stem-like structures bearing the individual flowers in an inflorescence are called **pedicels**. Each flower in an inflorescence is usually subtended by a bract. Some specialized inflorescences are typical of certain

taxa or groups, such as the umbel for the Umbelliferae (Apiaceae) and the spikelet found on the spikes, racemes, and panicles of the Graminae (Poaceae) (Figure 3-53). The highly membranous tough modified leaves known as **glumes, lemma,** and **palea** serve as the outer floral organs in such grass flowers. Like many other species of wind pollinated plants, these flowers tend to be inconspicuous and less showy.

There are two general types of inflorescences based on their pattern of floral opening: the **determinate**, in which the apical bud opens first, followed sequentially by those below and working downward (e.g., **cymes** of strawberry) or **indeterminate**, typical of many other common inflorescence types including **panicles** and **racemes**. Because botanists tend to think of fusion of floral organs and/or reduction in numbers as evolutionary advancements, species in the Compositae (Asteraceae) that have a **head** (**capitulum**) are considered more advanced than those species in which panicles are the norm. **Regular** flowers, radially symmetrical, are thought to be less advanced than irregular flowers, which are bilaterally symmetrical (Figure 3-54).

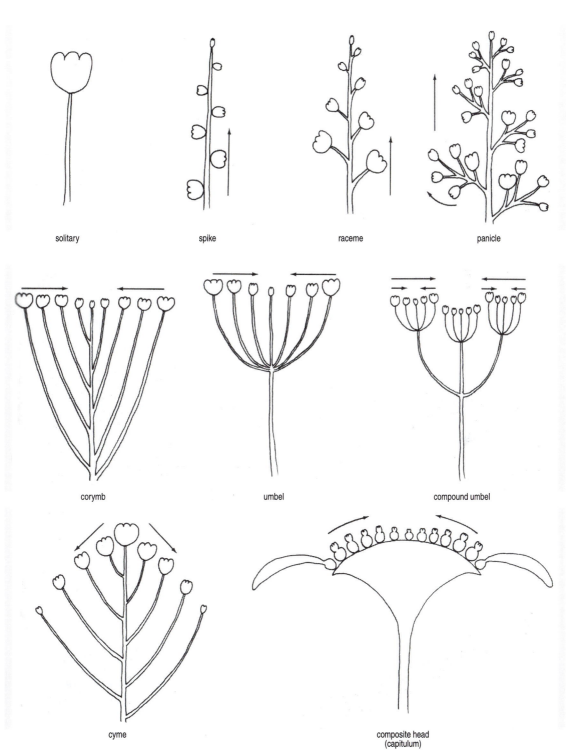

solitary   spike   raceme   panicle

corymb   umbel   compound umbel

cyme   composite head
(capitulum)

FIGURE 3-49. Some common inflorescence types. Flowers are represented schematically, with larger "flowers" indicating open flowers, and smaller structures unopened buds. Note that the first flower to open in the cyme is the apical, as in the "king flower" of strawberries, whereas the most proximal flowers open first and the apical last in the corymb. Arrows represent order of opening.

(a)

(b)

(c)

(d)

(e)

FIGURE 3-50. Commonly observed inflorescences. Umbels: (*a*) carrot (*Daucus carota*). (*b*) *Allium giganteum*. Heads: (*c*) *Zinnia elegans*. (*d*) Cockscomb (*Celosia argentea* var. *cristata*). (*e*) Tassel of sweet corn, a panicle.

FIGURE 3-51. Broccoli (*Brassica oleracea* var. *italica*), an edible inflorescence.

FIGURE 3-52. Interesting tubular-shaped flower of a cactus (*Mammillaria* sp.).

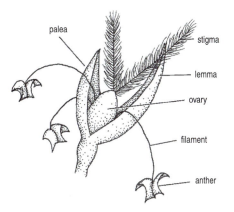

FIGURE 3-53. Structure of flowers representative of the grass family.

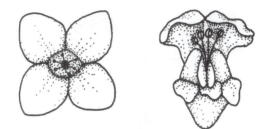

FIGURE 3-54. Regular flower, *left*, illustrating radial symmetry, and an irregular flower, *right*, which is asymmetrical.

FRUITS

The general public, horticulturists, and botanists often have different definitions of a fruit. A simple botanical definition asserts that a fruit is a ripened ovary and any associated parts. A horticultural, or usage, definition of a fruit does not include most vegetables that are correctly termed fruits botanically, but only those fruits that usually are consumed fresh or processed as a dessert food, (e.g., apples, strawberries, citrus). In this section, we will use the botanical definition for clarity and simplicity. In angiosperms, a fruit normally is formed following fertilization of the ovules and contains seeds resulting from such fertilization.

Fruits that form without fertilization are called **parthenocarpic** (**parthenos** = virgin and **karpos** = fruit). Examples are banana and navel orange (Figure 3-55). In some normally fertilized fruits such as apple and pear, it is known that hormones are produced by the developing seeds and that such hormones aid in fruit development (described further in Chapter 11). When the fruit ripens, the ovary wall is referred to as the **pericarp**. The pericarp usually consists of three parts, the outer layer, or **exocarp**; the middle layer, **mesocarp**; and the inner layer, the **endocarp**. The relative development of these layers is often different and distinctive for various fruits (Figure 3-56).

FIGURE 3-55. Parthenocarpic fruit (immature) of navel orange.

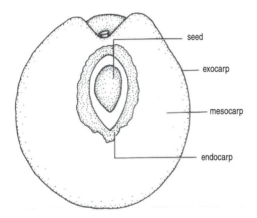

seed

exocarp

mesocarp

endocarp

FIGURE 3-56. Pericarp of a stone fruit (peach, *Prunus persica*). Note the three distinct layers—exocarp, mesocarp, and endocarp.

❖ TERMINOLOGY

MULTIPLE FRUITS. A multiple fruit is a fruit formed by the development of several flowers, which fuse during ripening. Usually such fruits have a common receptacle (e.g., the "core" of mulberry). Examples: pineapple (*Ananas comosus*), fig (*Ficus* spp.), mulberry (*Morus* spp.), and the "seed" of beet (*Beta vulgaris*) (Figure 3-57a).

AGGREGATE FRUITS. An aggregate fruit is one formed by the development of several ovaries produced by one flower. Examples: raspberry and blackberry (*Rubus* spp.). Note that in blackberry the receptacular tissue is part of the edible fruit at maturity, whereas

in raspberry the receptacle remains on the stem when the fruit is picked (Figure 3-57b).

SIMPLE FRUITS. A simple fruit is one that is formed by the development of a single pistil or ovary. Simple fruits are further sub-divided (Figures 3-57 and 3-58):

Fleshy Fruits. In fleshy fruits, the pericarp is usually soft, succulent, or fleshy when the fruit matures.

Berry. If the pericarp is fleshy throughout, the fruit is called a **berry**. Examples of berries are tomato, blueberry, eggplant, and cranberry. Two specialized types of berries also exist. The **pepo** has a thick, hard exocarp or rind at maturity. Examples of pepo fruits are most members of the Cucurbitaceae, such as squash, muskmelon, and watermelon. A **hesperidium** is a type of berry that has a leathery exocarp and mesocarp with a very juicy endocarp that has distinct segments. Examples include grapefruit, orange, and lime.

Drupe. A **drupe** is a fleshy, usually one-seeded fruit in which the seed is enclosed in a stony endocarp composed of masses of sclereids (the "pit").The exocarp is thin and the mesocarp is usually fleshy, constituting the edible portion of various species of *Prunus* (cherry, plum, and peach) and olive (*Olea europea*). In the case of almond (*Prunus dulcis*), the edible portion is the seed that is enclosed in the stony endocarp and the exocarp and mesocarp are not considered edible.

Pome. A pome is a fruit produced by a compound inferior ovary containing many seeds and which is composed primarily of fleshy perianth bases, exocarp, and mesocarp. The endocarp is often cartilaginous, encloses the seeds, and is called the "core." Apple, pear, and quince (*Cydonia oblonga*) are examples. In some classifications, pomes are referred to as **accessory** fruits.

Dry Fruits. The pericarp tissue of dry fruits is not moist and often is hard or brittle when the fruit is ripe.

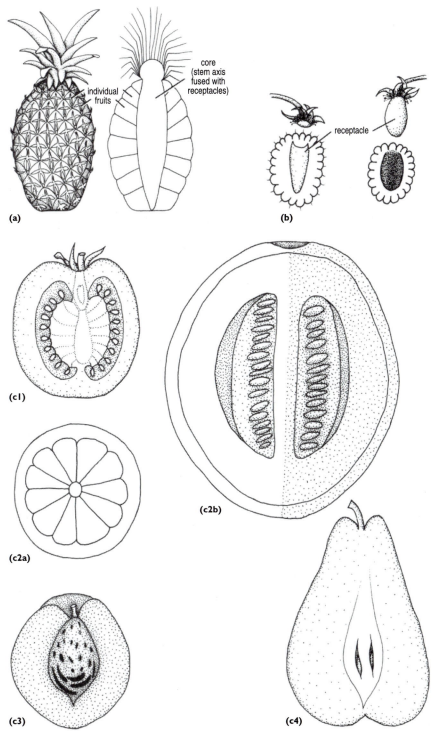

FIGURE 3-57. Fruit types. (*a*) Multiple: pineapple *left* and *right*. (*b*) Aggregate: blackberry (retains fleshy receptacle when harvested, *left*), and raspberry (does not include receptacle when harvested, *right*). (*c*) Simple fleshy fruits. (*c1*) Berry: tomato. (*c2*) Specialized types of berries: (*c2a*) Hesperidium: lime. (*c2b*) Pepo: muskmelon. (*c3*) Drupe: peach. (*c4*) Pome: pear.

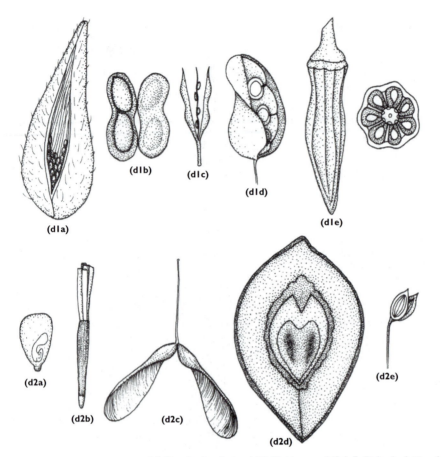

(d1b)

(d1c)

(d1d)

(d1e)

(d1a)

(d2a)

(d2b)

(d2c)

(d2d)

(d2e)

FIGURE 3-57. (continued) (*d*) Simple dry fruits: (*d1*) Dehiscent. (*d1a*) Follicle: *Asclepias*. (*d1b*) Legume: peanut. (*d1c*) Silique: radish. (*d1d*) Silicle: *Lunaria*. (*d1e*) Capsule: okra. (*d2*) Indehiscent. (*d2a*) Caryopsis: corn. (*d2b*) Achene: marigold. (*d2c*) Samara: maple. (*d2d*) Nut: pecan. (*d2e*) Schizocarp: carrot.

Dehiscent. The pericarp splits or **dehisces** along definite lines or sutures at maturity.

Follicle. A follicle is composed of a single carpel and dehisces along one suture at maturity. Examples: milkweed (*Asclepias* spp.), and *Delphinium*.

Legume. A legume is found only in the legume family (*Fabaceae/Leguminosae*) and is composed of a single carpel that dehisces along two sutures at maturity. Examples: bean (*Phaseolus* spp.), pea (*Pisum sativum*), and honeylocust (*Gleditsia triacanthos*).

Silique. A silique is composed of two carpels, which dehisce along two sutures and the fruit is divided lengthwise by a wall-like structure called a **replum**. A silique is usually longer than it is broad. Examples: Radish (*Rhaphanus sativus*), and mustard (*Brassica* spp.). A **silicle** is a modified silique that is usually as broad as it is long. Examples: sweet alyssum (*Lobularia maritima*) and honesty (*Lunaria annua*). Siliques and silicles are found only in the mustard family (Brassicaceae/Cruciferae).

Capsule. Capsule is a multi-carpelled fruit that dehisces along more than two sutures at maturity. Examples: poppy (*Papaver* spp.), okra (*Abelmoschus esculentus*), and azalea (*Rhododendron* spp.).

Indehiscent. The pericarp does not split or open at maturity; fruits usually have one or two seeds.

(a)

(b)

(c)

(d)

(e)

FIGURE 3-58. (*a*) Fruit of pear, a typical pome. (*b*) 'Gold Rush' squash. (*c*) 'Sunburst', squash. Both *b* and *c* are pepo-type fruits. Note the inferior ovary on flower of 'Sunburst'. (*d*) Okra 'Blondie', a capsule that is edible in the immature stage seen here. (*e*) 'How Sweet It Is (white)' sweet corn. Each kernel is a caryopsis.

Parts b through e, courtesy All-America Selections, Willowbrook, IL.

Caryopsis. A one-seeded fruit in which the pericarp and seed coat are fused. Examples: Kentucky bluegrass (*Poa pratensis*) and corn (*Zea Mays*).

Achene. One-seeded fruit in which the pericarp may be fairly easily separated from the seed coat; pericarp not fused to the seed coat. Examples: sunflower (*Helianthus annuus*), *Zinnia elegans*, *Dahlia* sp., and the "seed" found in strawberry fruits.

Samara. The samara is similar to an achene, but it has a winglike, somewhat membranous outgrowth of the pericarp. Examples: maple (*Acer* spp.), ash (*Fraxinus* spp.), and elm (*Ulmus* spp.).

Nut. This one-seeded fruit is somewhat like an achene, but possesses an extremely hardened pericarp. Examples: pecan (*Carya illinoensis*), walnut (*Juglans* spp.), and oak (*Quercus* spp.).

Schizocarp. A schizocarp is a compound fruit, usually composed of two single-seeded achene-like **mericarps** (half-schizocarps) that break apart readily at maturity. Examples: carrot (*Daucus carota*), parsley (*Petroselinum crispum*), dill (*Anethum graveolens*), and *Eryngium* spp.

It is useful for the horticulturist to understand fruit types and their characteristics. In some cases, the fruit itself is the valuable commodity, and in others the fruit is important because it bears the seeds.

Seeds

A seed develops as a result of the fertilization process described in detail in Chapter 4. It is composed of a developed embryo (the structure that will become the plant), enveloped by a protective **seed coat** or *testa*, which was formed by development (and usually hardening) of the outer ovule layers, or **integuments**. It generally contains stored food reserves, either in the endosperm or the cotyledons. If the stored food is primarily in the endosperm, the seed is an **albuminous** seed. If the food has been digested from the endosperm during the development of the seed and subsequently stored in the cotyledons, it is referred to as an **exalbuminous** seed.

In angiosperms, seeds develop inside the developing ovary, or fruit. However, in gymnosperms, the seeds are termed "naked seeds"; seed development takes place on the surface of a fertile leaf (the "scale" of cones of coniferous species) rather than in a developing ovary (fruit). A distinct taxonomic separation into subclasses of the angiosperms is based in part on cotyledon number; that is, plants having seeds with one cotyledon are monocots and those with two cotyledons are dicots. (Figure 3-59).

Two basic patterns of seed germination occur with respect to cotyledon position following germination. In **epigeous** germination, the hypocotyl elongates to bring the cotyledons above the soil line, but in **hypogeous** germination the hypocotyl does not elongate to any extent, so the cotyledons remain below the soil surface (Figure 3-60, 3-61).

Occasionally, seeds are formed without fertilization. Such seeds are termed **apomictic seeds** and are, therefore, asexual and a vegetative method of propagation. Such seeds may develop from cells of the nucellus or integuments (**adventitious apomixis**), directly from the ovule before meiosis takes place (**recurrent apomixis**) or

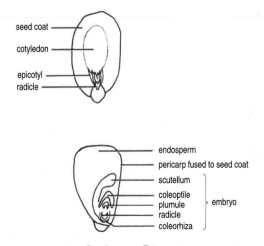

seed coat
cotyledon
epicotyl
radicle

endosperm
pericarp fused to seed coat
scutellum
coleoptile
plumule   } embryo
radicle
coleorhiza

FIGURE 3-59. Seed types. Dicot, *top*; monocot, *bottom*.

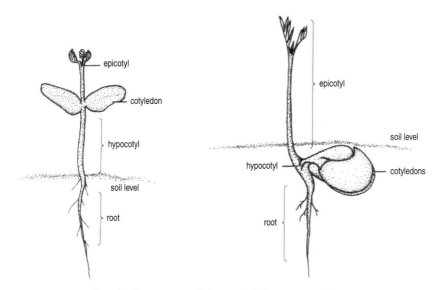

FIGURE 3-60. Germination patterns. Epigeous, *left*; hypogeous, *right*.

FIGURE 3-61. Germination of *Pinus*. Excavated seedlings at two stages of development, *left*. A germinated seedling in the nursery, *right*. (Note multiple cotyledons).

from the ovule after meiosis occurs (**non-recurrent apomixis**). Whether apomictic or normal sexual seeds are considered, it is readily apparent that seeds are an important propagation method.

It is clear that a knowledge of plant structure and function will enable the modern horticulturist to skillfully manipulate and control horticultural crops for the benefit of all people. Selected references are listed for readers interested in gaining greater knowledge in these important areas of study.

REFERENCES

Bailey, L. H., E. Z. Bailey, and the Staff of the L.H. Bailey Hortorium. 1976. *Hortus Third: A Concise Dictionary of Plants Cultivated in the United States and Canada.* MacMillan, New York.

Bailey, L. H. 1949. *Manual of Cultivated Plants Most Commonly Grown in the Continental United States and Canada*, (rev. ed.). MacMillan, New York.

Bidwell, R. G. S. 1979. *Plant Physiology.* (2nd ed.). MacMillan, New York.

Cronquist, A. 1982. *Basic Botany.* Harper & Row, New York.

Devlin, R. M. and F. H. Witham. 1983. *Plant Physiology.* (4th ed.). PWS, Boston.

Esau, K. 1977. *Anatomy of Seed Plants.* (2nd ed.). Wiley, New York.

Heyward, H. E. 1938. *The Structure of Economic Plants.* MacMillan, New York.

Lawrence, G. H. M. 1951. *Taxonomy of Vascular Plants.* MacMillan, New York.

Leyser, O. and S. Day. 2003. *Mechanisms in Plant Development.* Blackwell, Oxford, UK.

Mauseth, J. D. 1988. *Plant Anatomy.* Benjamin/ Cummings, Menlo Park, CA.

Norstog, K. and R. W. Long. 1976. *Plant Biology.* Saunders, Philadelphia.

Ray, P. M., T. A. Steeves and S. A. Fultz. 1983. *Botany.* Saunders, Philadelphia.

Salisbury, F. B. and C. W. Ross. 1978. *Plant Physiology.* Wadsworth, Belmont, California.

Walton, P. D. 1988. *Principles and Practices of Plant Science.* Prentice Hall, Englewood Cliffs, N.J.

Wilson, C. L. and W. E. Loomis. 1967. *Botany.* (4th ed.). Holt, Rinehart and Winston, New York.

CHAPTER 4

PLANT GENOTYPE

"AND HE GAVE IT FOR HIS OPINION, THAT WHOEVER COULD MAKE TWO EARS OF CORN OR TWO BLADES OF GRASS TO GROW UPON A SPOT WHERE ONLY ONE GREW BEFORE, WOULD DESERVE BETTER OF MANKIND, AND DO MORE ESSENTIAL SERVICE THAN A WHOLE RACE OF POLITICIANS PUT TOGETHER."

Johnathan Swift

For many horticultural crops, some of the most exciting advances have been made in the area of breeding, molecular biology, and selection. Most new cultivars that are introduced are a result of traditional breeding and frequently are improvements over previously available plants because they offer one or more of the following advantages: higher yields, earlier harvest, continuous fruit setting, greater vigor, altered plant growth habits, improved pest resistance, increased stress tolerance, improved flower colors, changed flower shape and size, improved foliar characteristics, increased nutritional value, better flavor, better fragrance, increased longevity, and novel plant types. Plant geneticists and breeders are continuing to improve these characteristics. Using newer biotechnological approaches of mutant selection and genetic transformation, scientists are developing plants with new and unique traits. Many of these new advances have resulted in plants with increased shelf life, and resistance to pests, viruses, and pesticides. Because genes can be moved from other species, there are now some plants available that are more nutritionally complete, such as yellow rice or have flowers with colors that previously have never been available, such as in Gerbera and Petunia. Another interesting use of gene transfer technology is what is known as molecular farming (sometimes called pharming) where plants are genetically engineered to produce useful substances, such as antibiotics, anticancer drugs, and other pharmaceuticals.

GENETIC IMPROVEMENTS

Genetic improvements resulting in greater yields have allowed for the steady increases that have occurred in the world population, while the number of hectares planted to horticultural crops has remained largely unchanged. High yields of fruit and vegetable crops combined with efficient farming techniques help feed an ever-growing world population with highly nutritious food.

Plants with increased vigor generally are more competitive for light, nutrients, and water. Thus with a rapid canopy closure they often can quickly shade out weeds. Vigorous plants establish themselves more rapidly than those with less vigor and quickly move out of the early stages of growth, where plants are often more sensitive to environmental stresses.

Plants with altered growth habits have facilitated the mechanical harvest of many crops. For example, bush-type vine crops, such as green bean can aid hand and mechanical harvests. The bush habit also has helped home gardeners conserve precious space.

Plants with genetically controlled pest and disease resistance can be an essential feature of integrated pest management programs. Not only does genetic resistance reduce reliance on pesticides, it also increases horticultural efficiency and allows producers to grow high quality plants on land where pests and diseases once precluded economical production.

Much research has been done on the development of plants that are tolerant to environmental stresses. For example, researchers have genetically engineered tomato plants that can grow in salty soils and still produce a crop. This is important because millions of hectares of land have been lost from increased salinity caused by irrigation. This could open up this land and previously idle land with high salt levels for the production of food, and may allow for irrigation with brackish water. Current research also includes the development of plants with tolerances to heat and cold, mineral element toxicities and deficiencies, drought stress, waterlogging, and air pollution.

Better and more flower colors are available in many perennial and annual garden flower species and most are a result of traditional breeding. Petunias with yellow flowers and large-flowered white marigolds now are available; both in the not-too-distant past were considered dreams. Recently, carnations, which lack blue pigments, have been genetically modified by inserting the blue pigment gene from petunia. The result is new carnation cultivars with flowers ranging from lavender to purple. In fact, the first genetically modified cut flower was released in 1997 and was the mauve 'Moondust' carnation. Likewise, changes in flower shape and size have provided the horticulturist with more double-flowered plants, such as petunias and roses, and the extremely small and large flowered cultivars found in marigolds. Regular consultation with seed catalogs, online information, and conducting small trials with new introductions are excellent ways to remain up-to-date in this area.

Breeding and biotechnology have also been used to modify the color of other parts of plants than flowers. For example, coleus plants are available with a wide variety of leaf colors and color patterns (**variegation**) and cabbages with red or green leaves add interest to salads. There are carrots on the market that are yellow, orange, red, or even purple. In fact, purple fleshed and skinned potatoes have become popular among many gourmet cooks.

Through breeding efforts turfgrasses have been developed with much better colors than those available in the past, and their growth habits are more desirable. These new cultivars are able to grow well either in full sun or in the shade, and they are better competitors because the canopy fills in rapidly and shades out the weeds. Some of the new turfgrasses, such as buffalograss are quite drought tolerant.

Breeding also has given us more nutritious and better-tasting foods. Tomatoes and watermelons are being bred to have higher levels of antioxidants that are important to health. Beans are being bred to accumulate more iron in their seeds because so many diets throughout the world lack sufficient amounts of this important micronutrient. Higher sugar content now is incorporated into many muskmelon cultivars. A 1% increase in the sugar content makes a big difference in flavor. Another example is that new apple cultivars are now available that are more flavorful than some of the currently popular cultivars.

Plant breeding and selection efforts have resulted in enhanced shelf-life with little loss of quality. Sugar enhanced sweet corn types will last longer on the grocer's shelf and still maintain good flavor and quality. New tomato cultivars have been developed with a longer shelf-life than many popular cultivars. When they receive the proper care, poinsettias now will keep into the spring and still retain their color.

By manipulating the phenomenon called **allelopathy**, breeders hope to develop plants that produce substance(s) that inhibit growth of other plants growing around them. In other words, some plants give off natural herbicides, most of which are phenolic compounds. A classic problem for horticulturists is the substances produced by walnut trees (thought by some to be a compound called juglone). One should not plant valuable ornamentals or vegetables under a walnut tree, since they will grow poorly and often die in a short period of time. Breeding work has been conducted to develop crops such as cucumbers and turfgrasses that will give off natural herbicides and thus control some species of weeds. This, if successful,

could reduce growers' dependence on herbicides and reduce labor required for weed control by organic farmers. Some researchers are studying in influence of planting cover crops that produce allelopathic chemicals. The cover crops generally are killed with herbicides and the desired crop seeds are sown directly into the soil through the cover crops. Any natural herbicides produced by the cover crops may then control weeds around the horticulturally important plants. Extensive trials are being conducted to identify cover crops that are not allelopathic to the main crop.

GENETIC TERMINOLOGY

To gain an understanding of how breeders and genetic engineers have developed such a wide variety of superior plants, a student must begin by learning about substances, structures, and the terminology of plant genetics. This section provides a brief review.

Plants are marvelously complex organisms that have the ability to reproduce themselves. They transmit information to their offspring, which leads to the tendency for offspring to resemble each other, their parents, and/or previous generations (**heredity**). The **deoxyribonucleic acid (DNA)** in the nucleus functions in the passing of heritable information from generation to generation. Much of the DNA in plants is nuclear, however, there also are extranuclear sources of DNA in plant cells, such as that present in plastids and mitochondria.

Plant cells contain another nucleic acid, **ribonucleic acid (RNA)**. RNA is found in the nucleus, ribosomes, plastids, mitochondria, and elsewhere in the cytoplasm. RNA is involved directly with the synthesis of proteins. DNA and RNA are similar because each is composed of four repeating subunits called **nucleotides**. Each nucleotide consists of three units: a five-carbon sugar, phosphoric acid, and an organic base that contains nitrogen. The two nucleic acid types differ in the structure of the sugar and do not contain all of the same bases. In ribonucleic acid (RNA) the sugar is **ribose** and in **deoxyribo**nucleic acid (DNA) the sugar has one oxygen less than ribose and therefore is called **deoxyribose** (Figure 4-1). DNA contains the bases adenine, cytosine, guanine, and thymine, whereas RNA has the same bases with uracil instead of thymine. Adenine and guanine are **purines** that structurally contain two rings made of carbon and nitrogen. Cytosine, thymine, and uracil are **pyrimidines** and are single ring structures (Figure 4-1).

The nucleotides chemically bond with each other via sugar-phosphate linkages and form long chains. RNA is usually a single-stranded helical structure. In DNA, each pyrimidine base in one chain bonds with a complementary purine base in another, i.e., guanine bonds with cytosine and adenine bonds with thymine. In order to form a stable molecule, the two chains must be oriented side by side in opposite directions. The two polynucleotide chains coil on each other in a spiral pattern, forming the double helix structure of DNA (Figure 4-2).

Purine bases

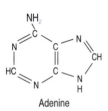
Adenine

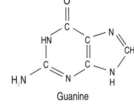
Guanine

Pyrimidine bases

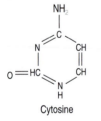
Cytosine

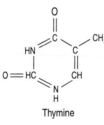
Thymine

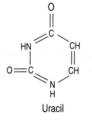
Uracil

Five-carbon (pentose) surgars

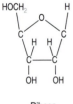
Ribose

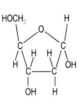
Deoxyribose

Phosphoric acid

FIGURE 4-1. The structure of the compounds that constitute DNA and RNA.

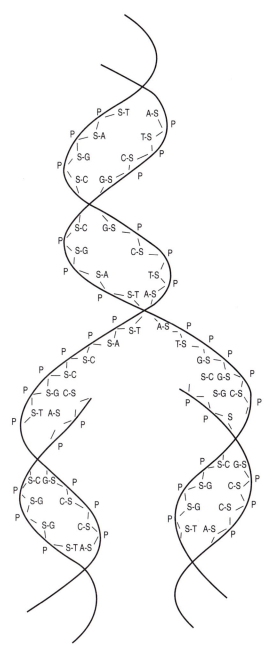

FIGURE 4-2. Replication of a DNA molecule. P = phosphate, S = sugar, A = adenine, C = cytosine, G = guanine, and T = thymine.

The order of the nucleotides is specific along a DNA molecule and forms a heritable code known as the **genetic code.** DNA is replicated by a complex of enzymes that first uncoil DNA, then a portion of the

molecule is split into single strands, and ultimately complementary deoxyribose nucleotides from the cell nucleotide pool are joined to each of the open areas of the single strands to eventually form two identical daughter DNA molecules (Figure 4-2). Each parental strand therefore serves as a **template** for subsequent chains of nucleotides. It is by this means that heritable genetic information is passed from parent to daughter cells.

The biosynthesis of RNA occurs in a manner similar to DNA replication (Figure 4-3). DNA directs RNA synthesis (**transcription**) by first uncoiling, then complementary ribose nucleotides are linked together and form the RNA polymer. Transcription is catalyzed by the enzyme RNA polymerase. There are three different forms of RNA: **messenger RNA (mRNA), ribosomal RNA (rRNA),** and **transfer RNA (tRNA).**

The three forms of RNA function together in the biosynthesis of proteins (**translation**). Proteins have many functions within a plant including catalyzing chemical reactions (**enzymes**), directing transport of substances across membranes, and serving as structural components of cells and tissues. Proteins are long chains of amino acids. Messenger RNA carries the code that directs the order in which the amino acids are placed. This code is dependent on the order of nucleotides in the mRNA. Groups of three nucleotides, called **codons,** each code for an amino acid. There are 64 different codons possible, which more than adequately account for the 20 distinct amino acids found in plants. Therefore, more than one codon may code for an amino acid; for example, four triplets— Guanine-Guanine-Guanine, Guanine-Guanine-Adenine, Guanine-Guanine-Cytosine, and Guanine-Guanine-Uracil—code for the amino acid glycine. Three of the 64 possible codons are called "nonsense codons" because they do not code for any amino acids; however, they are critical as stop signals in the translation process.

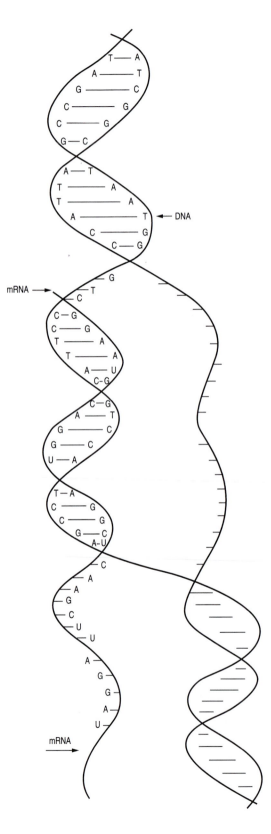

All three forms of RNA function in the assembling of amino acids into **polypeptides** (Figure 4-4). Polypeptides are chains of amino acids, and all proteins are polypeptides. Ribosomes contain three to five rRNA molecules and function in translation by "reading" the code on the mRNA, like a recorded tape. Each tRNA molecule conveys an amino acid to the site of protein synthesis and transfers it to the growing polypeptide. Attached to one end of a tRNA molecule is an amino acid, and on the opposite end is a nucleotide triplet known as an **anticodon**. Anticodons are complementary to, and recognize, specific codons. The codon sequence on an mRNA molecule, therefore, directs the order in which the tRNA molecules are bound and thereby governs the order in which the amino acids are transferred to the polypeptide. After the amino acid is added to the polypeptide, the tRNA is released and binds with another amino acid. Each tRNA molecule is specific for the transfer of a particular amino acid.

It is clear that genetic information flows from DNA molecule to DNA molecule during reproduction and from DNA to RNA to proteins. This is known as the **central dogma of molecular biology** and is the focus of much important, modern research in molecular genetics and **proteomics** (discovering and characterizing proteins).

Each living plant cell requires DNA and RNA transcription and for the passing of heritable information to daughter cells. **Chromatin** is a filamentous complex consisting primarily of DNA and proteins. **Chromosomes** are distinct granular bodies composed of chromatin that bear hereditary factors and are located in the plant cell's nucleus. **Genes** are specific sequences of nucleotide-pairs on a DNA molecule that code for a given polypeptide. Therefore, genes are the individual units of inheritance in the chromosome. The **genome** is one complete set of chromosomes, often represented by the letter x. Depending

FIGURE 4-3. Transcription (biosynthesis) of RNA that is directed by DNA.

of the species, this may be from one to more than 500 chromosomes. Generally one genome is represented by 3 to 50 chromosomes. The number of chromosomes in a genome is constant within a species, but varies among species, e.g., for potato (*Solanum tuberosum*), it is 12 chromosomes, and for banana, the number is 11. A cell with one set of chromosomes ($1x$) is called **monoploid** (from the Greek *monos* meaning single, and -*ploos* meaning -fold), so 12 is the monoploid number for potato and 11 for banana.

Vegetative cells of many plants contain two matching chromosome sets and are therefore said to be **diploid** ($2x$ or $2n$). In diploid cells, there are pairs of matching, or **homologous, chromosomes**. Each chromosome in a homologous pair contains genes for the same traits, and the genes occupy the same position in each chromosome. The gene pairs may be identical or somewhat different, depending on heritable changes that may have occurred in the past. The **locus** (plural, **loci**) is the location of a gene on a chromosome. **Alleles** are pairs or forms of a gene located at the same locus in homologous chromosomes.

Cells containing more than two complete sets of chromosomes are said to be **polyploid**. The commercial banana has three sets of chromosomes and is therefore **triploid** ($3x$), whereas most potato cultivars have four sets of chromosomes and are therefore **tetraploid** ($4x$). Higher levels of polyploidy also occur in the plant kingdom; dahlia and strawberry, for example, are $8x$. Polyploid plants are often more vigorous and productive than diploids (Figure 4-5) and are frequently preferred by horticulturists because their leaves and petals are often longer, thicker, and more intensively colored than those of diploids. Because some types of polyploid plants have poor fertility, most crops that are grown from seeds are diploids. For example, vegetatively propagated vegetables tend to be polyploid—Jerusalem artichoke (also known as sunchoke) is hexaploid (six sets of chromosomes), and potato is tetraploid—but seed-propagated vegetables—beet, carrot, lettuce, pea, and spinach—are all diploid.

There are two overlapping stages of cell division in plants called mitosis and cytokinesis. **Mitosis** is the process in which chromosomes replicate themselves as the nucleus divides; the result is two daughter cells that have the same characteristics of heredity as each other and as the parent cell. Mitotic cell divisions occur primarily in meristematic regions of the plant. The mitotic cycle has

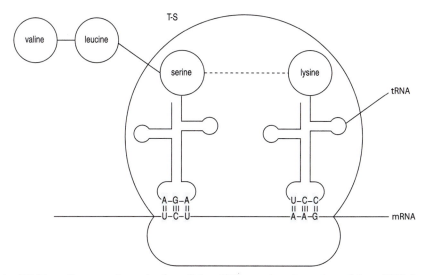

FIGURE 4-4. Within a ribosome the anticodon of the tRNA matches the codon of the mRNA; hence, the mRNA directs the order in which amino acids are placed in the process of polypeptide or protein translation.

FIGURE 4-5. The flower of a tetraploid daylily (*right*) is about double the size of the diploid flower (*left*).

organelles within the cell are replicated, chromosomes are duplicated, and spindle fibers are produced. During prophase, the chromosomes become thickened and condensed. Each identical half of the chromosome is known as a **chromatid**, and these halves are joined at one point, called the **centromere** (also known as the **kinetochore**). As the spindle fibers form around the chromosomes, the chromosomes move toward the midline or equator of the cell and the centromeres line up in the equatorial plane at metaphase. Full metaphase is when all of the chromosomes have moved to this center line, which puts them in a position to separate, each into single strands. It is during anaphase that the chromatids separate from each other as the centromere divides. This creates two daughter chromosomes that move away from each other toward the opposite poles of the

several phases: **interphase**, **prophase**, **metaphase**, **anaphase**, and **telophase** (Figure 4-6). Interphase is the nondividing stage between cell divisions. During this stage,

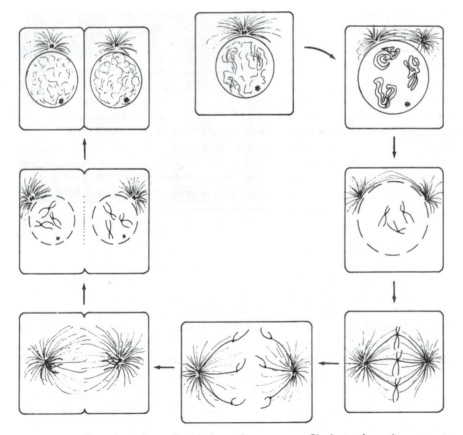

FIGURE 4-6. The mitosis in a cell with three chromosomes. Clockwise from the center: interphase, early prophase, prophase, metaphase, early anaphase, anaphase, telophase, and interphase.

cell as the spindle fibers that are connected to the centromeres shorten. This results in two identical sets of chromosomes at opposite ends of the cell. During telophase, nuclear membranes form around each of the two sets of chromosomes, creating "nuclear envelopes." The chromosomes elongate again into slender strands and the spindle fibers disappear. At this time, the two daughter nucleii enter interphase.

To complete the process of cell division, **cytokinesis** must occur. This is the formation of the cell plate, which divides the cell, approximately in the same area as the chromosomes were during metaphase. When the cell plate completes its growth, subdividing the cell, the two daughter cells deposit a cell wall on either side of the cell plate, thus the cell plate becomes the middle lamella (see Chapter 3).

Meiosis is another form of nuclear division in which the normal $2n$ chromosome number is reduced to $1n$ (**haploid**, $1n$ represents the chromosome number in the gametes). Meiosis ultimately leads to the development of **gametes**, or sex cells that have the capability of fusing together to form new plants. The **ovule** of a flower contains the female gamete that becomes a seed upon fertilization and ripening. The development of the female gamete occurs within an ovule; a diploid cell undergoes meiosis, and four large haploid cells, or megaspores, result. Only one of these four megaspores develops further; it undergoes three mitotic divisions and results in the formation of the **embryo sac**, which contains eight haploid nuclei. Six of these nuclei segregate into membrane-bound cells and one of the six becomes the **egg** (Figure 4-7). The two remaining nuclei become the polar nuclei. The male part of the flower, the anther, consists of four chambers in which numerous diploid cells undergo meiotic division, each yielding four haploid cells, or **microspores**. Within each microspore, each haploid nucleus divides mitotically, producing a cell containing both a vegetative and a generative nucleus. The cell wall subsequently thickens and the resulting structure is called a **pollen** grain.

The vegetative nucleus, or **tube nucleus**, is responsible for pollen tube growth and the **generative nucleus** divides again to form two **sperm nuclei**, usually during the growth of the pollen tube (Figure 4-8).

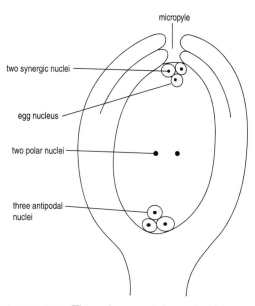

FIGURE 4-7. The embryo sac is located within an ovule as drawn.

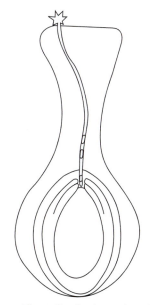

FIGURE 4-8. The pollen lands on the stigma and germinates and the pollen tube grows down through the style to the ovule, where fertilization will occur.

FERTILIZATION

Following the transfer of pollen from the anther to the stigma (pollination), the pollen grain germinates on the stigma, and its pollen tube grows down through the style and penetrates the ovule, usually through the micropyle (Figure 4-8). However, this can vary with the species. The pollen tube then enters one of the synergid cells and releases the two sperm nuclei. Both the pollen tube and synergid cell die and degenerate, whereas the sperm nuclei migrate independently within the embryo sac. One sperm nucleus unites with the egg nucleus to form a 2*n* zygote that eventually develops into the **embryo** of the seed. The other sperm nucleus fuses with the two polar nuclei, forming a 3*n* nucleus. The 3*n* nucleus divides mitotically and gives rise to the **endosperm**. Because there are two nuclear fusions, this process is known as **double fertilization** and is typical of the angiosperms.

With the exception of endosperm cells and gametes, the living cells of many plants are 2*n*. One set of chromosomes is contributed by the male parent and the other set by the female parent. Regardless of the ploidy number, the genetic makeup of an organism is called the **genotype**. The genotype is largely responsible for plant growth and developmental characteristics, but the environmental conditions under which plants are grown also greatly affect the plant's shape, structure, and physiology, collectively called the **phenotype**. A phenotype, therefore, is the **visible result of an interaction between the genotype and the environment.**

HERITABILITY

A monk named Gregor Mendel (1822–1884) conducted experiments at the Augustinian Monastery of St. Thomas in Brunn, Austria (now Brno, Czech Republic). Working with more than 10,000 garden pea plants (*Pisum sativum*) between 1856 and 1863, Mendel established a few simple principles of inheritance based on his observations. These principles gave us, for the first time, a scientific basis for plant breeding. Mendel's findings were first published in 1866; however, they were not widely noticed until about 1900. The science of genetics and plant breeding based on genetic principles was therefore developed mostly during the 20th century and continues to evolve in the 21st century. It is important to understand that Mendel knew nothing of the role of the nucleus and chromosomes in heredity but based his findings on laws of probability.

The garden pea was a good choice because it can be cross-fertilized although it is normally self-pollinated and self-fertilized. In addition, it is an annual and easy to grow. Although most of Mendel's plants were grown in the garden, a few were grown in pots in a greenhouse during flowering to escape insect problems in the garden.

Mendel was wise to choose pure and distinct traits and fortunate that each trait was controlled by one gene in a diploid plant. By initially considering only one trait at a time, he facilitated his understanding of basic plant genetics. Knowledge of Mendel's findings will give the student some understanding of plant genetics; however, many results of plant breeding cannot be interpreted in terms of simple Mendelian genetics (for example, plant yield and vigor are often more complex). For more complete coverage, readers are should consult the references listed at the end of this chapter.

The plants that Mendel used for **parents** (P) were from true-breeding stock because previous generations had been **inbred** (**self-fertilized**). Therefore, because these plants were self-fertilized, their offspring (progeny) were uniform and resembled the parents.

One trait that Mendel worked with was seed color (actually cotyledon color). Because cotyledons are part of the embryo of the seed and are not tissue produced by the maternal plant, their color represents the next generation. Therefore, when observing traits in the embryonic tissue, they can be seen as the seeds develop within the pods after sexual fertilization. Other traits, such as pod shape, distribution of flowers on the stem, and stem length are not possible to detect as the seeds develop. These traits require a second growing season to be evident and will become obvious after planting seeds that were produced by the F_1 hybrids and observing the seedlings as they grow and develop.

Mendel cross-pollinated yellow-seeded inbred parents with green-seeded inbred parents. To prevent self-fertilization in the female parents (seed parents) he **emasculated** their flowers by removing the anthers. To be sure that the traits were not simply influenced by the seed parent, he made **reciprocal** crosses. Reciprocal crosses were made in pairs. In one cross, the yellow-seeded pea was the seed parent and the green-seeded pea was the pollen parent, in the other cross, the green-seeded pea was the seed parent whereas the yellow-seeded pea was the pollen parent. The process of making a reciprocal cross is illustrated in Figure 4-9. The generation that results from crossing parents with different genotypes (**hybridizing**) is the **first filial** (F_1) generation; these plants are therefore called F_1 **hybrids**. When F_1 hybrids are self-fertilized or cross-fertilized with each other, the second filial (F_2) generation results, and when the F_2 generation is self-fertilized or randomly crossed with each other, the F_3 generation results. Each subsequent generation is numbered accordingly.

When Mendel made the reciprocal cross between the yellow-seeded and green-seeded parents, he observed that only yellow seeds were produced in the pods. These F_1 hybrid seeds were indistinguishable from seeds produced by self-fertilization of the yellow-seeded inbred line. This occurred regardless of the yellow seeded plant being the pollen or seed parent. The yellow seed color was **completely dominant** over the green seed color. The green seed color receded or disappeared and is therefore **recessive** to the yellow seed color. When Mendel planted these yellow F_2 seeds, however, he observed that their pods typically produced both yellow and green seeds, often within the same pod (Figure 4-10). When alternative classes of the observable traits occur within a filial generation, it is known as **phenotypic segregation**.

There were 258 plants in Mendel's F_2 generation that yielded 8023 seeds; 6022 were yellow and 2001 were green. This was a nearly perfect 3:1 phenotypic ratio.

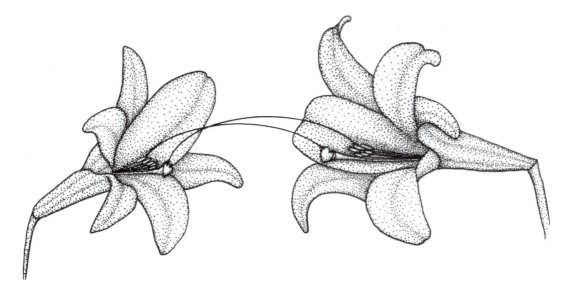

FIGURE 4-9. In a reciprocal cross, the sources of male and female gametes are reversed. This procedure helps determine whether a trait is sex-linked.

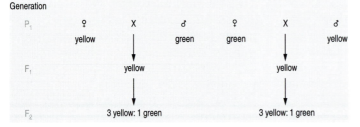

FIGURE 4-10. The phenotypes in the P₁ (parent) generation, F₁, and F₂ generations of the common garden pea, with a monohybrid cross. On the left, the yellow seeded pea was the female parent in the P₁ generation and on the right the female had green seeds; this represents a reciprocal cross. Phenotypic segregation occurred in the F₂ generation.

Mendel concluded that seed color was governed by a pair of hereditary factors, one that came from the seed parent and one from the pollen parent. It is now known that these hereditary factors are genes that occur as pairs of alleles.

Because of meiosis, each gamete contains one genome and, therefore, one allele for a given trait (assuming one gene for one trait). After fertilization, the resulting plant will have nuclei with like alleles on homologous chromosomes (**homozygous**) or unlike alleles (**heterozygous**). Mendel's P (parental) generations were homozygous, whereas his F₁ hybrids were heterozygous and his F₂ plants were either homozygous or heterozygous for seed color. This can be explained by using letters to symbolize the alleles. A capital letter, such as Y, is used to represent a dominant allele, and a lower case version of the same letter, y is used to represent a recessive allele. Therefore, in the pea seed color example, the dominant factor yellow would be represented by Y and the recessive factor green by y. The original yellow-seeded parent was homozygous dominant (YY) and the green-seeded parent was homozygous recessive (yy). Mendel's experiment can, therefore, be reconstructed as in Figure 4-11. Note that in writing crosses, the seed parent conventionally is listed first.

A **monohybrid** results from crossing two individuals that were homozygous for the alternate forms of *one* gene. In the monohybrid for seed color, the F₂ generation has a 3:1 phenotypic ratio but a 1:2:1 genotypic ratio (Figure 4-11).

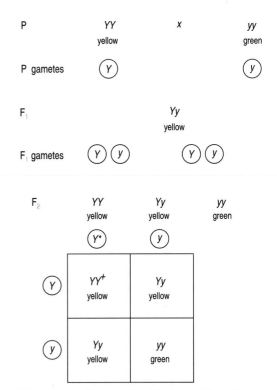

* F₁ gametes are listed outside of the box, and F₂ genotypes and phenotypes are listed within the box (frequently called Punnet Square).

+ note the 3:1: (yellow : green) phenotypic ratio and the 1:2:1 (YY:Yy:yy) geotypic ratio.

Figure 4-11. Reconstruction of Mendel's experiment on pea seed color showing phenotypic and genotypic results with a monohybrid cross.

To this point, the discussion has centered on one gene and one pair of alleles. However, thousands of other genes exist. For example, on a chromosome pair one might expect:

Yellow   Green
...YYBBCCddEE...   ...yyBBCCddEE...

A **dihybrid** results from crossing two individuals that were homozygous for the alternative forms of *two* genes, each coding for a different trait. For example, when cross-fertilizing two garden pea parents with two contrasting traits such as seed color and seed coat structure, a dihybrid is produced. The two seed colors yellow and green will be considered along with smooth and wrinkled seed coats. Smooth seed coats are dominant to wrinkled coats and yellow seeds are dominant to green. If W represents the smooth coat and *w* represents the wrinkled coat, the phenotypic and genotypic results can be represented as in Figure 4-12.

One can readily observe when comparing Figures 4-11 and 4-12 the increased

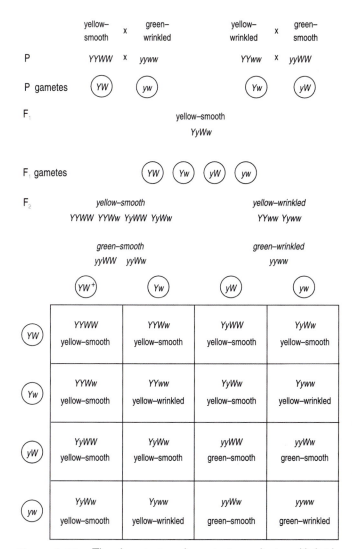

* Possible inbred parental combinations with different phenotypes for the 2 traits. Note the same F_1 results from each cross-fertilization.

† F_1 gametes are listed outside of the box; F_2 genotypes and phenotypes are listed within the box. Notes the 9:3:3:1 (yellow–smooth : yellow–wrinkled : green–smooth : green–wrinkled) phenotypic segregations ratio in the F_2 generation.

Figure 4-12. The phenotypic and genotypic results in a dihybrid cross are more complex than a monohybrid cross.

complexity of the results of a dihybrid cross fertilization compared to a monohybrid cross fertilization. The 9:3:3:1 phenotypic segregation of the dihybrid with 9 distinct genotypes reflects this complexity. In a **trihybrid** one deals with three pairs of heterozygous alleles; with complete dominance, the F_2 generation has 27 distinct genotypes and a phenotypic ratio of 27:9:9:9:3:3:3:1. Therefore, the more genes a plant breeder manipulates, the more complex identification and selection of plants with particular gene combinations become.

This knowledge has practical importance, not only in understanding principles behind plant breeding but also in learning the concept of **segregation**. F_1 hybrid cultivars are planted by horticulturists because they offer many advantages, as is discussed later in this chapter under the heading "heterosis." Seeds collected from F_1 hybrid plants begin the F_2 generation. Plants in the F_2 generation can be expected to segregate or not come "true" from seed. The variability generally occurs for several traits, and among the most noticeable is the difference in vigor among the seedlings. Some of the resulting plants may be tall and vigorous, but others may be short or even stunted. Horticulturally, the nonuniformity is a disadvantage and can lead to problems (e.g., growers experiencing difficulty in scheduling their crops or landscape horticulturists installing plants that are not meeting their needs). Therefore, if horticulturists collect or purchase F_2 seeds, they must be aware of segregating populations. In nature or in more primitive production systems, however, segregation can be an advantage. For example, if tree seedlings are too crowded, the most vigorous will out-compete those with less vigor and there will be a few strong individuals, rather than many weak ones. In nature, genetic variability can lead to some individuals in a population to survive adverse environmental conditions such as disease and insect infestations, thus ensuring survival of the species.

The plants of many species are highly heterozygous, which can lead to segregating seedling populations. Nursery plantings of seedling Colorado spruce (*Picea pungens*) typically segregate for the blue color; the forms with the most blue are selected because they are demanded in greater numbers by the landscape industry and can be sold at a higher price than green forms. Similarly, mugo pine (*Pinus mugho mughus*) seedlings generally segregate for plant size with the more dwarf forms considered more desirable (Figure 4-13).

CLONING PLANTS

Many crops are propagated asexually because of the uniformity among the resulting plants or **clones**. A clone is a population of plants that has been asexually propagated, can be traced back to one original individual, and is typically genetically uniform. An example of a clone is the commonly grown apple cultivar 'Golden

Figure 4-13. These seedling mugo pines are all the same age and are segregating for traits such as plant and needle size. The more dwarf plants on the right in each picture are considered more desirable.

Delicious,' which is one of the most popular apples for new plantings in the world. All 'Golden Delicious' trees can be traced back to one chance seedling that was growing on a hillside in West Virginia during the early 20th century. Methods of propagating clones include rooting of cuttings, layerage (forcing roots to form on a cutting before it is severed from the mother plant), grafting, and tissue culture (micropropagation). Potato is an example of a crop that generally is propagated vegetatively (by tubers) because seedlings of most of the important cultivars will not come true or will segregate for many traits. For details on cloning and other propagation methods, see Chapter 14.

PLANT BREEDING

Inheritance

Plant-breeding programs become increasingly complex because more than one gene can play a role in governing the phenotypic expression of one trait. For example, yield often is controlled by many genes, and rather than having only high- or low-yielding individuals, plants can differ in yield by only small amounts and can be placed into an array from low to high yield. Such an array often fits a bell-shaped curve, with most individuals intermediate in yield. Traits that fit such a continuous array between extremes are referred to as having continuous variation or **quantitative inheritance**. In quantitative inheritance, individual independent genes have a *cumulative* effect, and to maximize a trait, such genes must be brought together in an individual. In contrast, Mendel studied **qualitative characters**, or those traits with *discontinuous* variation (i.e., the two seed colors of yellow and green were distinct and exhibited no intermediates).

Dominance

The manifestation of one allele over another when heterozygous is called **dominance**.

A dominant allele can be **completely dominant** to the recessive allele as with the case of the pea seed color, where yellow was completely dominant to green. But this is not always the case; there are times when the heterozygote is intermediate between the two homozygotes (for the same alleles); often resulting from **incomplete dominance**. An example of incomplete dominance is a pink-flowered F_1 hybrid resulting from crossing a red flowered inbred with a white flowered inbred. The shade of pink depends on the degree of dominance. **Overdominance** occurs when the heterozygote is superior to either of the homozygotes with the same alleles, (i.e., a plant with the genotype *Dd* is superior to plants with *DD* or *dd*).

Heterosis

Heterosis is the technical term for **hybrid vigor** and is often exemplified by the increase in size or vigor of a progeny over its parents or over the average of its parents (Fig. 4-14). It is thought to result primarily from bringing together favorable dominant genes, although this explanation may not cover all cases. The following is an example of how the concentration of favorable dominant genes in one individual can lead to heterosis[1].

In the production of hybrid corn the theory works like this. For simplicity, let us assume that the dominant genes ABCDE are favorable for good yields. Inbred A has the genotype *AABBccddEE* (*ABE* dominant). Inbred B has the genotype of *aabbCCDDEE* (*CDE* dominant). The genotype of the F_1 hybrid is shown in the following:

Inbred A   Inbred B
AABBccddEE   ×   *aabbCCDDEE*

F_1 hybrid
AaBbCcDdEE

....
[1]From Poehlman, J.M. 1959. *Breeding field crops*. Henry Holt and Co., New York. p.35. Used with permission.

Figure 4-14. Hand emasculation (removal of anthers) of this tomato flower ensures that the pistil will not be pollinated by the male elements of the same flower (*top left*). Petunia stamens are being used to pollinate a previously-emasculated flower (*top right*). The results of controlled pollinations are hybrid plants that exhibit heterosis and are therefore often more vigorous than their parents, such as the snapdragons in the bottom photograph.
Photos courtesy of Goldsmith Seed Inc., Gilroy, CA.

In this example the F_1 hybrid contains dominant genes at all loci represented here (*ABCDE*) and would exhibit more vigor than either of the parent inbred lines.

In this example, heterosis is the result of partial dominance, complete dominance or over dominance at more loci than either parent. This F_1 hybrid would be more desirable for growers than either inbred line and would therefore command a higher price.

F_1 hybrids also are preferred because they have phenotypes with uniformity similar to that found within a clone. The uniformity facilitates crop scheduling.

F_1 hybrid seeds are often preferred by growers because the resulting plants exhibit heterosis. These plants may be high yielding, have superior flower color, large flower size, and exhibit great vigor and thus may be better competitors against weeds. The seeds are generally more expensive because they require more intensive production techniques than inbred lines, and this adds greatly to their expense. For seed production, monoecious and hermaphroditic plants generally require emasculation to prevent self-fertilization. Many species are hand-emasculated; although some crops such as sweet corn may be detasseled mechanically. Also male sterility has been bred into some breeding lines to avoid the effort involved with emasculation. Controlled cross-pollination in many crops requires humans. However, seed companies utilize other agents, such as the wind (on wind pollinated crops such as sweet corn) or insects, when possible, to increase efficiency and minimize costs. For example, common houseflies are used on carrots to carry pollen from one flower to another. Some seed companies raise flies specifically for the purpose of creating F_1 hybrid carrots. Likewise, hives of bees have been placed in cages with plants for the purpose of producing F_1 hybrids (Figure 4-15).

Inbreeding

Inbred lines are self-pollinated and are therefore cheaper seeds to produce. However, breeders prefer to produce hybrid cultivars to homozygous dominant cultivars in many species for several reasons: (1) Usually there are too many genes involved in a character such as yield to incorporate easily into one individual. (2) When breeding for homozygous dominance, advantages that might be gained from overdominance are lost. (3) F_1 hybrids provide an exclusive line that other breeders cannot produce unless they have access to the inbred parents. This is a form of cultivar protection—like a patent. (4) When inbreeding, the chances increase that homozygous recessive alleles will occur and can lead to a condition known as **inbreeding depression**. This condition is characterized by one or more of the following: small and weak plants with low vigor, reduced levels of fertility, and more defective plants. Some crops do tolerate inbreeding with few problems, however, and some commercially available cultivars, including some popular tomato cultivars, are inbred lines.

MOLECULAR BIOLOGY

Plant molecular biology is the study of molecular and biochemical processes in plants. These studies tend to focus on the central dogma of molecular biology of DNA replication, RNA transcription, and translation of proteins. This has practical applications for genome analysis and manipulation to increase production and quality of horticultural plants.

Figure 4-15. Bees are effective pollinators for a variety of horticultural crops (*left*) and their activity can result in excellent fruit set, such on the pear tree (*right*).

The study of **genomics**, the identification, function, and structure of many genes at the same time, and **proteomics**, the discovery and characterization of all proteins produced by cell types and organisms can be considered as part of molecular biology. As a result of genomic studies, the DNA has been sequenced (the exact order of nucleotides in the DNA of an organism) to some extent in thousands of species. The first plant species to have its DNA completely sequenced was a small weedy plant in the mustard family, mouse-ear cress (*Arabidopsis thaliana*).

Historically, the term **biotechnology** has meant to use organisms or biological systems to produce or modify processes or products such as yeasts to make beer or bread. However, many have come to accept the more restricted definition that biotechnology describes manipulating and transferring genes from one organism to another (**recombinant DNA technology**) and cloning of both plants and animals.

Gene transfer has been successful in many horticultural species. DNA can be isolated from virtually any living organism, cut into pieces using enzymes, replicated, and introduced into the genome of a desirable plant to change its phenotype. This process is known as **genetic transformation (genetic engineering)**. The result is a **genetically modified organism (GMO)**.

Many horticultural species have been genetically transformed and some are being grown commercially. Most of these **transgenic plants** have had foreign genes inserted that confer disease resistance, insect resistance, or herbicide tolerance. For example, growers are producing the transgenic 'UH Rainbow' papaya that is resistant to the papaya ringspot virus, which is the most serious disease problem in this crop. Yellow summer squash and zucchini squash are also being grown with transgenic resistance to viruses. A transgenic potato was available commercially that was resistant to the Colorado potato beetle, but was removed from the market by the company that sold it

because a new insecticide was introduced that controlled the same pests and farmers chose the insecticide rather than the transgenic potatoes.

There are different ways that foreign genes can be inserted into horticultural species. The most common is to utilize a soil-borne bacterium (*Agrobacterium tumefaciens*) that causes a disease known as crown gall. When this bacterium infects plants, it naturally genetically engineers the cells in the area of infection. A portion of its DNA, known as transfer DNA (tDNA) leaves the bacterial cell and migrates into the plant cell's nucleus and becomes incorporated into the chromosomal DNA and is expressed. Scientists have learned how to splice genes of interest into the *Agrobacterium* plasmid DNA (circles of DNA) and these genes are in turn incorporated into the target plant's DNA during the infection process. Another technique that has been successful is the use of a particle infiltration gun to shoot tiny DNA-coated tungsten or gold particles into plant cells. The DNA becomes incorporated into the plant DNA and is expressed.

There is a serious controversy about the production and sale of GMO plants and plant products. In most countries, GMO products are not labeled as such and consumers cannot readily distinguish between genetically modified plants or plant products and similar non-GMO plants. Some believe that GMO foods have not been proven to be safe. Part of this concern is that because GMO foods produce new and novel proteins, they could be allergens to some people and therefore unhealthy. There have been some studies that suggest that nontarget organisms will be killed or injured by ingesting the GMO plants or their pollen. Some other studies indicate that with some crop species, there can be interbreeding with weed or native species and that the transgenes will thus enter the wild population creating an environmental risk. This could confer herbicide resistance to weed species, making them more difficult to control. Another argument against the growing of GMO plants is that insects or disease-causing organisms will mutate and overcome the resistance conferred by the transgenes; rendering some pesticides ineffective.

Others tout the promise of growing GMO plants. If transgenic plants are resistant to insects and diseases, growers will use much less pesticide, which will be environmentally friendly. Farmers often choose to grow GMO crops because they can be much more economical to produce because there are fewer tractor trips over the fields applying pesticides and money is saved because less pesticide is purchased. In some poor countries, production of transgenic plants is attractive because they cannot afford the pest controls necessary to grow sufficient quantities of certain crops. In some cases, disease pressures can be extremely severe and there are no controls that are alternative to transgenic plants, therefore the only plants that can be grown successfully are genetically modified.

It is impossible to predict how the GMO controversy will be resolved. Since some consumers accept GMO foods and others do not, transgenic plants may find certain niches in horticultural production. It is doubtful that 100% of any crop will be transgenic. Consumer demand will probably result in clear choices between GMO and non-GMO foods.

CROSS-FERTILIZATION

Many plants have evolved so that cross-fertilization is favored and inbreeding avoided, thus reducing (or eliminating) the chances for **inbreeding depression** and a general weakening of the stock that could lead to extinction. In humans, inbreeding (incest) is taboo in most societies. Examples of inbreeding in humans have occurred in royal families as an attempt to preserve the pure or "blue blood" lines. When brothers and sisters or first cousins have intermarried,

however, the lines generally have been weakened rather than strengthened, resulting in cases of hemophilia, shortened lives, and greater incidence of disease. Humans, therefore insure cross-fertilization by making inbreeding socially unacceptable, whereas many plant species have physical or physiological means of reducing or preventing inbreeding. Plants employ the following means (among others) to ensure or favor cross-fertilization: dioecy, incompatibility, sterility, and dichogamy. These influence how we plant and grow many important horticultural species.

Dioecy

In dioecious species, the separate sexes are on different plants, thus assuring cross-fertilization. Many horticultural crops are dioecious, and this knowledge directs planting patterns and is used in the selection of plants. Hollies (*Ilex* spp.) are dioecious and often are grown for both their foliage and fruits. The distinctive red berries of American holly (*Ilex opaca*) form only on the female trees from flowers that have been fertilized by compatible pollen (Figure 4-16). Unless the hollies are planted in a region where they are native and there are ample staminant plants, horticulturists must be sure to provide a source of pollen. This is accomplished by planting a male tree in the vicinity of female trees, by planting a male tree and a female tree in the same hole and pruning the male to allow only minimal growth, or by grafting male branches onto female trees. Likewise, pistachio (*Pistacia vera*) trees must be planted at a ratio of one male to six females for profitable nut production.

The male forms of some other landscape trees are more desirable than females. The female ginkgo (*Ginkgo biloba*) produces plum-shaped, fleshy covered seeds that have a putrid odor and are very messy when they are ripe. Male forms, which produce no seeds, are therefore preferable to female trees. Ginkgos generally are asexually

Figure 4-16. The male American holly with its dried stamens (*left*) on the left is readily distinguishable from the female holly with fruit (*right*).

propagated by budding or grafting so horticulturists can be sure that they are planting male trees. Green ash (*Fraxinus pennsylvanica*) trees are also dioecious. Although the ash fruits are not obnoxious, they can be unsightly and messy. A male form known as 'Marshall's Seedless' is widely planted because of its good foliage color, growth habit, and lack of fruit. 'Marshall's Seedless' ash is asexually propagated.

Asparagus is a dioecious vegetable crop. Growers tend to select male plants because they generally outyield females and do not produce seedlings. Breeding work has focused on producing males with thicker spears because in traditional cultivars, female plants produce larger, more desirable shoots (spears) than male plants.

Incompatibility

Incompatibility is a genetically controlled biochemical hindrance to self-fertilization in some hermaphroditic plants, among members of a clone (self-incompatible), or between plants with a common incompatibility gene. The process of incompatibility can operate at any time following pollination and up to the point when fertilization would have occurred if there were compatibility. Under "normal" growing conditions, a self-incompatible plant is incapable of being fertilized by its own pollen. Sexually derived embryos therefore result only from cross-fertilization. Plants that serve as sources of compatible pollen (**pollinizers**) must therefore be included in a planting plan to result in seed or fruit production. Note that a **pollinator**, unlike a pollinizer, is a vector by which pollen is carried (e.g., bees or wind).

Horticulturally, self-incompatibility is important in fruit and nut production. In pomology, cultivars may fall into one of the following three categories: self-fruitful, partially self-fruitful, and self-sterile (often from self-incompatibility). Self-fruitful plants will set and produce a mature crop of either seeded or parthenocarpic fruit. If the cultivars are self-incompatible or only partially self-fruitful, compatible pollinizers must be included in a planting. Within a species, such as apple, some cultivars are self-sterile, such as 'Golden Delicious,' whereas some display at least a slight degree of self-fruitfulness, such as 'Wealthy' and 'McIntosh.' The following fruit or nut crops have at least some cultivars that display a degree of self-incompatibility: apple, pear, plum, sweet cherry (sour cherries are self-fruitful), hazelnut, and almond. Growers of these crops should therefore avoid planting only one cultivar, because poor yields will result. Figure 4-17 shows one planting plan for an orchard with two cultivars. Note that the main cultivar 'Jonathan,' is no more than two trees away from a 'Delicious' tree. When selecting or developing a planting

Jonathan	J	J	J	J	J		J	J	J	J
Jonathan	J	J	J	J	J		J	J	J	J
Delicious	D	D	D	D	D		D	D	D	D
Jonathan	J	J	J	J	J		J	J	J	J
Jonathan	J	J	J	J	J		J	J	J	J
Jonathan	J	J	J	J	J		J	J	J	J
Jonathan	J	J	J	J	J		J	J	J	J
Delicious	D	D	D	D	D		D	D	D	D
Jonathan	J	J	J	J	J		J	J	J	J
Jonathan	J	J	J	J	J		J	J	J	J

Figure 4-17. Planting plan of two cross-compatible apple cultivars where 'Jonathan' predominates. Cultivars are planted in solid rows to facilitate mechanical harvesting. There are many possible planting plans from which a grower may choose.

After N.F. Childers. 1975. Modern Fruit Science, Horticultural Publications, Gainesville, FL. Used with permission.

plan, a pomologist needs a knowledge of the cross-fruitfulness of cultivars and additionally must pick cultivars that will have marketable fruit in the future as consumers' tastes change.

Although planting plans are still used by growers, they have limitations and disadvantages. These include a lack of consistent and reliable bloom period overlap, a tendency for some cultivars to be biennial (blooming in alternate years), management problems including pickers mixing cultivars, and different spray schedules appropriate for each cultivar. Additionally, if the trees are managed as hedgerows, the bees generally do not go from row to row. Some modern orchardists are therefore planting ornamental crabapples between regularly spaced apple trees for use as pollinizers. The crabapples are often trained as slender poles so they will not be serious competitors for light, water, and nutrients with the main cultivar. Ornamental crabapples offer several advantages over other pollinizers. They produce flowers on both old and new wood resulting in profuse flowering and extended bloom time because the flowers on old wood often open earlier than on the new wood. They also can be pruned immediately after bloom, thus forcing new growth and ensuring bloom next year. It is recommended that an adequate number of trees each of three or four different ornamental crabapple cultivars be planted in an orchard

to ensure overlap of bloom between the main cultivar and at least one crabapple cultivar. Because bees have their maximum effectiveness as pollinators if the pollinizer is within 30 meters (90 feet), when using trees representing three or four different pollen donors, grower must carefully consider spacing.

Horticulturists often are called upon to counsel homeowners when problems result from self-incompatibility. A single apple tree that is planted 30 meters or more from another apple or crabapple tree may flower profusely each year and produce little, if any, fruit because of self-incompatibility. It is therefore wise to advise homeowners to plant at least two different cultivars in their yard for cross-pollination. Sometimes two or more cultivars will be grafted onto one tree to ensure good fruit production.

Many horticultural catalogs list pollinizers that are necessary if one wishes to produce a crop successfully. Sometimes, however, a pollinizer can be alternate bearing (biennial), in that it will only flower and produce a crop every other year. 'Red Delicious' is often alternate bearing, especially if the cultural practices are improper.

Sterility

Sterility in plants occurs when the male or female gametes are nonfunctional, which should not be confused with incompatibility where the gametes are viable. Male sterility, which results in nonviable pollen, appears only sporadically in nature. Breeders utilize male sterility, when available, by incorporating it into the female parents they wish to use in cross-fertilizations. Male sterility eliminates the need for emasculation of seed parents, thus reducing the labor requirement and cost of producing F_1 hybrid seeds.

Dichogamy

Some plants have ensured cross-fertilization by utilizing dichogamy, the shedding of pollen by the stamens at a time when the stigma is not receptive. There are two dichogamous methods in the plant kingdom called **protandry** and **protogyny**. Protandry is the maturation of the anthers before the pistils on the same flowers or on monoecious plants. Protogyny is the condition in which the pistils are receptive before the anthers on the same flowers or on the same plants shed their pollen. The monoecious nut species walnut and pecan both have some cultivars that are protandrous and others that are protogynous. Fertilization can be better ensured by planting two cultivars, whose pollen shed will coincide with pistil receptiveness. Similarly, when breeders choose inbred parents to use in a cross-fertilization, the time of pollen shedding by one parent and pistil receptiveness on the other must coincide closely, or a poor yield will result. An alternative is to collect pollen and store it for later use.

Xenia

Cross-fertilization between different types of corn, such as sweet corn and field corn, can lead to problems. This is because corn exhibits **xenia**, the effect of pollen on the endosperm and embryo. For example, if sweet corn and field corn are planted adjacently, the wind may cause cross-fertilization if both types of corn shed pollen simultaneously, and pistils (silk) are receptive. At harvest some kernels on an individual ear will look and taste like field corn, and some will resemble sweet corn. This will of course depend on the source of the pollen that landed on each individual silk. In extreme cases sweet corn, field corn, colored ornamental corn, and popcorn may be planted adjacent to one another. If they cross-fertilize, on each ear of sweet corn, some kernels will look like field corn, some will look like ornamental corn (colored), some like popcorn, and some like sweet corn. If good-tasting sweet corn, poppable popcorn, or pretty-colored ornamental corn is desired, one should maintain approximately 1 km (1 mile) between corn plantings.

Horticulturists are often asked, "If I plant squash near the cucumbers in my garden, will my squash taste like cucumbers because of cross-fertilization?" In other words, will a phenomenon such as xenia occur in cucurbits? The answer is *no*: although cross-fertilization occurs among some cucurbits, the fruit will not be affected. Fruit that does not taste as expected it is probably a result of environmental factors but is definitely not a result of cross-fertilization. Some cucurbits will cross-fertilize, such as members of the single species *Cucurbita pepo*, which includes zucchini, pumpkin, and yellow summer squash. Interplanting these vegetables will not influence fruit flavor, shape, or size; however, allowing the fruit to mature, collecting the seeds, and sowing them the next year might bring some interesting results. Some of the resulting plants may be sterile and take up valuable garden space, and some may produce unique-looking but generally useless fruit.

MUTATIONS

A **mutation** is a sudden heritable change in a gene or in chromosome structure that can produce single or compound effects, the latter being more common. Mutations occur spontaneously in nature in all living organisms. Mutations can also be induced by radiation or by chemical mutagens. Examples of forms of radiation that cause mutations are ultraviolet light, x-rays, gamma rays, and extreme temperatures. Examples of chemical mutagens are nitrogen and sulfur mustards, epoxides, peroxides, phenols, and alkaloids.

Mutation is a random and oft-repeated process; therefore, most mutations that are observed today probably occurred many times before. Because of selection, by both nature and humans, most alleles in existence are superior. New mutations generally are inferior and only a very few are worth saving. Many mutations are lethal—albino plants, for example, which produce no chlorophyll, are not able to photosynthesize, and ultimately die when the food reserves in their cotyledons or endosperm are exhausted. Although the observant horticulturist will see many mutations, most will not be worth propagating. The astute (and lucky) plantsperson may spot a superior mutant, propagate and patent it. This will add more diversity to our horticultural crops. Plant breeders, however, utilize mutations as a source of new traits and genetic diversity that they may wish to incorporate into a population. Few seed-propagated cultivars are a result of mutations. Many clonally propagated plants are mutants.

Chimeras

A **chimera** is an organism with two or more genetically different tissues. The term chimera originally was applied to a mythical beast with the head of a lion, body of a goat, and the tail of a dragon. In chimeral plants, there are genetically different cell layers in the apical meristem that divide and differentiate into the body of the plant. The arrangement of these genetically dissimilar meristematic cells is important to the stability of the chimera and the phenotype that is produced.

In higher vascular plants, the apical meristem is organized with one or more layers of surface cells known as the **tunica**. These cells divide in planes that are perpendicular to the meristematic surface (anticlinal divisions), so that each layer gives rise to cells that remain in that particular layer. The inner cells are called the **corpus**. They divide in various planes, and may consist of several cell layers. Monocots tend to have two or three layered meristems, and most dicots have three apical layers. These layers are numbered from the outside—inward, as L-I, L-II, L-III, etc. (Figure 4-18). If there are two layers, the outer is the tunica layer and the inner is the corpus. If there are three layers, the two outer layers are tunica layers and the innermost is the corpus.

If one or more entire layer of the apical meristem is genetically different from another entire layer, this conditions is known as a **periclinal chimera** (Figure 4-19). This is the most stable type of chimera because the tunica cells divide anticlinally and the daughter cells remain in their respective layer. That means as a shoot grows, there will be specific and stable

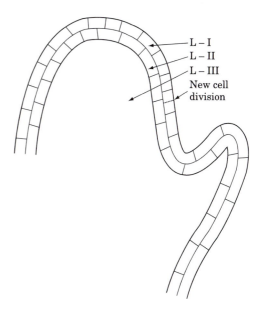

Figure 4-18. The tunica-corpus arrangement of a shoot apical meristem. L-I and L-II represent tunica layers, note their typical cell divisions perpendicular to the meristematic surface (anticlinal divisions). The L-III layer is the corpus.

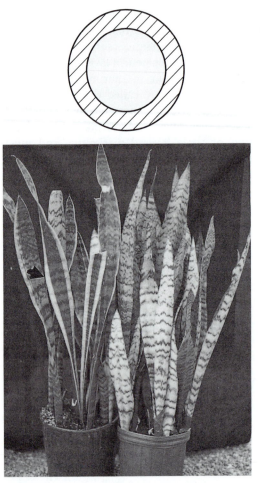

Figure 4-19. The shaded area represents a genotype that completely surrounds a different genotype in this plant organ (*top*); thus, this represents a **periclinal chimera**. The gold-banded *Sansevieria* on the left of the photograph (*bottom*) is an example of a periclinal chimera; compare with the solid green leaves on the plant on the right in the photograph.

patterns of the genetically different tissues. For example, variegation is where a plant organ shows two or more colors. The expression of these colors often is a result of the genetic differences between cells within a plant organ, such as a leaf or flower petal. The simplest example is with monocots. It is typical in many of these species to have leaves with green edges and white or yellow centers or white or yellow edges and green centers (Figure 4-19). This is seen in the variegation patterns of hostas, spider plants,

and other monocots. In these plants, the L-I cells divide and form the entire edge of the leaf blade, whereas the L-II cells give rise to the center of the leaf. If monocot leaves have white margins and green centers, they have the apical meristematic arrangement of WG (white for L-I and green for L-II layers). If the leaves have green margins and white centers, the arrangement is GW. These two stable variegation patterns are very common for monocots.

The situation is somewhat more complicated for dicots. Typically, the epidermis of the leaf is from the L-I, the L-II gives rise to the spongy mesophyll at the leaf margins as well as the palisade layer and the L-III does not contribute to the margins of the leaves, rather to the middle and upper spongy parenchyma layers of the leaf. The color of the leaf margins is usually contributed by the L-II layer. Therefore a dicot with a white leaf margin would have a GWG (L-I, L-II, L-III) arrangement of its apical meristematic layers. A WGW arrangement results in a plant with leaves showing an almost continuous green margin and white center. Another interesting periclinal chimera is on some thornless blackberries that have a thornless L-I layer surrounding thorny L-II and L-III layers.

Propagation

The method of propagation is critical for maintaining a periclinal chimera. Because embryos in seeds form from an initial zygotic cell, a chimera cannot be passed through a seed generation, therefore vegetative propagation is necessary to maintain a chimera. However, if the vegetative propagation technique results in adventitious shoot formation, the chimera will be lost in this new shoot. An **adventitious shoot** arises from a cell or group of cells that was not an apical or axillary meristem. For example, if an adventitious shoot forms from a leaf cutting or root cutting, it would typically originate from a specific tissue that would have been derived from one of the L-layers, and therefore would be genetically uniform throughout. The only way that a periclinal chimera can be maintained is when the new shoot growth is from preexisting apical meristems. This happens when stem cuttings are rooted, stems with buds are grafted, or clumps are divided, so that each shoot or group of shoots has its own apical and axillary buds. This maintains the integrity of the L-layers and therefore preserves the chimera. For example, if the gold banded *Sansevieria* in Figure 4-19 were propagated by leaf cuttings, adventitious shoots would form from the green tissue and the resulting plants would be solid green like the plant on the right in the photograph. The gold-banded character is maintained by dividing clumps of the plant. Likewise if chimeral thornless blackberries are propagated by root cuttings, the adventitious shoots form from the internal root tissues and are thorny. However, if shoot cuttings are rooted, the thornless chimera is retained.

Instability

Periclinal chimeras are the most stable type because of the tunica-corpus organization of the apical meristem and the anticlinal cell divisions in the tunica layers that maintain separation of the layers and the subsequent differentiated tissues of the plants. However, frequently in some species and infrequently in others, cell divisions in the tunica layers can be parallel to the meristematic surface. This results in two daughter cells layers within one L-layer. One of these daughter cells will remain in its appropriate layer, but the other will move between two cells in the adjacent outer or inner layer. Such "invasion" of a cell layer by a cell from a different layer can change the phenotype of the chimera. If a cell from an outer layer moves inward, it is called *replacement* and if a cell from an inner layer moves outward, it is called *displacement*.

Replacement and displacement will change the chimeral arrangement and this will often result in **bud sports** from chimeral plants (Figure 4-20). Additionally, if a cell mutates in an apical meristem, bud sports can result. If a portion, but not the whole, of one or more meristematic cell layers is genetically distinct, it is known as a **mericlinal chimera** (Figure 4-21). Sometimes the cells of one genotype will occupy a portion (or wedge) of all of the layers of an apical meristem. This chimeral arrangement is known as a **sectorial chimera** (Figure 4-22). Mericlinal chimeras and sectorial chimeras can be difficult to distinguish from each other and are both unstable compared to periclinal chimeras.

Bud sports, because of apical meristematic cell displacement and replacement, are more common in some species than others. An interesting example is *Spiraea japonica*

Figure 4-20. Coleus plant with a bud sport showing lighter-colored foliage.

Figure 4-21. The shaded area represents a genotype that is only 1 or a few cell layers thick and only partially surrounds a different genotype in this plant organ, this represents a **mericlinal chimera**.

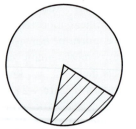

Figure 4-22. The shaded area represents a genotype that includes a portion of the surface and internal cells, and the light area is a different genotype in this organ. This represents a **sectorial chimera**.

'Anthony Waterer.' Its leaves are solid green, but it is a WGG periclinal chimera. This cultivar frequently sports to produce branches with white leaf margins and a green center (WWG) or pure white shoots (WWW), which lack chlorophyll and cannot photosynthesize. It is common in several variegated plants for them to sport and produce albino shoots, which are solid white (WWW) or chimeral (GWW) or produce green shoots that may be solid green (GGG) or chimeral (WGG).

If a bud sport is stable and will not continue to change or mutate, the branch can be propagated vegetatively and the resulting plant will display this new trait. Often these are patentable and have resulted in new cultivars. Bud sports have resulted in better fruit color in many apple cultivars; 'Connell Red' apple, for example, is a red sport of 'Fireside.' 'Connell Red' was found originally as one branch with bright red fruit on a 'Fireside' tree. It was clonally propagated by grafting and budding. The orchardist who patented this clone, gave it a cultivar name, in order to receive both royalties when it is propagated, plus the advertisement of having his name incorporated into the cultivar name. Over 150 red sports of 'Delicious' apple have been named and it is likely that more will continue to be introduced to the trade. Pomologists must know which cultivars are sports of other cultivars, however, because although fruit color may be different, sports may not be sexually compatible with their parent or related cultivars and will not be good pollinizers for each other.

In sports of some species, a small percentage of plants will revert to the original type, so growers must be aware of this phenomenon. Carnation is an example where the epidermal layer alone is responsible for flower color. 'White Sim' carnation is a WR periclinal chimera; 'Red Sim' is the internal tissue. 'White Sim' commonly sports from WR to RR, resulting in some red-flowered carnations in a greenhouse planting that is specializing in white-flowered carnations. This also must be considered in a breeding program when using 'White Sim.' Internal tissues give rise to the sexual parts, so 'White Sim' behaves like 'Red Sim.'

Another phenomenon of interest to horticulturists is the **witches' broom** (Figure 4-23) because it can be a new source of genetically dwarf plants. A witches' broom is the occurrence of many branches with short internodes growing from one point. This event generally occurs in trees, but sometimes may be observed in shrubs. Witches' brooms have many causes, including injury by insects and mites, viruses, invasion by dwarf mistletoe, and genetic mutation (bud sports) (See also Chapter 16). Pests, not bud sports, are generally the cause if there are many witches' brooms on one plant or in a given area. These will not result in dwarf plants if propagated. However, a single witches' broom in a tree may be a sport. Seeds can be collected from witches' brooms, and often, some will exhibit the dwarf habit. In others, cuttings or grafting wood can be harvested from the witches' broom for clonal propagation. Horticulturists covet genetic witches' brooms because they can serve as a source of new dwarf plants. These newer dwarf forms can add diversity to the landscape trade. Some nurseries, in fact, specialize in dwarf woody plants, many of which are members of clones derived from witches' brooms (Figure 4-23).

Figure 4-23. Witches' broom (*top*) on a Virginia pine tree that has much shorter internodes than the other branches. This witches' broom was discovered by Mr. James Youngers who generously shared its location with the authors. This 21 year old eastern white pine (*middle*) was 21 years old at the time and had been propagated from a witches' broom and planted at this site in 1963. This tree was approximately 1.3 meters tall with needles much shorter than normal. The sign (*bottom*) is a close-up of the sign in the top picture.

PEST RESISTANCE

Resistance to pests is only one of many characteristics to be considered in a breeding program. Genetic pest resistance can reduce a grower's use of pesticides and allow some crops to be more fully exploited in areas where pests have been a limiting factor. Traits such as yield and plant habit cannot be sacrificed for pest resistance, or growers will revert to pesticides for control and grow cultivars that have better economic returns. When all things are equal, however, added resistance to pests makes a crop or cultivar more desirable to grow. Some plants complete their life cycles at a time when the pest is absent (escape), or can withstand attack by the pest (tolerance) (see Chapter 16).

There are degrees of pest resistance in plants. Plants can be totally resistant (immune), partially resistant, or resistant only to certain races of a pest. A race is a group within a species that is different, but not so unique from other groups to be another species. Some races of pests cause more damage (are more virulent) than others. Although total resistance often may be the most desirable situation, partial resistance to a pest may greatly improve a crop and thus be economically important compared to the cultivation of extremely susceptible cultivars.

Inherent pest resistance in a plant results in reduced symptoms or even elimination of the effect of the pest, depending upon the degree of resistance (high, intermediate, or low). Resistance may arise as a response of the plant to the pest (**responsive** or **reactive resistance**). For example, a plant may produce a chemical that is active against fungi in response to a fungal attack. However, some plants possess the mechanism for resistance prior to pest attack (**passive resistance**). Examples of these types of resistance are the production of biochemicals or a surface structure, such as cuticle, that render the protection.

Pest resistance can be controlled by one gene (**monogenic**), a few genes (**digenic** or **trigenic**), many genes (**polygenic**), or by extranuclear DNA (**cytoplasmic**). Generally, the fewer genes that are involved, the easier it is to incorporate pest resistance into a crop.

Genetic resistance may exist for one or only a few genetic races of a pest, in which case it is considered transient, narrow, or vertical resistance. In such cases the pest can mutate, and the mutant races will be favored, because they can grow and reproduce on the host plant and thus the pest overcomes the resistance of the host plant. In other cases, the resistance can be broad and long-lasting if the plant is resistant to all races of the pest to which it is exposed. This broad (horizontal) resistance is most desirable but can be difficult to incorporate into a horticultural species because it is often polygenic.

Genetic engineering of plants has been used to add pest resistance to plants. Specific genes for resistance can be inserted without changing the rest of the genotype so cultivars will retain their desirable traits. Genes for resistance can also come from unrelated species. For example, genes that encode insecticidal crystal proteins from the bacterium *Bacillus thuriengensis* have been inserted into walnut, potato, and corn. Some of these transgenic plants are commercially available.

When purchasing seeds, a grower should be aware of pest resistance, and should utilize genetic resistance when it is practical. Tomato cultivars often are listed in catalogs with different letters following the cultivar names. These letters indicate pest resistance. Among the most common letters that horticulturists will note are *V*, *F*, and *N*, which represent verticillium wilt, fusarium wilt, and nematodes, respectively. Verticillium and fusarium wilts are fungal diseases caused by soil-borne pathogens. Nematodes are tiny round worms (see Chapter 16). All three pests can be difficult to control and can limit the productivity of a cultivar, so planting cultivars with genetic resistance to these pests may allow more successful production and profitable crops.

Pest Resistance to Pesticides

As stated previously, plant pests can mutate and circumvent pest resistance in the host. They can also mutate and develop races with resistance to pesticides. If a particular pesticide is used repeatedly because it is effective, those few individuals of the pest that do survive probably have some natural resistance to that pesticide. This may be the result of mutations or of selection for individuals in the population with inherent resistance to the pesticide. As the grower continues to spray with that pesticide, only the resistant pests will survive and reproduce. This may soon result in a new population of the pest with resistance to the pesticide, and this population has the potential to be devastating. Such repeated use of pesticides effectively breeds (selects for) resistance into pest populations, making specific pesticides obsolete by destroying efficacy. Therefore, pesticides from different classes of chemicals with different modes of action that control the same pest should be used in rotation. For example, a greenhouse applicator might use one pesticide for three or four months, a second pesticide for the next three or four months and finally a third pesticide for the following three or four months before starting the rotation over again. The chances of a pest population mutating and developing resistance to all three pesticides is quite low, especially if the three pesticides are from different chemical families. Wise use of pesticides, based on an understanding of their use, can lead to very effective pest control. Intelligent use of pesticides also can be important in an integrated pest management (IPM) program (see Chapter 16).

SUMMARY

Growth and yield characteristics (phenotype) are a result of the genetic makeup interacting with environmental conditions to which plants are exposed. A basic knowledge of plant genetics helps give a horticulturist a solid foundation on which to make decisions regarding many aspects of the culture of plants: knowing whether to purchase hybrid seeds, understanding seedling variability in segregating populations, choosing male or female plants, selecting cultivars with pest resistance, deciding how to propagate a plant, or developing a pest control program. The genotype has a dramatic influence on phenotype. In a field populated by one clone, some variability will be evident, but all plants will generally be similar. This variability is normally a result of environmental influences. Subsequent chapters will discuss environmental influences and manipulations important to horticulturists. After selecting the genotype, horticulture involves realizing the potential for growth determined by the genotype. This is accomplished by manipulating the plant and the conditions under which it is grown.

REFERENCES

Allard, R. W. 1999. *Principles of Plant Breeding.* (2nd ed.). Wiley, New York. 254 pp.

Banga, S. S. and S. K. Banga (Eds.). 1998. *Hybrid Cultivar Development.* Springer-Verlag, Berlin. 536 pp.

Chawla, H. S. 2002. *Introduction to Plant Biotechnology.* (2nd ed.). Science Publishers, Enfield, NH. 528 pp.

Chopra, V. L., V. S. Malik, and S. R. Bhat (Eds.). 1999. *Applied Plant Biotechnology.* Science Publishers, Enfield, NH. 384 pp.

Coors, J. G. and S. Pandey (Eds.). 1999. *Genetics and Exploitation of Heterosis in Crops.* American Society of Agronomy, Crop Science Society of America, Soil Science Society of America, Madison, WI. 524 pp.

De Nettancourt, D. 2001. *Incompatibility and Incongruity in Wild and Cultivated Plants.* (2nd ed.). Springer, Berlin. 322 pp.

Fairbanks, D. and W. R. Andersen. 1999. *Genetics the Continuity of Life.* Brooks/Cole, Wadsworth. Pacific Grove, CA. 848 pp.

Fu, T. J., G. Singh, and W. R. Curtis (Eds.). 1999. *Plant Cell and Tissue Culture for the Production of Food Ingredients.* Kluwer Academic, New York. 290 pp.

Geneve, R. L., J. E. Preece, and S. A. Merkle (Eds.). 1997. *Biotechnology of Ornamental Plants.* CAB International, Wallingford, UK. 402 pp.

Hrazdina, G. (Ed.). 2000. *Use of Agriculturally Important Genes in Biotechnology.* IOS Press, Amsterdam. 252 pp.

Henry, R. J. 1997. *Practical Applications of Plant Molecular Biology.* Chapman & Hall, London. 258 pp.

Janick, J. (Ed.). Annually. *Plant Breeding Reviews.* AVI, Westport, Conn.

Jensen. N. F. 1998. *Plant Breeding Methodology.* Wiley, New York. 676 pp.

Marcotrigiano, M. 1997. Chimeras and variegation: patterns of deceit. *HortScience* 32:773–784.

Owen, M. R. L. and J. Pen. 1996. *Transgenic Plants: A Production System for Industrial and Pharmaceutical Proteins.* Wiley, Chichester. 348 pp.

Poehlman, J. M. and D. A. Sleper. 1995. *Breeding Field Crops.* (4th ed.). Iowa State University Press, Ames. 494 pp.

Rajaskaran, K., T. J. Jacks, and J. W. Finley. (Eds.) 2002. *Crop Biotechnology.* American Chemical Society. Washington, D.C. 259 pp.

Shahidi, F., P. Kolodziejczyk, J. R. Whitaker, A. L. Munguia, and G. Fuller. (Eds.). 1999. *Chemicals via Higher Plant Bioengineering.* Kluwer Academic, New York. 280 pp.

Simmonds, N. W. and J. Smartt. 1999. *Principles of Crop Improvement.* (2nd ed.). Blackwell Science, Oxford, UK. 412 pp.

Snustad, D. P., and M. J. Simmons. 2003. *Principles of Genetics.* (3rd ed.). Wiley, New York. 840 pp.

THE AMBIENT ENVIRONMENT

LIGHT

All horticultural plants, except edible mushrooms, require light to complete their life cycles. Through the process of photosynthesis, plants convert light energy into chemical energy, which they use for growth, development, and the maintenance of life. Light is also important to plants for pigment (color) formation, plant growth habit, plant shape, plant size, flowering, fruiting, seed germination, onset of dormancy, onset of plant hardiness, leaf movements, formation of storage organs, autumn coloration, and defoliation of temperate zone trees.

Plants respond to light quality, quantity, and photoperiod. **Light quality** is perceived by the human eye as color and corresponds to a specific range of wavelengths. **Light quantity** refers either to the amount of light given off by a light source, such as the sun or a lamp, or to the amount of light that strikes an object, such as a leaf. The **photoperiod** refers to the duration of the lighted period (day length) and the relationship between the dark and lighted periods; the duration of darkness is critical to photoperiodic responses.

Light is electromagnetic radiation, some of which can be perceived visually. Visible light is only a small portion of the electromagnetic spectrum; it extends from about 380 to 770 nm (wavelength) (Figure 5-1). Light behaves both as waves and discrete particles of energy called **photons** or

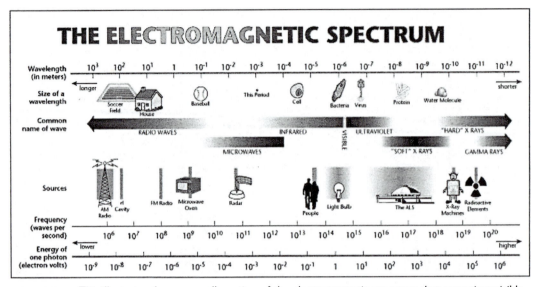

FIGURE 5-1. This illustrates the very small portion of the electromagnetic spectrum that comprises visible light. As the wavelengths increase, the energy of one photon also decreases.
Courtesy of the Advanced Light Source, Lawrence Berkeley National Laboratory.

quanta. The photons travel with specific associated wavelengths, expressed as lambda (λ) (Figure 5-2), which determine the color (**light quality**). Wavelength is measured in units called nanometers (nm), which were formerly called millimicrons (mμ). A nanometer is one billionth of a meter, or 10^{-9} meters. In the early literature wavelength was reported in angstroms Å, which equal one-tenth of a nanometer, or 10^{-10} meters. Short wavelengths indicate great energy. Colors such as violet and blue have short wavelengths (high energy), and red and far-red light have long wavelengths (low energy).

Horticulturists are concerned with light for proper plant growth and development. Because humans also perceive light, we tend to think that plants absorb light in the way that the human eye does, but that assumption is false. The eye is most sensitive to light of wavelengths in the region of 555 nm, which appears as a green or greenish-yellow color. Conversely, most plant leaves have a relatively high reflection of green light, which makes them appear green. Plant leaves absorb wavelengths at both ends of the visible spectrum, where the eye does not respond well. Therefore, one should not think of plants "seeing light"

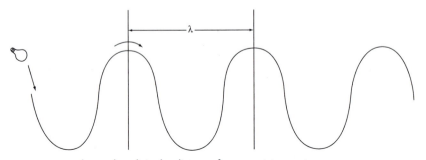

FIGURE 5-2. A wavelength is the distance from crest to crest.

(absorbing light) as humans see light. The importance of this will be clear in the following sections.

....

PHOTOSYNTHESIS

The process of photosynthesis is basic to the growth of higher plants and to life as we know it. It is through the process of photosynthesis that the sun's energy is incorporated into life. Therefore, it is among the most important chemical events on earth. This energy and the oxygen liberated during photosynthesis have enabled life to develop and flourish on our planet.

Photosynthesis requires the inputs of carbon dioxide (CO_2), water (H_2O), and light, and in the presence of chlorophyll, it culminates in the formation of glucose ($C_6H_{12}O_6$) or other carbohydrate and the liberation of oxygen (O_2). This is summarized simplistically in the following equation:

$$12H_2O + \text{light} + 6CO_2 \longrightarrow C_6H_{12}O_6 + 6O_2 + 6H_2O$$
$$\text{(in chloroplasts)}$$

Because glucose is food and has calories, and fertilizer mineral nutrients have no calories, it is true that we "feed plants with light, not with fertilizer." Through photosynthesis, plants incorporate light energy into chemical energy in the form of glucose. If plants are deprived of light for too long, they will starve to death, even with adequate fertilization.

Chloroplasts

Photosynthesis occurs in the subcellular organelles known as chloroplasts, which occur in large numbers in leaves (see Chapter 3 and Figures 3-1 and 5-3). Chloroplasts are bounded by two membranes separated by a narrow gap. Within the chloroplasts, there are layers of membranes that contain chlorophyll. Pairs of these membranes form flattened sacs called **thylakoids**. Thylakoids frequently stack in groups

in various locations within the chloroplasts. These thylakoid stacks resemble stacks of green flat dishes or pancakes and are known as **grana** (singular **granum**). The colorless matrix between the grana or outside the thylakoids is called the **stroma** (Figure 5-3). Both the grana and stroma have important roles in photosynthesis. Chloroplasts also contain their own DNA and RNA. Therefore, some protein synthesis can occur within chloroplasts. The DNA in the nucleus also codes for some of the chloroplast proteins.

Chloroplast Pigments

Within chloroplasts there are several different pigments. Of these, chlorophyll *a* is the most important for photosynthesis. Chlorophyll *b* is also important in higher plants, because it primarily transfers its energy to chlorophyll *a*. Chlorophyll *a* exists in two forms based on two minor absorption peaks: One has a specific light absorption peak at 703 nm and is known as

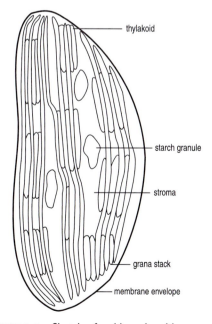

FIGURE 5-3. Sketch of a chloroplast. Note thylakoids, starch granules, grana stacks, stroma, and membrane envelope.

P_{700}. The other has an absorption peak at 682 nm and is known as P_{680}. Carotenes and xanthophylls also are present, and are discussed later in this chapter.

Chlorophyll *a* and chlorophyll *b* have major light absorption peaks between 400 and 500 nm (violet-blue) and between 600 and 700 nm (orange-red), with very little absorption between 500 and 600 nm (green and yellow-green) (Figure 5-4). Because of these specific absorption peaks, a measurement of **photosynthetically active radiation** (**PAR**) from 400–700 nm does include the important wavelengths. However, this can be considered too general because it also includes nonphotosynthetically active radiation.

The Process of Photosynthesis

The process of photosynthesis can be divided into two reactions based on the direct requirement of light: the light reactions and carbon dioxide fixation. These are discussed separately.

Light Reactions

The light reactions of photosynthesis are driven directly by light energy and occur in the grana of the chloroplasts. To understand the influence of light, one must first know the response of pigments (colored molecules) to light. Each chlorophyll and carotenoid pigment molecule can absorb one

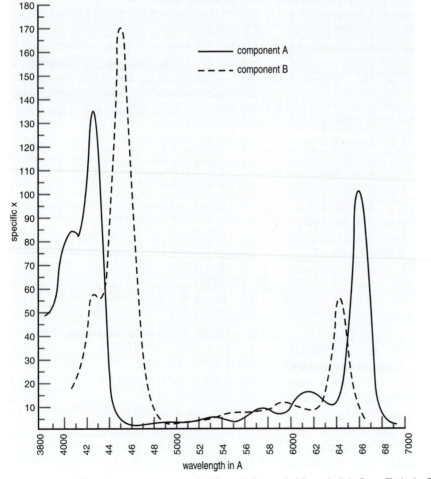

FIGURE 5-4. The absorption spectra of chlorophyll *a* and chlorophyll *b. From Zschiele, F. P. and C. I. Comar. 1941. Botanical Gazette 102(3):468. Published by University of Chicago Press.*

photon at a time; this will cause one electron in the molecule to become energized or to be in an *excited state*. This excitation energy can be lost as heat, emitted as radiation of lower energy or longer wavelength (fluorescence), or transferred to other molecules. The chlorophyll molecule loses excitation energy by heat loss and fluorescence. The latter is minimal in leaves, however, because the excitation energy can be used in photosynthesis. The carotenoids and chlorophyll *b* can transfer their excitation energy to chlorophyll *a*. Isolated chlorophyll solutions within vessels fluoresce as a dark red color when a beam of light is shined through. The red color can be seen from the side of the direction of the light source. Chlorophyll, like other fluorescent molecules, has the ability to change the wavelength of light that strikes it.

There are two photosystems in the light reactions of photosynthesis, photosystem I and photosystem II. In photosystem II, energy is received from chlorophyll *a* (absorption peak of 682 nm (P_{680})) and from accessory pigments including chlorophyll *b*. Photosystem I has P_{700} at its center and primarily utilizes energy from chlorophyll *a*. Associated with photosystem II is the **Hill reaction**, which is the light-dependent splitting, or **photolysis**, of water. A knowledge of the Hill reaction helps one understand that the oxygen that evolves during photosynthesis comes only from water. Accompanying the splitting of water are the release of hydrogen and electrons. These electrons are transported from molecule to molecule from photosystem II to photosystem I and energy is lost. Eventually the electrons result in reducing an important metabolic nucleotide, nicotinamide adenine dinucleotide phosphate NADP to make $NADPH_2$, which can be used in carbon dioxide fixation. In addition to the $NADPH_2$, about three high energy adenosine triphosphate (ATP) molecules are formed from adenosine diphosphate (ADP) and inorganic phosphate (Pi) through a process known as **photophosphorylation**.

Carbon Dioxide Fixation

The high energy ATPs and $NADPH_2$s that form in the light reaction of photosynthesis are used in many biochemical reactions, including the recycling of energy fixed during photosynthesis and CO_2 fixation (i.e., CO_2 fixation utilizes the energy from the light reactions).

Carbon dioxide enters the leaves through the stomates and is *fixed* (bonded) to another compound. A series of reactions follow, eventually resulting in the formation of glucose plus the original compound that fixed the CO_2. It takes six carbon dioxide molecules to generate one glucose molecule. From the glucose, the carbohydrate end products sucrose (table sugar) and starch can be made. Ultimately all organic components of the plant can be derived, starting with the glucose from photosynthesis.

There are two major pathways by which CO_2 may be fixed by plants. In one pathway, the molecule that fixes CO_2 splits to form two three-carbon molecules, hence this is called the **three-carbon pathway**. Plants that utilize only this pathway are referred to as C_3 **plants** and include most horticultural species. In the other pathway, a three-carbon compound fixes the CO_2 and a four-carbon compound results. A portion of this **four-carbon pathway** occurs in the leaf mesophyll cells and a portion occurs in the bundle sheath cells surrounding the vascular bundles in the leaf. In these C_4 plants, the bundle sheath cells are well developed, whereas in C_3 plants, the bundle sheath cells, if present, are much less distinct. In plants that utilize the four-carbon pathway, the three-carbon pathway also is employed, and there is a spatial relationship between the leaf mesophyll (C_4 enzymes) and the bundle sheath cells (C_3 enzymes). C_4 plants include horticulturally important species such as sweet corn and Bermudagrass.

In another group of plants that are variants of C_4 plants, the C_3 and C_4 pathways occur within the same cell. This system is known as **crassulacean acid metabolism**

(CAM). Plants that employ this mechanism are known as CAM plants; examples include many succulent plants such as moss rose (*Portulaca grandiflora*) and pineapple. CAM plants are unusual in that they open their stomates at night; thus CO_2 is primarily fixed at night. During darkness CAM plants fix the CO_2 with a compound called phosphoenolpyruvic acid (PEP) to form oxaloacetic acid. Oxaloacetic acid then is converted into malic and other acids that are stored in the vacuole. During daylight hours the acids leave the vacuole and are decarboxylated to yield CO_2 and PEP again. The CO_2 then is used by the CAM plants for the photosynthetic manufacture of sugars and starch. CAM plants include, but are not limited to, cacti and many succulents that are native to arid regions. Transpirational water loss is minimized when stomates are closed during the hot daylight hours and opened only after dark. See Table 5-1 and Figure 5-5 for some examples of C_3, C_4, and CAM plants.

The three-carbon pathway was elucidated by Melvin Calvin, Andrew A. Benson, and James A. Bassham. Dr. Calvin received the Nobel prize for work in this area and the three-carbon cycle is often called the **Calvin cycle**. The general scheme of the C_3 pathway is outlined in Figure 5-6. The four-carbon pathway often is referred to as the **Hatch-Slack pathway**, after two scientists who made great contributions to its clarification and its general scheme is outlined in Figure 5-7.

When both C_3 and C_4 plants are exposed to high light levels and relatively high temperatures, the C_4 plants have a more rapid photosynthetic rate on a leaf area basis. In fact, in full sunlight, at temperatures from 15–35°C (77–95°F) the C_4 plants are at least twice as efficient as C_3 plants in terms of the production of dry matter. One reason why C_4 plants are more efficient than C_3 plants is the fact that the enzyme that fixes CO_2 to PEP in the C_4 pathway has a higher affinity for CO_2 than the enzyme that fixes CO_2 to ribulose-1,5-bisphosphate (RuBP) in

TABLE 5-1. Examples of C_3, C_4 and CAM horticultural plants and common weeds

HORTICULTURAL PLANTS		
C_3 PLANTS	C_4 PLANTS	CAM PLANTS
Kentucky bluegrass	corn	*Agave americana*
beet	zoysiagrass	jade plant
spinach	Bermudagrass	pineapple
lettuce	Bahiagrass	Spanish moss
bean		*Sansevieria* spp.
carrot		
creeping bentgrass		

COMMON WEEDS	
C_3 WEEDS	C_4 WEEDS
quackgrass	crabgrass
lambsquarters	redroot pigweed
cocklebur	common Russian thistle
jimson weed	

FIGURE 5-5. Examples of C_3 (ornamental cabbage, *left*), C_4 (corn, *middle*), and CAM (jade plant, *Crassula argentea, right*) plants.

the C_3 pathway. In addition, light stimulates a process called **photorespiration**, which always accompanies the C_3 pathway. In this process, light causes the plant to use oxygen and release CO_2. In C_3 plants as much as one-half of the CO_2 that is fixed is reoxidized to CO_2 by photorespiration. This process occurs in cellular organelles called *peroxysomes*. In C_4 plants, photorespiration also accompanies the C_3 pathway; however, the C_3 pathway occurs deep within the leaf, in the bundle sheath cells, and the liberated CO_2 can be fixed with PEP or RuBP.

Because of the greater efficiency of C_4 plants in high light and high temperatures, C_4 plants often grow faster and better in the tropical regions or summer months of temperate regions. It is interesting that, as a rule, weeds that are C_3 species, such as lambsquarters, grow and flourish in the spring, whereas C_4 weeds, such as crabgrass, have a competitive advantage during the summer (Table 5-1). It also can be seen from Table 5-1, that many cool season crops are C_3 plants, whereas tropical and some warm season crops are C_4 and CAM plants.

SPACING AND ORIENTING PLANTS FOR LIGHT INTERCEPTION

Because photosynthesis is critical to plant life, light interception is always a consideration

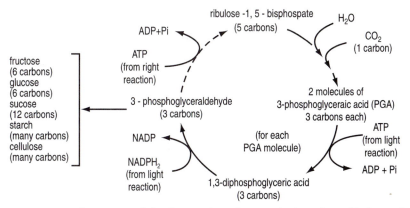

FIGURE 5-6. A summary of the three-carbon photosynthetic pathway (dark reactions). A broken line indicates several-step reactions in which products form 3-phosphoglyceraldehyde molecules combine to form multicarbon compounds. The number of carbon atoms in the molecule are indicated in parentheses.

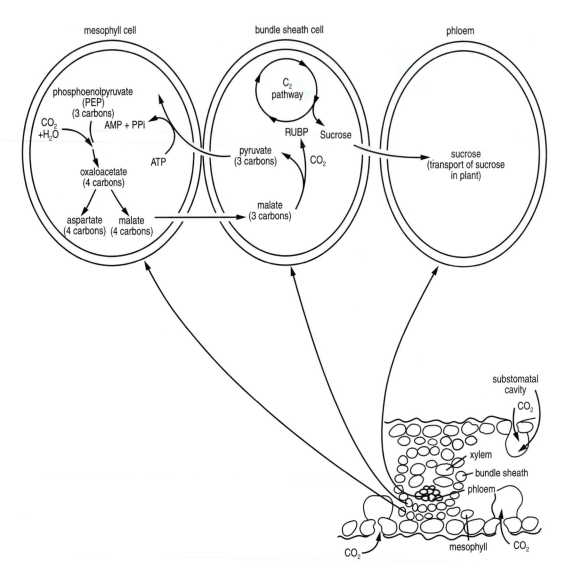

FIGURE 5-7. Schematic leaf cross section illustrating the relationships between the C_3 and C_4 pathways.

for horticulturists. Spacing and orienting plants to make the best use of available light is an important concern, but other factors also must be considered: desired plant quality, accessibility by the grower, air circulation, disease control, and profitability. Overcrowding plants can reduce quality because leaves will shade each other, internodes will elongate, and each plant will accumulate less dry matter than if the plants were spaced more widely.

Optimal Spacing

Field-grown plants frequently are planted at the ideal spacing for mature plants. Consequently, field space often is wasted when the plants are young and require less space. With some tree-fruit crops, growers will interplant other crops such as vegetables between the rows to obtain additional harvests until the trees reach a more mature size. This a common practice in many regions of the world, especially in countries with large populations, such as China. Alternatively, plants may be placed close together to maximize early yields, then thinned as they begin to crowd each other.

Proper spacing is critical for field-grown vegetables and other crops. If plants are planted too wide apart, space is not utilized efficiently, and yield per hectare is low. If plants are planted too densely, however, they shade each other and compete for nutrients and water, causing poor yields and reduced plant quality. When choosing proper spacings for plants, horticulturists should minimize shading for optimum photosynthesis and consider best plant growth habit.

There are times when uneven spacing of field-grown horticultural crops results from unplanned events. If seeds are sown that have an uneven germination rate or relatively low viability, for example, or if pests, diseases, or unfavorable environmental conditions are prevalent, an uneven or poorly spaced crop will result. Late germinating seeds often will not grow well because they will be shaded by plants that previously germinated. Therefore, when conditions are not optimal, competition for light can result in reduced yields and an uneven crop.

Weeds can compete with horticultural crops for light (see Chapter 16). Uncontrolled weeds will shade the horticulturally important plants, resulting in low quality plants and reduced yields. Weeds are controlled by using mulches, herbicides, cultivation, or mowing. Timing of weed control is critical to minimize their competition with the important plants.

Most growers of containerized plants consider spacing to be acceptable if the tips of leaves of adjacent plants are touching. Container-grown plants can be moved periodically as they grow to make maximum use of available space and still maintain quality. Some hydroponic growing systems automatically adjust spacing as the plants grow. Hydroponic troughs that contain lettuce seedlings, for example, are initially placed close to each other so that several thousand seedlings are in a relatively small

space. They remain at this spacing until leaves of adjacent plants touch or begin to overlap, at which time the troughs are moved farther apart until the plants grow and begin to crowd each other again. After a series of moves, the plants ultimately reach their final and optimum spacing for growing into high-quality, uniform plants ready to be marketed. Moving hydroponically grown lettuce or other containerized plants allows growers to make maximum use of available space while growing quality plants. Not all systems can be automated easily like the hydroponic systems, and hand labor may be necessary to move containerized plants. In these cases, growers must weigh the costs of moving plants against the advantages of early close spacings to determine if the initially close spacings are economically viable or if plants should be set at the final spacing.

Interplanting plants with upright habits and spreading plants within or between rows is efficient. In the field, rows of trees may be interplanted with shrubs. Some fruit and nut orchards are planted similarly, with upright and more spreading tree cultivars or species being interplanted. Container nursery stock is frequently placed in a similar pattern, alternating two or more cultivars or species with different growth habits. Whenever these planting patterns are used, spacing must be appropriate so the upright forms do not shed excessive shade onto spreading forms. It is also important that the branches of spreading plants do not intermingle with and shade the upright forms. Planting patterns and spacings must therefore be planned carefully to make best use of available space, while still allowing for optimum light interception by the plants for production of uniform and high quality plants.

Spacing of vegetables varies widely among commercial growers and between backyard gardeners and professional producers. Commercial growers consider the space required by the crop, total yield associated with different spacings, ultimate desirable size of the saleable product (crowding leads to smaller sizes), size and tire spacing of their equipment, and overall efficiency of their field operations. Spacing for the highest yield may be unacceptable because it may be difficult to control weeds and there may be poor air circulation, resulting in high disease incidence. Extreme crowding leads to competition for light and lowered yields while extra wide spacing will also result in unacceptably low yields for a field.

The spacings given on most seed packets are good if space is not limited and weeds are to be controlled mechanically. However, more intensive plantings can give acceptable yields for backyard gardeners with limited space. Rather than planting single rows, as in many commercial operations, double, triple or wide rows often utilize space more efficiently (see Figure 7-19). Planting small spaces that can be reached from the edges can maximize use of space, since rows for walking are not needed. For small yields needed to feed a family, planting groups of vegetables in blocks of various designs has proven efficient. However, even using these planting designs, individual plants must not be too crowded, or they will perform poorly because of competition for light, water, nutrients, and because of disease problems associated with poor air circulation.

Landscape plants must be properly spaced for light interception. In landscapes, plants of various species and cultivars that have different growth habits and growth rates are commonly interplanted. The beauty and function of these plants should not be the only considerations; light interception throughout the anticipated life of the landscape is also important. To allow viewing into a landscape planting, shorter statured plants, such as dwarf selections, short annuals and perennials, and low-growing spreading plants are usually located toward the front with taller individuals progressively located toward the rear. The amount of light and its direction must also be considered so that the plants that grow tallest will not shade out the shorter cultivars.

Plants that are spaced relatively closely often are pruned to maintain high quality and decrease competition for light (see Chapter 13). Alternately dwarf forms may be planted or cultivars may be grafted onto dwarfing rootstocks to reduce vegetative growth. This is common with apples, and dwarf trees can be spaced closely; yields per hectare are greater this way than when the same cultivars are on standard rootstocks. Smaller plants proportionally have a greater surface-area-to-volume ratio than larger plants. Because there is proportionally greater surface area and dwarf plants can be planted closer together, more leaf area is exposed for maximum interception of sunlight for photosynthesis than with standard-sized plants. Increased yields are subsequently reported for fields planted with dwarf fruit trees.

Applications of plant growth retardants are often made to keep closely spaced plants small and maintain quality (see Chapter 12). In controlled environments, size of many plants can be controlled by manipulating the differences (DIF) between night and day temperatures (see Chapter 6).

Orientation Issues

Row orientation to maximize light interception is an oft-debated subject. Many people believe that if the crop is uniform, the row orientation does not matter. Concerns, such as entry into the field, traffic patterns, and contour of the land often dictate the orientation of rows.

Greenhouse orientation is also discussed frequently. The type of greenhouse and latitude can influence orientation. A single greenhouse with a central ridge should be located in an east–west direction at latitudes above 40°N to minimize the time that the shadow from the ridge is on the plants during winter months. Below 40°N, a similar, single greenhouse should be oriented north–south because of the higher angle of the sun. Gutter-connected greenhouses may best be oriented in a north–south direction to minimize the shadow that would be cast by adjacent roofs and gutters. Prevailing winds should also be considered when planning greenhouse construction. During the winter, energy can be saved by having the smallest side, usually the end, facing the wind.

Experimentally, and for production in some regions, fruit trees are planted so their trunks are at an angle. These angles may alternate within a row. Additionally, various trellising systems are used on fruit crops. These are designed to expose a maximum amount of plant leaf surface area to sunlight. The increased light interception brings about more photosynthesis within each plant and thus higher yields. Sometimes these plants can be planted more densely than normal and can increase yields per hectare because of plant orientation or training.

Regardless of the orientation, it is wise to plan carefully when planting species that will grow to be different sizes, such as in a garden or landscape. If the planting is against a building or fence, the shortest plants must be placed away from the wall, those of intermediate height in the center, and the tallest plants close to the wall. This will maximize the effectiveness of the planting and minimize shading by the tallest plants.

In an open garden or other planting, the tallest plants should be located to the north and the shortest should be to the south. This also will minimize shading. Tall sweet corn or tomato plants can easily shade lower growing plants such as squash and thus reduce their yield.

Pruning Plants for Light Interception

Plants are pruned for a variety of reasons (see also Chapter 13). Among these reasons is the horticultural objective of improving the interception of light, which in turn enhances photosynthesis. Pruned apple trees generally outyield nonpruned trees, partly because of better light penetration and the increased surface area of the tree that is exposed to light. Consequently, fruit growers often make thinning-out pruning cuts where they remove entire branches. This admits more light into the center of the tree (Figures 5-8, 5-9, and 5-10). Thus productivity can be increased, and, because of better exposure to the light, fruit such as apricots, apples, and peaches will have increased red color (Figure 5-11). Additionally, some fruit trees, such as peach, are trained to have an open center (or open-vase system) to maximize the interception of light and thus enhance yield. The proper pruning of fruit trees and other plants often varies considerably with species and even cultivar, and appropriate references should be consulted before pruning any plant. A branch that is removed by pruning cannot be replaced.

Some horticultural crops, such as red raspberry are thinned by pruning to insure adequate cane-to-cane spacing and light interception. If not properly thinned, raspberry yields are reduced considerably. The distance between canes will vary with the planting and training system used by the grower.

FIGURE 5-8. Thinning-out cuts often involve removal of entire branches and allow better light penetration.

FIGURE 5-9. Peach trees are pruned so that they have an open center, and thus more leaf area is exposed for interception of light.

Ornamentals often are pruned to ensure good light interception to attain fuller, more attractive plants. An example is the proper pruning of a hedge. Hedges should be pruned so that they are wider on the bottom than the top so that the tops of the plants will not shade the bottoms. Fuller, more beautiful hedges result (Figure 5-12). Because of the sheared appearance of hedges, holes (Figure 5-13) or open bottoms stand out visually. Ornamentally interesting pruning techniques, including espalier, topiary, and bonsai must be done considering how the plant will intercept light so that it can grow and thrive (Figures 5-14 and 5-15).

FIGURE 5-10. This peach orchard has been subjected to a pruning system designed to maximize light interception.

FIGURE 5-12. A properly pruned hedge. Note that the bottom is wider than the top, facilitating good light interception by the plant's lower leaves.

FIGURE 5-11. Light is necessary for the formation of the red blush color on light-skinned peaches. The light yellow area being pointed out has captured the shape of the overlaying leaf that cast its shadow on this fruit.

FIGURE 5-13. An improperly pruned hedge. The resulting holes impair the desired visual effect.

FIGURE 5-15. Bonsai plants. *Top*: Common privet (*Ligustrum vulgarus*), approximately 20 years old, collected from a pasture in Louisiana. Height 42.5 cm. *Bottom*: Japanese garden juniper (*Juniperus procumbens nana*), approximately 20 years old. Height 57.5 cm. The art of bonsai, literally "tray planting," is said to have originated in China and has been highly developed by Japanese bonsai masters. Wiring of stems on bonsai plants is done to give special appearance to the plant, and light interception by these branches must be considered.

Photos courtesy Bob Hampel, Minnesota Bonsai Society.

FIGURE 5-14. Examples of special-purpose pruning techniques. *Top* and *middle*: Forms of **espalier** pruning, where fruit trees or ornamentals are trained against a wall. Espalier pruning can maximize light interception in these plants. *Bottom*: an example of **topiary** art, where plants are pruned into specific shapes, such as animals. If topiary pruning is done improperly, shading of lower portions will occur and dead spots will become evident.

ETIOLATION

Light is necessary for chlorophyll formation in most angiosperms but often not in gymnosperms, ferns, and many lower plants. Light is also a requisite for chloroplast division. Within limits, the number of chloroplasts in cells increases with the increasing light level in the blue and red wavelengths.

When plants or plant parts are covered to exclude light or are moved to a location devoid of light, the process is called **blanching** (from the French **blanche**, meaning white). **Etiolation** (from the French **étioler**, meaning to grow weak, or grow pale), however, means the development of a plant or plant part in the absence of light. Therefore, plants often become etiolated in response to a blanching treatment applied by the grower.

Etiolated plants often show characteristic symptoms including white to yellowish stems and leaves, spindly growth of stems, extreme elongation of internodes, suppression of leaf expansion in dicots, and formation of seedling hooks on dicot epicotyls or hypocotyls (Figure 5-16 and 5-17). The stems and leaves are white to yellowish because light is required for grana formation

and chlorophyll biosynthesis. In fact, plastids called **etioplasts**, which are characterized by a lack of grana, form in cells of plants grown in the absence of light. Etiolated plants often have a yellowish appearance because of the presence of pigments called **carotenoids**. Carotenes are discussed in more detail later in this chapter. Light inhibits internode elongation—hence the tall spindly growth so characteristic of etiolation. This is probably an adaptive mechanism that allows overcrowded plants to stretch for the light and thus become leggy. Although the leaves of etiolated dicots are small, those of etiolated monocots are not. Monocots generally produce either normal-sized leaves or sometimes produce larger leaves in the dark than in the light. The larger leaves are evident on many flowering bulbs, such as *Narcissus*, *Tulipa*, and *Lilium*, that are forced in the dark in coolers (Figure 5-18). The formation of the epicotyl or hypocotyl hook occurs in a dicot seedling when it is in the dark, below the soil surface. During the process of emergence though the soil surface, the hook emerges first (thus preventing the apical meristem from pushing through the soil first where it might be injured). Blue light causes a straightening of the hook. The apical growing point is thus lifted above the soil surface, receiving minimal damage; the shoot then grows upward toward the source of light (Figure 5-19).

Blanching is applied to several vegetable crops to increase the quality of the produce or to meet consumer demand. It is a common practice to blanch cauliflower heads to maintain a milky-white color and mild flavor; growers accomplish this by tying the outer leaves over the head to exclude light (Figure 5-20). Cauliflower plants grown without blanching, especially during the times of the year when the light levels are high, have yellowish heads of a lower quality. There are self-blanching cultivars in which the outer leaves bend to cover the heads and thus provide shade. Blanching of all or portions of other vegetables—such as green table onions, asparagus, and celery—can be accomplished

FIGURE 5-16. Corn (*top*) and pea (*bottom*) seedlings, note tall white etiolated shoots compared with normal light-grown plants.

FIGURE 5-17. Bare-root shrubs showing etiolated growth that occurred in relative darkness during cold storage. Bright sunlight and dry conditions could cause damage to these tender stems.

FIGURE 5-18. Etiolated lily plant that was forced in a dark cooler.

FIGURE 5-19. Honeylocust seedlings at several stages of development show the hypocotyl hook, which straightens upon exposure to blue light.

by ridging the soil along the rows to cover the portion of the plant in which an etiolated appearance is desirable. The lower portions of the table onions and celery may be blanched (Figure 5-21), and the asparagus must be harvested before emergence if white spears are desired. During the dormant season, rhubarb crowns and chicory roots are sometimes dug up and moved inside, where new growth is forced in the darkness. People in some regions demand high-quality, etiolated produce that results from forcing in the darkness.

sun-burned appearance. This also frequently occurs on grocers' shelves, especially when tubers are sold either in clear polyethylene bags or directly on the shelves. This is undesirable because the production of alkaloids accompanies greening of potato tubers. If eaten in enough quantity, such tubers may cause illness (see Chapter 15).

Etiolation also is important for the vegetative propagation of plants. It is well known that stem cuttings taken from etiolated stock plants often produce adventitious roots better than cuttings from stock

FIGURE 5-20. A cauliflower plant (*top*) with its leaves tied up to exclude light (blanching), and the resulting white curds (head) (*bottom*).

Soil or mulch often is mounded around potato plants so white (or nongreen) tubers can be harvested. Because tubers are actually stem tissue, portions of tubers that are exposed to the light will produce chlorophyll, causing a characteristic green or

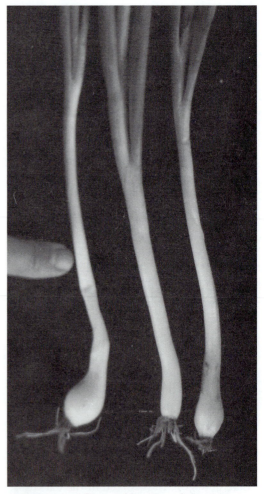

FIGURE 5-21. Green table onions (sometimes called scallions) with an etiolated portion produced by ridging the soil against the base of the plants as they grow.

plants that were not etiolated. This seems to be partly the result of light causing the destruction or inactivation of root-promoting factors in stems, including auxin and rooting cofactors. The fact that the rooting medium excludes light from the base of cuttings may be crucial to root initiation. In addition, the process of layering generally involves blanching of a plant part. Layering is a method of stimulating rooting while the propagule is still attached to the stock plant (Figure 5-22). With many types of layering, a portion of a stem is covered with soil and root formation occurs, partially as a result of the exclusion of light. The rooted stem can then be cut from the stock plant and handled like a rooted cutting (see Chapter 14).

EFFECT OF PIGMENTS

Carotenoids

Carotenoids are pigments that reflect light in the yellow to red-orange range and absorb blue light. When composed exclusively of carbon and hydrogen, they are orange compounds called **carotenes**. When carotenoids also contain oxygen they are yellow pigments termed **xanthophylls**. Carotenoids are responsible for the orange color of carrots; the red carotene, lycopene, is found in tomato and watermelon fruits and in other plants and plant parts. Carrots are said to aid vision because animals can, through digestion, split the β-carotene molecule into vitamin A, the lack of which is implicated in night blindness.

Although light is necessary for chlorophyll formation in angiosperms, it can also destroy chlorophyll through the process of photooxidation (also known as photobleaching). Carotenoids help protect chlorophyll from this photodestruction. It has been shown that some mutant plants that lack carotenoids will synthesize chlorophyll. However, that chlorophyll breaks down upon prolonged exposure to light. This

FIGURE 5-22. Tip layered black raspberry (*Rubus* sp.). The new plant formed where the primocane tip was buried in the ground.

suggests that some mutants that lack chlorophyll do so only because they do not contain carotenoids. Perhaps the same is true for some variegated plants. It is common to see the variegated or white portion of a

Dieffenbachia leaf turn a light green color when the plant is grown under low light and for this area to become white again when the plant is placed under a higher light level. This may be related to a lack of carotenoids, because the variegated portion of the leaf becomes white, not yellow. Carotenoids are produced by all normal photosynthetic cells and are important in protecting chlorophyll and transferring light energy to chlorophyll *a*.

Another function of carotenoids is the color that they impart to flowers, seeds, and fruits. This color attracts pollinators, as well as animals that dispense seeds, often after eating the seeds or fruits. For example, daffodil and dandelion flowers are yellow because of they contain the xanthophyll, violaxanthin in their chromoplasts (modified chloroplasts).

The green of chlorophyll masks the yellow to red of the carotenoids, so plant foliage generally appears green. Anything that results in chlorophyll being destroyed more quickly than it is produced allows the carotenoids to become visible. Such yellowing of leaves is called **chlorosis**, and yellow leaves are referred to as **chlorotic**. Chlorosis is the result of various disorders, including some fungal diseases and several nutrient deficiencies. Of particular note are those nutrients that comprise part of the chlorophyll molecules, such as nitrogen and magnesium, and those that are necessary for chlorophyll formation, such as iron, zinc, and manganese (see Chapters 8 and 9). The leaves of some species turn yellow in the autumn because chlorophyll destruction unmasks the carotenoids. Sometimes tannins (dark colored) will also be present, which, when present with carotenoids, produce a golden brown effect. The carotenoids decompose as winter approaches, leaving the tannins, and thus brown leaves grace the winter landscape.

Sometimes pigments will be present in abundant quantities and mask chlorophyll. Thus red or purple-leaved plants may result.

Cryptochromes and Phototropins

Blue light influences many plant responses, such as leaf area, petiole length, hypocotyl elongation, stem extension, and the bending of plants or plant parts toward the light. The **cryptochromes** and **phototropins** are proteins and are important blue light photoreceptors in plants. The cryptochromes are important in controlling stem elongation, the expansion of cotyledons and leaves, the daily rhythms of a plant (circadian rhythms), and flowering.

Both cryptochromes and phototropins can have similar influences. For example, phototropins are involved in the early stages of light inhibition of hypocotyl elongation during seedling development, whereas the cryptochromes mediate long-term inhibition of hypocotyls. When a seed germinates in the soil, it is dark and the seedling becomes etiolated as it pushes through the soil. When it emerges through the soil surface, blue light inhibits the growth thus de-etiolating the seedling and a sturdy, compact young plant results. Similarly, when plants have been grown in greenhouses under plastic films that filter out the blue light, they grow from 10 percent to 100 percent taller than plants grown in unfiltered sunlight.

Phototropins are important light receptors for plants bending toward a light source. This bending in one direction is known as **phototropism**. The hormone **auxin** is responsible for cell elongation on the side of a plant's stem that is away from the light source. This uneven elongation within the stem causes the bending toward the light that is evident when plants are grown in front of a window and is discussed in Chapter 11, and Figures 11-1 and 11-2.

Phytochromes

Plants irradiated with red (R) or far-red (FR) light or grown under different day lengths have more than 100 different growth and developmental responses (depending upon species). Important horticultural phenomena that relate to R and FR light, day length, or both include but are not limited to the following: gene expression, seed germination, branching, stem elongation, leaf expansion, chloroplast development, "sleep" movement of *Mimosa* leaves, flowering, storage organ (including bulb) formation, plantlet formation along the margins of some *Kalanchoë* leaves, fall coloration of leaves, leaf abscission, onset of cold hardiness, and onset of dormancy.

The understanding of these phenomena, has had important horticultural ramifications and was elucidated by a USDA research team headed by H. A. Borthwick and S. B. Hendricks. These researchers first confirmed that certain lettuce seeds, which would show little germination in the dark at 25°C (77°F) would germinate if briefly irradiated with red (660 nm) light. They further discovered that if the red was immediately followed by a brief irradiation with far-red (730 nm) light, germination was inhibited; the inhibition was reversed if the FR was immediately followed by a brief R light treatment. This was reversible for many repeated alternations and the germination percentage of lettuce seeds was dependent on the last light treatment received (Table 5-2).

Borthwick and Hendricks predicted that only one compound was functioning as the photoreceptor and this was alternately present as red and far-red light-absorbing forms. It was later discovered to be a family of blue-green (reflecting) protein pigments with at least some overlapping functions. To each of these pigments is attached a special, nonprotein, light-absorbing portion called a **chromophore**. These pigments are each known as **phytochrome** (P). Each is present in plants in two forms: One form has a maximum light absorbance at 660 nm (red) and is, therefore, called P_r or P_{660}; the other form has a maximum light absorbance at 730 nm (far-red) and is therefore referred to as P_{fr} or P_{730}. When P_r absorbs red light it is converted to P_{fr}. Conversely, when P_{fr} absorbs far-red light it is converted to P_r (Figure 5-23). P_r is the form of phytochrome that is synthesized by plants; the P_{fr} forms from P_r. In certain dicots, but apparently not in monocots, P_{fr} slowly reverts to P_r during times of darkness; the importance and practicality of this phenomenon is further discussed later in this chapter. Within plant tissues, P_r is a relatively stable compound, whereas P_{fr} is destroyed at different rates, depending on the type of phytochrome. This is independent of the presence or absence of light.

Phytochromes have been found in the highest amounts (in etiolated plants) in tissues that are responsible for plant development in relation to light treatment—that is, in meristematic tissues and their newly formed daughter cells, including shoot tips

TABLE 5-2. The effects of red and far-red light on lettuce seed germination

IRRADIATION	PERCENTAGE GERMINATION
R	70
R-FR	6
R-FR-R	74
R-FR-R-FR	6
R-FR-R-FR-R	76
R-FR-R-FR-R-FR	7
R-FR-R-FR-R-FR-R	81
R-FR-R-FR-R-FR-R-FR	7

From Borthwick, H. A., S. B. Hendricks, E. H. Toole, and V. K. Toole. 1954. *Botanical Gazette* 115:(3):216. Published by University of Chicago Press.

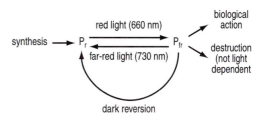

FIGURE 5-23. Phytochrome interconversions. P_r is the stable form of phytochrome, whereas P_{fr} is the unstable but active form.

and cambial tissue. Within these tissues, P_{fr} is the biologically active form. It seems to initiate a chain of events that culminates in the plant's phenotypic reaction to red or far-red light or to day length. The pool, or relative quantity, of P_{fr} is important for plant responses to light. In the lettuce seed example in Table 5-2, exposure to red light resulted in higher levels of P_{fr} in the seeds that ultimately led to germination; exposure to FR reduced the pool of P_{fr} and thus enhanced dormancy. It should be noted, however, that not all photoreversible plant phenomena are explainable with phytochrome. For example, blue light can inhibit lettuce seed germination. The effects of blue light are though cryptochromes, phototropins, and xanthophylls.

The level of light exposure has been used to classify plant responses to phytochrome into three requirements: very low fluence responses (VLFR), low fluence responses (LFR), and high irradiance responses (HIR). For example, some seeds that have VLFR can be initiated to grow with very short exposures of 0.1–100 nmol/m^2 of red light. This cannot be reversed by FR light. The most common response is the LFR, which is photoreversible and is typical of what is reported in Table 5-2. In contrast, the HIR is not photoreversible and is the result of a long exposure to FR light and not to many shorter FR light pulses. The synthesis of red anthocyanin pigments in apple peels and some dicot seedlings is an HIR.

Seed Dormancy

Seeds that are influenced by light are said to be **photoblastic**. This response may be stimulative (positive photoblastism) or inhibitory (negative photoblastism). Seeds often are most sensitive to light right after harvest and become less responsive during dry storage. Furthermore, those that are imbibed with water generally are more photosensitive than dry seeds. Phytochrome is involved intimately with the photosensitivity that is related to the maintenance or

breaking of seed dormancy. Examples of species in which germination is stimulated by light include: carrot (*Daucus carota*), rubber plant (*Ficus elastica*), gloxinia (*Gloxinia hybrida*), zoysiagrass (*Zoysia japonica*), and many weed species. In practice, rather than searching for a good source of red light, white light (mixed wave lengths) is used to overcome seed dormancy. Seeds of horticultural crops that require light to overcome dormancy are generally sown on the soil surface or are covered only lightly, because deep sowing will keep them in the darkness. The practice of tilling or otherwise disturbing soil brings positively photoblastic dormant weed seeds to the soil surface where they receive light and germinate. In their dormant condition, seeds of several important weed species can remain alive in the soil for many years; therefore, although tilling is an effective practice for controlling weeds, it can also be an excellent weed *propagation* tool!

Seed germination also can be *inhibited* by light, as it is in tomato and some lilies, or unaffected by light, as it is in peas and sweet corn. These species are therefore sown deeply to allow for a more uniform germination environment.

Branching and Stem Elongation

Irradiating plants with red light generally results in plants with relatively short internodes and vigorous axillary bud growth. This growth is typical of plants grown under cool white fluorescent lamps because they are high in red and low in far-red light (Figure 5-24). Plants irradiated with far-red light tend to show some of the symptoms of etiolation: long internodes, reduced axillary branching, and often smaller leaves. This tendency can be demonstrated by growing plants under incandescent lamps, which are higher in far-red than red wavelengths. It is logical that plants grown under far-red light resemble etiolated plants because in the darkness, P_{fr} is not formed, and in light, FR reduces the pool of P_{fr}. The P_{fr} that is

FIGURE 5-24. Plants growing under cool white fluorescent lamps.

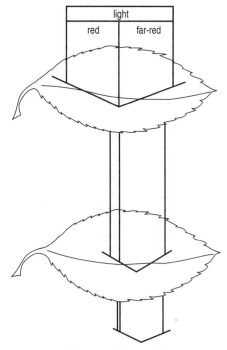

FIGURE 5-25. Drawing of leaves demonstrating preferential absorbance of R over FR.

present during darkness or when plants are grown under far-red light is biologically utilized, is destroyed, or reverts to P_r.

The control of stem elongation by R and FR light has important ecological significance. The sunlight that strikes the earth has approximately equal quantum flux of R and FR light. When this light strikes green leaves, because chlorophyll strongly absorbs R, approximately 90% of light at 660 nm is absorbed, whereas only about 2% of FR (730 nm) is absorbed; much of the remainder of FR passes through the leaf canopy (Figure 5-25). In addition, some sunlight is reflected by leaves. When plants grow densely, such as in a forest or a crowded flat, greenhouse bench, or nursery, the lower leaves and shorter plants will receive light that is much higher in FR than R. The shorter plants tend to stretch as a result of internode elongation. These plants can grow taller than their neighbors, and thus compete for valuable light needed for photosynthesis and growth. Tall, straight trees for lumber are produced in this way in forests. Horticulturally, however, stretched or leggy plants tend to be considered poor-quality plants because of their appearance or because their stems are weak. Therefore, although more plants can be grown in a smaller area by crowding of pots or by close

spacing, low-quality products often are produced as a consequence of this phytochrome-mediated response (Figure 5-26). In addition to stretching, crowding causes poor air circulation among closely spaced plants, allowing leaves and other plant parts to remain wet for prolonged periods of time and creating an ideal environment for plant infection by pathogenic fungi and bacteria (see Chapter 16).

Plants growing along the edge of a forest, field, bench, or flat receive radiation that is approximately equal in R and FR. The same is true for widely spaced plants, or solitary trees growing in a field or landscape. Because they receive a higher proportion of R than do crowded plants within a forest, field, bench, or flat, isolated plants tend to have shorter internodes and branch more freely (Figure 5-27). Therefore, wider spacings, to a point, often will result in higher quality horticultural plants. Wider spacings also allow more available light energy to strike the entire plant, permitting more

FIGURE 5-26. An end view of a flat of tomato transplants (*top*). Note the taller, leggy plant from the middle of the flat (*bottom center*), and the shorter plants on either side that were removed from the edge of the flat.

photosynthesis in uncrowded plants than in closely placed plants. Light is probably not the only important factor, because border plants also have less competition for nutrients and water than do crowded interior plants. Scientists correct for this "border effect" by planting extra border rows around their experimental plants; this way they can control a variable that could have a strong influence on research results.

There are greenhouse plastic films that specifically filter FR light. Plants grown under these films are shorter and can branch more than those grown under direct sunlight. Similarly, lamps can be selected for greenhouses that are high in red light, and thus reduce stretch of the plants. In floriculture, plant growth regulators are often used to keep plants small and in proportion to the pot size in which they are grown. These FR light-filtering films and lamps high in R light can be a substitute for chemical control of height growth.

FIGURE 5-27. Trees growing closely together in a forest (*top*) tend to grow in a straight and relatively unbranched pattern, whereas trees at the edge exhibit a much greater amount of branching (*bottom*).

Leaf Movements

Leaf movements of some plants such as sensitive plant (*Mimosa pudica*) and silk tree (*Albizzia julibrissin*) can be controlled by phytochrome. The closure or folding of the leaflets toward the midrib of the leaves of both species occurs at night ("sleep movement") and is generally controlled by light (Figure 5-28). It has been demonstrated that a brief irradiance with FR at the end of the day will inhibit leaflet closure; this can be reversed by following the FR with brief exposure to R. There are special cells called *pulvini* located at the points where leaflets are attached to the midrib. When the pulvini gain turgor—that is, become more filled with water because of a higher potassium ion content—the leaflets open; when they

lose turgor, the leaflets close. Therefore, phytochrome is thought to affect membrane permeability and ion movement and thus the loss or gain of turgor, which results in closing or opening of these leaves.

Anthocyanin Production

During periods of continuous light, many phytochrome-mediated responses are associated with high light levels. Very bright sunlight does not saturate this dependency, therefore the term *high irradiance reaction* (HIR) has been adopted for these phytochrome reactions. The production of anthocyanins (red pigments) is a typical HIR. The pigments generally form in the light from sugar that is produced during photosynthesis. The importance of light on anthocyanin production clearly is evident when observing that autumn leaves that are exposed to the sun are a brighter red color than those in more shaded locations. A wet fall season results in leaching of the sugars and anthocyanin pigments from those leaves and thus in a dull coloration.

The anthocyanin production in eggplant (Figure 5-29) and important fruit crops including apricot, apple, nectarine, peach, and pear is both horticulturally significant and light dependent. For example, apples and peaches on the exterior of a tree receive

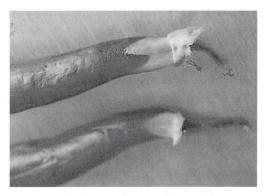

FIGURE 5-28. *Mimosa pudica* with leaves open (*top*) and closed (*bottom*). The closed leaves illustrate the sleep movement.

FIGURE 5-29. In many fruits, anthocyanin formation requires light. The calyxes of these eggplant fruit have been peeled back, exposing the shaded portion of the fruit with no anthocyanin pigment.

a greater level of light and thus have more red skin than those on the interior of the tree. Fruit growers will consequently try to control the canopy of an apple tree by selecting rootstocks and interstocks, by pruning both during the winter and summer, and by spreading the branches. Individual leaves lying on the surface of apple and peach fruit can cause sufficient shading to result in green patches, generally near the stem end (Figure 5-11). However, some fruit crops, such as blueberry, cherry, grape and plum do not require direct light for the production of anthocyanins and therefore color well even within the interior leaf canopy of the plant.

Anthocyanin production in species that require direct light appears to be related to FR light. In the HIR, FR stimulates anthocyanin production and results in a pool of P_{fr} that amounts to only three percent of the total phytochrome present. It should be noted, however, that other environmental factors may also affect fruit color on apples, including: temperature, nutrition, moisture, growth regulators, and certain pests. Weather conditions during ripening are critical to apple coloration and thus marketability.

PHOTOPERIODISM

During the early 20th century, two USDA scientists, Wrightman W. Garner and Henry A. Allard, discovered that the length of the daylight period influenced flowering in tobacco. They published their results in 1920 and after a suggestion of a colleague, A.O. Cook, named this response **photoperiodism**.

Photoperiodism is a response to day length—that is, the number of hours of light each day. Although the length of the light period is important, darkness is critical because many typical photoperiodic plant responses will be reversed if there is a brief light interruption of the darkness. Conversely, a brief period of darkness during the day will have no effect. Phytochrome is thought to be the major photoreceptor

related to photoperiod. In the absence of light, in some dicots, P_{fr} slowly reverts to P_r (see Figure 5-23); additionally P_{fr} is destroyed continuously, regardless of light. The pool of P_{fr} therefore is depleted during darkness. Irradiation with R during the middle of the night can negate the effects of darkness by replenishing the P_{fr}; in many plants this can be reversed by following the R with a brief period of FR.

Frequently, photoperiodism results from a cumulative effect of many days at a particular photoperiod. Each species has its own requirement for a minimum number of inductive photoperiods. Therefore, when one wishes to manipulate a particular photoperiodic response, plants must either be lighted to create long days or shaded to create short days for a number of weeks.

Flowering

An early, more or less distinct vegetative phase of growth, generally occurs, during which the plant will not flower, even if exposed to the proper stimulus. This growth stage is referred to as the **juvenile** phase, and the phenomenon is termed **juvenility**. Depending on the species, the length of the juvenility period may be from less than one month to many years. Following juvenility, many plants will continue to grow vegetatively until they receive environmental stimuli that will cause them to flower (adult phase). The process of flower initiation is when a shoot apical meristem changes from being vegetative to a floral meristem. When this occurs, the leaves that form from the floral meristem will differentiate as petals, sepals, and reproductive organs rather than the normal vegetative leaves typical of that species.

Related to the photoperiodic stimulation of flowering on vegetative plants, species can be grouped into three fairly distinct classes: day neutral plants, long-day plants, and short-day plants. P_{fr} seems to promote flowering in long day plants and inhibit flowering in short day plants.

Day Neutral Plants (DNP)

Plants such as cucumber, everbearing strawberry (*Fragaria chiloensis*), sweet corn, dandelion (*Taraxacum officinale*), and some tropical foliage plants can initiate flowers under a wide range of day lengths (Figure 5-30). These nonphotoperiodic species flower when they reach the adult phase of growth.

Some plants will produce more flowers or flower faster if they are exposed to long days or short days. Such plants are called quantitative long-day plants (short-night plants) or quantitative short-day plants (long-night plants), respectively. The category of quantitative long-day plants includes petunia, which initiates more flowers when grown under long days, and carnation, which flowers faster under long days. Quantitative short day plants include tomato, which may initiate more flowers under short days, and Rieger begonia, which flowers faster under short day conditions.

Long-Day Plants (LDP)

Many summer-flowering plants, including evening primrose (*Oenothera* spp.), spinach (*Spinacia oleracea*), and dill (*Anethum graveolens*), will flower if the light period is as long as or longer than a "critical day length" or if the nights are no longer than a

"critical night length" (Figure 5-31). The critical day length varies considerably among species, and sometimes among cultivars within a species. Spinach generally is not grown in the north in the mid-summer because it will bolt and thus become unmarketable. Therefore, spinach generally is grown under short-day conditions as a spring, fall, or winter crop.

Short-Day Plants (SDP)

Short day plants include several late summer-flowering or fall-flowering weeds and wild flowers such as ragweed (*Ambrosia elatior*), cocklebur (*Xanthium pennsylvanicum*), and goldenrod (*Solidago* spp.); and important floricultural crops such as chrysanthemum, kalanchoë (*Kalanchoë blossfeldiana*), and poinsettia (which has a

FIGURE 5-31. Spinach flowers under long-day conditions. This is often referred to as bolting.

FIGURE 5-30. The dandelion is an example of a day-neutral plant. It flowers when it reaches maturity, regardless of day length.

temperature and light interaction; Figure 5-32). Flowers are initiated in SDPs when the light period is shorter or no longer than a critical day length (i.e., the nights are as long or longer than a critical night length).

The discovery and understanding of phytochrome and its relationship to the photoperiodic control of flowering revolutionized the floriculture industry. For example, until the 1940s and 1950s, growers located at latitudes greater than 38° raised chrysanthemums only as a fall crop because this was when they flowered naturally. (At latitudes less than 38°, chrysanthemums will flower regardless of the season because of the daylength). Today, because of an understanding of photoperiod, chrysanthemums can be programmed to flower at any desired time throughout the year at any location. In particular, the crop is timed for holidays because of the great demand for flowers on these occasions.

Floriculturists usually want to control both the vegetative and reproductive growth cycles of their crops. They do not generally want to stimulate flowering immediately after propagation because the plants are small with minimal leaf area. Offering small plants would reduce the economic value of the crop. Larger plants with stronger stems

will result if the plants are first grown vegetatively and then stimulated to flower (Figure 5-33). To grow high-quality SDPs with the proper growth habit, growers must follow specified procedures in which they manipulate the critical day length either by controlling lighting or by providing darkness (shading) for the crop. The critical day lengths vary among crops and cultivars; for example, for hardy garden chrysanthemum it is 16 hours, for greenhouse forcing chrysanthemums it is 14.5 hours, for kalanchoë it is 12.5 hours, and for poinsettia it is 12 hours. Critical day lengths can differ among cultivars and can vary with the night temperatures under which the crops are grown. It may at first appear illogical to call species with critical day lengths of 16, 14.5, 12.5, and 12 hours short-day plants because in each case the lighted period can be one-half or more of the 24-hour day. These times, however, are approximately the longest light durations the plants can receive and still flower. Shorter days will still permit flowering, but longer days will keep plants in a vegetative state of growth.

Various books, manuals, industry bulletins, extension publications, and online sources are available that outline the protocol to be followed to achieve a flowering plant of a desirable size on any target date.

FIGURE 5-32. Poinsettia shown here as a landscape plant in a warm climate, is an example of a species that commercial floriculturists cause to flower by providing short day conditions (long nights). This enables the crop to be timed for the December holidays.

FIGURE 5-33. Vegetative growth on a chrysanthemum crop allows for adequately sized plants for sale, or the production of cuttings for vegetative propagation.

Information is given on crop, cultivar, propagation, medium, fertilizers, temperatures, and light. The following are some choices and information on lighting and shading systems for photoperiodic control of flowering; chrysanthemum is used as a representative SDP.

Short-day plants can be kept vegetative by extending the day length or interrupting the dark period. Either method results in an increased pool of P_{fr}, and because the dark periods are relatively short, the P_{fr} level does not diminish to a sufficiently low level to permit flowering. Because night interruption requires fewer hours of light than extension of the day, it is the preferred method of preventing flowering of SDP. The number of hours of photoperiodic lighting depends on the latitude and time of year. North of 40° latitude, for example, generally no lighting is required in the early summer, but four hours are required from late autumn to early spring. Generally, for chrysanthemums, if the periods of uninterrupted darkness are not more than seven hours, the plants will remain vegetative.

Incandescent lamps are commonly used for photoperiodic control (Figure 5-34). Numerous publications provide recommendations for the spacings among lamps, height above the plants, and levels of quanta that must strike the plants. The reader is referred to the greenhouse management books listed at the end of this chapter for specific details. Chrysanthemums usually are lighted from two to eight weeks, depending on the cultivar (Figure 5-35).

Night interruption with light can be accomplished by lighting for an extended time or with several brief periods throughout the night. Often the darkness will be divided into two short periods (less than seven hours each) by lighting for two to four hours in the middle of the night (e.g., lights will be turned on by timers from 10:00 P.M. to 2:00 A.M). This will vary with latitude and time of year. It also

has been shown that a series of short lighting periods will generally substitute for this longer time, resulting in substantial energy savings because the plants receive fewer minutes of light. Cyclic, intermittent, or flash lighting systems have been developed with the same overall light periods as just described, but the lights are timed to be on for only 6 out of every 30 minutes. Therefore, the time for which electric lamps are illuminated is reduced by 80 percent. Through the use of special time clocks, a series of timers, or computerized controls, different sections of greenhouse ranges can be lighted at different times. For example, a range could be divided into five sections, with each section lighted alternately for six minutes during each half-hour period for the full four hours from 10:00 P.M. to 2:00 A.M. This system brings about savings in electricity and requires less heavy-duty wiring, which provides additional cost savings during construction of the greenhouse.

Because momentary flashes of rather weak light can trigger the phytochrome response, growers have observed some interesting problems resulting in poor flowering. The light from a flashlight across a crop of an SDP, such as chrysanthemum, interferes with flowering; therefore, night watch personnel should be equipped with "safe green" lights. Safe green lights should emit light primarily in the green wavelengths that are generally reflected by plants. There also are reports of problems caused by flashes of light from oncoming automobile headlights across the crop, especially associated with newly constructed roads. Because there had not previously been a problem (the road was not there), the growers were unprepared to take preventative measures, and the crops were not ready for sale on time for the holidays. This problem can be prevented by placing shade cloth or black plastic over the crop to prevent this "light pollution" from striking the crop.

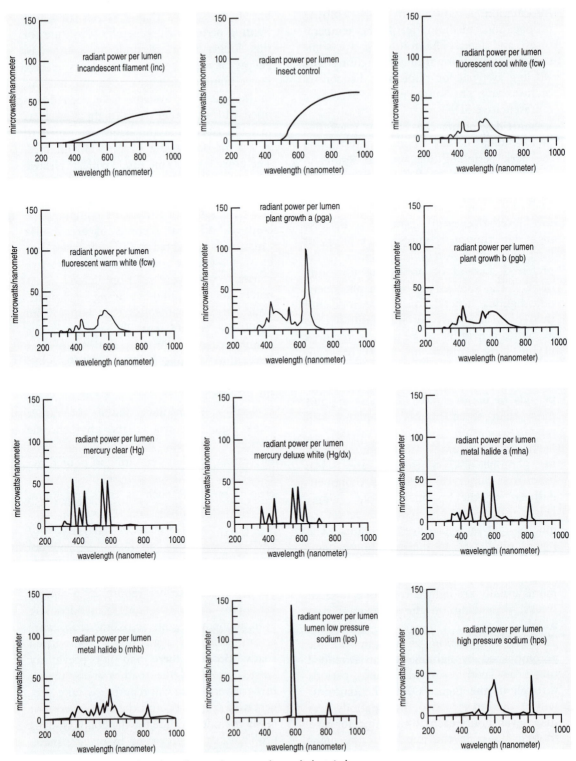

FIGURE 5-34. Spectral quality of several commonly used electric lamps.

From Campbell, L. E., R. W. Thimijan, and H. M. Cathey. 1975. Spectral radiant power of lamps used in horticulture. Transactions of the ASAE 18(5):952-956. Presented as ASAE paper No. 74-3025. Used with permission.

FIGURE 5-35. Light bulb with internal reflector used for night interruption on short-day plants, such as chrysanthemums.

FIGURE 5-36. Shade cloth is used for photoperiodic control of flowering for a chrysanthemum crop.

Following sufficient vegetative development, short-day floricultural crops must receive an adequate number of long nights (short days) to stimulate flower bud initiation. During the winter months, the nights are long enough to make special treatments unnecessary. During other times of the year, depending on the latitude, black cloth or plastic is pulled over the crop in the late afternoon or early evening and off again in the early morning (Figure 5-36). Growers must be careful about pulling the shading too early in the afternoon or late in the morning because heat built up under the material can result in delayed flowering and flower bud abortion. It is generally is recommended that 7:00 P.M. and 7:00 A.M. are the best times to pull black cloth or plastic every day for the appropriate number of weeks for each particular crop and cultivar; skipping only one night can lead to a slight delay in flowering, although it rarely causes major problems.

Black polyethylene (4 to 6 mil) is a relatively inexpensive shading material. Because it is not water permeable, however, drip from condensation can create a favorable environment for the development of plant diseases. Many growers choose not to use black polyethylene for this reason.

Black sateen cloth is a more expensive opaque material; because it is porous, water condensation is not a problem. Many growers use a cloth that is white or silver-colored on the outside and black on the inside. The outer white or silver color reflects light, and heating within the shaded area is less of a problem than with black cloth or plastic. Whether the material is cloth or polyethylene, it must be sufficiently dense to block light almost completely. Light leaks from holes, tears, or improper installation or pulling can result in chrysanthemums with **crown buds**, which are flower buds that are hollow or otherwise have incomplete development. It is not uncommon to see a more or less circular area on a bench of chrysanthemums showing crown buds. This alerts the grower to a pinpoint of light that entered the shaded area in a conical pattern. All holes should be repaired immediately.

Shade cloth or plastic can be pulled by hand or mechanically. Automatic pulling can be controlled manually, by timers, or by computer. The material can cover individual benches, or more efficiently, the entire inside of a greenhouse. Growers can use the same thermal curtains that provide insulation for heat conservation at night (see Chapter 6), for photoperiodic control. Although this option is initially expensive, it can pay for itself in a reasonably short period of time.

Alternating Day Lengths

Some plants require alternating day lengths for flower initiation. For example, the bellflower (*Campanula medium*) and creeping white clover (*Trifolium repens*) will flower only when a period of short days is followed by long days. *Kalanchoë* and some lilies, however, require a period of long days followed by short days. Therefore, the stimulus for flowering may be quite complex.

Formation of Storage Organs

In some species, storage organs such as tubers, bulbs, and tuberous roots form in response to photoperiod (Figure 5-37). The formation of tubers in potato, Jerusalem artichoke (*Helianthus tuberosus*), scarlet runner bean (*Phaseolus coccineus*), and aerial tubers in *Begonia* is stimulated by short days.

Bulbing in onion and early garlic is a long-day response. Although the bulb formation in Bermuda onions also is a long-day response, as in Spanish and other onion types, Bermuda onions have a shorter critical day length. They can therefore be produced in the southern United States, where they will form bulbs while the days are still relatively short and the temperatures are not too high. The days are too short for some of the other onion types to form bulbs at this time in the south. Bermuda onions are not produced in the north because they have such a short critical day length: By the time that it is warm enough and the soil sufficiently dry in the north for Bermuda onions to be planted, the plants would produce only a very few leaves before sensing long days. The plants would therefore produce small, nonmarketable bulbs because of insufficient leaf area for adequate photosynthesis.

The tuberous root formation of *Dahlia* can occur under any day length (Figure 5-38). Short photoperiods, however, will cause much greater tuberous root development, especially with large-flowered cultivars.

Storage organs can serve as propagules. They form either during the long summer days or as the days get shorter as autumn approaches, resulting in organs that are adapted for overwintering and will result in new plants the following spring. People often grow these crops in areas where overwintering temperatures are so low that storage structures are killed. In these areas, such structures (e.g., tuberous roots of *Dahlia*) must be dug in the fall and stored under cool and moderately dry conditions.

FIGURE 5-37. Potato tubers form in response to short days.

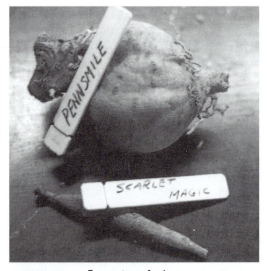

FIGURE 5-38. Formation of tuberous roots on dahlia is accelerated by short days.

Stem Elongation

Runner formation on plants such as strawberry and spider plant (*Chlorophytum*) is controlled partly by photoperiod (Figure 5-39). Strawberry runners form only under long days of 12, 14, or more hours of light. New plantlets form at specific nodes on these runners and root into the soil, allowing strawberries to spread vegetatively or be dug up and transplanted to new locations. Visible runners form more quickly on spider plants grown in day lengths of 12 or fewer hours than when the days are longer. The variegated spider plants tend to be more responsive than the all-green plants.

FIGURE 5-39. *Top* and *bottom*: Plantlet formation on runners of spider plant (*Chlorophytum*) is related to short days.

Temperate Trees and Woody Shrubs

The fall coloration of leaves, leaf fall (abscission), onset of cold hardiness, and onset of dormancy are all, in part, controlled by photoperiod. As summer progresses and autumn approaches in temperate climates, temperature declines, light levels decrease, and photoperiods become shorter. The role of lower light is not as clear, but decreasing temperatures and short days are important factors in determining the onset of leaf senescence, abscission, true dormancy, and cold acclimation in many temperate woody plants. Differences in sensitivity to day length have been shown between northern and southern races of woody species; the more southern races will continue stem elongation at shorter photoperiods than the more northern races. Similarly, it is common to see trees that are grown near street lights hold their leaves longer into the autumn when grown than those not receiving such extra light.

Phytochrome, especially the quantity of P_{fr} in plants, ultimately controls many different phenomena that are economically important in the production of horticultural crops. It was not until the basic knowledge of this pigment-protein was elucidated that horticulturists could better understand and grow their crops. This pigment impinges on all phases of horticulture, and by manipulating light quality (red and far-red), plant spacing, and photoperiod (either directly or by choosing planting dates), growers can produce high quality plants (Figure 5-40). For certain species growers can schedule their crops for the dates with the highest consumer demand.

LIGHT QUANTITY AND PLANT GROWTH

Plants fix CO_2 during the dark reactions of photosynthesis and generate CO_2 during the respiratory processes. At high levels of light, much more CO_2 is fixed by plants than is

FIGURE 5-40. Tiny, intact plantlets form on the margins of the leaves of *Kalanchoë* in response to long days.

respired; conversely, during darkness or at very low levels of light, respiration greatly exceeds photosynthesis and leads to a resulting loss in the dry matter of plants as the CO_2 is released by respiration. The level of light at which the CO_2 fixed by photosynthesis equals the amount generated by respiration is known as the **light compensation point**. This light level cannot support plant growth but is adequate for plant maintenance. The light compensation point varies among species, between plants of the same species that were grown under different light levels, and even among the leaves on an individual plant, depending on whether the leaves are on the exterior or interior of the plant. The lower interior leaves (shade leaves) on a tree, for example, have a lower light compensation point than those on the outer portions of the canopy (sun leaves).

At a certain point where photosynthesis is occurring at high light levels, and the supply of light energy absorbed by chlorophyll is no longer the rate-limiting factor. This is known as the **light saturation point** (light-saturated photosynthesis). The light-saturation point varies among species and between leaves that develop in the sun or shade.

Shade Tolerance

Outdoors, light levels vary from full sun to light shade to heavy shade. As light level decreases, it becomes increasingly difficult to find species that will grow well. This often presents problems for landscape horticulturists who are hired to establish plantings in dark courtyards or shaded areas between buildings. The light levels are often below the light compensation points of the available plant materials. Architects could probably achieve the desired effects better with plant materials around their buildings if they consulted with knowledgeable horticulturists during the design phase of their buildings.

The shade tolerances of all horticultural crops would make a lengthy list (Table 5-3), but some generalizations can be made. Nearly all vegetables and fruit crops grow best in full sun, although leafy vegetables, such as lettuce and cabbage can tolerate lower light levels than fruiting vegetables crops such as sweet corn, peppers, or squash. The latter crops should receive a minimum of six hours of direct sunlight each day. All fruit crops require full sunlight for production.

For optimum results, unless the site can be modified for light, the horticulturist must select the genotype of most commodities to fit the location (Figure 5-41). This selection should be made on the basis of utilization and adaptability of the plant material. One might keep in mind that many understory (low growing, shaded) plants in forests are either quite shade-tolerant or relatively intolerant of full sunlight. On the other hand, most weed species grow best in full sunlight, so as the canopy of field-grown plants, such as vegetables, flowers, or woody shrubs, fills in, it suppresses weed growth.

Many plants, including rhododendrons and various foliage plants, are produced under shading materials (Figure 5-42). Shading will not only result in lower amounts of light striking the plants but also will have

TABLE 5-3. Examples of horticultural plants that should be planted in shade or full sunlight.

COMMODITY GROUP	REQUIRE FULL SUNLIGHT	WILL GROW WELL IN SHADED LOCATIONS
Vegetables	tomato pepper sweet corn	lettuce cabbage spinach
Annual flowers	marigolds cosmos zinnia	ageratum lobelia sweet alyssum
Herbaceous perennials	peony iris dahlia	ferns hosta columbine
Trees	pines oaks maples	hemlocks flowering dogwood weeping fig
Shrubs	juniper lilac forsythia	barberries rhododendron yews
Turfgrasses	Kentucky bluegrass zoysiagrass bermudagrass	creeping red fescue St. Augustine grass

a considerable influence on lowering plant temperature and transpirational water loss. Shaded plants will thus require less frequent irrigation than plants grown in full sun. Many growers utilize wooden or aluminum lath to provide shade (Figure 5-43). Snow fencing is also frequently used as lath to provide shade for plants. Additionally, woven saran cloth is available that will result in specific percentages of shade for the plants.

FIGURE 5-41. Impatiens, *left*, and azaleas, *right*, are examples of plants that grow well and flower in shaded locations.

FIGURE 5-42. Saran, or other shade cloth, may be used to provide a shady environment.

FIGURE 5-43. Aluminum, *top*, and wood, *middle* and *bottom*, lath houses provide shade for plants requiring reduced light.

Sun and Shade Leaves

Leaves that develop in shaded areas, such as within a leaf canopy, are distinctly different from those grown in full sunlight (Table 5-4). Sun and shade leaves differ as a result of the direct effect of light on leaf morphology.

When plants are placed in relatively weak light, those that were grown previously in the shade are more productive than plants that had been grown in full sun because the photosynthetic apparatus of shade-grown plants is more efficient. Shade-grown plants often have larger, thinner leaves that have a deeper green color. The lighter green color of sun-grown plants is related partly to the fact that the high levels of light to which sun-grown plants are exposed causes photodecomposition of chlorophyll. Because shade-grown plants do not receive as much sunlight, more chlorophyll accumulates in the leaves, allowing for more efficient photosynthesis. In addition, the chloroplasts and grana are not stacked in the leaves of shade grown plants; therefore, all of the chloroplasts can receive more light, which is an advantage if the plants are to be grown in low light conditions.

Shade leaves have a lower light saturation point than sun leaves, and when shade-grown plants are moved to the bright sunlight, leaf scorching will occur because the chlorophyll is relatively unprotected inside the chloroplast. In Figure 5-44, leaf scorch is apparent on the *Hosta* grown in full sunlight. Another example of a problem of this type is encountered in transplants that are started under electric lamps in the home. With a species such as tomato, one can grow an apparently healthy-looking tomato transplant under electric lamps, but considerable damage will occur when it is moved into the garden. In fact, the plant may not survive because the high levels of sunlight will cause photodecomposition of the chlorophyll, and cells will collapse. Therefore, moving a plant from low to high light conditions should be done gradually, or the transplant will not perform very well.

TABLE 5-4. The relative comparisons of sun and shade leaves

SUN LEAVES	SHADE LEAVES
(Whole plant)	
1. smaller leaf area	1. larger leaf area
(Individual leaves)	
2. thicker leaf blades	2. thinner leaf blades
3. folding at the midrib (*Ficus benjamina*)	3. leaves flatter
4. more yellow because of less chlorophyll	4. greener because of more chlorophyll
5. fewer intercellular spaces	5. more intercellular spaces
6. increased stomatal density	6. fewer stomates per unit area of leaf
7. more well developed palisade layer or layers	7. more poorly developed palisade layer
8. elongated chloroplasts	8. round chloroplasts
9. chloroplasts aligned along side walls of cells so that they shade each other	9. more dispersed chloroplasts, many near upper surfaces of cells
10. more rudimentary grana	10. grana irregularly dispersed within chloroplasts, with larger grana stacks
11. most photosynthetically efficient in high light	11. most photosynthetically efficient in low light
12. do well in high light	12. high light damages photosynthetic apparatus
13. higher light compensation point	13. lower light compensation point
14. higher dark respiration rate	14. lower dark respiration rate

Leaf Area Index

Plant breeders continually strive to increase plant yields, sometimes by increasing photosynthetic efficiencies. However, these efficiencies never seem to exceed 22 percent of the absorbed photosynthetic light energy. Often only one to two percent of the photosynthetically active energy that strikes a crop during a growing season is converted and stored as carbohydrates.

In an effort to increase overall photosynthesis, breeders often strive to increase the **leaf area index (LAI)**. The LAI is the ratio of the area occupied by the upper sides of all of the leaves of a plant to the surface of the ground that the plant covers. Leaf area

FIGURE 5-44. *Left*: Leaf scorch is apparent on *Hosta*. *Right*: The center of the leaf is more chlorotic where the sunlight hit, whereas the base of this leaf is greener where it was shaded by another leaf.

index values vary with the crop, planting density, and amount of available light. A mature crop grown in full sunlight may have an LAI value of up to eight. Many of these leaves will be on the lower portions of the plant and will therefore be shade leaves. Under conditions of low light, many plants lose their lower leaves and thus have low leaf area indexes.

Although LAI is important, at a certain point, because of shading by upper leaves, increasing this value will not increase yield. Breeders therefore also focus on leaf orientation. Vertically oriented leaves will shade less than those that are horizontal, so breeding and selecting plants with erect leaves has resulted in greater yields.

Acclimatization of Foliage Plants

Foliage plants frequently are raised outdoors in the full sun of tropical and subtropical regions. If these plants are taken directly from the field and placed under the low light conditions of an office, shopping center, or home, problems will result that often lead to dissatisfaction. Field-grown plants moved directly indoors, under low light, are often subject to very rapid defoliation, poor growth, and death. Their sun leaves do not photosynthesize efficiently under these conditions, and thus cannot support the plant. Indoor foliage plants that were produced in full sun must therefore be acclimatized before being used for interiorscaping to ensure that they will maintain their aesthetic value (Figure 5-45).

During the process of acclimatization of indoor foliage plants, light, temperature, watering, and nutrition all are manipulated to improve the quality of plants moved indoors. Of these factors, light is the most important. Acclimatization can generally be accomplished by moving field-grown plants to lower light levels of between 40 percent and 80 percent shade for 5 to more than 10 weeks, depending on the species. Alternatively, plants can be grown under

FIGURE 5-45. Foliage plants, such as these offered for retail sale, make an attractive addition to the home. They must be properly acclimatized to prevent leaf-loss and maintain an attractive appearance.

approximately 40 percent shade, which eliminates the labor involved with moving the plants or placing shade cloth or some other material over the plants while they are growing. The temperature tends to be lower under shade than full sunlight; this also may play a role in acclimatization.

Plants should usually be fertilized and watered during acclimatization, but the needs vary among species, relating both to growth rate and subsequent leaf drop.

Generally, when foliage plants are placed indoors, the objective should be for plant maintenance rather than plant growth. Light is frequently the main limiting factor in many indoor locations but it can be provided by windows, skylights, or electric lamps (Figure 5-46). Light requirements among indoor foliage plants depend on the length of time that the plants will be maintained at a particular location. The light level should be at least at the light compensation point. If the plants can be replaced every two to four months, lower light levels probably will suffice because the plants can utilize stored carbohydrates, whereas higher light levels are necessary for longer term maintenance. Light levels as low as 15 to 30 $\mu mol \cdot m^{-2} \cdot s^{-1}$ photosynthetically active radiation are likely to be sufficient for a wide range of species.

FIGURE 5-46. Foliage plants often are included in the interiorscaping of a shopping center. Generally, these are located under skylights to provide for their photosynthetic needs and maintain a high-quality appearance for long periods of time.

LIGHT SOURCES AND THEIR USE

The sun is the most important source of light that strikes the planet Earth and is thus crucial to the production of the vast majority of horticultural crops. During times of the year when it is infeasible or impossible to produce unprotected plants and crops outdoors, horticulturists often utilize solar energy by growing plants under transparent coverings such as the glass or polyethylene of greenhouses, and polyethylene tunnels in fields. High value crops that are productive and adapted to intensive culture systems are generally selected by greenhouse growers and managers because of the expenses associated with greenhouse materials, construction, maintenance, heat control, and concentrated growing.

People also use sunlight when growing plants in windows of offices, businesses, or homes or maintaining plants under the skylights of large buildings, such as shopping malls. When plants are grown in windows, they generally are exposed to direct or indirect sunlight coming from one direction. Consequently, either whole plants or their parts (e.g., stems) bend toward the light; this phenomenon is called **phototropism**. Stems that bend toward a unidirectional light source are said to be **positively**

phototropic, whereas roots bend away from a light source and are **negatively phototropic**. Phototropism is associated with the near ultraviolet to blue wavelengths. The phototropin pigments are the photoreceptors that are responsible for phototropic bending. Phytochrome does not seem to be associated with phototropism. A person who does not like the appearance of plants bending toward the light can turn the plant periodically to give it a fuller appearance. However, rotating plants can cause problems related to sun and shade leaves and can lead to a plant that is not as aesthetically pleasing as one might desire.

Electric lamps frequently are used as either the sole light source for growing plants, as sunlight supplements during low light times of the year, or for photoperiodic controllers of certain flowering plants. The light that electric lamps produce is real light, so the term "artificial light" is inappropriate.

High-Intensity Discharge (HID) Lamps

HID lamps can be mercury (Hg) lamps, metal halide (MH) lamps, or sodium lamps. Sodium lamps are either high-pressure sodium (HPS) or low-pressure sodium (LPS) lamps; both have greater efficiency than Hg or MH lamps. The spectral quality of these lamps is given in Figure 5-34. HID lamps are rather large, with big ballasts and reflectors (Figure 5-47). They are generally available in sizes up to 2000 watts, but 400- and 1000-watt sizes of HPS lamps are used commonly in the United States. These lamps tend to be expensive compared to fluorescent or incandescent lamps but their life expectancy ranges from 10,000 hours to 24,000 hours, depending on the lamp. HID lamps are used sometimes in greenhouses to supplement sunlight on cloudy days during winter when the level of light is quite low, especially at high latitudes. Such supplemental lighting in greenhouses has been shown to increase rose production by as much as 240 percent.

FIGURE 5-47. HID lamps in greenhouses. The lamp on the left has the ballast attached directly above the reflector, whereas the center picture depicts ballasts that are separate from the lamps, against the greenhouse wall. Separating the ballasts results in less shading of the crop. Another type of HID lamp is illustrated at the right.

HID lamps also have been used for growing plants inside buildings with non-transparent coverings, such as warehouses. Because of savings in heating costs, such growing systems may have a more important horticultural impact in the future. The greatest successes have been with leafy vegetables such as lettuce and spinach, with the rooting of cuttings of crops such as chrysanthemum (to which photoperiodic treatments may be applied simultaneously), and with foliage plants. Plants can benefit from the addition of incandescent lamps along with the HPS lamps in these situations. The incandescent lamps provide additional light in the red wavelengths.

Fluorescent Lamps

The standard types of fluorescent lamps are the cool white and warm white tubes (see Figure 5-34). Cool white fluorescent lamps are used more commonly for growing plants. Although they are sometimes used in greenhouses, the reflectors take up valuable space and thus shade the crops; therefore, fluorescent lamps are used more frequently for seed germination and for producing plants in tissue culture systems. Hobbyists and home-owners also favor them. People often combine cool white fluorescent and incandescent lamps to increase the spectral distribution of the light striking the plants. This combination seems to be more important in the production of flowering plants than of plants grown primarily for their foliage. Some companies produce fluorescent "grow lamps" that emit a greater spectral distribution of light than warm or cool white tubes, but these lamps may be more expensive, less efficient, and less long-lasting than warm or cool white fluorescent lamps. Many fluorescent tubes decrease rapidly in efficiency over time, resulting in a lower level of light striking the plants. In trials with rooting of chrysanthemum cuttings, this inefficiency led to more stretching and a lower quality plant than desirable. These problems are solved by replacing the fluorescent lamps with HID lamps, which are much more efficient and produce more light to be intercepted by the plants (Figure 5-48).

Because of the lower quantity of light that strikes plants when using fluorescent lamps, the total number of photons striking a plant can be manipulated by extending the day length. A 16-hour photoperiod is recommended for growing plants under fluorescent lamps, if there is no photoperiodic effect to avoid.

FIGURE 5-48. (*a*) Various crops can be grown hydroponically under HID lamps indoors. This technology has been applied in the CELSS (Controlled Ecological Life Support System) by NASA scientists to study the growing of food crops for the International space station. (*b*) Experimentally, plants were grown in an environmentally controlled pressurized chamber. Crops such as wheat (*c*) and rice (*d*) have been grown for this purpose.

Incandescent Lamps

Incandescent lamps are used primarily for photoperiodic control because they emit too much heat to be used for growing plants. Furthermore, they are low in the blue end of the spectrum (Figure 5-34). Most plants grown exclusively under incandescent lamps exhibit tall, succulent growth characteristics.

Because incandescent lamps are small, provide little shade for the crop, and are high in light at the red end of the spectrum, and because the phytochrome reactions relating to photoperiod require only low light levels (beginning at 1.0 µmol m^{-2}), incandescent lamps are the major light source used for the photoperiodic control of floricultural crops. Incandescent lamps generally emit light in all directions, but external reflectors tend to shade the crop. Therefore, most growers purchase incandescent bulbs with internal reflectors (Figure 5-35). These internal reflectors cause less shading of the crop and reflect a large percentage of the light down onto the crop canopy.

SYSTEMS FOR MEASURING LIGHT

Much of the world uses the metric system for weights and measurements; however, some of the major developed countries utilize other systems, such as the English system. Because some countries such as the United States, are actively using both the metric and English systems, students should learn both systems and their relationship to each other. For example, most students are quite aware of the differences and similarities between measurements, such as a quart and a liter, a yard and a meter, and a gram and an ounce. However, there are three different systems for measuring light. It can be confusing if one reference gives light required for a plant growth in footcandles, lumens, or lux; and a second presents light as watts·cm^{-2}, and a third presents light as

µmol·m^{-2}·s^{-1}. The following sections define these units of light measurement and explain which ones are most appropriate for describing light in relation to plant growth.

❖ TERMINOLOGY

CALORIE (SMALL CALORIE). The amount of heat that is required to raise the temperature of one gram of water from 14.5°C to 15.5°C. In practice we generally assume it to apply to any 1°C temperature change. The term **calorie** that is used for food is actually the **kilogram calorie**, which equals 1000 small calories.

DYNE. A unit of force, which, when acting on something with a mass of 1 g, produces an acceleration of one centimeter per second per second (1 cm·s^{-2}). A steady speed differs from acceleration and is measured in distance per unit time, (e.g., kilometers per hour or miles per hour). During acceleration, speed increases over time. Therefore, kilometers or miles per hour increase with time, as in kilometers or miles per hour per hour, similar to the centimeter per second per second (centimeter per second squared). One dyne = 1 g·cm·s^{-2}.

EINSTEIN (E). A unit equivalent to one mole of light quanta (photons). It often is considered preferable to report light in moles rather than einsteins since the einstein is not an SI (Systéme International) unit.

ERG. A unit of work or energy that is accomplished by a force of one dyne moving through a distance of one centimeter. One erg = 1 dyne·cm^{-1}. An erg is an extremely small unit of energy, therefore the unit **joule** is used more frequently.

FLUX. The flow of a substance or the transfer of energy across a surface. When used in relation to light, it refers to the continued flow of light (energy) either from a source or onto a surface and is called radiant or quantum flux.

JOULE. A unit of work or energy that is done by a force of one newton moving through a distance of one meter. One joule = 10,000,000 (10^7) ergs = 1 newton meter.

LANGLEY. One langley equals one calorie per square centimeter, which is equivalent to 0.484 watts·m^{-2}. This unit often is used in meteorology but only rarely in literature on plants.

MOLE. The molecular weight of a substance expressed in grams (gram molecular weight). A mole of a substance contains Avogadro's number, or 6.02×10^{23} units of that substance.

NEWTON. A unit of force, which, when acting on something with a mass of one kilogram, produces an acceleration of one meter per second per second. One Newton = 1 kg·m·s^{-2} = 100,000 dynes.

PHOTON (QUANTUM). An individual particle of light, similar to an electron.

SI UNITS. Units adopted by the Systéme International d'Unités (SI). This is an international system that is based on the metric system. It was proposed in its present form in 1958 and was adopted by the signers of the Metre Convention in 1960 and by all countries that publish scientific journals. Its purpose is to improve communication among people within and among different countries.

WATT. A unit of power (work done per unit time) that is equal to one joule per second. This can be remembered easily by the student with the rhetorical question "Watts a joule per second?" (The answer: yes).

Photometric system

This system for measuring light measures radiant flux as light intensity. By definition, **light intensity** is the luminous flux *emitted* by a point source per unit solid angle. It is important to understand that this system does not measure the light intercepted by leaves, and that light intensity refers to light emitted by a source.

The **solid angle** also is known as the steradian (sr). To understand the steradian,

one must imagine a point source of light, such as a candle, from which light is emitted equally in all directions. Furthermore, one must imagine the flame of the candle as the center of a sphere (Figure 5-49) and a circular area on the inside of the sphere of a given area (e.g., $meter^2$ or $foot^2$). To illuminate that area on the sphere, light leaves the point source and spreads out, creating a cone. The steradian is the ratio of that circular area on the sphere that forms the angle of the cone (conical angle or solid angle) to the square of the radius of the sphere (length of one side of the cone).

Several units used in the photometric system that measure light intensity. A **footcandle**, equivalent to one **lumen**·ft^{-2} (lumen per square foot), is commonly used to measure light intensity. The lumen·ft^{-2} is defined as follows: There is a point source of light (Figure 5-49), or a standard candle, from which light leaves evenly in all directions. The light illuminates the inside surface of a sphere with a one foot radius. A lumen is the amount of light emitted that falls on 1 ft^2 of the inside surface of the sphere. The SI unit for intensity is the **candela** (cd), and 1 footcandle = lumen·ft^{-2} = cd·sr·ft^{-2}.

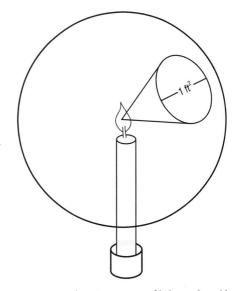

FIGURE 5-49. A point source of light enclosed by a sphere showing the unit solid angle.

Footcandles, lumens, and candelas are based on the English system of measurement, (e.g., ft^2) but another unit of intensity, the *lux*, is based on the metric system. One lux equals one lumen·m^{-2}. Therefore, as comparison with the footcandle in the preceding paragraph, with the lux, the radius of the sphere becomes one meter and the circular area on the surface of the sphere is one square meter. One lux is a lower light intensity than one lumen. The following relationship exists between photometric units: 1 lux = 1 lumen·m^{-2} = cd·sr·m^{-2}, and 1 footcandle = 10.76 lux. Light intensity also can be reported in terms of its energy content, (e.g., watts·sr^{-1}), however the number of lumens or watts varies with the light source.

The photometric system and its units deal specifically with the spectral sensitivity of the "standard human eye." In fact, the photometric units are based on the spectral responses of 52 pairs of American eyes – measured in 1923. The reader will recall that human eyes are most sensitive to light at 555 nm wavelength and that plants reflect much of the light that they intercept at that wavelength. Also, the amount of light intercepted by leaves is more relevant to plant growth and development than light that is emitted by a source. Therefore, the photometric system that reports illuminance or intensity in units of footcandles, lumens, candelas, and lux is not very useful when describing light for plants. Light meters or footcandle meters are adjusted so that their sensitivity matches the human eye, so a light meter is relatively insensitive to wavelengths that are most important for plant growth and development. For example, if an electric lamp provides light that has a high intensity, but is rich in green and yellow and poor in red and blue, the plants will grow poorly, even though the light meter measures adequate light levels. Thus the photometric system should not be used when describing light for plant growth.

Radiometric System

The radiometric system describes light in terms of its energy content. Radiant flux is measured as **irradiance**, which is analogous to **intensity** in the photometric system. Irradiance, however, is the light (radiant flux) *intercepted* per unit area. For plant growth and development the light that is intercepted by leaves, stems, and other parts is important. Measurements of irradiance are therefore much more appropriate than measurements of light intensity.

Irradiance should be reported as the amount of energy that strikes a surface per unit area. The SI unit for energy is the **joule**, so it is appropriate to report irradiance as joules·s^{-1}·m^{-2} (joules per second per square meter). Because a watt is one joule per second, irradiance is frequently reported as watt·m^{-2}. For example, the peak solar radiation level that strikes temperate regions of the Earth at sea level is approximately 1000 watts·m^{-2}. Irradiance also is reported as ergs (10^{-7} joules), gram calories (4.19 joules), or langleys (4.19 joules per square centimeter or 0.484 watts·m^{-2}). Fortunately, each unit is easily converted to joules.

Although the energy intercepted per unit time per unit area is useful, the wavelength region also must be specified. The waveband can be narrow, or, more commonly, spanning 400 to 700 nm. It is best to describe light in more narrow wavelength bands. The band from 400 to 700 nm is usually referred to as **photosynthetically active radiation (PAR)**. For example, irradiance might be reported as 10 mW·cm^{-2} (milliwatts per square centimeter) PAR at plant level, which describes the quantity and quality of light that is striking the plants. However, expressing light in this way does not explain how the light is distributed within the PAR band. This can be somewhat clarified if the light source is stated, because the spectral distribution is known and readily available for most lamps (Figure 5-34). The 10 mW·cm^{-2} PAR would be different for a metal halide lamp than for a cool white fluorescent lamp.

Instruments known as **radiometers** are used by researchers to measure irradiance levels. Radiometers are quite variable in cost and sensitivity. **Spectroradiometers** are excellent instruments for measuring irradiance. These instruments can scan one or more nanometers at a time and can graphically plot the irradiance continually across a wide spectrum, which provides an excellent method to evaluate lamps for plant growth. Other radiometers are equipped only with wide-band sensors (e.g., PAR) and unfortunately do not respond to all wavelengths within these bands with equal sensitivity. Such radiometers are superior to light meters (photometric) but are inferior to spectroradiometers. Wide-band sensor-equipped radiometers may be purchased for only a small fraction of the cost of spectroradiometers and are therefore used more commonly. A quality spectroradiometer with accessories can cost about as much as a new medium-sized automobile.

Photon (Quantum) Flux

The primary photochemical act in photosynthesis depends on the number of photons that are trapped. Therefore, it is logical to express incident light and PAR as the number of photons per unit area per unit time. This measurement is referred to as **photon (quantum) flux**. The number of photons generally is reported as moles. Because the einstein (E) is one mole of photons, **photon flux (PF)** is often reported as einsteins per unit area per unit time $(E \cdot m^{-2} \cdot s^{-1})$ or microeinsteins per square meter per second $(\mu E \cdot m^{-2} \cdot s^{-1})$. The einstein is not an SI unit, so it is preferable to report PF as moles of photons per unit area per unit time $(mol \cdot m^{-2} \cdot s^{-1}$ or $\mu mol \cdot m^{-2} \cdot s^{-1})$.

Reporting only PF has the same limitation as was discussed for the radiometric system—the quality of the light is not known. The wavelength bands also must be specified, and the more narrow these bands, the better the information. However, **photosynthetic photon flux (PPF)** generally is reported, and it refers to the wavelength band from 400 to 700 nm. It is possible to convert from footcandles to PPF if the electric lamp is known, for example with cool white fluorescent lamps footcandles should be multiplied by 0.15, with incandescent lamps footcandles should be multiplied by 0.22, and with low pressure sodium lamps footcandles should be multiplied by 0.10. The different conversion constants reflect the differences among these common lamps.

Quantum (photon) light meters are available that measure PF. Various filters and qualities of meters are available to scan various wavelengths. Like radiometers, photon light meters vary both in sensitivity and price.

SUMMARY

Light is crucial for the growth and development of higher plants. This is true not only because of the necessity of light for photosynthesis, but also because of photomorphogenesis, the effect of light on the development of the structure of the plant. Plants have other important pigment systems in addition to chlorophyll. Horticulturists manipulate plants through phytochrome-mediated responses; this knowledge is the basis for the production of photoperiodic crops, such as chrysanthemum. By choosing the light source and sometimes by employing shading, horticulturists can govern light quality, quantity, and photoperiod and thereby achieve the phenotype that is most desirable, and often most profitable.

REFERENCES

Bickford, E. D. and S. Dunn. 1972. *Lighting for Plant Growth*. The Kent State University Press, 221 pp.

Cathey, H. M. and L. E. Campbell. 1980. Light and lighting systems for horticultural plants. *Horticultural Reviews* 2:491–540.

Gould, K. S. and D. W. Lee (Eds.). 2002. *Advances in Botanical Research Incorporating Advances in Plant Pathology Anthocyanins in Leaves*. Vol. 37. Academic Press, Amsterdam. 212 pp.

Hanan, J. J. 1984. *Plant Environmental Measurement*. Bookmakers Guild, Longmont, Colorado. 326 pp.

Incoll, L. D., S. P. Long, and M. R. Ashmore. 1977. SI units in publications in plant science. *Current Advances in Plant Science* 28: 331–343.

Kendrick, R. E. and G. H. M. Kronenberg (Eds.). 1994. *Photomorphogenesis in Plants*, 2nd ed. Kluwer, Dordrecht. 828 pp.

Langhans, R. W. (Ed.). 1978. *A Growth Chamber Manual--Environmental Control for Plants*, Comstock, Ithaca, New York. 222 pp.

Leyser, O. and S. Day. 2003. *Mechanisms in Plant Development*. Blackwell, Oxford, UK. 241 pp.

Lumsden, P. J. and A. J. Millar (Eds.). 1998. *Biological Rhythms and Photoperiodism in Plants*. Bios Scientific Publishers, Oxford.

Nelson, P. V. 2003. *Greenhouse Operation and Management*, 6th ed., Prentice-Hall, Upper Saddle River, New Jersey. 692 pp.

Raghavendra, A. S. (Ed.). 1998. *Photosynthesis A Comprehensive Treatise*. Cambridge University Press, Cambridge, UK. 376 pp.

Sage, R. F. and R. K. Monson (Eds.). 1999. C_4 *Plant Biology*. Academic Press, San Diego. 596 pp.

Smith, H. 1975. *Phytochrome and Photomorphogenesis*. McGraw Hill, London. 235 pp.

Smith, H. (Ed.). 1981. *Plants and the Daylight Spectrum*. Academic Press, London. 508 pp.

Taiz, L. and E. Zeiger. 2002. *Plant Physiology*, 3rd ed., Sinauer Associates, Sunderland, MA. 690 pp.

Thomas, B. and D. Vince-Prue. 1977. *Photoperiodism in Plants*, 2nd ed., Academic Press, San Diego. 428 pp.

CHAPTER 6

TEMPERATURE

"WHO LOVES A GARDEN,

LOVES A GREENHOUSE

TOO."

William Cowper

Temperature may be the most important environmental factor that limits the distribution of plants on the planet. In the northern hemisphere, low temperature extremes limit the northern distribution of plants and summer heat and lack of chilling in warm climates limits their southern distribution. Horticulturists manipulate temperature by growing plants in controlled environments such as greenhouses. In the field and landscape, sites are chosen for plants considering temperature averages, fluctuations, and extremes throughout the year. This can determine whether plants survive, thrive, or produce acceptable yields.

There are many factors, other than plant survival and yield that are determined by both day and night temperatures. Plant growth, development, flowering, dormancy, propagation, color, yield, life, and death are all greatly influenced by temperature. We cool and warm plants and protect them from temperature extremes. We consider root zone and air temperatures, and temperature fluctuations when we grow plants. We choose cool season plants for some times of the year and warm season plants for other times. Horticulturists must be acutely aware of temperatures, how they can be manipulated and how plants can be protected.

141

Plants have evolved to survive and function effectively under various temperature regimes. This is an important reason why different species are found in tropical and temperate climates. Higher plants grow in diverse locations such as the Sahara Desert and Death Valley and in the severe cold of the Arctic and Antarctica. The biological activities of most plants, however, are limited primarily to the narrow temperature range of approximately 0° to 50°C (32–122°F), and within this range, during the growing season, most horticultural crops respond best to temperatures of 10° to 30°C (50–85°F). For many species, biological activity slows or stops at approximately 0°C (32°F), the melting point of ice. At temperatures above 50°C (122°F), proteins are irreversibly destroyed (denatured), and plants are injured or killed.

The different temperature requirements of horticultural species not only determine the climate in which they are best produced,

but also are factors to consider when selecting the best season to grow a crop (e.g., spring vs. summer plantings). Because of the seasonal variability for planting times, horticultural crops may be classified into three broad categories, based on their temperature requirements: cool season crops, intermediate season crops, and warm season crops (Table 6-1).

TEMPERATURE AND PLANT GROWTH

Temperature has a direct influence on the rate of many chemical reactions, including those that are catalyzed by enzymes. In many of these reactions, substrate is utilized or consumed and energy is liberated. This process is known as **respiration** and occurs in the mitochondria of living plant cells throughout the plant (see Chapter 3). Carbohydrates, lipids, and proteins are used as

TABLE 6-1. Some examples of cool, intermediate, and warm season crops.

COMMODITY	COOL SEASON CROPS	INTERMEDIATE SEASON CROPS	WARM SEASON CROPS
Vegetables	Pea Radish Carrot Cole crops (cabbage, broccoli, etc.)	Tomato Potato	Cucurbits (squash, pumpkins, melons) Corn Sweet potato
Turfgrasses	Kentucky bluegrass Creeping red fescue Creeping bentgrass	Zoysiagrass	Bermudagrass St. Augustine grass
Fruits	Apple Pear Plum Cherry	Peach Blackberry	Banana Citrus Coffee
Floriculture	Carnation Snapdragon Freesia	Poinsettia Rose	Gloxinia foliage species
Woody ornamentals	Lilac Norway spruce	Holly Rhododendron	Norfolk Island pine Rubber plant

substrates for respiration. This process proceeds continuously in plants, during periods of darkness and light. In fact, because of photorespiration (which occurs within plant photosynthetic cells in peroxisome microbodies and mitochondria working together), total respiration of C_3 plants during the light usually is two to three times greater than during darkness.

The basic respiration reaction, commonly is given with glucose as the substrate, is as follows:

$$C_6H_{12}O_6 + 6O_2 + 6 H_2O \rightarrow 6CO_2 + 12H_2O + energy$$
glucose oxygen water   carbon water
                              dioxide

This equation briefly summarizes many chemical reactions that lead to the oxidation of glucose.

The energy that is liberated principally is heat that is lost to the air or soil. Additionally, energy is trapped in high-energy molecules, including adenosine triphosphate (ATP). This chemical energy is necessary for many aspects of life, including maintaining cellular integrity, plant growth, and nutrient uptake.

Many intermediate chemicals are produced during the respiratory cycles, chemicals that may also have important biological functions within plants. Therefore the liberation of energy is not the only important consequence of plant respiration.

Plant growth and respiration are results of many enzymatic reactions, and their rates are correlated with temperature. The rate of a reaction or even of a complex process such as plant growth can be assessed in relation to each 10°C (18°F) increase in temperature. The factor that measures the ratio of the reaction rate at one temperature to the rate at a temperature 10°C (18°F) higher is called the **temperature coefficient**, or Q_{10}.

The Q_{10} values in biological systems generally range from 1.3 to 5 but are often close to 2 (Figure 6-1). A Q_{10} value of 2 means that for each 10°C rise in temperature, the rate of a reaction doubles; a Q_{10} of 3 means that for each 10°C rise in temperature, the reaction rate triples. Within temperature limits, the Q_{10} value for plant respiration is approximately 2. Within temperature ranges, growers can increase the growth rate of many crops by increasing the temperature, or slow the growth by reducing temperature (Figure 6-2). The following plant processes show a quantitative relationship to temperature: respiration, part of the photosynthesis reaction, maturation, ripening, germination, and growth in general.

A. $Q_{10} = \dfrac{R_2}{R_1}^{\left(\frac{10}{T_2 - T_1}\right)}$

or

B. $\text{Log } Q_{10} = \left(\dfrac{10}{T_2 - T_1}\right) \log \dfrac{R_2}{R_1}$

or when $T_2 - T_1 = 10$

C. $Q_{10} = \dfrac{R_2}{R_1}$

Where T_1 = the lower temperature
          T_2 = the higher temperature
          R_1 = the rate at the lower temperature
          R_2 = the rate at the higher temperature

FIGURE 6-1. The general equation defining Q_{10} or temperature coefficient.

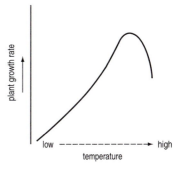

FIGURE 6-2. The relative relationship between plant growth rate and temperature.

In a "typical plant," growth rate varies with the portion of the temperature range being considered. From 0° to 15°C (32–59°F), growth rate increases rapidly, whereas it increases steadily from 15 to 30°C (59°–86°F). Above 30°C (86°F), growth rate usually declines because processes such as respiration utilize food substances as fast or faster than they are manufactured by photosynthesis. At approximately 50°C (122°F), thermal-induced death occurs. The 20° to 30°C (68–86°F) temperature range is the optimum for many plants. It should be kept in mind that *optimum* means best productivity, which is not necessarily maximum growth; in horticulture, the demand is for quality plants or produce, not necessarily the largest or fastest growing plants. Consumers often demand small, compact ornamentals, or food that is within a certain size range. A potato that is 1 kg (2.2 lbs) is generally considered too large.

By knowing the relationship between plant growth rate and temperature, a grower often can speed up a crop by raising daytime temperatures or slow the rate of growth and development by reducing daytime temperatures. However, this capability varies with the crop and stage of development. Within limits, for example, daytime temperature manipulation can control lily growth, but it has relatively little effect on chrysanthemums. This information may be very useful to a floriculturist who sees a responsive crop developing either too rapidly or too slowly for sale on a particular holiday

Thermoperiod and DIF

The relationship between temperature and plant growth and development involves both day and night temperatures. Most plant growth is thought to occur at night because the stomates of most plants close during the dark periods and turgor increases, causing cells to swell and thus enlarge. Although photosynthesis stops after dark, other temperature-sensitive metabolic

processes, including respiration, continue during the night. For this reason the lowering of nighttime temperatures in a production greenhouse has been a common commercial practice. Lowering nighttime temperatures reduces the losses of the of the products of photosynthesis caused by respiration, so food is conserved for use in growth and other metabolic processes. Plants grown under lower night temperatures than day temperatures also can have higher levels of growth-promoting hormones, such as gibberellins (see Chapter 11).

It has been shown that plants can grow more satisfactorily in alternating day and night temperatures than in a constant temperature environment. This response is called **thermoperiodicity**. It was reported as early as 1944 that when mature tomato plants were grown at a constant temperature of 26.5°C (80°F), the growth of the fruit was reduced dramatically, and stem growth was generally less than when the plants were grown with a day temperature of 26.5° C (80°F) and a night temperature of 20°C (68°F).

With many floricultural crops, compact plants are considered attractive and desirable. Such compact plants can be produced by using chemical growth retardants (see Chapter 12) or by manipulating thermoperiod. Some crops, such as lilies, poinsettias, and chrysanthemums, will grow taller if the night temperature is held constant and the day temperature is increased. Conversely, shorter plants will result if the day temperature is held constant and the *night* temperature is increased. The difference between day and night temperature (called **DIF**) is important for these crops. DIF is calculated by subtracting the night temperature from the day temperature. For example if the day temperature is 25°C (77°F) and the night temperature is 20°C (68°F), the result is a *positive* DIF of 5°C (9°F); however, if the day temperature is 20°C and the night temperature is 25°C, the result is a *negative* DIF of 5°C (written as −5°C). If the day and night temperatures are equal, the result is *zero* DIF. The more positive the value of DIF the taller the plant. DIF controls plant height by changing internode length but not the number of nodes produced by the plant.

With some species, such as *Kalanchoë blossfeldiana* a constant temperature (zero DIF) results in shorter plants than either positive or negative DIF. With some species, there is a greater reduction in height growth changing from positive to zero DIF than from zero to negative DIF.

A two-hour drop in temperature at sunrise can be as effective as cool days for some crops. This expands the use of DIF into warm climates or during parts of the warmer times of the year in temperate zones. This temperature drop is known as Drop or DIP.

Plants should be exposed to DIF each day. If they are growing too rapidly, they can be slowed by changing to zero or negative DIF. Conversely, if plants are growing too slowly, they can be switched to a positive DIF.

The most effective time to use DIF is when plant stems are growing rapidly. For many floricultural species with determinate growth, such as chrysanthemum and poinsettia, stem growth is most rapid during the middle of the production period. Stem growth of these plants slows near the time of flowering so DIF will have less effect at this time.

With some crops, use of far-red light filtering greenhouse coverings, which raise the red light:far red light ratio will enhance the growth retarding effects of zero or negative DIF. Lighting with far-red light or incandescent lamps (low red light:far-red light ratio) can also negate the effects of negative DIF.

There can be some undesirable plant responses under strongly negative DIF, such as yellow (chlorotic) leaves and Easter lilies with down-turned leaves. Fortunately, this can be reversed by changing to a positive DIF.

A grower should therefore carefully monitor both day and night temperatures for greenhouse crops; there are various temperature recorders available for collecting such information 24 hours per day. Alarms are often set to sound at a grower's residence in response to night temperatures becoming too high or too low. It should be noted that temperature recommendations for many greenhouse crops list only night and not day temperatures, since night temperatures are considered more critical.

Graphical tracking (crop modeling) has been developed for many crops, especially in floriculture. This is accomplished by measuring plant height and plotting it on a computerized graph that compares the crop height to the expected height on a predetermined growth curve. If the crop is too tall, the grower can begin negative DIF or Drop, or apply a growth retardant, and if it is too short, positive DIF can begin so that the crop will reach the predicted height range.

For most rapid growth potential, on cloudy days it is recommended that the day temperature be maintained 3°C (5°F) higher than the night temperature. However, on sunny days, it is recommended that the day temperature be maintained 5° to 7°C (10–15°F) higher than the night temperature because of the greater photosynthetic potential during periods of increased light.

Growing Degree Days

The development and reproduction of living organisms, such as plants and insects that cannot maintain a constant internal temperature, is highly dependent on environmental temperature and time. Predictions of when these organisms will develop from one point to another in their life cycles can be calculated in time and temperature units known as **growing degree days** (heat units). Growing degree days are calculated to determine approximate time to harvest and to estimate when fields and plants should be scouted for damaging insect populations as part of an integrated pest management

(IPM) program. Vegetable processing companies, in particular, calculate growing degree days to estimate the appropriate time to send harvesting equipment to fields. Growers often combine the use of degree days with other methods, such as measuring the sugar content of the fruit with a refractometer (Figure 6-3) or the internal level of ethylene to determine optimum timing of harvest.

As discussed earlier in this chapter, organisms grow within a range of temperatures. The minimum temperature at which growth occurs is known as the **minimum developmental threshold** (minimum threshold, base temperature) and the maximum temperature at which growth will take place is called the **maximum developmental threshold** (maximum threshold). A single growing degree day is 24 hours at one degree above the minimum developmental threshold. For example, one growing degree day is accumulated if the minimum developmental threshold is 10°C and the temperature remains 11°C for 24 hours.

Growing degree days can be calculated in degrees Celsius or degrees Fahrenheit. Because a Fahrenheit degree is smaller than a Celsius degree, five Celsius growing degree days equals nine Fahrenheit degree days. Therefore it is important that temperature measurements be standardized when calculating growing degree days.

The oldest and easiest way to calculate growing degree days is the simple or rectangle method. There are other different methods for calculating growing degree days, including the cosine, Huber's, sine, and triangle methods. These measure areas under a curve between the minimum and maximum developmental thresholds and can be complicated, so are often calculated by computers.

With the rectangle method, growing degree days are calculated according to the following equation:

$$\text{Degree days} = \frac{\text{daily maximum air temperature} + \text{daily minimum air temperature}}{2} - \text{minimum threshold temperature}$$

Because growth does not occur below the minimum or above the maximum threshold temperatures, if the daily maximum air temperature exceeds the maximum threshold temperature, the maximum threshold temperature must be used as the daily maximum air temperature. Likewise, if the daily minimum air temperature is less than the minimum threshold, the minimum threshold temperature must be used as the daily minimum air temperature.

Accumulation of growing degree days begins with a **biofix date**. This biofix date varies depending on species and the use of the growing degree day calculations. The biofix date for most seed-propagated plants begins with seedling emergence, although it can begin with the planting date. For cranberry, the biofix date is when the ice is off the plants and bog and for apple scab, it is the date of bud break. Insects may have calendar dates as biofix dates, or the date may be based on the first catch in a trap, or the first occurrence of a pest or eggs.

Minimum and maximum threshold temperatures also differ with species. For example, the minimum threshold temperatures are 2°C for onions, 4°C for canning peas, and 10°C for sweet corn, snap beans, and tomatoes, and fruit crops such as apples and grapes. For various pests and pathogens

FIGURE 6-3. Sugar content of fruits such as blueberries and strawberries is often measured with a refractometer.

that cause disease, the minimum threshold temperatures are 0°C for development of apple scab, 6°C for cabbage maggot, and 10°C for common asparagus beetle, flea-beetle, imported cabbageworm, and tree and shrub pests. Some pests have no known maximum threshold temperature, including apple scab, cabbage maggot, common asparagus beetle, fleabeetles, and imported cabbageworm. However, cranberry, sweet corn, and tree and shrub pests have a maximum threshold temperature of 30°C.

Examples of calculations for growing degree days for sweet corn (10°C and 30°C for minimum and maximum threshold temperatures) are one day with a temperature range from 8° to 20°C, another day with a range from 12° to 24°C and a third day with a range from 24° to 35°C. The degree days for each of these three days would be calculated as follows (recall that when the daily temperature is lower than the minimum or higher than the maximum threshold temperatures, the threshold temperature must be used as the appropriate low or high temperature for that day):

$$\text{degree days} = \frac{10 + 20}{2} - 10 = 5 \text{ degree days}$$

$$\text{degree days} = \frac{12 + 24}{2} - 10 = 8 \text{ degree days}$$

$$\text{degree days} = \frac{24 + 30}{2} - 10 = 17 \text{ degree days}$$

If these were three days in a row, there would be an accumulation of a total of 30 growing degree days during this time period. The producer will make this calculation daily until the total number of degree days required by that crop to reach optimum maturity for harvest has accumulated. These numbers can be fed automatically into a computer and the producer can very quickly get an estimate on the relative state of maturity of each field and crop. The number of degree days required for peas is approximately 1000 to 1200, depending on cultivar and location. Grapes require from 120 to 170 days from bloom to harvest or 1955 to 3375 degree days and

apples take from 70 to 170 days from bloom to harvest and require between 1400 and 2800 degree days.

There are various websites, ranging from the United States National Weather Service to various university extension sites that publish growing degree day accumulation charts throughout the growing season. This is often on a county-by-county basis. Other sites are available where threshold temperatures and time periods can be specified and the growing degree days are calculated instantly. Some of these sites also offer growers advice on how to use the growing degree days information and when they should be scouting their crops.

One must exercise caution, however, when relying solely on growing degree days. Although air temperature has a great influence on plant growth and development, other environmental factors such as soil temperature, moisture status of the crop, solar radiation and photoperiod also play important roles. Therefore, growing degree days should be used as a tool to help alert the producer when to either scout the fields or to otherwise check the crop. It is by intelligent use of a variety of methods that the highest quality products will be produced and harvested at their optimum maturities.

SUGAR AND STARCH CONVERSIONS

Starch is a carbohydrate polymer that is abundant in the plant kingdom and consists of chains of glucose (sugar) molecules. Starch is a major form in which most plants store energy, and it is often found in large quantities in seeds, stems, and roots.

Temperature influences the conversion of sugar to starch and starch to sugar, according to the following equation:

$$\text{sugar} \underset{\text{cool}}{\overset{\text{warm}}{\rightleftharpoons}} \text{starch} + \text{water}$$

Under warm conditions, sugar in plants is converted to starch and water, and under lower temperature regimes, starch and water are converted to sugar.

Knowledge of the preceding equation has important implications for the quality of horticultural food crops in relation to the time of year and location where they are grown and their handling after harvest. Asparagus (*Asparagus officinalis*) should be harvested early in the spring, for example, because the lower temperatures enhance the flavor. The asparagus will have a higher sugar content and will therefore taste sweeter compared to harvests made later in the spring when the temperatures are higher. Likewise, Brussels sprouts (*Brassica oleracea* var. *gemmifera*) are best planted in the summer for a late fall harvest. Such plants will produce sprouts that are sweeter because of the higher sugar content related to the lower temperatures experienced in the fall. Similarly, when sweet corn is harvested, it should be cooled immediately (or eaten at once) to retain its sweetness ("supersweet" corn cultivars that contain the SH_2 gene are the exception because they do not readily convert their sugar to starch). In sweet corn with the normal sugary *su* gene or with the sugary enhancer *se* gene, the sugar is very rapidly converted to starch, particularly under higher temperature conditions. This is important for canning and freezing companies, many of which will hydro-cool (pass through cool water bath or spray) the ears on the trucks shortly after harvest to reduce the temperature of the kernels quickly, thus maintaining the high sugar levels and sweet-flavored corn. Peas respond in the same fashion as sweet corn and are therefore handled similarly. Asparagus, sweet corn, and peas are also often harvested in the early morning to avoid heating of the product, a practice that aids in hydro-cooling and reduces the conversion of sugar to starch.

Potato tubers, on the other hand, will become undesirably sugary for chipping when stored between the temperatures of 0° and 5°C (32° and 41°F). This process can be reversed if tubers are conditioned at higher temperatures. High sugar levels in potato tubers cause problems for the potato chipping industry, because during the cooking of the chips, the sugar will be converted to caramel (burnt sugar) during cooking. Caramel will make the chips a darker brown, which is considered undesirable by the industry because of consumer preference for light-colored chips (Figure 6-4). A knowledge of the influence of temperature on the conversion of starch to sugar and sugar to starch is important, particularly in postharvest handling of many horticultural crops. This is especially true for vegetables, but also plays an important role with fruit crops and some flowers (see Chapter 15).

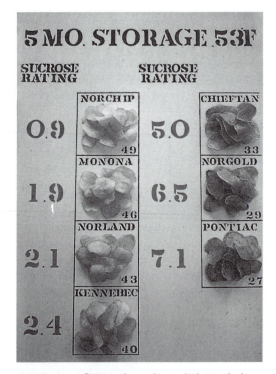

FIGURE 6-4. Potato chip color is darker at higher sucrose contents. The sugar level that is accumulated at a given storage temperature varies among cultivars, thus influencing choice of cultivar to chip from cool storage.

Courtesy of J. Sowokinos, Red River Valley Potato Research Laboratory, MN.

This principle is also illustrated by the practice of reducing greenhouse temperatures to cause brighter flower or bract color (e.g., in poinsettia). The class of red pigments known as anthocyanins is formed from sugars, so any practice responsible for causing an accumulation of sugars may result in enhanced pigmentation. Because they often attempt to hasten development of a crop by raising temperatures or to slow such development by lowering temperatures, growers need to consider the potential impact on crop quality when pursuing such practices. This knowledge also helps explain why red-skinned apples have deeper pigmentation when grown in climates with cooler night temperatures, such as at higher elevations or latitudes distant from the equator. Related to this is when the color fades to greenish on many red leafed trees in temperature zones with high night temperatures during the summer. The red color is much more intense in cooler climates on many plants, including red leafed maples and colored beeches.

The intensity of autumnal red leaf color is another example of the effect of temperature on coloration. Cool nights (and bright days) cause greater anthocyanin pigment production, probably because of greater accumulations of sugars on cool nights. Rainy weather, in addition to reducing photosynthesis, contributes to a more drab coloration because some of the sugars are leached from the leaves. Therefore, the beauty of the autumn season is greatly dependent on the weather.

....

SOIL TEMPERATURE

Soil temperature influences many plant processes including seed germination, root growth, plant growth and development, water uptake and disease susceptibility. Generally, optimum soil temperatures range from 15° to 30°C (59°–86°F), depending on the species. Cool season crops tend to grow best at lower soil temperatures and warm

season crops usually grow best at the upper end of the range.

Because soil temperatures influence seed germination, the timing of sowing seeds is critical for obtaining an acceptable seedling stand. Low soil temperatures associated with winter or early spring plantings can lead to poor or no seedling emergence. Seed metabolism is slowed by the cold soil, and disease-causing organisms, such as the water mold *Pythium* (see Chapter 16), can grow in cooler soils. Therefore, seeds planted too early in the spring may germinate so slowly that they often decay in the soil either prior to or just after emergence through the soil surface (pre-emergence or postemergence damping-off, in this case caused by *Pythium* spp.) (Figure 6-5). Attempts to obtain an earlier crop through early seed sowing often result in either a stunted, nonuniform crop or total crop failure. In the field, clear or dark-colored mulches, low tunnels, or high tunnels are used to warm the soil to allow for earlier sowing to help time harvests for premium prices.

One should sow seeds of cool season crops in cool but not cold soils and warm season crops in warm soils. The maximum soil temperature at which cool season seeds will germinate is about 25°C. Sometimes cool season crops are planted in late summer for a fall crop, but if the soil temperature is too high, the seeds of cool

FIGURE 6-5. Postemergence damping-off of a marigold seedling.

season crops will fail to germinate, or the seedlings will not grow normally. Lettuce seeds (a cool season crop) may be planted late in the day so sprouting will occur during cooler night hours. This may avoid high-temperature-induced dormancy. The minimum soil temperature at which warm season seeds will germinate is about 10°C, but this temperature may still be too low for some crops. It is often advisable to dust the seeds with a suitable fungicide to protect against damping-off organisms (see Chapter 16).

Root growth is similarly influenced by soil temperature. When soil temperature is above or below the optimum for a particular crop, root growth slows or stops. Generally, the roots of cool season crops grow better at lower soil temperatures than do roots of warm season crops. Conversely, warm season crop root growth continues at higher soil temperatures, whereas root growth of cool season crops often will stop. When strawberries are grown in hot soils, for example, the root system becomes reduced to the extent that an insufficient amount of water is provided to the tops, resulting in wilting and reduced growth.

Root resistance to water uptake is greater in low soil temperatures than in warm soils. Therefore, the tops of the plants with roots in cold soil are often not provided with adequate amounts of water. If plants are watered with cold water, the tops often will wilt, particularly with warm season crops. In wet, heavy soils, watering with cold water is not a great problem, because wet soil has a fairly high heat capacity (see Chapter 7). Plants grown hydroponically or on sandy soils are not buffered against the sudden temperature change of irrigating with cold water; such an irrigation, because of increased root resistance to water uptake, may cause the crop to wilt (Figure 6-6). The plants generally will recover gradually as the root zone warms. Low soil temperature can also result in decreased nutrient uptake because of increased root resistance to uptake.

FIGURE 6-6. This coleus plant is wilting because of the introduction of cold water (*pitcher*). On this sunny, warm day, the plant wilted within a few minutes, even though the medium was saturated.

Many greenhouse growers go to the extra expense of heating all of their irrigation water to maintain higher soil temperatures. The use of such tempered water may cause less shock to the plants during watering, and thus better growth. Growers believe that they notice fewer problems with those crops, but experimental evidence with tempered water is inconclusive.

Controlled-environment growing affords growers several means of maintaining optimum soil temperatures. Historically, manure and straw were mixed together and placed under the soil in hot beds so that the exothermic bacteria decomposing the organic matter would provide heat for the soil. This system is now used less commonly because growers cannot accurately control soil temperature. Growers currently have many methods for maintaining high soil temperatures, in addition to the use of tempered water. Electrically controlled heating pads and heating cables placed in greenhouse benches allow the grower to control the temperature of the root zone thermostatically. Hot water or steam pipes are commonly found under greenhouse benches or under the floor of the greenhouse. When the pipes are located under a bench, growers often hang a polyethylene "apron" or "skirt" from the edges of the bench to the floor to hold in valuable bottom heat (Figure 6-7).

FIGURE 6-8. These PVC pipes are part of a closed hot water system used to provide bottom heat. The pipes are placed at intervals in this ground bed to disperse the heat evenly.

FIGURE 6-7. A plastic apron is used to conserve heat from heating pipes under a bench used for propagating cuttings. A view from under the bench *top*; a close-up of a heat pipe showing "fins" that provide greater surface area for radiating heat, *center*; and an overview of a bottom-heated bench filled with hardwood evergreen cuttings, *bottom*.

Systems in which a continuous series of plastic pipes or flexible rubber-plastic tubing are laid out in loops to conduct hot water can be constructed or are commercially available (Figure 6-8). The flexible tubes are generally placed less than 20 cm (8 inches) apart in ground beds, on benches, or in the floor of a greenhouse. A suitable medium—flats, pots, plastic or fibrous mats, or the concrete of a greenhouse floor—may be placed directly on top of the pipes or tubes. A boiler or hot water heater is used to supply the hot water and a pump circulates the water through the system at a rate that will maintain good heat transfer. The temperature can be controlled thermostatically, and a temperature probe should be inserted into the growing medium and monitored regularly. Water generally leaves the boiler at 40° to 42°C (104°–108°F). Some systems utilize antifreeze in the tubes in case of a boiler failure in the greenhouse. The lines of these systems must be bled of air to function properly. Such systems offer a means of effective and economical control of the temperature of the growing medium.

Bottom heat also has been used in attempts to conserve valuable energy resources. Although their results are somewhat inconclusive, researchers have studied the influence of keeping roots warm at night while further reducing night air temperatures. It appears, however, that soil temperature will not wholly substitute for night air temperature.

Many sources of waste heat available, including that produced by electrical power plants (Figure 6-9). The waste hot water is commonly circulated through a network of pipes that runs under a greenhouse or range of controlled-environment facilities. This allows for the maintenance of high soil temperatures, greatly reducing heating costs. An alternative approach is to run the waste hot water through heat exchangers for use in heating the greenhouse air, either as the primary heat source or as a supplement to soil heating. A limitation on this system is the temperature of the waste hot water. If it is not sufficiently hot, it may not be a valuable resource.

Regulating soil temperatures also is important in the vegetative propagation of horticultural plants by cuttings. The ideal air temperature for rooting cuttings of many species is 21° to 27°C (70°–80°F) during the

(a)

(b)

(c)

(d)

(e)

FIGURE 6-9. Electric power plants (*a*) produce waste heat in the form of hot water that normally is cooled in cooling towers (*b*). Greenhouses (*c*) also can utilize this energy to heat the air and soil. This heat exchanger (*d*) removes heat from the hot water and distributes the warm air through convection tubes within the greenhouse (*e*) in which various crops are grown.

day and 15°C (59°F) at night. The temperature of the rooting medium should be maintained at least at 15°C (59°F), although higher root zone temperatures often are employed by propagators who use bottom-heating methods. These growers usually provide rooting zone temperatures from 21° to 27°C (70° to 80°F) to initiate roots on cuttings. This practice results in faster rooting of cuttings and often a higher rooting percentage compared to cuttings rooted in a cooler medium. Best rooting generally occurs if the temperature of the rooting medium is maintained higher than air temperature. The optimum temperature for a rooting medium for root *initiation* is approximately 5°–7°C (10°–15°F) higher than the optimum temperature for root *elongation*. It is therefore recommended that once cuttings root, the root zone temperature be lowered to encourage root elongation. Occasionally problems can arise with heating the rooting medium excessively. The soil temperature can become so hot that the soil surface heat damages or burn the plants.

In the field, the soil surface also can heat to levels that will cause plant damage or burning, which is particularly a problem with young, tender seedlings. This phenomenon often occurs on black muck soils and other dark colored soils because the reflection of sunlight is low and the absorption is high, causing the soil surface to heat to high levels. Onion seedlings exhibit this problem in the spring before they become large enough to shade the soil and reduce the incident radiation that strikes the soil surface. As a result the onion seedlings will exhibit symptoms similar to those of postemergence damping-off because heat causes the collapse of cells of the seedling near the soil line. Such symptoms can also occur where black polyethylene touches the hypocotyl of young seedling grown with a black plastic mulch. This phenomenon also occurs in some hydroponic situations, particularly the nutrient film technique (NFT) (Figure 6-10). If the nutrient film lines or gullies are constructed of black polyethylene, the black polyethylene can (and often does) touch the

FIGURE 6-10. Heat damage (*arrow*) on these tomato seedlings exhibits symptoms similar to postemergence damping-off.

stem of the young seedlings. During periods of high incident radiation, such as hot summer days, heat burning will occur and will cause damping-off symptoms.

Mulches also influence soil temperature (see Chapter 10). Clear and infrared transmitting polyethylene mulches are often used for warm season crops such as curcurbits to aid in warming the soil. The solar radiation strikes the soil and is converted to long wave radiation (heat) and the dead air space trapped between the soil and the mulch insulates, thus reducing convective heat loss. Additionally, the mulch prevents the mixing of cooler air with the hot air beneath the mulch. Growers are able to produce better and earlier crops of some species by using these mulching materials.

Some mulches, especially organic materials such as straw and bark, also insulate the ground and reduce the temperature fluctuations in soils. The soil may then remain cooler in the spring and stay warmer in the fall. The lower spring soil temperature can result in delayed bud break on perennial species such as trees and shrubs, or later planting dates for annual crops, and delayed cooling of the soil in the fall can force new succulent growth on some perennials, which may be more susceptible to freezing damage. Therefore, the influence of mulches on soil temperature can have positive or negative consequences.

TEMPERATURE MONITORING AND CONTROL

As previously mentioned, the temperature of both the growing medium or soil and the air (night and day) are important to plant growth. Although controlled environments may be easier to manipulate, the grower who produces outdoor crops must be acutely aware of plant, soil, and air temperatures as well as atmospheric and weather conditions.

Growers therefore often monitor the temperature in fields, nurseries, or orchards by the judicious placement of several minimum/maximum thermometers or other temperature recording devices such as recording chart thermometers throughout the field or orchard. Similar to minimum/maximum thermometers, recording chart thermometers give the daily low and high temperatures, and they also alert the producer to temperature fluctuations and the length of time at each temperature. Hygrothermographs are also used to record temperatures and leaf wetness (for disease monitoring) over a period of time, but cost more than recording chart thermometers and minimum/maximum thermometers. An even more expensive alternative is the use of data logger weather units. These are associated with computer programs and provide growers with temperature information, as well as when conditions are conducive to the development of disease-causing organisms. In-field computerized weather stations are also available to help growers monitor conditions related to temperature and development of disease-causing organisms, such as *Botrytis* and those that cause downy mildew and powdery mildew.

Thermistors are often used in computerized systems for sensing temperature. These solid-state chips function by changing the voltage output as the temperature fluctuates. Thermistors provide more accurate temperature sensing than thermometers and are therefore gaining popularity among growers.

Field temperature monitors should be in wooden shelters and may be distributed according to the terrain, with focus on high and low spots. Often one thermometer or thermistor per 2 hectares (or one per five acres) can be adequate if the terrain is relatively uniform.

Some growers hang thermometers from branches, however, this may result in inaccurately low readings on cool, clear nights and are not representative of the temperatures on and in plant parts. Electronic thermocouples are two wires of different metals that can be inserted into buds or other plant parts to measure the current flow through the junction of the wires. These wires are connected to electronic monitoring equipment that converts these current flow rates into temperatures. Thermocouples record accurate temperatures *within* plant parts. This monitoring is valuable for managing freeze protection. Although thermocouples provide accurate and valuable information and are relatively inexpensive, they have not been widely adopted by growers of field-grown horticultural crops.

An infrared thermometer is a line-of-sight instrument that is aimed at the surface of a plant (or other object) and detects the thermal energy wavelength from that surface without touching the plant. It gives the surface temperature of the part of the plant at which it is aimed. Surface temperature can be closely related to heat or cold stress on the plant.

Although the temperature of plants is more difficult to control outdoors, it must be closely monitored by horticulturists so that they can react to temperature extremes. Knowing the symptoms of temperature stress enables growers to read their crops using a basic understanding of plant biology and the environmental factors that affect plant growth, and to react accordingly. This reading is crucial because leaf and fruit temperatures can be different than air temperature and may not be measured accurately without thermocouples.

In a controlled-environment situation, the grower can govern the environmental variables that influence plant growth. The temperature in a greenhouse can be monitored by a variety of methods, including use of thermometers, thermostats, and thermistors. Temperature readings in a greenhouse may be inaccurate if care is not taken in placement of sensors, because both hot and cold spots, as well as shaded and sunny areas, are present. Locating a thermometer, thermostat, or thermistor in a hot or cold spot or in the direct sunlight will result in inaccurate temperature readings and thus poor control of the heat in a greenhouse, resulting in erratic plant growth, or the production of crops with less-than-desirable quality.

Aspirating (passing air by) and shading the temperature sensor is the most accurate way to monitor the air temperature in a greenhouse (Figure 6-11). If the sensor tip is exposed to direct sunlight or is near a heat source, it will give an inaccurately high temperature reading. Therefore, temperature sensors should be housed in a white or light-colored box or tube to provide shade. The box is aspirated by an exhaust fan located at one end to pull air through the unit. It is considered best to use the fan to exhaust the box rather than to take in air because the heat from the blower motor could give inaccurate readings if blown into the box. The air flow rate of the fan can range widely, which has been interpreted to mean that the fact that air moves in probably more important than the rate of movement. Many growers have used a 1/500 HP fan rated at approximately 3000 RPM.

The average temperature at crop height should be monitored so temperatures can be regulated to produce optimum growing conditions. Locating the aspirated temperature sensor at crop height with a louvered box, or using an adjustable, flexible intake hose, such as a clothes dryer hose, will accomplish this. The intake can then be adjusted to plant height as the crop grows or placed among the different crops that are grown throughout the year. The box also should

FIGURE 6-11. Views of the outside (*top*) and inside (*bottom*) of an aspirated thermostat.

not be located in any hot or cold spots in the greenhouse.

Aspirated or other thermostats or thermistors in a greenhouse are connected directly or by computer to switching systems controlling heat sources, vents, and cooling systems. It is by automatically controlling the greenhouse temperatures that one can accurately control the environment in these growing facilities.

COOLING CONTROLLED ENVIRONMENTS

It gets hot on sunny days in greenhouses, automobiles, and under clear and infrared-transmitting mulches. This heating up is commonly known as the **greenhouse effect** making greenhouse cooling essential for maintaining temperatures conducive to plant growth. Greenhouses trap air, prevent this air from mixing with the outside air, and characteristically lack convection (movement of heated air) compared to outdoors. Sunlight that passes through the greenhouse covering is absorbed by plants and objects within and loses energy. This energy is re-radiated as long wave radiation (heat) that will accumulate. The air within the greenhouse insulates, the covering is less transparent to long wave radiation than to short wave radiation (light), and the covering prevents the mixing of cooler air from outside with the warm air within, all contributing to the warming. Temperatures can get very hot within, leading to plants drying quickly and suffering heat stress. To prevent thermal-induced injury or death of plants, growers exhaust excessively warm air by vents or fans, use greenhouse structures with retractable roofs or the whole greenhouse is retractable, cool the air, reduce the amount of sunlight entering the greenhouse, or use a combination of these.

To understand both the cooling (during summer) and prevention of heat loss (during winter) from a greenhouse, one must first understand how heat is transferred. Heat can flow through a material by molecular action, which is known as **conduction**. Materials available for greenhouse construction differ in their ability to conduct heat. As a general rule, dense compounds such as metals conduct heat well, whereas less dense materials such as wood do not conduct heat as well. Glass conducts better than air-inflated double polyethylene film and has consequently lost much popularity as a greenhouse covering.

Heat also is transferred by moving air, or **convection**. When heat moves along an air stream it is called **forced convection**, whereas **free convection** is when warm air rises and cold air settles into lower areas.

For cooling greenhouses, growers utilize the principle of convection when they exhaust hot air. In glass, some other rigidly covered greenhouses, and those with retractable roofs, vents located at the tops

are opened when the cooling requirements are not great, such as times of the year when the sun is not high in the sky or during moderately overcast periods (Figure 6-12). The free convection of warm air rising through the vents and the entry and settling of cooler air effectively reduce greenhouse temperatures. Within limits, the degree of cooling can be governed by the distance the vents are opened, which often is controlled thermostatically. Sidewall vents also can be opened in conjunction with the top vents to allow the entry of cooler outside air into the greenhouse to replace the warm air that freely convects through the top vent. The wind may cause forced convection through those vents.

In greenhouses that do not have top vents, such as many quonset-shaped or gutter-connected double polyethylene-covered structures, the hot air is exhausted by fans. The fans are located at the end or along the side of the greenhouses and are thermostatically controlled to turn on when the air gets too warm. Fans create air currents and thus results in forced convection so cooler air enters the greenhouse at a rate sufficient to replace the exhausted warm air. These exhaust fans are usually capable of one complete air change each minute.

During the winter it is often necessary to cool greenhouses, but problems can result from the influx of the cold outside air (e.g., cold spots and plant damage can occur). Therefore, the exhaust fans may be supplemented by a louvered inlet at the opposite end of the greenhouse, which is often equipped with a fan of equal capacity to that of the exhaust fan (to prevents cold spots). The inlet fan blows the cooler air into a convection (or distribution) tube, which extends the length of the greenhouse. Along the sides of the convection tube are holes through which the cooler air is distributed. This setup provides a good mix of the cold and warm air and allows for uniform temperatures throughout the greenhouse (Figure 6-13). The convection tube can also be used when heating in order to distribute warm air uniformly.

FIGURE 6-12. Vents (*left*) are a common feature of many greenhouses. Double-polyethylene–covered greenhouse (*right top*) with the side rolled up to provide ventilation. Side view of greenhouse (*right bottom*) with sections of the roof retracted as well as having its sides rolled up to provide good air mixing with cooler outdoor air.

During the summer, the use of vents, exhaust fans, or both may not sufficiently cool the air, and temperatures can climb to damaging levels within a greenhouse because the outside air is too warm to cool the inside of the growing facility. Cooling systems therefore are necessary to maintain optimum greenhouse growing temperatures.

Cooling can be accomplished with air conditioning units, often in conjunction with exhaust fans (Figure 6-14). Air conditioners generally are quite expensive to

FIGURE 6-13. Inflated (*left*) and collapsed (*right*) convection tubes in a greenhouse. Holes along the sides of the tube facilitate uniform air distribution.

FIGURE 6-14. Window air conditioners sometimes are used to cool greenhouses.

install and operate, however, so their use in greenhouses has been limited although they have been used in other controlled-environment growing facilities such as warehouses that are equipped with lights for growing plants indoors. Air conditioning also is the primary cooling source for plants in interiorscaping situations such as a shopping malls or office buildings.

Evaporative cooling systems are widely used in greenhouses. These systems utilize the principle of the heat of vaporization of water (see Chapter 7). The most common evaporative cooling systems are called **fan and pad cooling**. Along one wall in a greenhouse, pads are located either horizontally or vertically. Water passes through the pads,

keeping them wet. The traditional material for pads has been shredded aspen (excelsior), but this material generally has been replaced by a cross-fluted cellulose material impregnated with anti-rot salts, wetting agents, and fungicides (Figure 6-15). This newer material resembles corrugated cardboard in appearance and lasts longer than excelsior.

Exhaust fans are mounted on the wall opposite the pads (Figures 6-16, 6-17). These fans remove hot air from the greenhouse and draw air through the pads. The process of evaporation absorbs heat from the incoming air (heat of vaporization) and thus cools the air. The air warms as it moves by forced convection across the greenhouse, where it is then exhausted by the outlet fans. To be effective, there are specific relationships between pad area and fan sizes. The greenhouse management books listed as references at the end of this chapter provide further information.

Other evaporative cooling systems utilized by greenhouse operators are forced mist or fogging units. These inject a fine mist or fog into the greenhouse that will cool the air. The efficiency of fan and pad or a fog system depends on the relative humidity of the air: the lower the relative humidity, the more efficient the evaporative cooling system.

FIGURE 6-15. A view of excelsior cooling pads on the outside of a greenhouse (*left*) and cross-fluted cellulose (*right*) from the inside of a greenhouse.

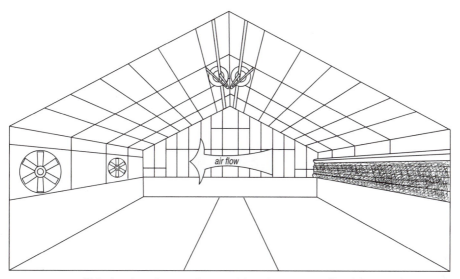

FIGURE 6-16. This drawing of a greenhouse with pads on one wall and fans on the opposite wall illustrates the concept of a fan and pad cooling system.

FIGURE 6-17. The insulation panel (*arrow*) will be placed over the fan during winter nights to prevent substantial heat loss.

Because sunlight is the major source of heat during times when greenhouses must be cooled, reducing the quantity of radiation that enters the greenhouse effectively reduces the heat load within the structure. Therefore, partial shading is important from spring to fall. There are several ways to shade a greenhouse, including the construction of lath over a greenhouse, use of various shade cloth materials, or most commonly the application of a shading compound directly to the greenhouse covering (Figure 6-18). Growers may use one of several commercial shading compounds available or some growers simply use whitewash materials. White shading compounds are recommended because they reflect more sunlight than colored materials can. These shading materials reduce the quantity of light that reaches the plants and thus reduce photosynthesis. Therefore, using too much shading compound may cause poor plant growth. In the fall, the shading compound must be removed promptly to allow more light into the greenhouse.

Some growers cool their greenhouses by maintaining a film of water on the glass or

FIGURE 6-18. Application of a shading material reduces the amount of light entering the greenhouse, thus reducing potential buildup of heat.

between the layers of the covering material (e.g., double polyethylene). The water effectively absorbs infrared radiation and thus reduces the temperature within the greenhouse. Systems employing this approach store the warmed water during the day and the absorbed heat subsequently is released during times of demand for heat during the night.

A newer approach is to use an infrared (heat)-reflecting plastic film to cover the greenhouse. These materials contain a pigment that selectively filters infrared radiation and can reduce greenhouse temperatures by 5°–8°C (9°–14°F).

··

HEATING AND ENERGY CONSERVATION IN CONTROLLED ENVIRONMENTS

In many areas where greenhouses and other controlled-environment growing facilities are located, the ambient air temperature is too cool for the crop. Solar energy does not always provide enough heat during daylight hours, and unless the energy is efficiently stored, it becomes too cool at night. Heating is therefore a major expense for operators of controlled-environment growing facilities.

Many factors contribute to heat loss: the covering material, the condition of the

covering material, insulation, shape and size of the structure, the ability to store energy, and the size of the area that is heated. The source of energy for heating, the units that provide the heat, and the efficiency of the operation of these units all contribute to the expense of heating a greenhouse or other plant-growing area.

When selecting a greenhouse covering material, one should consider such variables as initial expense, availability, durability, flammability, taxes, and heat loss characteristics. Many commonly used greenhouse coverings have relatively poor insulating qualities. These include such materials as: a single layer of either glass, polyethylene, or fiberglass (Table 6-2). Because of the higher R values (hr·°F·sq ft/Btu) of air-inflated translucent double-layer polyethylene, double acrylic, or twin-wall polycarbonate (Table 6-2), these materials frequently are used for construction of energy-efficient

TABLE 6-2. The resistance to heat flow (R value) of some common greenhouse construction materials.

MATERIAL	R VALUE[a] (HR·°F·FT^2/BTU)
Glass, single layer	0.88
Clear polyethylene, 2 or 4 ml single layer	0.87
Clear polyethylene, double layer, air separated	1.43
Fiberglass	1.00
Double acrylic	1.82
Softwood, 2.5 cm thick	1.79
Polystyrene (styrofoam) 2.5 cm thick	4.00
Concrete block, 20 cm thick plus 2.5 cm foamed urethane	7.69

Source: Blom, T., J. Hughes, and F. Ingratta. *Energy Conservation in Ontario Greenhouses*. Ontario Ministry of Agriculture and Food. Publication 65.

[a]A higher R value indicates better insulating quality.

greenhouses. Most greenhouse coverings and combinations that effectively transmit light have R values below 2. Some people believe that greenhouses someday may be too expensive to heat and thus will become obsolete. Therefore, some controlled-environment growing facilities in which plants are grown are insulated buildings with electric lamps. Note the R value of 7.69 for 20 cm thick concrete blocks insulated with 2.5 cm foamed urethane compared to 0.88 for a single layer of glass in Table 6-2. Some of these new building are so well insulated that the electric lamps alone provide sufficient heat to grow the crop, even during the winter.

In greenhouses the condition of the covering material also can have a great influence on the energy usage. For example, if panes are missing or if spaces exist between the panes of glass, heat will escape very rapidly, adding greatly to heating expenses. Overlapping panes of glass and applying a lapseal (essentially sealing the area between overlapped panes) may bring about substantial reductions in energy usage. Likewise, air spaces around doors, fans, and vents can be major avenues for heat to escape, but using weather stripping and caulking can reduce heat loss (Figure 6-19).

Hanging a single layer of polyethylene inside of a glass-covered, single layer fiberglass-reinforced, plastic-covered, or single layer polycarbonate-covered greenhouse can

create a dead air space, thus adding insulation to a greenhouse (Figure 6-20). Some growers utilize a double layer of polyethylene on the outside of their glass covered greenhouses and thus reduce heating costs. One should be aware, however, that with each layer added to a greenhouse covering there is a corresponding reduction in the amount of light striking the plants, which may in turn cause poor plant growth.

A direct relationship exists between greenhouse design and heat loss. Proportionally, the greater the surface area, in relation to the volume, the greater the heat loss. For example, several freestanding greenhouses will have more surface area (more sidewalls through which to lose heat) than the same number and size of gutter-connected

FIGURE 6-20. Application of a single layer of polyethylene on the inside of a glass-covered greenhouse (*top*) or two air-separated layers of polyethylene over existing glass glazing (*bottom*), can substantially improve heating efficiency. However, this will reduce the amount of light reaching the plants.

FIGURE 6-19. The caulking of a greenhouse can result in energy cost savings.

greenhouses (Figure 6-21). Although free-standing houses allow more house-to-house temperature variability for different crops, their heating costs are greater than gutter-connected greenhouses. Growers report that by hanging vertical sheets of polyethylene between different areas of gutter-connected houses, they are able to grow quality crops that have different temperature requirements. They are also able to reduce the heat of unused areas by hanging the vertical sheets.

In the northern hemisphere, because the sun is southerly, some greenhouses are constructed with a solid north wall (Figure 6-22). This wall usually is insulated and painted white on the inside to reflect light. It is constructed at an angle so that the light will be reflected down onto the crop. A greenhouse with a solid north wall is better insulated and thus more energy efficient than a greenhouse with conventional coverings on this wall.

There are many different choices of fuels and sources of energy available for heating a controlled-environment growing facility. The price and availability must be considered when making decisions about the

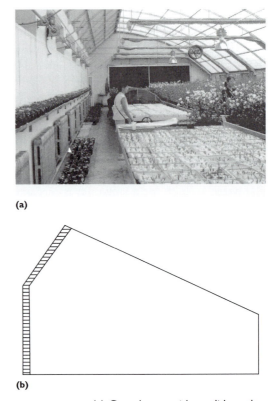

(a)

(b)

FIGURE 6-22. (*a*) Greenhouse with a solid north wall to the left side. (*b*) This wall is typically designed so that the top of the wall is angled to reflect light down onto the crop.

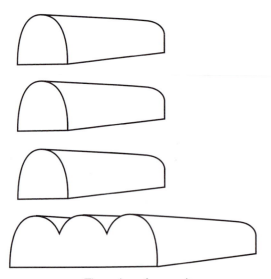

FIGURE 6-21. These three freestanding greenhouses have the same growing area as the gutter-connected house, but a much greater exposed surface area.

source of fuel. There is no universally correct statement as to which fuel is cheapest and which is most expensive. For example, natural gas is often cheaper when there is a locally available pipeline, but not all areas have this supply. Some commonly used and more conventional fuel sources are fuel oil, propane, coal, and electricity. Other energy sources—wood and wood residues, waste heat from power plants and industry, geothermal energy, heat from mine shafts, and solar energy—may offer substantial savings over such conventional fuels. These energy sources also must be considered for meeting modern greenhouse heating requirements if they are readily available and may be utilized efficiently.

If fuels are not used efficiently, their cost can be prohibitively expensive. Clean, well maintained heaters such as boilers will use fuel much more efficiently than if they are dirty or in disrepair.

Energy also will be used more efficiently in the heating of smaller areas. It is necessary to heat the area where the plants are growing, or only the plants themselves, and not necessarily the air. This limited area heating may be accomplished in a number of ways. Growers may move convection tubes used for heating down to plant level and pull **thermal blankets** (Figure 6-23) over their crops at night. This practice allows them to heat only the area where the plants are located and not the air near the top of the greenhouse. On snowy nights, however, the thermal blankets should not be drawn, because it is necessary for the roof to stay warm and melt off the snow. Properly installed thermal blankets can be closed manually or automatically, and result in considerable cost savings in heating costs at night. Additionally, thermal blankets have made achieving negative DIF affordable.

Electrically powered infrared-radiant heating also is reported to be economical because the radiation is absorbed by the crop, pots, benches, walks, and other objects, which in turn is re-radiated and warms the surrounding air. Therefore, only the crop and area immediately around the crop are kept warm; the remainder of the greenhouse can be quite cool, and good growth of uniform crops can result. Some growers have not had good uniformity among their plants using infrared radiation, possibly because they were unaccustomed to the mechanics of implementing this radiant heating system.

Some growers close their greenhouses during the coldest months of the year because of high energy costs. Other growers have switched to crops that require lower temperatures (e.g., substituting lettuce for tomatoes or freesias and alstroemerias for chrysanthemums).

Although the cost of heating is a major consideration and conservation measures are extremely important when producing plants in controlled environments, one must also plan and schedule crops carefully to use the available space efficiently. Utilizing all heights in a greenhouse by creating different levels on benches for various crops, using the area under benches for shade-tolerant crops, and growing in hanging baskets are all methods of maximizing available growing space (Figure 6-24). Use of movable benches can minimize wasted aisle space and provide a higher return for every square (or cubic) meter of growing space that is being heated or cooled. Speeding up crops and maximizing production (e.g., producing twice as many rose stems per square meter of greenhouse space) are often more important in the economics of production than reducing energy input and usage. Therefore, although energy conservation is important, there is no substitute for wise and efficient management. Because horticultural crops require intensive production to gain the highest returns on an investment, the successful grower must carefully plan production schedules and operations and leave nothing to chance.

Other Protected Cultivation

Growers can save heating and cooling costs, extend the growing season, and overwinter plants by growing in unheated coverings

FIGURE 6-23. Thermal blanket that also serves for photoperiod control. The silvery reflective surface is located on the underside of the blanket.

FIGURE 6-24. The owner of the greenhouse at left has made efficient use of space by placing plants at several heights. Use of hanging baskets also can increase greenhouse production (*right*).

and structures. Vegetable and small fruit growers are able to obtain earlier harvests and extend the growing season by several methods that modify the microclimate in which they are growing their crops. This can be as simple as using a milk jug with the bottom removed or paper **hot caps** to cover plants (Figure 6-25). Commonly, **floating row covers** are placed over the tops of plants like a bed sheet (Figure 6-26 a–d). These are made of lightweight spunbonded polyester or polypropylene, similar to the material used to cover the underside of a box spring. These provide a few degrees of protection and will also exclude insect and animal pests. Wind can cause rubbing and plant damage so some growers place wire hoops under the row cover to keep it off the plants. **Low tunnels and slitted row covers** are clear or white polyethylene materials supported by wire hoops that provide similar protection to floating row covers. Such tunnels are modeled after the older glass **cloches** (Figures 6-27) that were commonly used in European gardens as early as the

Victorian era. Low tunnels can be difficult to ventilate, therefore polyethylene row covers that are slitted at the top are used to provide ventilation.

FIGURE 6-25. A milk jug (bottom removed) helps protect this young tomato plant from a late spring frost.

(a)

(c)

(b)

FIGURE 6-26. (*a*) Polyethylene row covers (low tunnels) in an experimental field act as miniature greenhouses and allow for earlier planting dates and yield and can protect plants into or through the autumn months. (*b*) Slitted row covers provide less insulation at night but offer some ventilation during the day and allow plants such as these tomatoes to grow through. (*c*) These lettuce plants were given an initial boost because of the spun-bonded polyester row covers that are now pulled off and on the ground. (*d*) Saran row covers over pine seedlings provide shade, and reduce transpiration.

(d)

Unheated polyethylene-covered, quonset-shaped hoophouses (greenhouses) have been used in nurseries for years to over-winter herbaceous perennials and woody plants. Similar structures, known as **high tunnels** (Figure 6-28) are used for both early production and extension of the growing season into the fall and winter months of small fruits, vegetables, and flowers. These have no permanent heating system or electrically powered ventilation and their size is typically 4.3 × 29.3 meters (14 × 96 feet). Sides are rolled up and ends can be removed to provide ventilation. High tunnels allow for early and late season production when prices are high. For example, tomatoes are about one month earlier when produced in high tunnels than

FIGURE 6-27. Cloches function like miniature greenhouses, trapping heat in the vicinity of the young plants and warming the soil. Cloches have been popular in European gardens.

FIGURE 6-28. The growing season is being extended for these tomato plants by use of a polyethylene-covered high tunnel.
Photograph courtesy of S. Alan Walters, Southern Illinois University Carbondale.

directly in fields. The higher price commanded for early tomatoes can pay for the high tunnel in the first year.

Bedding plants and other species are frequently grown, held, or hardened-off in small structures, which either rely solely on solar energy (**cold frames**), or have a source of supplemental heat (**hotbeds**) (Figures 6-29 and 6-30). Their use is also discussed in Chapter 14. Cold frames, hot beds, and low tunnels are economical because they maintain or provide heat only in the area where the plants are grown.

TEMPERATURE AND SITE SELECTION

The geographical areas in which specific horticultural crops are grown are governed to a large extent by both the average temperature and by extremes of temperature. Factors such as air drainage, orientation of a slope, altitude, and location near large bodies of water are all important when selecting a site. Latitude is also a major factor in site selection, and this is discussed later in this chapter.

Cool air is more dense than warm air and is therefore heavier. Because it is heavier, cool air flows downhill and warm air rises

uphill by free convection (Figures 6-31 and 6-32). Horticultural crops such as fruit orchards are generally located on slopes with good air drainage. Similarly, fruit trees should be located on slopes in the home landscape, avoiding low spots. Because cool air will settle into low areas or pockets, it moves away from the fruit plants, reducing or eliminating injury caused by low temperatures. Low-lying areas or "frost pockets" are therefore more subject to frosts and hard winter freezes than the hillsides. Conversely, hilltops and ridgetops frequently are avoided as planting sites because of high winds and eroded soil.

FIGURE 6-29. A cold frame constructed out of wood. Clear polyethylene serves as a covering material.

FIGURE 6-31. The vineyards pictured here are located in a valley (Napa Valley, California). The cool night air drains and settles into the valley, increasing the sugar content and enhancing the organic acid components of the grapes; the result is the production of high-quality wines.

Obstructions, such as fences or hedgerows that are perpendicular to the slope of a hill can create artificial frost pockets (Figure 6-32) and are more common in home landscapes than in commercial orchards. The cool air that drains down a slope can be trapped by an obstruction and cause a localized frost or freeze on the uphill side. Such factors must be considered when selecting a site or diagnosing plant injury that appears in localized areas.

When possible, one should both avoid frost pockets and consider the direction that the slope faces. North-facing slopes will remain cooler in the spring than south-facing slopes, and buds generally open later on fruit trees and grape vines located on north-facing slopes. Because frosts are often still likely in the early spring, the delay of bloom on trees located on north-facing slopes helps those plants avoid injury. On the other hand, planting on south-facing slopes can result in earlier maturing fruit than on north-facing slopes; therefore if no spring frosts occur, this earliness could be an important factor in obtaining a higher price for the crop.

Altitude also affects the location of horticultural crops. For every 300 meter

FIGURE 6-30. A window sash is frequently used to cover cold frames and hotbeds (*top*). Hot water is circulated through this pipe to provide supplemental heat in the hotbed (*center*). A thermometer should be placed inside of a hotbed so temperatures can be monitored (*bottom*).

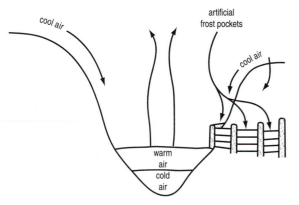

FIGURE 6-32. Diagram of air drainage and frost pockets.

(1000 foot) rise in elevation, the temperature drops 5°C (9°F). Higher elevations have more of a rhythm of cold nights and warm days (thermoperiodicity) than do lower elevations. This characteristic causes problems in growing staple crops in some areas of the world, even in locations near the equator, such as in the Andes region of South America. There it is not uncommon for plants to receive a frost at any time throughout the growing season because of the high altitude. A strong research effort has been initiated on the cold hardiness of crops, including some of the tender potato species, in order for them also to be grown at these higher elevations. Many cold hardy potato species are native to this area.

Many of the green foliage plant species that are used for interiorscaping have been selected from the general location of the equator and are thus tropical. Near the equator the temperature close to sea level is very uniform year-round and humidity is high; whereas the temperature within homes and commercial buildings may fluctuate and humidity is low. Consequently, tropical plants selected near sea level often do not survive well indoors because of temperature fluctuations and humidity changes. For this reason, many of the green plants selected for indoor use are chosen from the higher altitude regions of the tropics where plants have evolved to be more hardy.

Water has a high heat capacity (see Chapter 7), and land located near large bodies of water usually has fewer temperature fluctuations and a more moderate climate than inland regions. Many horticultural commodities, particularly fruit crops, are produced near major bodies of water, such as the Great Lakes of North America. The primary fruit producing region of Michigan is located along the eastern shore of Lake Michigan, and in New York significant amounts of land are devoted to grapes and apples along Lake Ontario and Lake Erie. Many nurseries are located near Lake Erie in Ohio. Therefore, one should consider large bodies of water as temperature moderators and also as sources for irrigation.

VERNALIZATION

Vernalization is the effect of low temperature on flower induction. The phenomenon of flowering usually is separated into three stages or events: **flower induction**, when a vegetative apical meristem is stimulated to begin to change preparatory to forming a flower; **flower initiation**, wherein floral organs begin to differentiate; and **flower development**, the phase when such organs undergo necessary physiological and anatomical changes to become a mature flower.

Many biennials require vernalization before they will flower: Examples include beet, cabbage, celery, onion, foxglove (*Digitalis purpurea*), honesty or moneywort (*Lunaria annua*), and the old garden favorite, hollyhock (*Althea rosea*).

The vernalization stimulus is perceived in regions of active cell division in the plant, especially the apical meristem. Experiments have been conducted in which a vernalized apical meristem was grafted onto a nonvernalized plant, and the portion that arose from this apical meristem flowered. Therefore, not only does the apical meristem perceive the stimulus, but it is altered to the point where it continues to have the capability of responding after removal of the stimulus.

Plant hormones, especially gibberellins (see Chapter 12) have been implicated in the vernalization phenomenon, but their roles are not yet fully understood. It has been shown that the application of auxins, and more commonly gibberellins, will substitute for the low-temperature requirement of some plants. Curiously, as shown by grafting and decapitation experiments, the stimulus to flower is not translocated out of the meristem.

Vernalization is utilized commercially to stimulate the flowering of the florist's stock (*Mathiola incana*), which is grown as a cut flower. This plant normally requires at least 3 weeks at 10°C (50°F) to initiate flower buds. The plants should have 10 or more fully developed leaves prior to vernalization for effective flower induction. As few as six hours per day of temperatures greater than 18°C (65°F) during this low-temperature period will prevent flowering.

High temperatures can reverse low-temperature vernalization; this is called **devernalization**. Onion bulbs that are to be used for sets are often stored at temperatures near freezing to retard spoilage. Sometimes in grocery stores, onion sets will be stored in the cooler. Such sets, when planted in the field, will flower because they have been vernalized. Flowering interferes with the bulbing process, however, and prevents further leaf production, thus reducing photosynthetic potential and causing small bulb size. It is recommended that the bulbs be exposed to temperatures greater than 27°C (80°F) for two to three weeks following such cold storage. This exposure devernalizes the onion sets and causes them to remain vegetative. However, if onion sets (or even some transplants, if past the juvenile stage) are planted too early in the spring, they can also be vernalized in the field by low ambient temperatures and will consequently flower. The bulb from an onion plant that has flowered will not store well and breaks down rapidly because it is a true biennial that has completed its life cycle.

Most spring-flowering bulbs, such as hyacinths, narcissus, and tulips, require a low temperature treatment before they will flower. This is *not* a vernalization phenomenon, however, because these species already possess flower buds that formed during the previous growing season. Therefore, the low temperature provided by winter when such bulbs are planted outdoors, or by floriculturists when forcing them for winter pot crop production, simply overcomes dormancy and is necessary for flower stem elongation.

DORMANCY

Dormancy applies to seeds and buds and is the condition where their germination or growth is inhibited by their own physiology (some textbooks call this **rest**). A seed or bud is considered to be dormant when, although quite alive, it will not germinate or grow, even though the environmental conditions of temperature, water, oxygen, and pH are all favorable. However, when these processes are inhibited solely by unfavorable environmental conditions, the buds or seeds are considered to be **quiescent**. Examples of such conditions that maintain buds or seeds in a quiescent state are the lack of water or extremes of temperature. In this book, the terms dormancy and quiescence are used. However, many scientists studying dormancy use the terms: **ecodormancy**, **endodormancy**, and **paradormancy**. Ecodormancy is synonymous with quiescence, endodormancy is synonymous with dormancy as defined above, and paradormancy is also known as apical dominance where the apical bud inhibits growth of lateral buds.

Dormancy is an adaptive mechanism that dictates that bud break or seed germination will occur when conditions generally will remain favorable for further growth and development (Figure 6-33). Plants would not survive if, for example, the buds on temperate trees began active growth during a January thaw, or if the seeds of desert plants germinated during a small rainstorm during the dry season. In both cases, the plants would be subject to injury or death if dormancy did not prevent growth or germination. Dormancy, therefore, ensures that species survive in many ecological niches. Knowledge of a plant's native ecology offers direction to the horticulturist in choosing treatments to induce or overcome dormancy.

The induction of bud dormancy is probably of more concern to horticulturists than induction of seed dormancy. This is because seeds generally go dormant as a part of their normal ontogeny, whereas buds, or the plants containing the buds, generally must receive the proper environmental stimuli to induce dormancy. Bud dormancy on many temperate woody plants and herbaceous perennials generally occurs as a response to changing environmental conditions such as photoperiod and low temperatures. Photoperiod plays a major role, especially if temperatures remain high as autumn approaches. It is interesting to note that an increase in the cold hardiness of buds on temperate woody plants generally accompanies the beginning of dormancy. The onset of dormancy is usually accelerated during dry conditions and when mineral nutrients (especially nitrogen) are withheld. Late summer and early fall fertilizations with materials high in nitrogen will delay dormancy and stimulate soft and succulent new growth, which may then be subject to freezing injury. For this reason, it is recommended that one should *not* apply fertilizer to temperate woody plants late in the growing season (see also Chapter 9).

When terminal buds first set in the summer on many temperate woody plants, they normally will not expand. However, these buds are considered to be predormant rather than truly dormant, because defoliation (removal of mature leaves) by people or pests often will result in terminal bud outgrowth. This mechanism allows many temperate woody perennials to survive severe foliar infestations by insects; for example, the gypsy moth defoliated many trees in the northeastern areas of the United States but because the buds were not truly dormant, the trees put out additional flushes of growth. This later growth resulted in adequate leaf area for photosynthesis to provide food for overwintering and growth the following spring. Lateral buds, likewise, exhibit predormancy and their outgrowth is inhibited by mature leaves and the terminal bud. Pruning will therefore force the outgrowth of those buds; if pruning is done in late summer, this new growth may not properly harden and may be damaged by low winter temperatures.

FIGURE 6-33. Dormancy in temperate species allows them to withstand harsh winter conditions.

When buds no longer respond to pruning or defoliation, they are said to be truly dormant. At this time, even if a plant is brought into a greenhouse, the buds will not begin growth.

Bud dormancy of many woody and herbaceous plants is overcome by an exposure to the chilling temperatures (0°–7°C or 32°–45°F) of winter (photoperiod may also be used to overcome bud dormancy in some species). Temperatures below freezing, however, are generally far less effective than those just above freezing. Dormancy gradually decreases throughout the winter. In the northern hemisphere, some southern regions do not consistently receive chilling winter temperatures sufficient to meet the dormancy requirements of many plants. This is the reason for the southern limit of some fruit crops such as apples and peaches. Such species require a minimum number of hours of chilling temperatures (Table 6-3) that vary not only among the species, but also among cultivars within a species. In fact, apples and peaches are bred so that the number of hours of cold that are required to break bud dormancy are reduced in order to extend these crops into warm locations (e.g., peaches grown in Florida and Texas). These southern states have been mapped for the number of hours of chilling that they receive during the winter months and divided into **chilling zones** (Figure 6-34). If

a cultivar is planted in a chilling zone that receives an inadequate number of chilling hours, there can be poor or slow leaf development, flower buds may abscise resulting in poor or no cropping, and fruits can have irregular development. The poor leafing-out results in inadequate shading of the main framework of the tree and therefore sunscald damage. Likewise, herbaceous perennials such as members of the genus *Hosta* that are planted out of their chilling zone will have delayed emergence and weak growth.

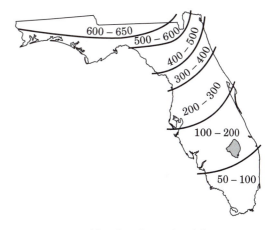

FIGURE 6-34. Map that shows the chilling zones within Florida. The numbers represent the number of hours below 7° C (45° F) that are received to February 10 during 75% of the winters.

Figure courtesy of University of Florida, Institute of Food and Agricultural Sciences and by J.G. Williamson and T.E. Crocker.

TABLE 6-3. Winter chilling temperature range required by some fruit crops to break bud dormancy.

Fruit species	Hours below 45°F (7°C) needed	Fruit species	Hours below 45°F (7°C) needed
Apple	250–1700	Peach (Texas)	350–950
Walnut	400–1500	Peach (general)	800–1200
Pear	200–1500	Domestic plum	900–1700
Cherry, sour	600–1400	Japanese plum	300–1200
Cherry, sweet	500–1300	Apricot	300–900
Peach (FL)	50–400	Almond	100–400
		Fig	0–300

Source: After N.F. Childers, *Modern Fruit Science* (9ᵗʰ ed.). Horticultural Publications, Gainesville, FL.

A few genera of nursery stock that are overwintered bareroot under refrigerated conditions remain dormant even after being planted in the landscape or field; these eventually die. Examples are black gum (*Nyssa sylvatica*), hackberry (*Celtis occidentalis*), and some oaks. These can be forced into growth by **sweating** them before transplanting. This is accomplished when it is sufficiently warm for planting outdoors in the spring by first soaking the roots for several hours, covering the floor and plants with moist burlap or straw for three to four days. Buds will then swell and the plants must be planted immediately because growth has started. Many large nurseries recommend sweating their bareroot stock.

The seeds of many species exhibit dormancy that is overcome by **stratification**, a moist-chilling treatment. Additionally, dormancy of some seeds, such as lettuce are brought about by exposure to high temperatures. This is called **thermodormancy**. If seeds of lettuce are planted during the hot summer months for a fall harvest, the high temperatures may induce thermodormancy and poor stands result. Hormonal relationships in seed dormancy are discussed in detail in Chapter 14.

Other perennials go dormant during the summer months and resume growth during the cooler times of the year. An example is the perennial *Arum italicum* (Italian arum). This plant does well in the shade and sprouts in the fall and remains in leaf during the winter and spring. Its leaves die down in June as it enters dormancy for the summer months. Likewise, during hot, dry months of the summer, Kentucky bluegrass will turn brown and become dormant if it is not watered. It can remain dormant for months until it is watered. It will then green up and resume growth.

After the dormancy requirements are met, buds and seeds will grow or germinate normally when conditions become favorable (Figure 6-35). During periods in which the growing conditions remain unfavorable, however, such as cold or dry spells, the buds or seeds will not develop or germinate and are therefore termed *quiescent*.

FIGURE 6-35. Following completion of its dormancy requirements, this hardy azalea greets spring with gorgeous flowers.

TEMPERATURE STRESS

A **stress** is a condition that can potentially cause either reversible or irreversible injury. This section is limited to stresses caused by either low or high temperatures or by temperature fluctuation.

LOW-TEMPERATURE EFFECTS

Chilling Injury

Chilling injury occurs when plants are damaged by low temperatures and ice crystallization has not occurred. Generally, the lower the temperature, the greater the injury; however, chilling injury has been observed on some plants exposed to temperatures as high as 10° to 15°C (50°–59°F). Although temperate zone plants can be injured by chilling temperatures, plants native to the tropics are often much more chilling sensitive. In horticulture, chilling injury can occur in the field or during storage (e.g., produce in a grocery store is vulnerable). Banana fruits, for example, turn black after being stored in a refrigerator (Figure 6-36) or when otherwise exposed to low temperatures (e.g., during transport to or from market in the fall, winter, or spring). Plants may exhibit differences in tolerance to chilling temperatures at different phases of their life cycles depending if they were cold acclimated (Figure 6-37).

FIGURE 6-36. Chilling injury in banana. The black fruit was stored in a refrigerator, whereas the yellow fruit was stored at room temperature.

The plasma membranes of the cells are commonly implicated as the site of chilling injury. When the plasma membrane of the cell is injured, it becomes leaky and injured cells will first look "water soaked." It is when the plant is rewarmed that most symptoms are manifested. Typically symptoms of chilling injury include lesions on organs, discoloration, death of older leaves, partial or complete defoliation, wilting, poor growth, poor keeping quality (produce), increased susceptibility to organisms that cause rot, and plant death.

To avoid chilling injury, it is best to store some warm season fruits and plants at room temperature and not under refrigeration. Examples include avocados, bananas, potted tropical plants, tropical flowers, and tomatoes. Chilling injury has a detrimental effect on tomato flavor and can be part of the reason for poor tasting fruit during the winter. Purchasing from a store that does not refrigerate its tomatoes and not storing them in the refrigerator at home results in more flavorful tomatoes year-round.

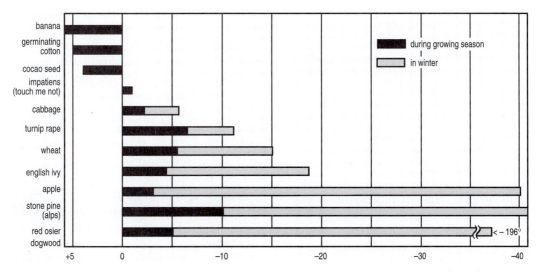

FIGURE 6-37. Hardy plants are killed at a higher temperature during the growing season if they have not cold acclimated. Thus, if a frost occurs during the period of active growth it can cause considerable damage of many species.

Figure 19-2 from Botany by P. M. Ray, T. A. Steeves, and S. A. Fultz. Copyright 1983 by Saunders College Publishing, a division of Holt, Rinehart and Winston, Inc., reprinted by permission of the publisher.

Freezing Injury

Low-temperature stress that is a direct result of the freezing of water in plant tissues is called **freezing stress**. There are differences among plant genotypes, within a plant at different times of the year, and between different cells and tissues within the same plant for their susceptibility to freezing stress. For example a normally hardy temperate plant, such as an apple tree, will not tolerate the same low temperatures during the growing season that it will during the winter after it has acclimated to the cold (Figure 6-37). Rhododendrons, may have their flower primordia within the buds killed by freezing and the leaves will survive without injury indicating differences in hardiness between the two organs.

In relation to the cellular makeup of the plant, water can either freeze within the cells (**intracellular freezing**) or outside of the cell protoplasts in cell walls, the intercellular spaces, within the xylem vessels, or on the surface of plant tissues (**extracellular freezing**). Intracellular freezing causes much more plant injury and death than extracellular freezing. In laboratory situations, intracellular freezing generally accompanies rapid drops in temperature. Although intracellular freezing has not been conclusively documented in nature, it is thought to result in rupture of the plasmalemma, thus causing cell death. Conversely, extracellular freezing can be tolerated to a great extent by hardy, acclimated plants. In fact, because of the heat of crystallization of water, when water freezes it warms the cells.

Extremely rapid rates of freezing in laboratory conditions leads to the formation of ice nuclei within cells as the water becomes glasslike or vitrified. True vitrification does not result in ice crystal formation, just ice nuclei, which can be tolerated. This system is used to preserve plant germplasm and is known as **cryopreservation**. Even tender tropical plants that will not cold harden can survive this condition. Therefore, it is the formation of ice crystals (freezing) not the low temperatures that damages or kills plants.

Cold Acclimation

Plants in an active growth stage are generally more susceptible to environmental stresses, such as low or high temperature stresses than when they are dormant. For example, during the winter many evergreens can endure extremely low temperatures but in the summer a light frost can cause injury on these same plants. Many plants therefore can harden or acclimate to withstand temperature extremes. The degree to which plants acclimate varies with the species or cultivar and the season of the year. For example, more hardy apple cultivars (e.g., 'Jonathan') show greater cold hardiness in the fall than less hardy cultivars (e.g., 'Duchess'). Often a freeze in the late fall or early winter will be more damaging than one in midwinter because plants have not fully hardened when they are exposed to the stress. Such knowledge is useful to horticulturists who attempt to understand and take steps to try to reduce temperature-related plant damage.

Plants cold acclimate in response to environmental stimuli including photoperiod, temperature, and water. Light and photosynthesis are necessary for cold acclimation and during shorter photoperiods plants acclimate to survive the cold winter conditions. The freeze hardiness of plants increases with exposure to low temperatures, especially the first frosts, and decreases with exposure to high temperatures. Such hardening generally does not occur above 5° to 10°C (41°–50°F). Cold acclimation accompanies the onset of dormancy, but involves more than growth ceasing. It also is known that some degree of water stress also plays a role in cold acclimation. A water stress may increase cold hardiness, a phenomenon observed when water is temporarily withheld from vegetable transplants. Hardening involves many chemical changes and genes being activated within the plant that are necessary for winter survival.

Winter Hardiness

To survive winter's freezing temperatures, plants have adapted different mechanisms. These differences account, to some extent, for the latitudinal limits of the ecological distributions of species. Some plants have adapted **freeze tolerance** mechanisms, whereas some species utilize **freeze avoidance** mechanisms.

Freeze Tolerance

When temperatures slowly drop below freezing, ice crystals will grow in the extracellular areas, beginning in the cell walls, because water is purer there and the cell wall microfibrils serve to nucleate the water to allow for ice crystal growth. However, within plant cells, there are solutes that depress the freezing point a few degrees, allowing the water to remain liquid. Many plants can tolerate extracellular freezing.

Although ice is frozen water, as a solid it is dry (no longer liquid). This creates a water potential gradient within the plant (see Chapter 7) between the moist conditions inside the cells and the dry conditions (ice) outside of the cells. Water moves with a gradient across membranes by osmosis. The water that leaves the cells freezes in the extracellular areas causing freezing-induced cell dehydration. Many hardy woody plants can tolerate extreme dehydration of their cells, even to the point where no freezable water remains in the cells. Acclimated red-osier dogwood (*Cornus sericea*), paper birch (*Betula papyrifera*), and trembling aspen (*Populus tremuloides*) utilize this mechanism and can survive at least to the temperature of liquid nitrogen (−196°C or −321°F). This is why these plants can survive in the arctic zones (Figure 6-38). Some plants, however lose the stability of their plasma membranes during freezing-induced cell dehydration, leading to cellular death. In plants that can tolerate such dehydration, various genes are activated, often by cold stress, that contribute to membrane stability by influencing the calcium and other factors in the membranes.

Freeze Avoidance

In some species, freezing-induced cell dehydration is not as complete as with the more hardy species, such as red-osier dogwood. As water is drawn out of the cell, however, the solute concentration becomes higher, and an increase of solute in a solution lowers the freezing point. This is similar to what occurs with salts used for deicing roads in the winter. Freezing point depression offers some protection for plants, but seldom more than −4°C (25°F). Therefore, this phenomenon does not contribute dramatically to plant winter hardiness, but offers some freeze avoidance.

Zero degrees centigrade (32°F) often is called the freezing point of water; it should more correctly be called the melting point of ice because **supercooled** (undercooled) water can remain liquid below 0°C (32°F). There is some supercooling of water prior to extracellular freezing beginning within plants. However, small volumes of water within cells and tissues can remain liquid independent of the ice in adjacent tissues. Such water is not in equilibrium with the ice and this phenomenon is called **deep supercooling** and occurs in overwintering buds and xylem ray parenchyma cells of many woody perennials. Because supercooled water remains liquid plants avoid freezing and its associated injuries.

Water initially freezes when a microscopic ice crystal known as a nucleus forms. Nucleation is followed by rapid crystal growth, which is the actual freezing phenomenon that can cause plant injury or death. Ice can nucleate around dust particles, cell walls, or bacteria (*Pseudomonas syringae*). In fact, there are natural strains of *Pseudomonas fluorescens* and *P. syringae* that do not nucleate ice and may be sprayed on plants early in the growing season to displace, prevent, or reduce the growth of ice-nucleating bacteria. Additionally, pesticides have been used to kill ice-nucleating bacteria, resulting in supercooled water on the plant and better plant survival.

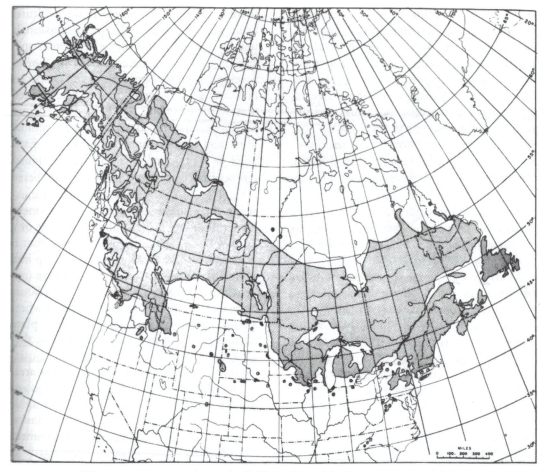

FIGURE 6-38. The geographic range of paper birch. Plants that exhibit freeze-induced cell dehydration must be able to withstand cellular desiccation. The tolerance of the dehydration can be critical to winter survival; therefore, the most hardy plants can best withstand cellular dehydration.
USDA Forest Service Photo.

When supercooled water freezes within plants, the intensity of the freezing process is so great that death of tissues occurs. Pure water can remain liquid (or supercooled) down to a minimum temperature of approximately −38°C (−36°F) when in solution. Although water has been shown to supercool in plants down to −47°C (−53°F), it may only supercool a few degrees before it freezes, depending on the species, organ and environmental conditions. It has been shown that even within an azalea flower bud, the different flower primordia supercool to different temperatures (i.e., they are killed at different temperatures). The killing temperatures are those at which the supercooled water freezes. Because of this phenomenon there may be sparse flowering on azaleas following a cold winter.

Many temperate tree species have a northern limit (in the northern hemisphere) in which they can be grown, and this limit is nearly equivalent to the line where winter temperatures stay above −40° (Figure 6-39). The −40° minimum temperature is roughly equivalent to the U.S.–Canada border. Supercooling is thus a mechanism employed by many winter hardy plants to avoid freezing.

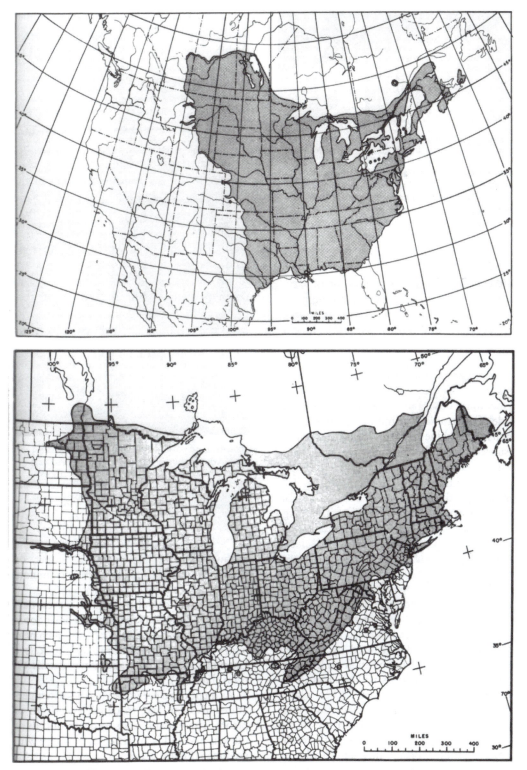

FIGURE 6-39. The range of green ash (*top*), and the range of American basswood (*bottom*).
USDA Forest Service Photo.

Examples of Winter Injury

Freezing injury in plants not only is related to the minimum temperature, but also to how rapidly the temperature drops, the duration of the low temperature levels, and the rate of thawing. Much more plant injury is observed when temperatures drop rapidly than when they drop slowly. Therefore, areas that commonly experience wide and rapid temperature fluctuations (e.g., high altitudes) are unsuitable for many horticultural crops. The influence of the length of time during which the plant tissues are frozen is not clearly related to plant injury. Evidence indicates that there is an interaction between time and temperature for some species (i.e., the lower the temperature, the shorter the length of time that some plants can survive), but with many species, the length of time frozen has little or no effect on survival. Although rapid thawing can increase freezing injury, it is probably not a critical as rapid freezing.

Rapid temperature drops do occur in plants growing outdoors in temperate regions, leading to a common winter injury problem called **southwest injury** (sometimes referred to as **sunscald**). Note that this discussion refers to a different disorder than sunburn (sometimes called sunscald) of fruits during the growing season. Young, thin-barked trees such as maple and apple are particularly sensitive to this problem. During the winter, while the air temperatures are below freezing, the sun's rays can strike young tree trunks and warm them as much as 25° to 30°C (45°–52°F) higher than the shaded side. Extracellular ice crystals may then melt, causing the cells in the thawed tissue to imbibe the water and become active. When the sun is then obscured by a cloud or blocked by a building or other object, or after sunset on a cold day, the temperature of the thawed side of the trunk drops rapidly. This is thought to cause intracellular freezing that ruptures the plasmalemma and kills the cells. As a result, in the following growing season the dead bark dries, splits, and pulls away from the wood (xylem) (Figure 6-40), leading to entry of disease-causing organisms, poor growth, branch die-back, or even the death of young trees. In temperate regions of the northern hemisphere, the sun is in the southern part of the sky during the winter. Therefore, the sun strikes the south

FIGURE 6-40. Southwest side of a young tree with southwest injury (sunscald) (*left*). This injury had occurred one or more growing seasons before the picture was taken because there is callus growth along the margins of the injured area. This tree had its trunk wrapped to avoid southwest injury (*center*). If some types of wraps are not removed, they may damage or even girdle the tree (*right*).

or west side of the trunk, causing sunscald symptoms to be only present on these sides (which explains why this disorder is often called southwest injury). Using protective wraps, shading the trunk, or painting the trunk with white latex paint can prevent southwest injury (Figure 6-40). Some wrapping materials serve the dual purpose of protecting the trunk from the sun's rays and from rodent (vole) or equipment damage. These protective wraps are not without problems and therefore must be checked frequently during the growing season to be adjusted so that they will not girdle the trunk. There tends to be a greater frequency of insect borers in wrapped trunks and the increased moisture under the wraps can favor the development of fungal and bacterial diseases. Additionally, the trunk can overheat under these wraps during the summer leading to cambial damage and splitting or cracking of the bark. Leaving the wraps on for only the winter months and removing them as it warms in the spring will protect from southwest injury and reduce the incidence of these problems.

Another common disorder that affects trees is called **frost cracking**; the symptoms of which are seen when the trunk or large branches of a tree split longitudinally. Frost cracks occur during very cold winter periods. Because plants cells contract when water leaves them and freezes in the extracellular areas, stresses can develop and the trunks can suddenly split. This splitting makes a loud sound similar to the sound of a gunshot. If frost cracking occurs year after year, the crack then reopens annually. With repeated healing, a longitudinal ridge forms, which is called a "frost rib" (Figure 6-41).

Southwest injury and frost cracking are two types of freezing injury that can occur to plants. There are many other disorders that are associated with freezing. Plant death is the most severe freezing injury and is the primary reason that we commonly grow many perennial species as annuals

FIGURE 6-41. A typical frost rib in the trunk of a tree.

(e.g. tomato and geranium). Freezing injury ultimately can lead to death of plant parts, such as growing points, buds, cambium, roots, flowers, leaves, or fruits.

It is well known that the above-ground portions of temperate woody plants can withstand lower temperatures than buried parts, such as roots. It is fortunate that the roots normally are insulated by the soil and often by snow cover. Shallow-rooted plants often are injured during years of no show cover. Root freezing and the resulting plant injury or death also is a major concern when overwintering container-grown plants (Figure 6-42). An additional type of freezing injury is less direct: When water freezes in the soil, it causes frost heaving, which can damage roots and crowns of plants.

FIGURE 6-42. This normally hardy shrub was killed by winter temperatures because the cold-sensitive roots were above-ground in a raised planter, thus lacking the insulating properties of the soil.

Cold Hardiness Zones

Knowledge of cold hardiness is critical to success in many branches of horticulture. Fruit growers and landscape professionals will be unsuccessful if they plant tender species or cultivars of perennial plants. Fortunately, the winter hardiness of many species and cultivars is well known. The United States, for example, has been divided into hardiness zones based on the average annual minimum temperatures (Figure 6-43). Each zone is a range of 10°F (5.6°C) minimum temperatures. These zones are numbered from 1 to 11 (north to south) with Zone 1 being the coldest and Zone 11 the warmest. Generally, the northern hardiness limit of a species or cultivar is the information that is published (e.g., honeylocust is hardy up into Zone 4a). Because there are relatively small areas (microclimates) with different temperature zones, there are always exceptions to where plants can be grown. Landscapers may place certain plants in protected locations so that they might enjoy species that are not considered to be cold hardy in their particular area.

Another factor to consider when choosing plants is their seed source. The geographic origin of seeds is known as the **provenance**. Plants from seeds collected from southern provenances often are not as cold hardy as plants of the same species grown from seeds collected from northern locations (northern hemisphere). Therefore, when ordering

woody plant seedlings, the wise horticulturist who plans to plant in the north should be sure that the seed source is from a location with sufficiently long and harsh winters.

Protecting Plants to Avoid Freeze Damage

An **advective freeze** is when a mass of cold air moves into an area blown by moderate to strong winds that will drop below 0°C for a day or more. These are difficult to protect plants against, but some natural and human means have been successful. A **radiational frost** occurs on clear nights with calm winds as the heat radiates from the surface into the atmosphere. These are easier to protect plants than with an advective freeze. Covering plants can help with both types of conditions.

Some plants avoid freeze damage by having a low growth habit. Snow cover provides insulation to low growing plants and thus helps many species avoid both types of freezing (Figure 6-44). The snow cover offers protection because it traps air and creates dead air spaces, which minimizes convective heat loss and insulates the ground, low growing plants, and lower portions of larger plants (Figure 6-45). Some species can therefore be grown successfully in temperate climates if the snow cover is adequate. In some areas, plants such as forsythia may flower only on branches that were located below the snowline the previous winter because the flower buds possess limited cold hardiness. A practice in some northern areas (northern hemisphere) is to dig shallow trenches or holes and actually cover or bury the stems of plants such as hybrid tea roses and tender European grapes for the winter to avoid freezing injury. In sufficiently cold climates during autumn, plants such as butterfly bush (*Buddleia davidii*) are cut back to near the ground line and mulched deeply to survive the winter. Mounding with straw or leaves can allow some marginally hardy plants to survive harsh winters. Simply locating marginally hardy plants in protected areas of the landscape can increase their survival.

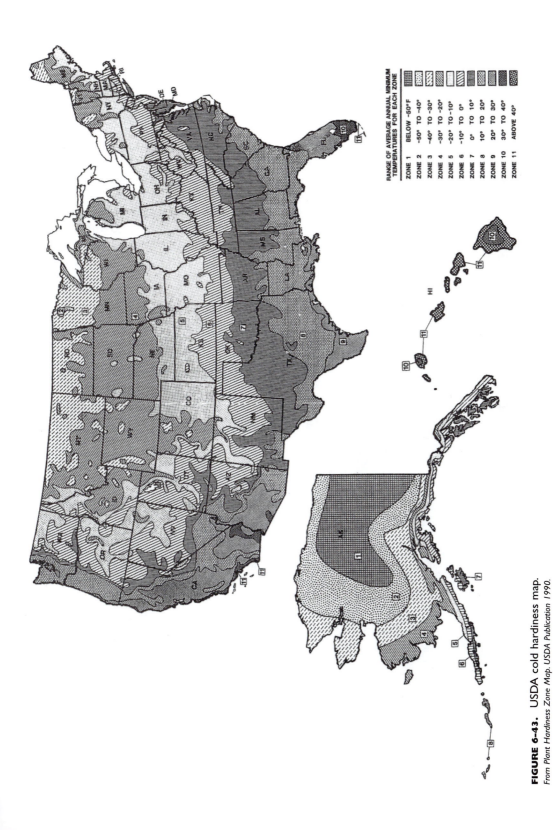

FIGURE 6-43. USDA cold hardiness map.
From Plant Hardiness Zone Map, USDA Publication 1990.

FIGURE 6-44. Some plants exhibit freeze avoidance as a result of their low spreading habit, which allows the snow cover to insulate their branches.

There are other means by which people protect more tender plants in the landscape for the winter. Hybrid tea roses are often covered with Styrofoam "rose cones" for

FIGURE 6-45. The top of this plant was killed when winter temperatures were too low. The lower branches avoided the low temperatures because they were covered by snow. The snow line at the time of the critical winter temperature is readily apparent.

protection. There are series of tubes filled with water that are used around plants for protection, as well as various other protective sleeves and covers (Figure 6-46).

Protection From Radiational Frosts

Growers know that clouds in the sky will serve as a blanket and hold in heat, therefore lower temperatures can be expected on clear nights, when **radiational cooling** will occur. Radiational cooling takes place when heat is emitted (or radiated) from the earth upward into the atmosphere, with few or no clouds to hold it in. It results in a cooling of plant parts and the air around them. On still, dry, long nights with clear skies, the air near the ground becomes cooler than the air above it. The condition of dense, cool air near the ground with warm air above it is called a **temperature inversion**. Radiational cooling has caused chilling, frost, or freezing injury in many crops and usually results in earlier frosts than when an advective freeze (cold front) moves into an area.

To minimize crop damage caused by radiational cooling, growers may utilize overhead irrigation to cause ice to form over the plants, thereby taking advantage of the heat of crystallization of water (see Chapter 7). Some growers burn fuels such as kerosene to provide local heating of the air around the crop when frosts or freezes are likely (Figure 6-47). Older practices such as burning tires or smudge pots are banned in many areas because of the smoke pollution. Wind machines (Figure 6-48) and even helicopters are used to mix the warm air from above with cold air below to avoid frost or freeze injury during temperature inversions.

Although row covers accomplish this in commercial fields, in a home landscape or gardening situation, about 1° to 3°C protection is offered by covering plants with woven cloth, such as a bed sheet before a clear, cold night. Paper or plastic sheeting can also be used, but they do not have

FIGURE 6-46. The young trunks of citrus are sensitive to the cold. Mounding sand around the trunk, or similarly using black plastic sleeves will offer some protection (*left* and *center*). The clear corrugated plastic on the right (*top* and *bottom*) offers protection to the whole young orange tree.

nearly the insulating properties of woven materials. Paper hot caps or milk jugs with the bottoms removed are also used to cover tender plant and protect them from frosts (Figure 6-25).

FIGURE 6-47. Return-stack burners are often used to prevent damage to tender grapes from radiational frost.

....

HIGH-TEMPERATURE EFFECTS

Heat Stress

As in low-temperature stress, there are differences in the ways plant species respond to high temperatures. Additionally, plants or plant parts in a state of rest (quiescence or dormancy) can generally tolerate temperature extremes better than actively growing plants. The high temperature to which plants are exposed and the length of time of exposure are important in causing plant injury or death. A relationship exists between high temperatures and time of exposure: In general, higher temperatures require shorter exposure times before plant death.

Some commonly occurring disorders of plants are associated with high temperatures. One injury occurs to the south or southwest side of the trunk of thin-barked

FIGURE 6-48. Wind machines mix the warm air from above with the cooler air down at plant level to prevent low-temperature injury in the case of a temperature inversion.

FIGURE 6-49. Sunburn on a bell pepper fruit caused by poor foliage cover that allows exposed cells to reach lethal temperatures.

trees and is indistinguishable from the southwest injury (sunscald) described previously (Figure 6-40). The sun's rays are thought to overheat the trunk, leading to death of the cambium. The bark then dries and peels away from the trunk leaving a dead cambium. Shading, wrapping, or painting will help reduce this injury, as it does with southwest injury.

The fruits of crops such as apple, pear, tomato, pepper, and eggplant often exhibit solar injury on the sides exposed to the sunlight, especially on the south or west sides of the plant (Figure 6-49). This disorder is known as sunburn (sometimes also called sunscald) and is primarily caused by localized overheating in the fruit, although ultraviolet radiation may also be involved. With good leaf canopy coverage most fruits will be shaded, resulting in minimal sunburn. When plants have few leaves or sparse foliage, or if branches bend over with heavy fruit loads, however, many fruits will be exposed directly to the sun's rays and serious damage will result. The symptoms typically range from off-color or yellowish, white, or tan areas, to water-soaked regions, to dead areas of fruit tissue, to total death of individual fruits. Overhead sprinkler irrigation cools the fruit through evaporative cooling and has been shown to reduce sunburn. A less expensive, newer method is to spray light-colored reflective particles on the plants. A reflective, processed kaolin clay particle film material has been used successfully to reduce sunburn of fruit. This material has an added benefit of controlling or suppressing certain insect pests and spider mites. In some cases, yields are improved with proper applications of the kaolin material.

Heat injury can also occur on leaves, particularly when plants are moved from indoors or from a greenhouse to outdoors. Such heat injury can be a cause of transplant shock.

As mentioned previously, high soil temperatures also can lead to damage or death of plants. The hot soil leads to death of cells at the soil line and is a particular problem with seedlings (Figure 6-10).

Heat Tolerance

Some plants can tolerate higher temperatures than others. Higher transpiration rates (see Chapter 7) at higher temperatures may make it difficult to separate heat and drought stress. Plants can gradually harden, however, and thus are better able to withstand drought stress. High temperature hardening is controlled by environmental stimuli such as temperature, water, and light. High temperatures and low moisture increase the heat tolerance of many plants. Light tends to enhance heat hardening, therefore, many species are more heat tolerant when grown in full sun than under electric lamps. When hardened, differences in heat tolerance among plants also have been noted. When properly hardened, the warm season grass, Bermudagrass, is much more heat tolerant than the cool season grass, Kentucky bluegrass.

Heat Avoidance

When trees develop thick barks on their trunks, they generally are less susceptible to southwest injury. This is true regardless of whether injury is caused by temperature fluctuations in the winter or by high summer temperatures. This protection is probably related to the greater insulating value of a thick bark, thus leading to less heating in the cambial region. Some species reflect light very efficiently and thus do not heat up as much as other species. Desert plants reflect a relatively high amount (60%) of visible radiation; therefore, to some extent, they tend to avoid overheating. The spines on the Saguaro cactus (*Carnegiea gigantea*) not only offer protection but also provide shade to the green stem, thus helping to prevent overheating on hot summer days in the desert. Plants also have the ability to cool themselves. This is a function of transpiration (see Chapter 7). It therefore becomes clear that soil–water relations can play an important role in heat avoidance in plants.

TEMPERATURE FLUCTUATIONS

It is well known that rapid temperature fluctuations can cause general or localized plant injury. A classic example is the effect of cold water on African violet (*Saintpaulia ionantha*) leaves. African violets are typically grown in relatively low light conditions and their leaves assume a temperature roughly equivalent to or slightly above that of the air. When cold water lands on African violet leaves, a temperature shock occurs and chloroplasts in the palisade layer break down, resulting in the appearance of yellow rings or spots on the foliage (Figure 6-50). This result is a temperature effect, not a water effect, because warm water on the leaves will not cause this type of damage. In fact, placing a cold metal rod or an ice cube in a plastic cup on African violet leaves will cause identical symptoms.

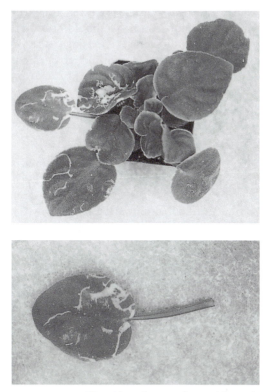

FIGURE 6-50. The chlorotic spots (*top* and *bottom*) are examples of cold water damage on these African violet leaves.

SUMMARY

An understanding of temperature is crucial for the care and management of horticultural plants. Planting the correct species (genotypes) at the proper time with respect to averages and extremes of temperature in each area will play an important role in the production of high-quality plants. The appearance (phenotype) of two genetically identical plants can differ dramatically if they if they are grown at two different temperature regimes. Although temperature extremes or rapid fluctuations can destroy crops, growers do have some means to moderate plant temperatures; alternatively they may select proper planting sites, planting dates, crops, and cultivars. One way to avoid temperature extremes is to use a controlled-environment facility, such as a greenhouse, although to be economical, high value crops that would not perform as well in the field must be selected for such enterprises. Horticulturists should study all options carefully before choosing either a growing system or a method to protect crops from temperature extremes.

REFERENCES

Bigras, F. J. and S. J. Colombo (Eds.). 2001. *Conifer Cold Hardiness*. Kluwer Adacemic. Dordrecht. 596 pp.

Bonan, G. B. 2002. *Ecological Climatology Concepts and Applications*. Cambridge University Press, Cambridge. 678 pp.

Dirr, M. A. 2002. *Dirr's Trees and Shrubs for Warm Climates: An Illustrated Encyclopedia*. Timber Press, Portland, OR. 446 pp.

Dirr, M. 1998. *Manual of Woody Landscape Plants: Their Identification, Ornamental Characteristics, Culture, Propagation and Uses*, 5th ed. Sipes, Champaign. 1187 pp.

Levett, J. 1980. *Responses of Plants to Environmental Stresses*, 2nd ed. Volume I. *Chilling, Freezing, and High Temperature Stresses*. Academic Press, New York. 497 pp.

Li, P. H. and T. H. H. Chen (Eds.). 1997. *Plant Cold Hardiness Molecular Biology, Biochemistry, and Physiology*. Plenum Press, New York. 368 pp.

Li, P. H. and E. T. Palva (Eds.). 2002. *Plant Cold Hardiness Gene Regulation and Genetic Engineering*. Kluwer Academic. New York. 294 pp.

Margesin, R. and F. Schinner (Eds.). 1999. *Cold-Adapted Organisms Ecology, Physiology, Enzymology and Molecular Biology*. Springer, Berlin. 416 pp.

Nelson, P. V. 2003. *Greenhouse Operation and Management*, 6th ed. Prentice Hall, Upper Saddle River, NJ. 692 pp.

Swiader, J. M. and G. W. Ware. 2002. *Producing Vegetable Crops*, 5th ed. Interstate Publishers, Danville, IL. 658 pp.

Viémont, J.-D. and J. Crabbeé. 2000. *Dormancy in Plants From Whole Plant Behaviour to Cellular Control*. CABI, Wallingford. 385 pp.

Westwood, M. N. 1993. *Temperate-Zone Pomology: Physiology and Culture*, 3rd ed. Timber Press, Portland, OR. 523 pp.

PART III

THE RHIZOSPHERE
(ROOT ZONE)

CHAPTER 7

WATER

Water! It covers 70 percent of the earth's surface and yet it is the single most limiting factor for plant growth in vast areas of the world, especially in desert areas and regions of low rainfall, where drought is commonplace (Figures 7-0, 7-1). The location of early civilizations was determined by the availability of an abundant source of fresh water. The so-called Cradles of Civilization—China, Babylon, Egypt, and Peru—are examples where humanity first flourished, and all are located near readily available, high-quality sources of fresh water.

Water is the major constituent of a plant. Herbaceous plants contain 80 to 90 percent water by weight, and even woody plants are made up of more than 50 percent water. In important horticultural commodities, such as lettuce and tomatoes, the edible part is composed of approximately 95 percent water. Water is used in tremendous quantities for crop growth; in fact, just over 200 g (for millet, *Panicum* spp.) to over 1000 g (alfalfa, lucerne; *Medicago sativa*) of water are required to produce 1 g of dry matter.

Abundant water is clearly required for horticultural crop growth and development. This chapter discusses how water becomes available to the plant, how it is involved in growth and development, and how horticulturists can manage or manipulate water and water loss to maximize plant production, even when conditions are dry.

190

Water, so abundant in lushly vegetated sites, is critically limiting to plant growth in much of the world.

PROPERTIES OF WATER

Water has many unique and interesting chemical and physical characteristics, many of which greatly influence plant production. In fact, every plant process is affected by or dependent on water, which has great ecological significance, as evidenced by the observation that water, along with temperature, has been the primary determinant of the geographic origin of plant species. A water-deficient plant, for example, has reduced yield potential and will die if the deficiency is not remedied (Figure 7-2).

Water has a higher specific heat than most other substances, with the notable exception of liquid ammonia. This means that it takes a large amount of heat to raise the temperature of water. Thus, large bodies of water have a tremendous ability to stabilize temperatures, as evidenced in the moderate climates of islands and land near large lakes, seas, and oceans. This fact is important to the horticulturist who is considering site selection. Perhaps equally important is the fact that the high specific heat of water

annual precipitation (mm)

- > 2000
- 1500–2000
- 1000–1500
- 500–1000
- 250–500
- < 250

FIGURE 7-1. A global picture of areas of water deficit.
Photograph courtesy USDA.

FIGURE 7-2. A plant wilts as a response to insufficient water. This aluminum plant has already had its growth potential reduced, although it will recover if watered at this stage.

plays a critical role in stabilizing internal plant temperatures (in cells) on a short-term basis.

Water is a substrate (or base) for many plant processes, including photosynthesis, the process in which it is combined with carbon dioxide to form simple sugars. Photosynthesis is discussed in more detail in Chapter 5.

Water is often referred to as the "universal solvent," partly because it has a high boiling point and a low freezing point. This property explains its ability to dissolve fertilizer salts. Much of the subsequent discussions about water's function and movement in the plant will therefore also be true for fertilizer salts moving in solution. In part of this chapter we will discuss how water and nutrients move through the soil, into and through the roots, up the stem and through the branches, and into the leaves, with the water exiting via the stomates.

The formation and maintenance of the shape or structure of most plants rely on water; wilted leaves, flowers, and fruits look quite different from fully turgid organs. Turgidity of cells is dependent on water, with the crucial function of the stomatal guard cells being a particularly pertinent example. If the stomates cannot open, carbon dioxide cannot enter the leaf and photosynthesis

cannot occur. Cell enlargement, if the cell walls are sufficiently elastic, is caused by the engorgement of the cell by water, thus stretching the cell to a larger size. This rather simple process is an essential feature of the physical increase in size of most plant parts. In fact, in many fruits, the increase in size depends entirely on cell enlargement because no new cells are produced in the period commencing shortly after fruit set. Growers sometimes retard plant growth simply by withholding the driving force for growth—water.

Physical Properties

Water is a simple molecule consisting of two hydrogen atoms and one oxygen atom and is written chemically as H_2O (Figure 7-3). The hydrogen atoms have a specific relationship to each other and are always located on one side of the oxygen atom. The position of the electrons in the two hydrogen atoms to one side and the oxygen atom on the other side of the molecule causes a partial positive charge on the side of the molecule where the hydrogen ions are located and a partial negative charge on the side where the oxygen atom is situated. The partial positive is indicated by the Greek letter gamma (γ) and a plus ($+$) sign adjacent to the hydrogen atoms, and the partial negative charge associated with the oxygen atom is denoted by a gamma negative sign ($\gamma-$). These charges account for many of water's unique properties. Figure 7-4 illustrates water's loose self-attraction (shown by the

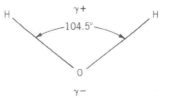

FIGURE 7-3. Schematic representation of the water molecule. The 0 represents an oxygen atom and the H signifies hydrogen. The $+$ and $-$ symbols denote electrical charges and the γ indicates that these are partial charges.

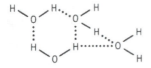

FIGURE 7-4. Diagrammatic illustration of hydrogen bonding in ice. Dotted lines indicate hydrogen bonding, creating the lattice (crystal) form.

broken lines); this is known as **hydrogen bonding** and results in a more or less ordered structure that represents the currently accepted explanation of water's unique properties. This configuration apparently accounts for the open crystalline structure when water is in its solid form (ice). However, individual water molecules are not associated with one another during the vapor phase. When in its liquid form, water is composed of molecules with associations intermediate between the ordered structure of ice and the lack of association that is characteristic of water vapor. Because liquid water has this intermediate, somewhat crystalline structure with hydrogen bonds possessing a half-life of 10^{-11} seconds, it is sometimes called a "flickering iceberg." For a summary of phenomena associated with hydrogen bonding, see Table 7-1.

When a salt such as potassium chloride is dissolved in water, because of the polar nature of water, the negatively charged chloride ion will migrate to the side of the water molecule where the hydrogen atoms are located and will become associated with the gamma positive charge of the hydrogens. In a similar manner, the positively charged potassium ion will migrate to the side of water molecule where the oxygen is located, and will be attracted to the gamma negative charge of the oxygen. This example helps demonstrate the good solvent properties of water; many substances will dissolve in water in this manner. This characteristic is not only important in the soil for dissolving nutrients and subsequent uptake and translocation, but it is also essential for the translocation of various metabolites throughout the plant (e.g., sugars and hormones). When water moves, it pulls itself along because of the hydrogen bonds (Figure 7-4). Many water molecules moving in the same direction in a mass, such as water flowing in a stream, is referred to as **mass flow,** and is an important means of water and **solute/ nutrient** movement throughout the plant.

Another important phenomenon associated with water and its movement is called

TABLE 7-1. Hydrogen bonds and measurable (sensible) heat related to phase changes in water.

	PHASE CHANGES			
RESULTS	WATER FREEZES	WATER VAPOR CONDENSES	ICE MELTS	WATER EVAPORATES
Hydrogen bonds form	Yes	Yes	No	No
Releases heat (exothermic)	Yes	Yes	No	No
Warms plants and atmosphere	Yes	Yes	No	No
Hydrogen bonds break	No	No	Yes	Yes
Requires heat (endothermic)	No	No	Yes	Yes
Cools plants and atmosphere	No	No	Yes	Yes

surface tension, which is the tendency for water molecules to cling to each other (cohesion) and to surfaces they touch (adhesion). Because of this, water molecules on the surface are attracted into the body of the liquid and thus are like a stretched skin. This is why it is possible to fill a glass of water slightly above the lip without spilling. When the water is poured out of the glass, some of it will cling to the sides and the lip of the glass. The same thing will happen when very thin glass tubing is placed in water; if the diameter of the channel in the tubing is small enough, the water will actually rise up in the tubing because of surface tension. It has been demonstrated for a vessel or tube with an inside diameter of 0.2 mm that water will rise to a height of 30 cm. Similarly, in a tube 40 μm in diameter, the water will rise to a height of 75 cm, and a tube 1.0 μm in diameter will draw water to a height of 30 meters. Many plants can have vessels of such a small diameter in their vascular system. When the surface tension of the water in a tube or vessel is greater than the pull of gravity, the water rises by capillary action. An additional pull on water occurs as plant organs elongate.

Transpiration

But how can water continue to be transported to the leaves at the top of a tree that is taller than 30 meters? Trees, like most plants, have an additional pull on the water caused by **transpiration,** which is the loss of water from a plant in the form of vapor, (essentially an evaporation process). Most transpiration occurs through the stomates and many plant scientists view this process as a necessary evil because it results in large amounts of water loss and has the potential to cause injury from desiccation. It has been estimated that a single corn plant in midsummer may lose over two liters of water per day. In fact, 95 percent or more of the water absorbed by plants is lost by transpiration, so that if there were no water lost by transpiration, a single rain or irrigation might suffice to provide enough water for the growth of an entire crop. Some plant scientists feel that transpiration is necessary to get sufficient nutrient elements into the plant. Most nutrient elements go into a plant in solution through the young roots with the water (see Figure 3-15), but just the amount of water needed for normal plant growth could bring in sufficient nutrients to meet the plants' requirements. Therefore, the huge quantities of water used in transpiration are probably not necessary to bring nutrients into the plant.

It is often argued that transpiration is necessary because it tends to have a cooling effect on the plant. This results from water's high **heat of vaporization;** that is, in order for water to change from the liquid state to the vapor form it must absorb 540 calories of energy per gram of water from its surroundings. Thus, when liquid water is vaporized, large amounts of energy are required, and a net cooling effect occurs. The problem with this argument is that stomates of many plants are often closed during the hottest part of the day, so there is no cooling when it is most needed. Additionally, leaves are rarely injured by overheating. Although it is apparent that plants transpire, and in a sense waste great amounts of water, perhaps in the future scientists will learn how to better control transpiration to enable more efficient utilization of that tremendously important and limited resource, water.

Evapotranspiration is the evaporative loss of water from plants and the soil. Transpiration and evapotranspiration are influenced by the amount of water vapor in the air. To better appreciate this concept, it is necessary to understand **humidity.** The **specific humidity** of the air is the amount of water the air contains per unit weight (i.e., grams of water per kilogram of air). The **relative humidity** is expressed as a percentage and is the amount of water vapor in the air relative to the maximum amount of water the air can hold at a given temperature. The higher the temperature of the air, the more water vapor the air can hold. This is why the relative humidity is often so low indoors in the winter. When cold air (which can hold only small amounts of water) is brought inside a heated building, it still contains the same amount of total moisture as it did outside (i.e., the specific humidity is essentially the same inside and outside). However, the warmer air indoors can hold more water vapor than it could at a lower temperature outdoors. The relative humidity, expressed as the amount of water vapor the air can hold at that temperature, is therefore lower inside the building. Under conditions of high relative humidity, there is less of a gradient (**vapor pressure deficit**) between the amount of water inside a leaf (usually 100 percent relative humidity) and the amount of water in the air, which results in a reduction in transpiration rate. In other words, greater amounts of water in the external air will lower the transpiration rate. Generally speaking, vapor pressure deficit increases as ambient temperature increases, resulting in higher evapotranspiration. Vapor pressure tables can be helpful in estimating potential evapotranspiration.

Diffusion results when a substance in fluids moves in response to a gradient, that is, it goes from a higher to a lower concentration until an equilibrium is reached, as in transpiration. An example of a diffusion phenomenon is illustrated by the fact that water moves with a **water potential** gradient. When water potential is measured at two points in a system—for example, between roots and leaves, or soil and plant—direction of water flow along this gradient can be readily discerned. Water potential, expressed as ψ_w, has been historically measured in **bars**, and one bar is approximately equal to one atmosphere of pressure. Water potential is now more often measured in the SI unit, the megapascal (Mpa). One bar is equal to 10^5 Pa or 0.1 Mpa. The Mpa is considered to be the preferable unit for reporting water potential. Water potential, ψ_w, is the sum of its component potentials, that is, ψ_w is composed

of osmotic or solute potential (ψ_π), plus turgor or pressure potential (ψ_p) and gravitational potential (ψ_g). ψ_π is primarily related to dissolved solutes,[1] the turgor potential (ψ_p) is based on xylem tension and positive pressure of water pressing against cell walls, and the gravitational potential (ψ_g) is related to plant height. This latter component varies at a rate of 0.1 Mpa (or one bar) for each 10 meters of plant height.

Although the utility of water potential measurements has been somewhat controversial, most plant/water relations scientists accept them as useful. Measurement of water potential may be accomplished in several ways; one of the most common is the pressure chamber. The pressure chamber technique employs a pressure safe vessel attached to a pressure gauge. An excised plant sample such as a petiole or rachis is subjected to slowly increasing pressure until xylem sap is observed exuding from the sample. The gauge pressure is then noted, and this reading can be related to the water potential of the leaf's cell walls. For a more detailed discussion of water potential, see the references at the end of this chapter.

A negative water potential lowers the concentration of water (or raises the concentration of solute) and therefore increases the attraction for water. An example of diffusion caused by a difference in water potential is osmosis, where a difference exists on two sides of a membrane, generally because of differing solute concentrations.

The **permanent wilting point** of a plant is the point of wilting beyond which the plant cannot recover, even if it is placed in a saturated atmosphere. This point occurs when what little water is left in the soil is held so tightly by the soil particles that it becomes unavailable to the plant. Negative 15 bars (−15 bars or 1.5 Mpa) water potential in

....
[1]Some researchers separate out a further component, matrix potential (ψ_m), related to water activity effected by charged surfaces.

the soil is considered to be at or beyond the usual permanent wilting point for many plants.

Incipient wilting occurs when a plant wilts temporarily but then recovers even without additions of water. This condition often occurs when a plant loses turgor briefly under conditions in which the absorption of water cannot keep pace with the transpiration demand. Incipient wilting may occur during summer afternoons for many plants. The stomates then close, cutting transpiration loss, although the plant may remain wilted. As air temperatures drop, water loss is less and uptake catches up with transpiration demand, allowing the plant to recover. Stomates may then re-open until dark. Several horticultural crops exhibit incipient wilting in stress periods, especially tomato. This wilting should not cause alarm as long as soil moisture levels are adequate, and it is quite common. If plants remain wilted even if the soil is moist, the horticulturist should be concerned about root damage from diseases or other soil problems such as high soluble salts that interfere with water absorption.

The movement of water by diffusion across a differentially permeable membrane is termed **osmosis.** Water is generally taken up from the soil solution by osmosis. To understand better how water uptake and movement within the roots occur, one should realize that certain parts of the root are hydrophobic. **Hydrophobic** means a substance repels water; the opposite is **hydrophilic,** meaning that the substance is water-loving or attracts water. Examples of hydrophobic substances are fats, oils and most lipids. Paper towels, salts, and water are examples of hydrophilic substances. Cells of the root endodermis (see Chapter 3) contain the Casparian strip, which is composed of bands of suberin. This substance is hydrophobic and represents a potential barrier to water movement into the interior of the root (Figure 7-5). However, plasmodesmata extend through numerous pits in the endodermal cell walls,

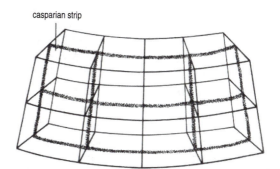

casparian strip

FIGURE 7-5. The Casparian strip, illustrated here in a schematic of endodermal cells, is an anatomical feature important to uptake of water and minerals by roots.

connecting the internal portions (protoplasts) of endodermal cells with those of the adjacent cells to the outside (cortex cells) and inside (pericycle cells); this facilitates movement of water and nutrients within the cellular membranes and is called **symplastic** movement.

Root Tips

As explained in Chapter 3 (Figure 3-15), much water uptake occurs in and immediately behind the root hair zone where the roots are nonsuberized (Figure 7-6). Horticulturists consider white root tips desirable because they appear healthy, and are an indication of root surface area for the uptake of water and nutrients. Water and nutrients move readily, with little or no resistance, into the root tip and through the epidermis and cortex near the root tip until they encounter endodermis cells with their well-developed Casparian strip. Movement of water and nutrients in the cells to the outside of the endodermis may occur in one of three ways: through the cell walls; across the plasmalemma and into the symplast where they go from cell to cell via the plasmodesmata that connect protoplasts of adjacent cells; or across the vacuoles (Figure 7-7). The cells of the endodermi less and less permeable to wate mature, that is, near the root apex no Casparian strip. The Caspa

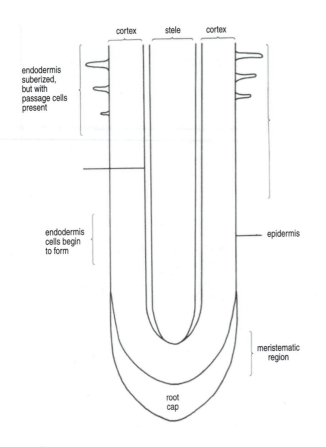

cortex   stele   cortex

endodermis
suberized,
but with
passage cells
present

endodermis
cells begin
to form

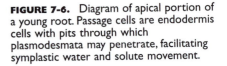
— epidermis

meristematic
region

root
cap

FIGURE 7-6. Diagram of apical portion of a young root. Passage cells are endodermis cells with pits through which plasmodesmata may penetrate, facilitating symplastic water and solute movement.

develops farther from the root tip and the endodermal cells become considerably less permeable as the cell walls thicken and further suberize. Water uptake near the root apex is not of major concern, however, because the vascular tissue has not differentiated in this young region of the root. It is apparent, therefore, that for water and nutrients to enter the stele, they must cross the cell membrane (plasmalemma) into the cell and move through the symplast. Because water and nutrients must cross the membrane, the plant, through special properties of the membrane, is selective in the materials that it takes up.

Membranes

A membrane is a lipid bilayer with associated proteins. The generally accepted structure of a lipid is a molecule containing a polar hydrophilic end and a nonpolar hydrophobic end (Figure 7-8). A schematic representation of why a drop of oil (lipid) remains intact in a vessel of water is illustrated in Figure 7-9. The hydrophilic ends of the lipid molecules are attracted outwardly to the water and the hydrophobic ends are repelled by water and are therefore attracted to themselves, forming the center of this drop of oil. The structure of a membrane is illustrated in Figure 7-10. Note the classic fluid mosaic model and lipid protein bilayer pattern. The fluid mosaic model illustrates the constant motion of the component lipid and protein molecules, accounting for the property of a membrane to "mend" itself (e.g., when it is pierced by a needle).

Plant membranes are differentially permeable, meaning that water can freely pass through such membranes by osmosis, but many nutrients cannot pass freely through them. Absorption of such nutrients therefore

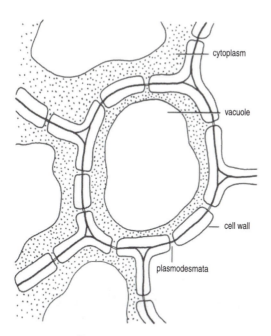

FIGURE 7-7. Diagram of cortical parenchyma cells showing pathways for radial water and salt movement: through the cell walls, through the symplast via the plasmodesmata, and across the vacuoles.

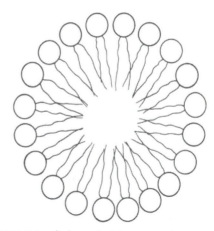

FIGURE 7-9. A drop of oil in water, schematically showing the hydrophilic ends of the lipids attracted to the surrounding water and the hydrophobic ends attracted to themselves and forming the center of the drop of oil.

requires energy, necessitating aerobic respiration to enable uptake of both cations and anions. This fact underscores the absolute necessity of an adequate supply of oxygen to the roots. Soil oxygen level is influenced by soil type, compaction, and amount of water in the soil. Another pertinent feature of the differential permeability of plant membranes is the fact that once nutrients are absorbed, they will not leak out.

Plant scientists recognize that ions are selectively absorbed. As an example, consider the two monovalent cations, sodium and potassium. Although both elements are in Group I in the Periodic Table of Elements and potassium (39) has a greater atomic weight than sodium (23), potassium is readily absorbed by plant roots and sodium is absorbed to a much lesser degree. This selection process usually occurs at the plasmalemma. It is there that water and

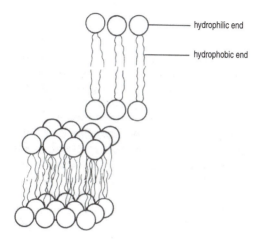

FIGURE 7-8. Structure of typical lipids.

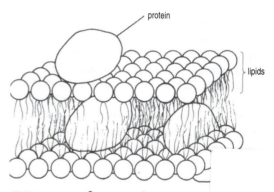

FIGURE 7-10. Structure of a membrane.

nutrients must cross the membrane in order to traverse the Casparian strip and enter the stele for subsequent translocation to the rest of the plant. Thus plants can, to some extent, exclude certain toxic substances and accumulate other substances (e.g., acids in lemon).

Water Uptake

Water moves by mass flow, diffusion, and capillary action through the soil and into the root. It can then move through the cell wall area (through the cellulose and the microfibrils) of the epidermis and cortex. However, when the water reaches the Casparian strip, it can go no further into the root unless it passes through a membrane, the plasmalemma, and enters the symplast. The water generally crosses the membrane, entering the symplast in the cortex or endodermis. The Casparian strip is the inner limit of water movement through the cell walls directly, so the water must cross the plasmalemma at this point. Once it has crossed the plasmalemma, water moves from within one cell to the next cell via the plasmodesmata until it reaches the xylem, where it passes back out through the membrane into the xylem vessels. At this point most of the water taken up by the plant moves in response to a gradient, from the soil through the plant and into the air. The transpiration demand is the main pulling force, although growth itself—together with mass flow, capillary action, and diffusion—is also a force involved in water movement in a plant (Figure 7-11). Nutrients, too, move in solution along such a pathway, although they may be exchanged on and off of the xylem cell walls. This is a function of the negative charges associated with the cellulose that comprise the major part of xylem cells (which, of course, are no longer living). Therefore, such cells are often thought of as pipelines, or water and nutrient conduits, but because of the aforementioned charges, it is clear that they are more dynamic in function than mere pipelines.

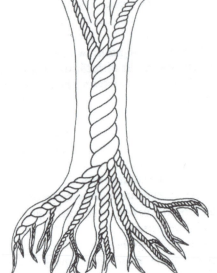

FIGURE 7-11. Movement of water in plants. Water moves in the plant along the pathway illustrated by this braided rope diagram. Component ropes lead to main branches and branch roots and eventually the fibers lead to the leaves (or small branches) and small branch roots.

HORTICULTURAL APPLICATIONS AND WATER MANAGEMENT

Water and Temperature Control

Misting or syringing

The practice of applying fine sprays of water is sometimes useful for some important horticultural species in field situations. An example is the practice of syringing closely mowed, high-value turf plantings such as those used for golf greens, bowling greens, and tennis courts. In northern temperate regions, the species often used for these purposes is creeping bentgrass (*Agrostis stolonifera*), which is usually mowed to a height of 1.25 to 2.5 centimeters (0.5–1 inch). On hot, sunny days in the summertime, from late morning until late afternoon, this closely clipped grass will go into stress. This is a combination of heat stress and water stress brought about by the inability of the root systems to absorb water at a pace equal to the rate of transpirational loss. The grass consequently turns a dull blue-green color, signaling the person caring for the turf (e.g., the golf course superintendent), that the turf is being stressed. To keep the turf healthy, it is therefore a good practice to bring it out of stress by syringing with water. This usually is accomplished by turning on the irrigation sprinklers for a short period of time for example, for about ten minutes—or by having workers spray the turf with water from hoses (Figure 7-12). The objective is just to wet the foliage, which is then cooled directly because the water is cooler than the grass and indirectly because of the evaporation of the water. This latter mechanism takes advantage of water's high heat of vaporization and is an example of evaporative cooling. Heat stress and water stress are both reduced because the increase in relative humidity around the leaves of the turf slows transpiration losses. If the grass is not cooled in such a situation, the plants could die, or the leaf tips could desiccate and turn brown. The latter effect is unsightly and reduces the value of such sports turf. In addition, stresses may weaken plants, causing them to be more susceptible to disease organisms (see Chapter 16). The water applied in this manner normally does not remain on the foliage long enough for fungal spores to germinate or bacteria to grow and lead to disease problems. This is because the sun and wind will dry the outdoor plants; misted indoor plants usually do not benefit from such air movement.

Fan and pad cooling

Horticulturists also use the benefits of water's high heat of vaporization in fan and pad cooling systems for greenhouses (Figure 7-13). Even in temperate climates, cooling is commonly required during the bright days of summer because of the ability of a greenhouse to trap and accumulate heat (see Chapter 6). Fans alone can exhaust the hot air, but such systems can only cool the greenhouse interior to approximate ambient external temperatures.

FIGURE 7-12. *Left*, syringing a golf green by hand to cool this fine turf and reduce transpiration on a hot day. Close-up of a bentgrass green, *right*. Note the fine character of this grass species (penny in the photo for scale). Automatic pop-up systems such as illustrated in Figure 7-25 are also used for this purpose.

FIGURE 7-13. Crossfluted evaporative cooling pads for two adjacent greenhouses served by a common water source and catchment tank. This system illustrates efficient use of resources.

However, replacing all or part of the wall opposite the exhaust fans with a continuously moistened fibrous wall, or pad, can significantly reduce the temperature as the interior air is exhausted by the fans and outside air is pulled in through the pad. This reduction occurs because as the air flows through the moist pad, it picks up water, which is subsequently evaporated (a process which, as we have already seen, requires heat). The cooled air is then pulled through the greenhouse where it replaces the warmer greenhouse air that is exhausted to the outside. Efficiency of this system decreases as the relative humidity of the external air increases, but even in quite humid environments some cooling will be accomplished. Under ideal conditions (high temperature and low relative humidity) interior temperature can be reduced 10°C or more below that of the exterior ambient air. It is important to note that the pad must be composed of a material providing a high surface-area-to-volume ratio and that the fan must be large enough to achieve one air change per minute. Historically, shredded aspen (*Populus tremuloides*), generally referred to as excelsior was the fibrous material used. Because this material tends to settle and may rot, some newer, more stable materials

have been developed. The most commonly adopted of these materials are crossfluted cellulose materials that have been extruded into a rigid and rippled, or corrugated appearing, shape. Excelsior is still popular with many greenhouse operators, however probably because of its proven performance, lower cost, and availability. Regardless of pad material selected, water must be dripped uniformly over and through the fibrous material so that a uniform, continuous flow of moisture-laden air will be introduced into the greenhouse to facilitate the evaporative cooling process.

Fans in the United States are rated in cubic feet per minute (cfm), so the appropriately sized fan can be determined by selecting a fan that can exhaust approximately 8 cfm per square foot of floor space in the greenhouse to be cooled. This figure is an approximate one and may vary with elevation of the land upon which the greenhouse is located, with light levels and with crop requirements.

Frost Protection

Irrigation for frost protection is an example of using water's property of heat of crystallization (heat of fusion) to protect horticultural crops from damage by frost or temperatures slightly below 0°C (32°F). In this method, low-volume, frequently cycled, overhead irrigation is applied to the crop beginning just before the onset of predicted frosts or freezes. Special low-volume heads may be used for this purpose. As the temperature drops, gradual and continuous freezing of ice commences on the plant surfaces (Figure 7-14). The irrigation must be continued until the ice melts completely following the return of higher temperatures. For water to freeze, 80 calories per gram of water must be removed after the water has reached the freezing point. As the heat is given up by the water the immediately adjacent plant part is warmed, thus preventing the tissues from freezing. This technique is commonly employed by growers of strawberry, citrus, and other fruits, and by producers of early vegetables during seasons of high frost probability. However, it is important to remember the limitations to the utilization of this approach. Firstly, it generally is only successful if the freeze is of short duration and only a few degrees below 0°C. Second, the irrigation must continue to be applied until the ice melts completely as a result of the return of the air to higher temperatures, combined with the temperatures of the irrigation water itself. If the irrigation is turned off before the ice is completely melted, severe plant damage may result, because for ice to melt, the ice must receive 80 calories per gram of water from its surroundings. Without continued irrigation, part of this heat will be obtained from the plant tissue, thus lowering its temperature below a critical level and causing tissue damage. It therefore behooves the prudent grower to monitor the plants carefully, turning off the water only when the ice has melted and the danger period is past. It should also be noted that the plants can break from the weight of ice accumulated over a lengthy period of irrigation, especially during windy weather. Other aspects of irrigation, besides this specialized use, are discussed later in the chapter.

FIGURE 7-14. Branch of peach encased in ice from irrigation applied for frost control. These flowers survived −4°C (25°F) and set a good crop of fruit.

Heat Capacity of Water and Site Selection

Another important property of water is that it has a high heat capacity (see Chapter 6). This means that the amount of heat required to change the temperature of water is higher than that to change the temperature of a similar volume of air. Large bodies of water, such as oceans and large lakes therefore have a moderating effect on nearby land masses. As a result, land near large bodies of water warms more slowly in the spring and tends to stay warmer later into the fall. Many perennial fruit crops benefit from this tendency because flowering in the spring may be delayed until after frost dangers have passed. Sudden temperature fluctuations, which could also damage flowers and early crops, are less likely near large bodies of water. Temperature extremes are not as likely near large bodies of water as they are for a more continental climate location: summers are cooler and winters are warmer. Furthermore, the likelihood of early frosts in the fall is reduced because, once warmed, water tends to cool off slowly, thus moderating temperatures of adjacent land masses. For these reasons, many fruit production areas are associated with large bodies of water. Examples include the location of grape, sweet cherry, peach, and other fruit plantings along the coasts of the Great Lakes in North America. These lakes represent huge volumes of water capable of exerting strong moderating effects on adjacent land.

This phenomenon also helps to explain the moderate climates of many island locations. One example of this is the highly favorable temperatures, horticulturally speaking, of the British Isles. Relatively mild temperatures occur although the isles are located at a more northerly latitude than most of the United States. Many species can be grown in Ireland and the United Kingdom that are not sufficiently cold hardy in the northern U.S. because of the moderating effects of the Gulf Stream and

Atlantic Ocean. In the southern hemisphere, Tasmania and New Zealand are similar examples. Large bodies of water often affect the location of the horticultural industry in many areas of the world because of the moderating effect that such bodies have on air temperatures of the surrounding areas (Figure 7-15).

Reducing Transpiration

Plant Factors

The plant itself possesses many characteristics and adaptations enabling the plant to reduce transpiration. These include leaf size, shape, orientation, and surface characteristics, in addition to the water conserving nature of the relatively impervious corky layer of bark found on the stems of most woody species. It is generally believed that leaf orientation that is favorable for photosynthesis will also promote higher

transpiration. Plants with small leaf areas normally transpire less than those with large leaf areas. Indeed, some woody plants such as buckeye drop all or some of their leaves during periods of severe moisture stress. Other species such as sweet corn and other grasses reduce exposed leaf area by the curling or rolling of their leaves. The ratio of leaf area to root area is even more important than leaf area alone, because plant water deficits will develop if absorption fails to keep pace with transpiration. Therefore, plants with deep greatly branched root systems will be less likely to be damaged by drought than plants with less extensive root systems.

Leaf surface characteristics are also important in exerting an influence on transpiration. The cuticle, composed of wax and cutin and cemented to the epidermis by pectin, often covers the surface of leaves, fruits, and other plant parts. As noted in

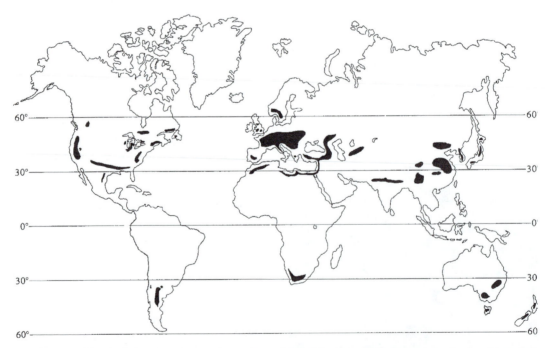

FIGURE 7-15. The shaded areas represent the major regions of deciduous fruit and nut production. Note that regardless of hemisphere, these areas lie mostly between 30° to 50° latitude, but are extended to higher latitudes by proximity to large bodies of water and into lower latitudes by the cooling influence of higher elevations.

Temperate-zone Pomology by Melvin N. Westwood. Copyright 1978 by W. H. Freeman and Company. Reprinted with permission.

Chapter 3, the waxy cuticle forms the "bloom" found on some fruit and leaf surfaces. Amount, absence, or presence of cuticle can be an important factor in reducing water loss, which is important to postharvest handling of fruits, vegetables, flowers, and other horticultural commodities (see Chapter 15). Thick layers of epidermal hairs (trichomes) may cover the leaves of some species resulting in a significant increase of leaf boundary layer resistance and densely pubescent leaves of such plants as lamb's ear (*Stachys byzantina*) reflect a substantial amount of incident radiation, which contributes to a reduction in transpiration. As noted in Chapter 5, sun leaves are smaller and thicker than are shade leaves; these characteristics can further influence transpirational water loss.

Although stomatal aperture controls transpirational water loss, the relative depth of stomata is also important. Stomata are recessed in pine needles, creating a greater boundary layer and a lower vapor pressure deficit. Therefore pine trees have leaves adapted for winter conditions when the soil is frozen and water uptake is limited. Conversely, aquatic species have relatively high stomatal ridges and high rates of transpiration. The fact that the majority of stomata are located on the lower leaf surfaces helps control water loss because it is more likely that the sun will strike the tops of the leaves, contributing to a greater vapor pressure deficit. Even with these adaptations to reduce transpiration, drought stress is a major reason for crop losses in horticulture.

Antitranspirants

Because water conservation is of utmost importance in increasing the efficiency of crop production, it has been proposed that a reduction in water lost through transpiration could lead to more efficient water utilization, if photosynthesis were not concomitantly reduced. Antitranspirants are compounds that horticulturists and researchers have applied to plants in an attempt to achieve this goal. Antitranspirants fall into three main classes: chemicals that influence the stomatal closure apparatus, compounds that seal leaves or plug stomata, and substances that reflect light from the leaves. Phenylmercuric acetate (PMA) is an example of a chemical that appears to close stomata, but to date it has not been widely used, in part because it may have a negative effect on photosynthesis. Even if effective, PMA is unlikely to be used because it would release mercury into the environment. Likewise, the waxlike substances used as sealants have also been less than satisfactory, partly because the sealing mechanism excludes CO_2 which is necessary for photosynthesis. Substances such as kaolinite clay reflect sunlight and thus cool the leaves, which reduces transpiration. Because of the white appearance they make plants unsightly, however, and can reduce photosynthesis. It should also be pointed out that some antitranspirants (primarily the sealant type) are used commercially by nursery growers and landscape contractors in an effort to reduce transplanting shock when establishing leafy plants in the field and to lessen winter "burn" of transplanted evergreens. Reports of efficacy of such approaches are inconclusive. Nevertheless, researchers continue to explore this avenue in hopes of achieving more efficient water use by reducing transpiration losses.

Misting House Plants

Many horticultural practices have been developed in order to capitalize on the unique properties of water and its functions in plants. In the home, many plant enthusiasts diligently apply water to the plant foliage with a mist bottle. The objective of such an exercise is to increase the relative humidity in the immediate vicinity of the leaf lamina in an attempt to reduce water stress (Figure 7-16). Such stress could be caused by the rapid transpiration that might be engendered by high temperatures and low relative humidities commonly found in

FIGURE 7-16. (*a*) In general, misting of houseplants is discouraged because it leaves plants vulnerable to diseases. (*b*) Houseplants may be grouped to increase relative humidity.

Photo courtesy Deborah Brown, University of Minnesota.

many modern homes. However, use of the mist bottle is usually discouraged because it is not particularly effective in increasing relative humidity for a useful length of time and the presence of water on the leaves often leads to disease problems (see Chapter 16). Alternative procedures—such as placing potted plants on a tray of moist gravel or by locating plants in close proximity to each other—are recommended. In the gravel tray method, the water level is allowed to come only to within about one centimeter (0.4 in.) of the surface. This enables evaporation of the water from the moist gravel and free water surfaces, increasing the relative humidity above the gravel, but it prevents water movement into the pots by capillary action, which could happen if the water level were at the

interface of the pot bottom and the gravel. Increasing the relative humidity in the vicinity of the plants is also the objective of grouping plants close together. However, here it is achieved by direct increase in the water content of the air resulting from evapotranspiration from the leaves and growing medium of the group of plants. High humidity is also employed to prevent root desiccation in nursery storage facilities (Figure 7-17).

IRRIGATION

Meeting water requirements of crops by irrigation is as old as civilization itself. The fabled lost continent of Atlantis is reputed to have possessed an elaborate system of irrigation canals, and if such a civilization

FIGURE 7-17. View of commercial bare-root plants in a nursery storage facility. High humidity provided by fog nozzles helps keep the roots from desiccating.

actually existed, irrigation would date from 12,500 years ago! More verifiable records exist that document the use of irrigation in China over 5000 years ago, and the Egyptian ruler Menes (2700 B.C.) turned the course of the Nile in order to flood higher ground for crop production. The development of the Mesopotamian Empire made use of the waters of the Tigris and Euphrates rivers for irrigation, while the Inca and other early western hemisphere civilizations employed elaborate irrigation systems long before the arrival of Europeans in the New World.

One of the oldest and simplest methods of remedying insufficient water supply to plants is hand watering, or manual irrigation (Figure 7-18). Manual irrigation is relatively simple, is low in material costs, and generally uses only moderate amounts of water. However, it does represent a high investment in labor inputs and can be highly variable in amount of water applied. Hand watering is also impractical for large-scale field applications and inefficient for many greenhouse and other intensive horticultural production systems. Therefore, let us examine some of the commonly used irrigation methods and systems appropriate to horticultural production.

Surface Irrigation

There are several kinds of irrigation systems in use, each with inherent advantages and disadvantages. One of the oldest systems that has been widely used for thousands of years is **surface irrigation**. Two main types exist, *furrow* and *flood* irrigation Figure 7-19).

The furrow system and its modifications require the construction of usually shallow ditches between the crop rows or beds, which are filled from supply ditches by opening gates or by the use of siphon pipes or tubes (Figure 7-20). The beds often are raised several inches above the original soil level to provide good drainage and aeration in the root zone. Furrow irrigation methods

FIGURE 7-18. Watering transplants with a hose. Some growers will periodically water their plants by hand because it enables them to inspect plants more closely.

FIGURE 7-19. Furrow irrigation for vegetables in China, (*top*). Note raised beds employed to keep roots above the high water table. Efficient land use is apparent because the green onions will be harvested before the interplanted Chinese cabbage is large enough to compete with them (*bottom*).

(a)

(b)

(c)

FIGURE 7-20. (*a*) Perforated pipe for filling the furrows from the main water line in furrow irrigation of this pepper crop. (*b*) and (*c*) Siphon tubes also are used to deliver water from the main channel for this furrow-irrigated crop of tomatoes.

are essentially limited to fields that are levelled by machine to slope slightly from one end to the other, and they are rarely successful on sandy or well-drained soils that hold little water. Soil type is therefore a critical consideration for furrow irrigation

applications (see Chapter 8). An advantage of furrow irrigation is the fact that it wets very little of the foliage, which causes less predisposition to foliar disease problems. This type of irrigation requires very large quantities of water and is labor intensive, however, requiring constant attention to manage water flow. Much water is also lost to evaporation because of the great surface area exposed to the air. Another potential drawback results if the irrigation water is high in soluble salts, because large amounts of salts will build up in the soil. If furrow irrigation is used over an extended period of time, the salt levels could become high enough to seriously reduce yields and in severe cases may even render the land unsuitable for crop production.

Flood irrigation differs only slightly from furrow irrigation, primarily in the fact that flood irrigation covers most or all of the field soil surface rather than just the surfaces between the rows or beds. Although flood irrigation requires less labor, the advantages and disadvantages are essentially the same as those for furrow irrigation. However, flood irrigation is generally considered to be one of the least efficient irrigation systems in terms of water and labor use. *Basin* irrigation is a type of flood irrigation commonly employed in arid regions to concentrate the water in the immediate vicinity of a specific tree (Figure 7-21). It also is used by landscape contractors on a small scale when transplanting trees and shrubs in order to reduce transplanting shock and to increase chances for survival.

Sprinkler Irrigation

Sprinkler irrigation is widely used for a great number of crops. It is commonly called overhead irrigation because the water is applied above the plant or crop canopy, generally wetting both the foliage and the soil in the process. It is most often employed for outdoor plantings and crops, including fruit and vegetable crops, turf, and nursery

FIGURE 7-21. Basin-type flood irrigation for individual trees (*top* and *bottom*). The soil dike holds the water within the drip line, providing for efficient water use.

plantings. It is also useful outdoors for containerized nursery stock and indoors for greenhouse flower, nursery, and vegetable production. In greenhouse applications, good air circulation is essential to avoid diseases, and it may also be necessary to institute a disease control program that may require judicious use of appropriate pesticides (see Chapter 16).

Two modified uses of sprinkler irrigation discussed elsewhere involve application of

water for frost control and use of a fine spray or mist to encourage rooting of cuttings. Water-soluble fertilizers may also be applied effectively with sprinkler irrigation systems; this practice is often referred to as "fertigation." Timing of such fertilizer applications may be better tailored to the needs of the plant than when application is by more conventional methods. More information on the subject of applying fertilizer through the irrigation system is provided in Chapter 9. Because it requires no special land preparation and little or no attention during application, the annual cost of labor for most sprinkler irrigation systems will be less than that for surface irrigation methods. Of course, initial installation costs may be high both in capital and labor and the amount will depend on the system selected.

One of the most expensive systems is the pop-up solid-set sprinkler design frequently installed for high-value turfgrasses such as golf greens. Once installed, permanent sprinkler irrigation systems require only small labor inputs for maintenance of the system. Nonpermanent sprinkler irrigation systems usually rely on portable aluminum pipe, however, and moving the pipes is tedious and labor intensive. Other potential disadvantages of sprinkler irrigation include distortion of the irrigation pattern by wind, which causes uneven distribution; high power requirements for pumping; a constant, high-quality water supply requirement; and inefficient irrigation of heavy soils in hot windy climates. The difficulty with heavy soils occurs because in order to supply an adequate amount of water to heavy soils without run-off losses, the irrigation rate must be very slow, which allows an excessive loss of water from evaporation.

Although perforated pipe systems and stationary, or fixed head sprinklers have been used to some degree, the most common field uses of sprinkler irrigation employ a portable system using double or single nozzle revolving head sprinklers (see Figures 7-22, 7-23, and 7-24). Such heads may be mounted at various heights above the supply

FIGURE 7-22. A riser for sprinkler irrigation with a double nozzle revolving head in use on nursery crops under lath.

pipe on smaller pipes called *risers*. For low crops such as strawberries, the risers may be only a few centimeters tall, usually 15 to 30 cm (6–12 in.). They are generally one to two meters tall for most row crops, but risers several meters tall may be installed for frost protection and irrigation of tree fruits (Figure 7-23).

FIGURE 7-24. A solid set irrigation system in use in a tree seedling nursery.

FIGURE 7-23. Risers several meters tall (*left*) may be used for crops such as citrus (*right*), both for irrigation and for frost control.

If necessary because of water quality or source, filters or debris screens are commonly installed at the source or in the water lines to prevent trash or particles that might clog the nozzle orifices. Such screens also catch weed seeds, thus preventing dissemination of weeds when surface water sources are used. Freedom from particulate matter is even more important for most of the low-volume irrigation systems discussed later.

Types of Sprinkler Irrigation

Numerous systems for sprinkler irrigation are available for field-scale fruit and vegetable production. Additionally, several modifications are used primarily for home garden and small-scale enterprises. The components of all sprinkler systems for field use are essentially similar. They include a pump to provide the required pressure, main pipelines (**mains**), lateral lines (**laterals**), and the nozzles. Smaller scale sprinkler systems for home use differ only slightly in that hoses are substituted for the mains and laterals and the pressure is provided by the home water supply. The laterals and nozzles on the laterals must be spaced closely enough to enable sprinkler pattern overlap sufficient to provide complete coverage. To ensure that the irrigation system is delivering adequate and uniform coverage, several straight-sided containers should be placed at intervals in the area being irrigated. The amounts of water found in the containers should be similar if the irrigation system is delivering a uniform coverage. Although rain gauges are often used for this purpose, any straight sided containers such as coffee cans may be substituted. Field mains may be permanent or portable. The permanent pipes are used most effectively for systems that will remain in place for the entire season or for several years and often are buried well below the soil surface. Portable mains are employed in cases where the system must be used for several fields during a growing season.

The term **solid set** is used to describe systems that have laterals placed close enough together that they will enable complete field coverage without requiring any moving or repositioning of the laterals. Such systems are particularly appropriate where soil-water conditions require irrigation frequencies of less than 5 to 7 days. They also are well-adapted for situations requiring low volume applications, for soils that are very slowly permeable to water infiltration, and for frost control or crop cooling. A variation of the solid set system is one in which the mains and laterals are placed underground with only the sprinklers and risers above ground. This type of system is often simply called a **permanent system**; the aforementioned pop-up system used for high value turfgrass is perhaps the most sophisticated example (Figure 7-25).

FIGURE 7-25. Pop-up sprinkler systems are permanent irrigation installations used for turf and other high value crops. A pop-up nozzle (*left*), and several such sprinklers in action (*right*). Note that they direct the water away from the sidewalks.

The **center-pivot** continuous move system is probably one of the most widely adapted types of irrigation systems. It can operate continuously, moving slowly through the field. Figure 7-26 shows a section of a center-pivot irrigation system in use. The tower and lateral assemblies are driven through the field on wheels or tracks by hydraulic, pneumatic, or electric drive motors. Although not as suitable as solid-set for heavy soils of low permeability, center-pivot systems are well-suited for many large-scale field applications. Such systems usually employ one relatively long lateral ranging from 200 meters (660 feet) to over 915 meters (3000 feet) in length. A center-pivot system with a 400 meter (1300 foot) lateral can irrigate approximately 50 to 55 hectares (135–145 acres) out of a 65 hectare (160 acre) field. Because this long lateral pivots around a central supply point, the center pivot system irrigates a circular area (Figure 7-27). This leaves the corners unirrigated, but some modifications have been made to enable modern center pivot systems to irrigate most of the corners. One method in use for irrigating corners with a center-pivot system employs an extra tower that folds back along the rest of the lateral, opening up and sprinkling when the corner section of the field is reached.

FIGURE 7-27. Circular pattern commonly obtained with center-pivot irrigation systems.
Photo courtesy of Valmont Industries, Inc., Valley, Nebraska.

Another method uses an extra sprinkler of large capacity, also activated as the corner is encountered. Other growers may opt to place a solid-set system in each of the corners to provide coverage of the parts missed by the center-pivot system. Uniformity of coverage is achieved by using either different sized sprinkler nozzles spaced evenly along the lateral or by employing nozzles of the same size spaced differently along the lateral. In the latter case nozzles are spaced farther apart near the pivot point and increasingly closer together as they become more distant from the pivot (Figure 7-28). As with other sprinkler systems, the size of the nozzle and of the orifice determine in part the output per sprinkler head. Pumping pressure and length and diameter of the lateral also markedly affect delivery rates from the sprinkler nozzle. Therefore, pressures required to provide necessary outputs for center-pivot systems are high. A range of water pressures at the pivot point from 310 kPa (45 lb/in.^{2}) to 690 kPa (100 lb/in.^{2}) is needed to operate center-pivot systems, depending on type.

Center-pivot systems require larger initial investments, and are expensive to operate because of high water demand and pumping power requirements. The potential return per hectare must therefore be relatively high in order to justify such a sizable investment.

FIGURE 7-26. Section of a center-pivot lateral. Note individual towers and several flex points or joints, which prevent breakage due to rough ground surfaces or other stresses. The end is fitted with a gun type sprinkler to extend coverage.

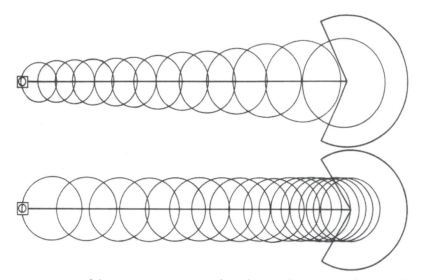

FIGURE 7-28. Schematic representation of evenly spaced, varying sized nozzles (*top*) and same-sized, variably spaced nozzles (*bottom*) along the lateral of a center-pivot irrigation system. Pivot center and water source are represented at left by the tiny circle enclosed by a square.

However, in many cases, the efficiencies of scale achieved through the ability to irrigate a large number of hectares provide handsome economic returns.

Other lateral irrigation systems that can move continuously also are employed for relatively large areas, although some are more suitable than are center-pivot systems for smaller fields. An additional advantage of such systems, when compared with center-pivot, is that they move linearly in a straight line (hence the name, **straight line linear move**). This enables them to provide complete coverage for square and rectangular fields. The drive mechanisms and power sources are often similar to those used for center-pivot systems. However, the directional "tracking," or control, is provided differently, because there is no anchoring effect of the pivot for straight line linear move systems. Historically, a common tracking device for such systems was a simple ditch or groove for wheels or rollers to follow. More recently a number of electronically guided systems, relying on various sensing methods, have been the norm. In greenhouses, a modification of this system, sometimes referred to as a "boom system," is used for overhead irrigation or misting cuttings (see Chapter 14).

Other traveling irrigation systems include **hose-pull reel** types, **cable-tow,** and other slight modifications of these systems. Such systems rely on motor-driven winches, cables, and a flexible, large-diameter hose for the water supply. The actual irrigation device may ride on a cart or sled and often consists of a boom-type sprinkler or a "big-gun" type of high volume sprinkler (Figure 7-29). Numerous other types of

FIGURE 7-29. This big gun, large volume irrigator operates under high pressures, with a large nozzle and orifice. It is frequently used with cable-tow or hose-pull (as illustrated here) movable irrigation systems.

traveling irrigation systems exist, each with its own special features and uses (Figures 7-30 and 7-31). The books listed at the end of the chapter provide more detailed treatment of this subject.

FIGURE 7-30. Irrigation pump installed to take advantage of the water available from this pond. When using such water sources, care must be taken to use screens to prevent debris from fouling the pump.

Problems With High Volume Irrigation

There are several problems associated with high-volume irrigation systems. They include:

1. Large evaporative losses. Furrow and flood irrigation systems expose large surface areas of water, and sprinkler systems place droplets directly into the air; both systems lead to significant losses by evaporation.

2. High cost of labor. Placement of siphon tubes for furrow irrigation and moving pipe for sprinkler systems involve large inputs of labor.

3. Expensive initial establishment. Initial costs of landscaping for furrow or flood system and the components of a center-pivot system are two examples of high establishment costs.

4. Power costs are high. Pumping large volumes of water requires significant inputs of energy, resulting in high fuel or electricity costs to the grower.

5. Water and soil quality deterioration. If the irrigation water is high in soluble salts, salts may build up in the soil and in the runoff water.

6. Costs related to treatment or recycling of runoff water. Reduction or prevention of contamination of surface and ground water is often required by law and is an

FIGURE 7-31. Types of irrigation devices employed for home horticultural uses. Most sprinklers of this sort deliver approximately 0.6 cm (1/4 inch) of water per hour.

essential, environmentally sound, practice necessary to protect water supplies. Built-up salts, pesticides, or other contaminants must be reduced by chemical treatment, filtration, settling, and recycling. This can be an especially serious problem for high density horticultural enterprises such as container nurseries and greenhouses (see Figure 7-32).

Low Volume Irrigation

In recent years a great deal of interest has developed in low-volume types of irrigation, frequently referred to as trickle or drip systems. We will use all three terms interchangeably. One obvious reason for the intense interest in drip irrigation is the increasing world-wide concern for conservation of one of the earth's most precious

FIGURE 7-32. A large nursery in a densely populated area (*top left*) must recycle its water. Runoff from irrigation applied to the plants (*top right*) flows into channels that lead to a collection lagoon. It is then pumped to a holding pond (*middle left*), tested for pH, soluble salts, specific nutrient content and contaminants (*middle right*). It is them filtered and treated to eliminate undesirable organisms (*bottom left*), topped up with fresh water to replace losses to evaporation and plant growth, and pumped to a holding tank (*bottom right*) from which it will be used again for irrigation of the nursery.

resources, water. Researchers have also shown that highly efficient crop production can be achieved when the water is delivered directly to the root zone of the target plant. In some ways, certain low-volume field irrigation methods may be thought of as simulating what occurs with similar "spaghetti" tube irrigation of pot crops in a greenhouse (Figure 7-33). It will be seen later that a volume of soil immediately encompassing the root zone can be watered from such a point source (Figure 7-34), but the rate of penetration and pattern of distribution is influenced by such factors as soil type and by barriers, such as clay pans and hard pans (see Chapter 8).

Just what is trickle irrigation? This low-volume method of irrigation is the most recently developed commercial irrigation approach being employed in horticulture and involves frequent, slow application of

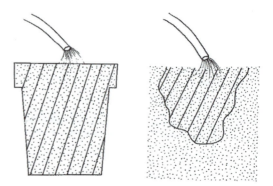

FIGURE 7-34. Watering a soil from a point source, where the medium is contained in a pot and thoroughly wetted (*left*), and the pattern of penetration in a field soil (*right*).

water at specified points along the water lines through emitters or applicators (Figure 7-35). The pressure in the water distribution system is reduced to a trickle by various emitter characteristics, including orifice design and size, vortexes, and long or twisting flow paths. This design makes it possible to discharge a limited volume of water under very low pressure. The movement through the soil of the water so emitted will be primarily by unsaturated flow, and the distribution pattern will usually be elliptical. The characteristics of the soil will influence the extent and pattern of distribution and, together with the target plant's root system characteristics, will dictate the spacing required to adequately wet the root zone.

Greenhouse watering systems developed in the United Kingdom in the late 1940s are considered to be the progenitors of trickle or drip systems of irrigation for field use. The development of economical plastic pipe and tubing enabled such watering methods to be widely accepted in the greenhouse industry. Blass is generally credited with adapting such technology for field use in Israel in the early 1960s. The drip irrigation approach essentially discards the idea of using the soil for storage of water and instead employs frequent low-volume applications to keep up with evapotranspiration losses.

FIGURE 7-33. Low volume irrigation system for greenhouse pot crops. The delivery tubes and spaghetti microtubing are illustrated at *top*. Note the lead weights facilitating insertion of the tube into the pot (*bottom*).

2. Weed growth is reduced because less soil surface is wetted.

3. Limited soil wetting may permit less interruption of field cultural practices.

4. Foliar disease problems will be reduced because foliage is not wetted. This advantage is especially pertinent in greenhouse installations, where foliar diseases such as those caused by *Botrytis* and certain other fungi may be a problem.

5. The uniform water application prevents plant stresses and enhances yield.

6. Fertilizers can be injected (long a common practice for automated greenhouse systems) (Figure 7-36).

FIGURE 7-35. Irrigation timing may be controlled by use of a series of time clocks, one for each greenhouse or location (*top*), or may be programmed by a single device with settings for location, day, hour and duration of application (*bottom*).

Advantages of Drip Irrigation

Low-volume irrigation has several advantages over more conventional, high-volume irrigation systems:

1. Less water is used, reducing costs and conserving water resources.

FIGURE 7-36. Fertilizer injector for greenhouse trickle irrigation system. A tube siphons a measured amount of concentrated fertilizer solution from the stock tank in the foreground and meters it into the supply line for delivery to the crop.

7. Daily or frequent applications of water keep salts in the soil solution more dilute and leached to the outer portion of the wetted soil, enabling use of somewhat saline water sources.

8. Water is placed where it is needed most.

Disadvantages of Drip Irrigation

By contrast, there are only a few disadvantages inherent to low-volume irrigation systems. The most serious concern is the clogging of emitters, microtubing, and orifices. This clogging may be caused by particulate matter in the water, which must therefore be eliminated by use of efficient filtration devices. Microorganisms, which may cause clogging, can generally be controlled by chemical treatment of the water. For these low-volume systems to function effectively, *high quality water is essential*. To determine acceptability of water quality, the services of a water testing laboratory should be employed and appropriate remedial measures taken, if necessary. Damage to the tubes and delivery lines by rodents and other animals such as coyotes and dogs is another problem common to field installations. Another major disadvantage is that high plant densities may cause high costs, thereby making such systems economically infeasible for some crops. Furthermore, in regions where saline soils predominate, the low volume of water applied may lead to buildup of salt levels in the root zone.

Low-volume irrigation systems may be placed on or above the soil surface or may be buried (Figure 7-37). Although the earliest attempts to develop such systems occurred in Germany in 1869 and involved the use of buried clay pipes, surface systems are now more prevalent. Advantages of the subsurface installations include less evaporative loss and reduced damage by equipment and animals. However, subsurface systems are more difficult to observe and repair, less easily moved, and therefore not as versatile as systems placed on the surface.

Components of Low-Volume Systems

Regardless of whether low-volume systems are used in field or greenhouse, several components are common to all drip irrigation systems. As with most modern irrigation methods, high quality (clean) water must be delivered to the orchard, field, bed, or bench to be irrigated, and various pumps, headers, lines, valves, and gauges are employed in this process. In addition, highly selective filters are required for drip irrigation because of the aforementioned water quality problems. Various **line source applicators**, laterals, hoses, or pipes may be the vehicle of delivery, or an assortment of emitters may be part of these lines. The spaghetti microtubing commonly used for greenhouse crops (Figure 7-33) and several other types of emitters are also available. Several types of

FIGURE 7-37. Irrigation with above-ground drip irrigation in grape vineyards. General view *left*; closer views of the line and emitters center and *right*.

line source applicators are in common use, including soaker hoses, double-wall pipes, and porous plastic tubes. One popular "biwall" type of pipe is made of flexible polyethylene and consists of one tube within another. Each tube has holes in it, but they are not aligned, thus pressure is reduced so the water trickles out slowly, which helps avoid damage to soil structure. Another type of line source applicator is a tube composed of a porous paper and plastic material. The water flows or oozes through the walls of the tube by mass flow and capillary action. Either type of tube may be placed along the greenhouse bench or adjacent to the crop row in the field, enabling a gradual and gentle soaking action to take place.

Subirrigation

Subirrigation involves application of water to the soil medium from a source below the root zone. Low-volume irrigation methods are readily adaptable for various methods of subirrigation. Although line source applicators such as rigid types of bi-wall tubes can

be placed underground alongside the crop row or around or adjacent to a tree in an orchard, all types can be used as a means of delivering water to **capillary mats** for greenhouse pot crop production (Figure. 7-38). Capillary watering is not a new concept; historically pots were placed on beds of sand, the sand was kept moist and the pots were watered by capillary movement of the water into the pots. In most modern systems, the sand has been replaced by mats composed of fibrous material, often resembling thin carpeting or a blanket. Capillary mats can also be wetted by placing one end of the mat in the water source; the water will move up the mat by mass flow and capillary action, the water then flows through the mat and up into the pot. Such a system provides a more or less continuous water supply to the potted plants and like other low-volume irrigation methods, does not wet the foliage. As mentioned earlier and in Chapter 16, keeping the foliage dry is an effective method that can help control foliar diseases.

FIGURE 7-38. *Top*: Diagram of capillary mat being wetted by placing one end of the mat in a water supply. Arrows represent water flow from water source through capillary mat. *Bottom*: A line source applicator (ooze-tube type) supplying water to the pots by wetting a capillary mat.

Closed Systems of Irrigation

Most irrigation or watering systems are open systems where water is applied to the soil or medium surface and the drainage water is not reused. In many greenhouse production schemes, however, **closed systems** are rapidly gaining favor because the drainage water is reused or **recycled**, thus reducing environmental contamination. An example of such a closed system is the ebb-and-flood system (formerly referred to as "ebb-and-flow"); essentially it is a closed subirrigation system used for containerized bedding plants and pot plants. The pots or flats are placed in a watertight bench into which nutrient solution is pumped, usually to a depth of about 20 mm ($1/2$ to $3/4$ in.), where it is pulled into the potting medium by capillary action. It usually takes about 10 to 15 minutes to thoroughly wet the potting medium, after which the solution drains into a holding tank where it is held until the next watering. The solution normally contains fertilizer and the solution is routinely tested, adjusted as necessary, and recycled, often for the duration of the crop cycle.

Although the ebb-an-flood system is relatively expensive to install, it offers potential long-term financial advantages:

1. Fertilizer is not lost into surface or ground water.
2. Labor costs are less than many systems because watering is automated.
3. Both water and fertilizer are used in lesser amounts.
4. No special adaptations are required when changing crops or pot size.

One disadvantage of the ebb-and-flood system often cited is the potential for high humidity to accumulate in the crop canopy, but this problem is easily overcome by good management techniques such as encouraging good air circulation through the crop.

A modification of the ebb-and-flood system is the flood-floor system. This is another example of a closed system of irrigation where a watertight floor, often lined in concrete, with a lip around the edge and a central drain, is employed for growing pot crops or bedding plants requiring minimal maintenance. Pumping and recycling of the nutrient solution is implemented in a fashion similar to the ebb-and-flood system.

Although hydroponic systems are somewhat similar to the aforementioned closed systems, most hydroponic systems are not truly closed and are a specialized nutrient delivery system, so they are discussed in Chapter 9, Mineral Nutrition.

Subirrigation can cause problems because water is constantly rising to the soil surface, where it evaporates. In a capillary mat system, the water is taken up through the drainage holes in the bottom of the pot and rises to the surface of the medium along a diffusion gradient. Along with the water, any dissolved or soluble salts will also move to the soil surface; when the water evaporates the salts will accumulate on the soil surface. This salt buildup may dictate that the grower should periodically flush or leach the medium by watering from above. Such leaching should reduce the salt buildup at the soil surface and avoid soluble salts damage ("burn") to the plants.

A tremendous upsurge in implementing drip irrigation is evidenced by the rapid increase in field use from virtually zero in 1970 to approximately 175,000 hectares planted in 1980 in the United States alone. The rate of usage has continued to increase, and rapid adoption of drip or trickle systems has continued at a rapid pace in Israel and numerous other countries. Clearly, low volume irrigation will continue to have a significant impact on horticulture and agriculture world-wide.

TIMING AND AMOUNT OF WATER

It has often been pointed out that crops of the greatest value suffer the most when water is deficient. Although this is not universally true, it points out the fact that horticultural crops (where *quality*, as well as yield, is crucial) demand great attention to their water requirements.

Water deficiency may be evidenced by many symptoms. Slight water stress can cause stomatal guard cells to become less turgid and the stomates to close. Loss of turgidity leads to limited cell expansion, which results in a reduction in growth. Furthermore, closure of the stomates reduces uptake of CO_2, drastically reducing rate of photosynthesis. Production of substrates (food) necessary for growth is diminished, generally resulting in smaller leaves and fruits, shorter internodes, and overall smaller plants. Thus, a slight moisture stress can ultimately lead to slight, or relatively large, yield reductions.

It should be noted however, that in some cases smaller statured plants are desirable and growers of certain pot crops (e.g., chrysanthemums and lilies) may withhold water as one method of preventing plants from becoming too tall or "leggy." (Internode elongation may also be suppressed by the application of plant growth regulating chemicals. This approach is discussed in detail in Chapter 12). Another important application of withholding water is to "harden" seedlings or cuttings prior to transplanting. Using this method in conjunction with lower temperatures and reduced or withheld fertilizer may cause tissues of such plants to become physically harder and increase their ability to withstand cold and drought stress inherent to the transplanting process.

On the other hand, slight excesses of water may cause the plant cells to become unduly stretched, resulting in lush or luxuriant growth. Stems of such plants may become soft, weak, and leggy. They are then prone to physical damage and possibly more susceptible to insect or disease attack.

Severe water deficiencies or prolonged water deficits may cause wilting; drying of leaf tips and margins; yellowing of older leaves; death of leaves; leaf, flower, and fruit abscission; and even death of the plant. It is interesting to note that many of these same deficiency symptoms may result from severe or prolonged water excesses. This is in part a result of an oxygen deficiency inherent in overwatered, or waterlogged, soils. To avoid problems associated with water deficiencies and excesses, it behooves the horticulturist to understand the principles of proper watering of plants. It is also important to note that a large number of the problems associated with growing plants indoors are water related. Therefore, a good knowledge of how to water plants is essential for one to grow healthy, attractive and useful plants.

Soil should generally be watered thoroughly to wet the soil completely in the root zone, and it should be allowed to dry between waterings. Wetting the entire root zone and thus encouraging a deep root system to form is important. If only the surface few centimeters of soil or growing medium are watered, only the upper part of the root zone is wetted. This can be an especially serious problem in container grown plants. Because water in such containers moves vertically as a response to gravity, partial watering results in a layer of thoroughly wetted soil over dry soil, rather than partially wetting the entire soil volume. Therefore, the roots will tend to accumulate in this shallow zone. Since the soil surface dries out first, roots concentrated in this zone will be susceptible to desiccation and death. Certain disease problems may also be accentuated by this problem. Although this problem may be more prevalent in greenhouse crops, it may also occur in field situations.

Watering Container-Grown Plants

Potted plants are best grown in containers with drainage holes located on the bottom. Many decorative containers have no drainage holes, so it is difficult to know how much water is being applied, and overwatering plants in such containers is therefore common. Overwatering results in poor aeration of the medium, which may also lead to disease problems. For such decorative containers, double potting—placing the plant in a pot with drainage holes and putting that pot in the decorative container—is recommended (Figure 7-39). Some pots have holes on the sides of the pot near the bottom instead of on the bottom. Such pots are preferable if the pot is to be placed on a surface that has poor drainage, such as a clay soil or plastic sheets, as in many nurseries specializing in container grown trees and shrubs (Figure 7-40).

The proper way to water container grown plants from above is to irrigate sufficiently so that 10 to 20 percent of the water applied trickles out of the drainage holes in the bottom of the pot. This will ensure that the entire column of medium is wet. However, this rule can be misleading with media that contain a high percentage of sphagnum moss peat, because peats tend to shrink and pull away from the edge of the container as they dry. When watering, the water can run off the surface of the medium and down the sides of the container, never sufficiently wetting the root zone. If this happens, because of the hydrophobic nature of the peat, several successive irrigations may be necessary. Excessive applications of water will leach nutrients, and this too should be avoided. Overwatering generally occurs from watering too frequently, not from applying too much water at one time.

FIGURE 7-39. Double potting of ornamentals in China. The interior pot has drainage holes, thus circumventing the problem of the lack of drainage holes in the outer decorative pot.

FIGURE 7-40. Container grown plants placed on plastic sheeting, which helps keep contamination to a minimum. Note that drainage holes are on the lower sides of the pots.

Watering Field Grown Plants

Horticultural species vary greatly in their potential rooting depth and spread. Although deciduous fruit crops and alfalfa are irrigated to a depth of up to two meters, some vegetable crops may only require irrigation to wet a root zone of only one-half meter (1.5 feet) or less. Generally, roots of tree species are assumed to spread beyond the dripline. An understanding of root penetration and spread for a given species and of the soil characteristics is necessary to properly plan irrigation schemes requisite for the attainment of optimum yield and quality. A common rule of thumb is to apply 2.5 cm (one inch) of water at one time per 5 to 7 day period for sandy loam soils. This approach is only a guide, however, and numerous other factors should be taken into account, including crop species, soil type, evapotranspiration rate and ambient rainfall.

SOIL MOISTURE MEASUREMENT

Accurate measurement of soil moisture status, although sometimes difficult to achieve precisely and economically, is a desirable parameter to consider in determining irrigation timing and rate. Several kinds of soil moisture meters and measurement devices are available, with the following types in most common use.

Gravimetric Method

This is considered to be the most accurate moisture measurement method because with it moist soil samples are weighed, then oven-dried and weighed again. The weight difference is expressed as the percent of dry soil weight. This information can then be compared with the available water (i.e., the difference between the field capacity and permanent wilting point of the soil in question), and decisions regarding water application can then be made. However, this method is labor-intensive and time-consuming, so it is not considered feasible for most practical applications.

Gypsum Moisture Blocks

Blocks of gypsum and a porous material are placed in a tube in the soil and when equilibrium with the soil moisture is reached, the block is removed and weighed. This method, although not as tedious as the gravimetric method, is not used a great deal because of its relative inconvenience and high degree of variability.

Tensiometers

Tensiometers measure tension or condition of the water in the soil. They usually consist of a porous cup or bulb filled with water which is attached to a vacuum gauge or manometer by a connecting tube (Figure 7-41). When placed in the soil at a critical root depth, the relationship of soil moisture to the moisture in the porous cup is translated to a needle in the gauge. The gauge

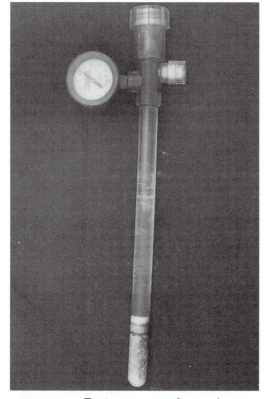 

FIGURE 7-41. Tensiometers are often used to measure soil moisture. One commonly used type, *left*, and being used in greenhouse medium, *right*.

reading is calibrated to give an approximation of the available water. Tensiometers or modifications are among the most commonly used moisture measurement devices for estimating moisture needs of a crop.

Electrical Conductivity Methods

Electrical moisture measurement methods usually depend on electrical resistivity or conductivity. Measurement may be accomplished by one or more types of conductivity/resistivity electrode probes that are connected to a meter. The major disadvantage of such methods, although they are rapid, is the fact that readings are dramatically altered by levels of salts in the soil solution. **Bouyoucos blocks**, developed by G. J. Bouyoucos in 1940, partially solved this problem by combining some of the best characteristics of the gypsum block with electrical resistance methods. Two electrodes on a stainless steel 20-mesh screen are imbedded in a block of plaster of paris, and the block, which can be attached to an electrical source and gauge, is buried in the soil to be tested. The amount of gypsum in solution is reflective of soil moisture and can be read on the gauge. Bouyoucos blocks, or modifications, have been used for years in soil moisture and irrigation research. They can also be modified to help program automatic irrigation systems.

Neutron Probes

These methods depend on equipment that is capable of measuring neutron scattering, a phenomenon that received a great deal of interest when radioactive materials became available. In essence, this method reflects the ability of hydrogen nuclei to reduce the energy of "fast" neutrons. Therefore the amount of hydrogen atoms in the soil will be indicated by the number of slowed neutrons measured by the probe. Problems with inaccuracies and high costs have prevented wide use of neutron probes for measuring soil moisture.

Evapotranspiration and Solar Radiation Measurement

Soil moisture has often been described as a bank account; precipitation and irrigation add to the account and evapotranspiration makes withdrawals. This relationship ultimately determines growth and yield, so it is important to consider evapotranspiration measurement as an aid in irrigation scheduling. **Lysimeters** are large soil tanks in which crops are grown in a representative field situation. Weighing-type lysimeters are considered to provide the most accurate evapotranspiration data. Such direct field measurements are expensive and difficult to maintain, so lysimeters are used primarily by research stations to provide a guide for estimating evapotranspiration from climatic data. **Net radiation** is calculated from the solar radiation measured by various devices and, with other climatic data such as wind velocity and relative humidity, is taken into account in estimating evapotranspiration. Evapotranspiration information and climatological data are valuable aids in estimating appropriate crop irrigation timing and amounts.

Crop Requirements for Water

Total crop needs for water vary considerably among species. In arid regions, some fruit crops may require 5 to 10 times the amount of water needed by a low-moisture-requiring crop such as sorghum. Short-season vegetable crops, although requiring a large amount of water per gram of dry matter produced, do not have a high requirement for a given crop season because they are in the ground for such a short time (e.g., radishes for 21–25 days).

In general, water requirements for horticultural crops are high. A steady or nearly constant amount of available water is essential for many vegetable crops, especially where the economic part is vegetative, e.g., spinach, cabbage, and lettuce. In other crops, moderate fluctuations in available water are not serious during parts of their growth period, but they may have critical periods (which may be relatively short) where a water stress will result in crop failure or extreme reduction in quality of the product. Classic examples of this are seen in crops such as sweet corn, snap beans, and tomatoes. For corn the most crucial time is during pollination and the period immediately following pollination. If water is limiting during pollination, ears will be small and kernel number per ear will be greatly reduced, thus diminishing both quality and overall yield. In snap beans, water deficits followed by irrigation or excess water during pod development stages will result in a tearing apart of the newly formed internal cells, which causes cavities or air spaces between the ovary walls (**interlocular cavitation**, Figure 7-42).

Similarly, **hollow heart** of potatoes may occur as a result of uneven or very high moisture levels during tuber development. Carpel separation in cucumbers is another disorder that may occur in this fashion. Cracking of tomatoes, growth cracks, and second-growth of potatoes, and splitting of cabbage heads also are phenomena related to water availability. In each case, if water availability is low for the period approaching maturity, organ growth stops. Then, if water availability abruptly increases, a renewed spurt of growth occurs

causing rupture of the skin of the tomato fruit, for example. In the case of potato, either a growth crack or second growth ("knobby" potatoes) of the tubers, or both, may result (Figure 7-43). A somewhat related problem is the physiological disorder in tomatoes known as **blossom-end rot** characterized by cell breakdown at the junction of ovary and style (Figure 7-44). Blossom-end rot is thought to be related to

FIGURE 7-43. Disorders related to inconsistent water supply: radial cracking in tomato, *top*; growth crack in a carrot root, *middle*; and second growth in potato, *bottom*.

FIGURE 7-42. Cell tearing (*interlocular cavitation*) in snap bean pods caused by improper water amounts during early pod development.

FIGURE 7-44. Early stage of blossom-end rot of tomatoes. Note water soaked area at blossom end of the fruit.

both water availability and its involvement with nutrient availability, notably calcium. However, it has been shown for lemons, peaches, and tomatoes that the leaves draw water from the fruit during periods of water stress. This, therefore, may in part contribute to blossom-end rot incidence in tomatoes and other fruits.

Soluble Salt Problems

World-wide, one of the most severe problems in crop production is caused by excess salts in the soil (field). A similar problem is frequently encountered in media for container-grown nursery and greenhouse crops. Soluble salts problems may be the result of an inherently high level of salts in the soil, use of saline irrigation water, application of excessive amounts of fertilizer, or all three causes. If the water application rate is less than optimum and high levels of fertility exist, the soluble salt problem is accentuated and symptoms soon become apparent. Excessive fertilizer levels may result from a variety of causes, including mistakes in calculations of fertilizer requirements, incorrect calibration or malfunction of equipment used in applying fertilizers (e.g., spreaders in the field, injectors in greenhouses), or simply fertilizing too frequently

or at times when plants are unable to adequately utilize the nutrients. Soluble salts levels are measured by an instrument called a **solubridge**. This instrument measures electrical conductivity in millimhos per centimeter (the opposite of ohms of electrical resistance). Optimum levels vary with the crop and dilution method employed by the soil test laboratory. Each laboratory will normally provide a chart that enables interpretation of its method.

Symptoms of Soluble Salts Injury

Whole plant symptoms include cupping of leaves; wilting; drying of the edges of the leaves; leaf, flower, and fruit abscission; and in extreme cases, death of the plant. It is readily apparent that these symptoms are nearly identical to those described earlier for insufficient water. Indeed they should seem similar, since the most common result of high soluble salts in the soil solution is to cause the plant to be unable to absorb water.

In fact, in most cases of soluble salt problems, the root cells lose water to the surrounding soil solution. This water loss occurs because the water is in a higher concentration (lower solute concentration) inside the cell than outside the cell; water is not taken up by the roots, but in fact is lost to the surrounding soil solution by osmosis. The plasmalemma then pulls away from the cell wall resulting in a condition known as **plasmolysis**. The plasmodesmata may break, causing a loss of connection from cell to cell. This means that water cannot be translocated readily from cell to cell and cannot cross the endodermis with its Casparian strip. Thus, water is not taken into the symplast. A compounding problem is that as the water is drawn out of the cells, it will cause the salts that are already present in the cells to become more concentrated. This can cause death of the cells, resulting in what is often called fertilizer burn, such as might be observed when fertilizer or salt is spilled on a lawn. The practical results of high soluble salts in the soil solution are the symptoms of a drought-stricken plant; the plant may show wilting, browning of leaf margins, organ abscission, and other damage even though its soil feels wet to the touch.

The problem of soluble salts injury may be accentuated by growing plants in a limited soil volume, so it is often observed in container-grown plants. Potted plants watered from below frequently are subject to a concentration of soluble salts at or near the soil surface. A related problem frequently observed is the deposition of salts on the rim of a pot, leading to direct phyto-toxic effects or burn of leaves or petioles. Houseplants watered from underneath and greenhouse crops grown on capillary mats are particularly prone to this malady unless they are periodically **leached**. Leaching simply involves applying copious quantities of water from above so that the water moving through the soil profile will dissolve and carry away the excess salts. Low-volume irrigation that targets a plant's root zone is another means of dealing with potentially high soluble salts under field conditions. Again, periodic leaching may be required to prevent salt buildup.

Irrigation water that is high in salts can be used with low-volume systems. However, if measures are not taken to reduce salt buildup in soils repeatedly irrigated with saline water, yields will be reduced and the land may eventually have to be removed from production. Indeed, a lesson from history suggests the serious nature of this problem. Flood irrigation was common in Roman times, and studies suggest that a resulting salt accumulation caused low food yields and may have been a major factor in bringing the Roman Empire to its knees!

Another serious soluble salt problem is seen in climates where salt is used for de-icing roads. Grasses, wildflowers, trees, and shrubs all may suffer from the salt spray splashed onto plant surfaces from passing vehicles and from the accumulation of the salts in the soil. Limiting the use of salt for this purpose, selecting a less toxic salt, or planting more salt-tolerant species in vulnerable areas may help with this problem. Indeed, some horticulturists are actively working on selecting and breeding more salt tolerant plants for locations or planting sites adjacent to highways.

Xeriscaping

It is important to plan landscapes to efficiently address the water requirements of the plants composing the landscape design. Many parts of the world are arid or semi-arid and cannot sustain classic landscape designs employing high water-use plants. The concept of xeriscaping addresses this concern. Credit for creating the term xeriscape is generally attributed to the Denver, Colorado Water Department, which first used the concept in 1981 as a response to prevailing drought conditions. Xeriscaping is based on attention to seven principles:

1. Designing for efficient water use.
2. Selection of drought tolerant plants or plants that exhibit low water requirements.
3. Irrigation equipment and design for efficient water use.
4. Relatively small lawn areas.
5. Use of water conservation measures (see also Chapter ten, Mulches).
6. Employment of water harvesting techniques.
7. Use of appropriate maintenance practices.

Zones

A basic concept of xeriscaping is the use of **zoning**.

Zone I

A mini zone near the house is referred to as the **oasis** zone, where components such as potted plants, flowerbeds, vegetable gardens, and small lawn or water feature may be used. Drip irrigation is preferred in this zone.

Zone 2

Zone 2 is the **transitional** zone. It includes plants that use very little water, but need occasional deep watering during drought periods. Berms and swales are often included in this zone to facilitate the catching and directing of rain water for subsequent irrigation (a type of water harvesting).

Zone 3

Zone 3 is the arid zone; this is the area of the xeriscape design that employs plants needing little or no irrigation. The goal for this zone is to use plants that will survive on rainfall alone; often these are plants native to the area. Again, berms and swales are used to enhance harvesting of rainwater for irrigation.

Xeriscaping does not mean elimination of lawns, nor is it exclusively use of xerophytic plants and rocks, gravel, or sand. Instead, it is a highly specialized approach to landscape design. Its goal is to conserve and efficiently use available water while creating a well-planned and beautiful landscape that is appropriate for the region.

FIGURE 7-45. Water drop striking the soil. The impact of the water drop on bare soil caused soil particles to scatter, thus beginning the processes of soil compaction and erosion.
(USDA photo).

SUMMARY

Water is the single most limiting factor for production of horticultural crops in many areas of the world. It is required for virtually all important plant processes—from photosynthesis and translocation to cell enlargement and growth—and is crucial to crop yield and quality. Therefore it behooves the prudent horticulturist to become thoroughly grounded in the properties of water and its functions in growing high-quality profitable horticultural crops. One might reflect thoughtfully on Stewart's comment from more than 100 years ago, "five or ten acres well cultivated and supplemented with abundant water will yield as much profit as 50 or 100."

REFERENCES

American Society of Agricultural Engineers. 1981. *Irrigation Scheduling for Water and Energy Conservation in the 80's.* ASAE Publication 23–81, St. Joseph, Michigan.

Borghetti, M., J. Grace, and A. Raschi. 1993. *Water Transport in Plants Under Climatic Stress.* Cambridge University Press.

Cantor, Leonard M. 1967. *A World Geography of Irrigation.* Oliver and Boyd, London. 252 pp.

Doorenbos, J., and W. O. Pruitt. 1977. *Guidelines for Predicting Crop Water Requirements.* FAO Irrigation and Drainage Paper 24. Food & Agriculture Organization of the United Nations, Rome. 144 pp.

Elfving, D. C. 1982. "Crop response to trickle irrigation." In: *Horticultural Reviews* 4:1–48.

Hansen, V. E., O. W. Israelson, and G. E. Stringham. 1980. *Irrigation Principles and Practices,* 4th ed., John Wiley & Sons, New York. 417 pp.

Hillel, D. (Ed.) 1982. *Advances in Irrigation,* vol. 1. Academic Press, New York.

Hillel, D. 1997. *Small-scale Irrigation for Arid Zones: Principles and Options.* FAO Development Series 2, Food and Agricultural Organization of the United Nations, Rome.

Jensen, M. E. (Ed.). 1980. *Design and Operation of Farm Irrigation Systems.* American Society of Agricultural Engineers Monograph No. 3, St. Joseph, Michigan.

Kozlowski, T. T. and S. G. Pallardy. 1997. *Physiology of Woody Plants* (2nd ed.). Academic Press, San Diego.

Kramer, P. J. 1983. *Water Relations in Plants.* Academic Press, New York.

Meidner, H. and D. W. Sheriff. 1976. *Water and Plants.* Halstead Press, John Wiley & Sons, New York.

Nelson, Paul V. 2003. *Greenhouse Operation and Management* (6th ed.). Prentice Hall, Upper Saddle River, N.J.

Northington, D. K. and E. L. Schneider. 1996. *The Botanical World* (2nd ed.). Wm. C. Brown, Pub., A Times Mirror Co., Dubuque, Iowa.

Pair, C. H., W. H. Hinz, K. R. Frost, R. E. Sneed, and T. J. Schiltz. 1983. *Irrigation* (5th ed.). The Irrigation Association, Silver Spring, Maryland.

Shalhevet, J., A. Mantell, H. Bielorai, and D. Shimshi, (Eds.). 1979. *Irrigation of Field and Orchard Crops Under Semi-Arid Conditions* (rev. ed.) Int. Irrigation Information Center, Bet Dagan, Israel.

Stefferud, A. (Ed.). 1955. *Water. The Yearbook of Agriculture.* USDA, Washington, D. C.

Stewart, H. 1889. *Irrigation for the Farm, Garden and Orchard.* Orange Judd, New York.

U.S. Dept. of Agriculture. 1977. *Gardening for Food and Fun. The Yearbook of Agriculture.* USDA, Washington, D.C. 392 pp.

Wilcox, L. 1895. *Irrigation Farming.* Orange Judd, New York.

Yaron, D. (Ed.). 1981. *Salinity in Irrigation and Water Resources.* Marcel Dekker, New York.

CHAPTER 8

SOILS AND SOIL MANAGEMENT

What is soil? Although there seem to be almost as many definitions of soil as sources giving the definition, it is clear that soil as we know it did indeed have its origin in rock. Weathering of the rock has produced tiny rock fragments which, when combined with remains of plants and animals in various stages of decomposition, constitute the solid phase of the soil. Soils also contain pores and spaces that contain air and water; the relative proportions of solids, water and gases often govern the suitability of a particular soil for plant growth. Soils experts distinguish soil from the underlying material (rock) by the presence of several criteria: (1) soils contain a relatively high organic matter content; (2) soil contains organisms and roots of higher plants; (3) soils have been more intensely weathered; and (4) soils have characteristic horizontal layers. If a section is cut downward through the soil, several horizontal layers can be distinguished. Viewed from the side, such a section is a soil *profile*, and the layers are referred to as *horizons* (Figure 8-1).

The rock from which a soil was derived is called the **parent material**. The type of parent material,—such as limestone, granite, sedimentary, or volcanic rock—has a profound effect on the characteristics of the soil it produces. Climatic conditions of the area, especially rainfall, also

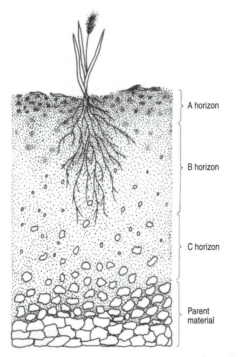

FIGURE 8-1. A schematic representation of a soil profile. Note that the surface soil (A Horizon) has the most organic matter and organisms.

A horizon

B horizon

C horizon

Parent material

CHARACTERISTICS OF SOIL

The living and dead organisms present in a soil have a great influence on soil characteristics. Although macroscopic animals such as rodents, insects, and earthworms are readily apparent, a fertile soil contains a much greater number and weight of microscopic organisms. In fact, a productive soil is teeming with living organisms, including bacteria, fungi, nematodes, actinomycetes, and protozoa. One source estimates that a teaspoon of soil contains 5 million bacteria, 20 million actinomycetes, 1 million protozoa, and 200,000 algae and fungi.

The soil serves several functions for plants. It provides mechanical support for plants and anchorage for the roots. It is the source of all chemical elements necessary for plant growth except carbon. The soil also serves as a reservoir of water and oxygen to the roots. If a soil is well-drained, good air exchange will be possible, thus preventing the buildup of toxic levels of gases, such as the CO_2 that is produced by the respiration of living organisms and decomposition of organic matter.

Characteristics of Mineral Soils

A majority of agriculturally important soils are termed **mineral** soils. Mineral soils are soils that consist primarily of mineral matter and contain less than 20 percent organic matter. Such soils may have a surface layer up to 30 cm thick, which may contain varying amounts of organic matter, often less than 5 percent.

Textural Characteristics

The **soil class** refers to the textural name of a soil. A soil **texture** is determined by the relative proportions of **sand, silt,** and **clay** in a soil. The sand, silt, and clay particles are actually small rock fragments and in a soil test are separated mechanically based

dramatically influence soil properties. Soils in regions of high rainfall usually lose nutrients through the leaching action of the rain and they frequently are more acidic, characterized by a low pH value (see later discussion of pH). Conversely, soils developed in regions of low rainfall are typically high in salt content, are less acid and have a high pH.

Although soil is not absolutely *necessary* for plants to grow, it is abundantly clear that fertile, productive soils represent a tremendous resource. Indeed, as was true for great historical civilizations (such as the Nile Valley and Mesopotamia), today's most powerful and prosperous nations derive much of their strength from an agriculture based on an abundant and fertile soil resource. Soils are highly variable from one location to another, and they are dynamic because they constantly change in response to their environment.

on their size and shape. They therefore are referred to as **soil separates**. Their characteristics are as follows:

Sand mineral particles ranging in size from 0.05 mm to 2.0 mm in diameter. (Particles larger than 2.0 mm are designated as gravel.)

Silt mineral particles between 0.05 mm and 0.002 mm in size.

Clay These are the tiniest rock fragments, being less than 0.002 mm in diameter, and flat in shape. There are two basic types of clays, the **kaolinite** or 1:1 type and the **montmorillonite** or 2:1 type. The 2:1 type clays include several subtypes including illite, vermiculite, and montmorrillonite. The 1:1 clays consist of one layer of silica and one layer of aluminum, while 2:1 clays

are made up of two silica layers and one aluminum layer. These clays are the major colloids of a mineral soil and contribute greatly to the **cation exchange capacity** (CEC) of such soils. The cation exchange capacity is the ability of a soil colloid to hold cations and is important to the fertility of a soil. Cation exchange capacity is therefore discussed further in Chapter 9.

Soil textural names, or soil classes, may be determined by mechanically analyzing the percentages of sand, silt, and clay particles and then applying these percentages to the chart shown in Figure 8-2. For example, a soil containing 40 percent silt, 35 percent clay, and 25 percent sand is a clay loam, but a soil containing 40 percent silt, 25 percent

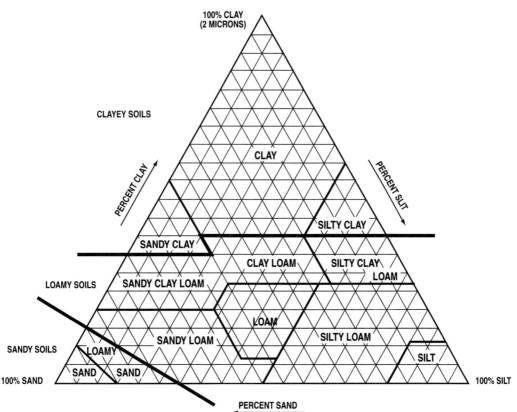

GUIDE FOR TEXTURAL CLASSIFICATION

FIGURE 8-2. Proportions of sand, silt, and clay for different textural classes. The compartment containing the point of intersection of specific percentages of sand, silt, and clay designates the appropriate soil class.

clay, and 35 percent sand is classified as a loam. Characteristics of some horticulturally important mineral soil classes are as follows:

Sand

Soils classified as sands are often subject to drought. They tend to be excessively drained and usually require frequent overhead irrigation to be productive (see Chapter 7). Soluble nutrients leach readily through the profile because sands are very porous and have proportions of silt and clay that are too low to hold water well. The are also usually low in organic matter, thus further reducing their ability to hold water.

Sandy loam

A sandy loam is a soil that is predominantly composed of sand particles but which contains enough silt and clay to hold water and nutrients better than sands. Such soils warm up quickly in the spring, are easily sprinkler irrigated, and are therefore often favored by producers of vegetables and fruits—such as muskmelons and strawberries intended for early market. Some tree fruits must also be grown on sandy soils in order to sustain good growth.

Loam

The combination of sand, silt, and clay in a loam is such that the characteristics of one do not predominate over the others. The sand component imparts good drainage and aeration characteristics, and the silt and clay give a loam its characteristic "mellow" feel—slightly plastic, yet somewhat gritty— and good water and nutrient-holding abilities. Loam soils are often highly productive and are sought by growers of many horticultural commodities.

Silt Loam

Silt loams contain over 50% silt particles, usually fine sands and low amounts of clay

particles. When pulverized, such soils seem smooth and feel almost like flour. Although silt loam soils may tend to hold large amounts of water, they are frequently good soils for apple production and may be profitably furrow irrigated.

Clay

A clay soil is one which is very high in this smallest mineral fraction. Typically it is poorly aerated and poorly drained. Clay soils are easily compacted, and they warm up and dry out very slowly in the spring, thus making them less desirable soils for most horticultural purposes.

Soils with a high percent of clay particles historically have been referred to as "heavy" and sandy soils as "light." This terminology has nothing to do with the actual weight of these soils; rather the soils were so named because of the energy required to plow them. *Fine textured* and *coarse textured* are therefore more appropriate terms.

Bulk density of a soil is the mass (weight) of a given *volume* of dry soil, which takes into account both solids and pores. It is a useful measurement for evaluating a soil for possible inclusion in container soil mixes.

Understanding soil texture is key to understanding water movement in the soil. Water is attracted to charged particles (adhesion) and to itself (cohesion) and thus moves by mass flow and capillary action through the soil (see Chapter 7). In other words, water moves as a film through the soil—coating the sand, silt, clay, and organic matter particles—being pulled by gravity, and pulling itself along. The maximum amount of water that a soil can hold against the pull of gravity is said to be the **field capacity** of that soil. Abrupt changes in the soil texture, or the existence of large pores will break the continuous film of water, and stop downward drainage at this location. Water will continue to accumulate above this change in texture and will fill all of the pore spaces, resulting in a **perched water table**.

A perched water table will exist in the soil column above the interface of the two textures. Perched water tables are not conducive to root growth, because of the lack of oxygen for root metabolism. The longer the column of soil of a similar texture, the better the drainage and the less the problem with perched water tables. A similar phenomenon exists in pots. With shallow pots the perched water table is a much more serious problem than it is with longer or deeper pots.

Perched water tables can be problematic when soil is placed on top of gravel. Because of the macropores in gravel, when a bucket of water is poured directly onto gravel, the water goes through the gravel quite rapidly. A different condition exists when soil is placed on top of gravel. When the film of water moves through the soil and encounters the gravel with its macropores, a break in the film will cause a perched water table. Therefore putting gravel under soil is not recommended because unless the soil column is long, the perched water table can cause serious root problems and in extreme cases, plant death. A similar problem occurs when an organic soil, compost, or topsoil is added preparatory to installation of turf or landscape plantings. Such materials must be blended with the existing soil by rototilling or similar methods.

This logic also applies to transplanting a plant into the soil. Many people will dig a hole for a tree or shrub and set in the root ball and fill in with a potting soil, moss peat, or similar "superior" medium. Because there is a change in soil texture at the amendment/soil interface, watering or rainfall may cause a perched water table. In fact the hole will serve as a basin and will hold water, blocking sufficient oxygen from reaching the roots. In addition, the roots of a plant will not cross a soil texture interface readily. Thus they will remain in the original hole, rather than growing out into the surrounding soil. Poor anchorage, poor growth, and sometimes plant death will result. In transplanting plants, then, it is recommended that the hole should be back-filled with the native soil. If the native soil is poor, it can be improved greatly by incorporating organic matter throughout, not just in the planting hole.

Soil Structure

Soil structure is described as the arrangement of primary soil particles (sand, silt, clay) into secondary units called **peds**, that often function in the same fashion as the primary soil particle. This grouping together is known as **aggregation**, and results in clusters such as crumbs, granules, blocks, or prisms. The aggregation of clay particles into crumb or granular structures generally improves porosity, bringing about improved drainage and aeration and making a heavy soil more horticulturally desirable. Formation of granules is influenced in nature by several factors. These include freezing and thawing, wetting and drying, physical activities of soil animals and plant roots, decay of organic matter, and microorganisms and adsorbed cations. Tillage activities can further modify soil aggregation; this can be especially destructive of the fragile granular and crumb structure of a soil if tillage activities take place when the soil is too wet (Figure 8-3).

FIGURE 8-3. Handfuls of two soil classes that have been squeezed, showing relative differences in ability of the ball to remain intact. The "squeeze test" is helpful in determining a soil's readiness for cultivation. The soil ball should break apart when dropped if the soil is not too wet (ball at right).

Improvement of Soil Structure

Many soils benefit from tailoring cultural practices to maximize soil particle aggregation. This practice improves aeration and drainage in heavy soils and it often increases water-holding ability in sandy soils. Organic matter derivatives bind soil particles together into granular aggregates and tend to lighten and expand them also. This organic matter "glue" is sometimes referred to as **humus** or **humic** material. Humus binds soil particles together, and it and the tiny colloidal clay particles impart electrochemical properties to the aggregates. These electrochemical properties help to stabilize the aggregates and also increase the ability of these particles to hold nutrients and water. It therefore behooves the practical horticulturist to manage field soils in a manner designed to maintain a high level of organic matter in the soil. It should be noted that it is extremely difficult to *increase* organic matter content of a field soil under continuous production, primarily because a well-aerated soil will provide conditions conducive to oxidation (breakdown) of the organic fraction of the soil. However, it is possible to *maintain* a soil's organic matter content at or near its optimum levels by cultural practices that add organic matter back into the soil, such as adding manure or using green manure or cover crops. Nevertheless, it is a common practice to amend a landscape planting area with organic material such as compost and tilling it into the existing soil prior to planting. (For more detailed discussion of composts, see Chapter 9.)

Organic Soils

Although they constitute a small percentage of the world's important agricultural soils, organic soils can be extremely productive and therefore are worthy of attention. An organic soil is one which contains more than 20 percent organic matter. If such a soil has been highly decomposed, it is generally termed a **muck** soil (Figure 8-4). If

FIGURE 8-4. A crop of onions growing on a muck soil. Because muck soils are made up of substantial amounts of highly decomposed organic matter, they are ideal soils for high-value crops that require large amounts of nitrogen, although muck soils may be very low in potassium.

the organic particles are mostly not decomposed or only slightly decomposed, the soil is said to be a **peat**. Peat soils often contain 50 to 100 percent organic matter and usually have developed under conditions of excess moisture; frequently such soils have lain under water for all or large portions of the year for hundreds of years. Certain peat soils, such as those derived from mosses in the genus *Sphagnum*, are much desired for a number of horticultural purposes, especially as components of soil mixes for container grown plants (Figure 8-5).

FIGURE 8-5. Close-up view of a piece of sphagnum moss peat. Note that the fibers are still relatively intact and distinguishable.

Peat soils so derived are properly referred to as **sphagnum peats** or **sphagnum moss peats**, although in the trade, they are sold as "peat moss." Peats are also derived from other parent materials such as reeds, sedges, woody plants, or combinations, and are named for the parent plant material, (e.g., *sedge* peat, *woody-sedge* peat). Reed-sedge peats and woody peats, those derived from forest tree species, are often used as mulches, but hold much less water than do sphagnum peats. For this reason, sphagnum peats are preferred as components of potting soils. Peat deposits are abundant in colder climates, with the greatest number of hectares being found in Russia, Sweden, Canada, and Alaska. In addition to their horticultural uses, peat deposits have been used extensively for fuel for homes and for industrial purposes such as fuel for power plants. Historically, peat was harvested by cutting blocks from the peat bog and letting them dry in the sun, but now modern peat harvesting machinery is used (Figure 8-6).

Soil Reaction (pH)

An important characteristic of soils, or more specifically of the soil solution, is its **reaction**. The reaction of the soil solution refers to whether it is **acid, neutral,** or **alkaline.** Pure water has a neutral reaction

FIGURE 8-6. Commercial peat harvester in a peat bog in Ireland.

(pH). In pure water a few of the water molecules dissociate into cations (hydrogen ions, H^+) and anions (hydroxyl ions, OH^-). In acid soil solutions, H^+ ions predominate over OH^- ions but in alkaline soils, the OH^- ions are more numerous than the H^+ ions. If the H^+ ions and OH^- ions are present in equal concentrations, the soil reaction is considered to be neutral.

This relationship is measured in terms of the H^+ ion concentration and is referred to as the pH of the soil solution. Specifically, the pH value of a solution is the logarithm of the reciprocal of the H^+ ion concentration (pH = $\log^{-1}$; H^+). Sometimes this equation is stated as the negative logarithm of the H^+ ion concentration. A pH of 7.0, the point at which the H+ ions and OH^- ions are essentially equal, is said to be neutral. At a pH of 6.0, the H^+ ion concentration is ten times greater than the H^+ ion concentration at pH 7.0, while the OH^- ion level is ten times less than the level at pH 7.0. Therefore at pH 6.0, there are 100 times more H^+ ions than OH^- ions and the soil solution is considered acid. The reverse is true at a pH of 8.0. There are 100 times as many OH^- ions as H^+ ions and the soil solution is said to be alkaline. The range of pH values generally goes from a low of 1.0 (acid), to a high of 14.0, (alkaline). Most mineral soil solutions fall into a range from 3.5 to 10, with 5 to 7 the most common in soils of humid regions and 7 to 9 in arid zone soils. A few peat soils may fall below a pH of 3.0, and some alkaline soils may reach a pH of 11 or more.

Soil pH is not constant; rather it is a dynamic, changing phenomenon. As discussed in Chapter 9, addition of fertilizers may either acidify a soil or increase its pH. Weathering and leaching in regions of high rainfall tend to naturally acidify the soil. Similarly, frequent and heavy irrigation will leach cations associated with alkalinity (e.g., calcium and magnesium), leading to a predominance of hydrogen ions and thus an acidifying effect. If precipitation or irrigation is less than evaporation, the soils of the region usually have a high, or alkaline, pH because the basic cations are not leached and replaced by hydrogen ions.

Horticultural Importance of pH

The soil reaction influences higher plant growth primarily in three ways: by an *indirect* effect on the availability of nutrients and toxic ions, by an *indirect* effect on the activity of soil microorganisms and by a *direct* effect of the hydrogen ion. Direct damage attributable to the hydrogen or hydroxyl (OH⁻) ion concentration is rare and occurs only at extremely low or high pH values. In general, most higher plants can tolerate a wide range of hydrogen ion concentrations. For this reason, the indirect effects hold much greater horticultural significance.

Nutrient availability is greatly influenced by pH. Some nutrients are sensitive to fluctuations of pH, but others remain available over a broad range of pH values (Figure 8-7). For example, several of the metallic **micronutrients**[1] such as iron, zinc, manganese and copper are very soluble at low pH and become less available as the pH approaches 7.0. Likewise, the toxic element, aluminum, is more soluble at low pH and thus becomes phytotoxic to many species in highly acid soils. Conversely, molybdenum (a micronutrient) and several **macronutrients**—notably nitrogen, potassium, calcium, sulfur and magnesium—become more readily available as pH increases from very acid levels to slightly acid and higher pH values. Phosphorus falls into neither of these categories, and its availability often dictates the soil pH of

6.2 to 6.8 requisite for successful production of many horticultural crops. This is a satisfactory range because all other essential nutrients are relatively available between these pH values. As pH drops below 6.0, phosphorus begins to be precipitated by the soluble metallic ions, especially aluminum, iron and manganese. Phosphorus thus becomes less and less available in very acid soil solutions; it is also complexed by calcium at slightly alkaline pH levels. Boron follows a somewhat similar pattern to that of phosphorus, becoming less available in very acid and moderately alkaline soils.

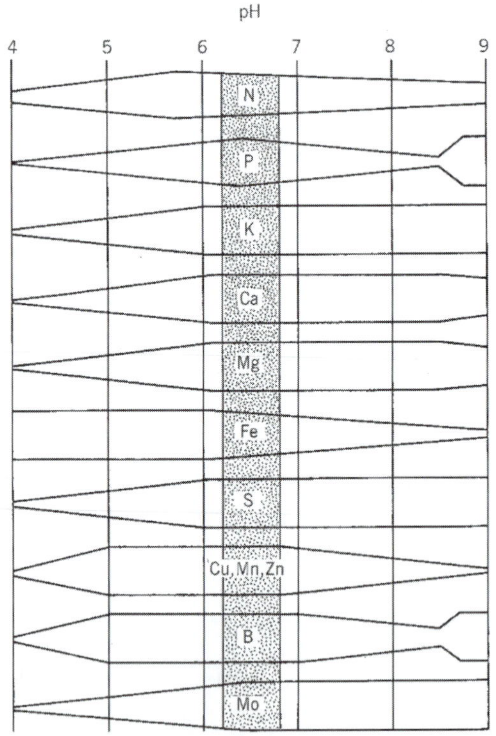

FIGURE 8-7. Diagrammatic representation of the relative solubility (availability) of nutrients at different pH values of the soil solution. Note that most important horticultural crops are best grown in a range from 6.2 to 6.8. A few species grow better at a distinctly acid pH, such as several members of the Ericaceae (e.g., blueberries, cranberries, azaleas, and rhododendrons), which do best at pH 4.0 to 5.5. Potatoes are grown at an acid pH to escape the potato scab organism.

....
[1] *Micronutrients* refers to essential nutrient elements required in very small quantities for normal plant growth. *Macronutrients* are elements required in relatively large quantities. These terms are preferred over *major nutrients* and *minor nutrients* because all required elements are essential to plant growth, and none of their roles should be considered minor.

Acid Loving Plants

Many members of the heath family (Ericaceae) evolved in peat bogs or similar acidic environments. In the process of their evolution, they developed adaptations which enabled them to tolerate high levels of aluminum, iron, and manganese and to take up phosphorus that is normally unavailable under such acid conditions. One such adaptation is their symbiotic relationship with mycorrhizal fungi that aid the roots in absorbing nutrients such as phosphorus. Examples of horticulturally important plants that fall into this category include blueberries, cranberries, and lingonberries (*Vaccinium* spp.); azaleas and rhododendrons (*Rhododendron* spp.); and heaths and heathers (*Erica* and *Calluna* spp.).

Disease/pH Interaction

A few commercially important horticultural crops are grown at specific pH levels because certain serious disease organisms are suppressed at these levels (see Chapter 16). Perhaps the most important example is potato (*Solanum tuberosum*), which must be grown at a relatively low pH to escape the ravages of potato scab. The potato scab organism, *Streptomyces scabies*, is an actinomycete that is especially virulent within a pH range of 5.2 to 8.0, but its development falls off rapidly at pH levels below 5.2. Because the potato is the third or fourth most important food crop in the world, the impact of this relationship is immense. In contrast, the fungus that causes clubroot of crucifers (e.g., cabbage, broccoli, and cauliflower), *Plasmodiophora brassicae*, is most pathogenic at pH 5.7. Its severity drops off sharply as the pH is increased to 6.2. A knowledge of such phenomena enables the horticulturist to select soils of appropriate pH for specific crops or to modify soils accordingly.

Erosion and Soil Management

The loss of precious soil resources to soil *erosion*, the wearing away of the land by water, wind and other geological agents, is one of the most serious problems facing humankind. Soil erosion causes a direct loss of valuable topsoil and nutrient resources that can drastically reduce crop production and, in extreme cases, lead to famine and significant loss of human life. Furthermore, the movement of soil and nutrients into lakes and streams causes pollution of water supplies essential for human and agricultural consumption.

Water **infiltration** is controlled in part by the soil characteristics. Sandy soils absorb water more rapidly than do heavy clay soils. Water can also move more rapidly through sandy soils, which can contribute to losses of soluble nutrients such as nitrogen and potassium. This movement through the soil profile of dissolved nutrients is termed **leaching**. Leaching in soils high in clay content is much slower, however and may be virtually non-existent. Because water moves slowly through such soils, they are more suitable for furrow irrigation and less suitable for sprinkler irrigation, which is preferred for sandy soils.

If rainwater or irrigation water is not absorbed by the soil, run-off will occur. The degree of erosion caused by such run-off is influenced by several factors, including the rate and amount of water landing on the soil, the slope of the land and the vegetation covering the soil. Soil organic matter content may also modify water uptake and infiltration rate, further affecting run-off and leaching losses.

The planting of **cover crops** is an effective technique for reducing loss of soil to wind or water erosion. The above-ground plant parts interrupt and reduce wind and water movement along the soil surface and reduce the impact of rain and irrigation droplets striking the soil. The plant roots further aid in reducing soil loss by physically holding the soil in place. **Wind breaks** are larger plantings of trees or shrubs that are also designed to interrupt the wind and reduce its potential damage to crops and soil (Figure 8-8).

FIGURE 8-8. Windbreaks are helpful in reducing wind velocity and thus help suppress wind erosion. *Pinus resinosa* and *P. banksiana* are used adjacent to this new planting of cabbage transplants in Minnesota.

FIGURE 8-10. Alternation of two crop types, or *strip cropping*. Note that the grain crop will intercept run-off water and reduce its erosion potential.

Contour planting, strip cropping, terracing, minimum tillage, and **mulching** are also practices helpful in prevention of soil erosion. Contour planting or contour tillage involves conducting all cultivation practices along the contour lines. The contour line is the imaginary line between two points of equal elevation. When a crop is planted "on the contour," the field boundaries tend to curve rather than follow a geometric pattern as with a typical rectangular field (Figure 8-9). Strip cropping is the alternation of two types of crops requiring different tillage practices, such as a strip of row crop alternated with a sod-forming crop (Figure 8-10). Strip cropping may be practiced by planting alternately along contour lines or across the prevailing wind or slope.

Terraces are usually level or nearly level strips of land constructed across the slope or on the contour. They are intended to facilitate cultivation of difficult, often highly sloping sites while minimizing erosion (Figure 8-11). Terraces catch the rainfall and irrigation water and allow it to soak into the soil or be conducted away on a grade likely to minimize soil erosion. Developing countries frequently use terracing techniques.

Minimum tillage practices reduce the erosion losses and damage to soil structure resulting from weathering and tillage practices. **No-tillage** agriculture involves no plowing or other soil-disturbing cultivation practices, with the exception of the planting operation. In no-tillage systems, the crop is

FIGURE 8-9. A contour planting of horticultural crops helps retard erosion. Strawberries, *left*, and Taxus, *right*.

FIGURE 8-11. Terraces cut into a hillside to facilitate cropping of a slope that would otherwise be impossible to cultivate without causing severe erosion. Grapevines are planted in this fashion in many of the world's important grape producing regions.

FIGURE 8-12. No-tillage corn.
Photo courtesy Donald Elkins, Southern Illinois University.

planted directly into the existing vegetation, usually sod or crop residue (Figure 8-12). Weeds or the cover crop are generally controlled in such systems by the use of herbicides used alone or in combination with mowing or other mechanical and chemical controls. Other variations of minimum tillage involve the occasional use of cultivation equipment, but considerably less than in conventional agricultural systems. Some minimum tillage systems actually require more use of pesticides than would be used in conventional tillage systems. However, an additional interesting characteristic of reduced tillage systems is that pesticides are more rapidly inactivated and degraded. This quality and the lessening of run-off minimizes potential pollution of the environment by pesticides and fertilizers. Reduced tillage systems are also thought to increase or help maintain soil organic matter levels, which results in an increased water-and nutrient-holding ability. Leaching losses are thus also reduced. Because minimum tillage research is still relatively new, there are many potential problems on which research is needed, such as "volunteer" seedlings from the previous year's crop (e.g., tomato seedlings from seeds that overwintered in the soil) and build-up of insect populations and disease inocula.

Mulches The use of mulches is addressed in Chapter 10. Minimum tillage systems are in essence modified mulching practices. In minimum tillage systems "living mulches," or cover crops, are employed, or the remains of the previous crop function as a mulch. These crop residues provide many of the same benefits discussed in Chapter 10.

AERATION AND DRAINAGE

Numerous practices may be used to increase aeration and to improve drainage of soils that tend to hold too much water. A common approach is the use of various kinds of **drainage** improvement methods. Digging ditches in or around a field to lower the water table is an ancient practice. This approach works reasonably well in relatively free draining soils where the existing water table is near the surface and in situations where the water can be easily conducted away from the field. In some cases of unusually high water tables, such as in Florida and the south of China, the use of **raised beds** enables the crop to be planted further above the free water to improve aeration (Figure 8-13). Installation of clay tile drains ("tiling") is another ancient practice used to reduce water-logging of soils. Tiling involves digging ditches, inserting cylindrical lengths of tile into them, and then refilling the ditch with soil to cover the tile. The tile is laid at a slight slope so that it functions as a conduit to remove the water that seeps into the tile through the joints. In recent years, perforated plastic piping has supplanted clay tile to a large extent. Intervals between lines of tile (or plastic) are determined primarily by soil type, which governs permeability and rate of water flow through the soil and into the drains. As would be expected, the spacing of tiles for a clay loam soil, for example, must be much closer than for a sandy soil.

The **mole drain** is a relatively temporary drainage method achieved by pulling a special plow equipped with a projectile-shaped metal "mole," which makes an un-walled cylindrical channel through the soil. Impermeable or slowly permeable distinct layers of hardened soil, usually containing high amounts of clay particles that may be cemented tightly together by organic matter or other substances, are referred to as **hardpans. Clay pans** and **plow pans** are similar to hardpans, and like hardpans, are generally found near the zone of transition from the A horizon to the B horizon. The plow plan is thought to be caused by repeated plowing of the soil, because it is a compact layer found at the depth of the plow slice. Implements used for making mole drains, or blades without the mole, are often pulled through the soil to break through such pans to improve drainage. The power requirements for mole drainage are great, necessitating use of a very large tractor. See also Figure 8-14 for another aeration method.

FIGURE 8-14. *Top:* Equipment used for aerifying high value turf plantings such as golf greens; resulting holes, with cores removed and raked aside (*bottom*).

FIGURE 8-13. Raised beds, such as these near Guangzhou, Peoples Republic of China, can be used to enable crops to be grown in locations where the water table is near the surface.

Media for Containers

Numerous types of important horticultural plants are grown in containers. A wide range of containers—pots, packs, flats, and so on—are available for holding the medium in which these plants are grown (Figures 8-15 and 16). Although in some cases the container and medium are one and the same (Figure 8-17), generally an appropriate medium must be placed in the container to facilitate growth of the desired plants. This medium must perform the same functions requisite of a good field soil for plant growth—that is, provide anchorage and support for the plant, serve as a source of nutrients, and make a proper balance of water and air available to the plant roots. Furthermore, a good container medium must be free of toxic substances and undesirable organisms such as weeds, weed seeds, insects and insect eggs, nematodes, and pathogenic fungi and bacteria. Disinfestation of container media is discussed in Chapter 16. It is also desirable that the medium be uniform and reproducible, readily available and relatively inexpensive. For some purposes, especially seed germination and rooting of cuttings, a light weight medium may be desirable, but a heavier medium may be preferable for tall plants

(a)

(b)

(c)

(d)

FIGURE 8-15. Containers for growing plants. (*a*) Pots may be of various sizes, shapes and composition, such as clay, polyethylene, and other plastics. (*b*) Packs and flats, historically of wood construction but now mostly plastic. (*c*) Multiple pots and flats. (*d*) Flats for seedling production in strips (*left*) and plugs (*center* and *right*).

FIGURE 8-16. A standard pot (*left*) is roughly as tall as the inside diameter of its top. Hence, a 15-cm standard pot is approximately 15 cm (6 inches) deep and the inside diameter at the lip is 15 cm. An azalea (or 3/4) pot (*center*) is three fourths as tall as its diameter, and a *bulb pan,* (*right*) is about one half as tall as its diameter.

that might tend to become top-heavy. Less physical work is required for handling light-weight media, and a reduced weight has obvious advantages for shipping plants in pots or other containers.

Most field soils are totally inappropriate for use as container media without modification. This is true for two reasons: the afore-mentioned need for disinfestation, and the fact that most field soils become badly compacted when placed in containers and watered repeatedly. This compaction occurs because a water table exists immediately above the container bottom and the capillary depth is short, causing inadequate pore space for water and air. It has subsequently become common practice to modify field soils intended for container use by adding one or

FIGURE 8-17. Young tomato plants in different medium-container combinations, in which the plant-and-container combination is planted as an intact unit.

more amendments. Physical properties of several materials useful as soil amendments or as constituents of soilless media are presented in Table 8-1. Sand, perlite, calcined clay, and polystyrene beads are among the materials most frequently added to a field soil or soilless mixture to create a mixture with improved drainage and aeration characteristics. Sand and calcined clays also add to the bulk density of a medium, whereas perlite, styrofoam, and vermiculite reduce it.

Peat is the material most commonly used to improve water retention in an amended field soil and concurrently to improve drainage and aeration. It is also frequently used as a component of soilless media. Sphagnum moss peats are preferred for most such uses because of their long fibers and slow rate of decomposition. The fibrous structure enables peat to hold a great amount of water because of its large surface area. It is important to avoid excessive grinding of peats, because the good aeration properties are lost when peat aggregates are broken down to individual fibers. Hypnum moss peats are also used as soil amendments and in soilless media, but they have shorter fibers, a higher pH, and a tendency to break down faster than sphagnum peats. Other peat types such as those derived from reeds and sedges, are common agricultural soils; they are more highly decomposed and thus compact more when incorporated into a medium. This feature makes them less desirable than sphagnum or hypnum peats.

Barks have become popular in recent years as a major ingredient of many soilless mixes for container use. They tend to break down slowly and have many of the beneficial characteristics of peats. Barks usually are composted or allowed to decompose partially before use in media. The composting process reduces the content of inherently phytotoxic phenolic compounds commonly found in some hardwood and softwood barks and also increases nutrient- and water-holding ability. Barks often are relatively inexpensive because they are by-products of the lumber industry.

TABLE 8-1. Physical properties of selected materials used as soil amendments or as medium constituents.

SAND

❖ Rock fragments 0.05–2.0 mm in diameter
❖ The pH varies with the source
❖ Contains few or no nutrient elements
❖ Weight(s) 1360 kg/cubic meter (100 lbs./cubic foot) when dry
❖ Good drainage, if clean (coarse sands especially)

MILLED SPHAGNUM MOSS

❖ Good buffering properties on a weight basis
❖ pH of approximately 3.5
❖ Contains only very small amounts of mineral nutrients
❖ Holds 10–20 times its dry weight in water
❖ Contains a fungistatic substance that inhibits damping off

SPHAGNUM MOSS PEAT

❖ Up to 50 percent decomposed
❖ Good buffering properties on a weight basis
❖ pH of 3.8–4.5
❖ Contains only small amounts of available mineral nutrients, depending on the state of decomposition
❖ Holds 10 times its dry weight in water
❖ Light weight

VERMICULITE

❖ Expanded mica that has been heated to 1100°C (approximately 2000°F)
❖ Good buffering properties
❖ pH of 7, sometimes higher
❖ Contains enough Mg and K to supply most plants for a short time
❖ Weighs 80–135 kg/cubic meter (6–10 lbs/cubic foot) when dry
❖ Well-drained
❖ Holds water well, about 1 liter of water per three liters of vermiculite (3–4 gal/ft^3)
❖ The lattice flakes apart readily
❖ Four size grades: No. 1, insulation grade, largest size
　　　　　　　　　　No. 2, regular horticultural grade
　　　　　　　　　　No. 3, Intermediate size
　　　　　　　　　　No. 4, Smallest size, component of many commercial mixes for seed germination

PERLITE

❖ Siliceous material of volcanic origin that has been heated to 760–870°C (approximately 1400–1600°F)
❖ No buffering capacity
❖ pH of 6–8
❖ Can contain up to 20 parts per million (ppm)fluoride
❖ Well-drained
❖ Weighs 80 kg/cubic meter (6 lbs/cubic foot) when dry

(Continued)

TABLE 8-1. Physical properties of selected materials used as soil amendments or as medium constituents (*continued*).

CALCINED CLAY

❖ A montmorillonite clay processed at high temperatures
❖ Has some buffering capacity
❖ Has no fertilizer value, but can absorb nutrients
❖ Porous
❖ Resistant to breakdown
❖ Holds water
❖ Weighs 400–540 kg/cubic meter (30–40 lbs/cubic foot)

COMPOSTED BARK

❖ Composting destroys injurious/inhibiting compounds
❖ Composting increases cation exchange capacity
❖ Composting eliminates insects, disease organisms, nematodes and weed seeds
❖ Wide C:N ratio (300:1)
❖ Decomposes slowly
❖ Good water and nutrient retention
❖ Particle sizes of 10 mm or less preferred for pot-plant mixes
❖ Particle sizes of 10 to 19 mm preferred for addition to greenhouse beds for cut flowers
❖ Particle sizes over 20 mm used for landscape mulches
❖ Less expensive substitute for sphagnum moss peat

ROCKWOOL

❖ Fibers spun from melted limestone, basalt, and slag following melting at 1600° C (approximately 2900° F)
❖ Little to no nutrients or nutrient-holding ability
❖ pH ranges from 7.0–8.5
❖ Horticultural grade must be used (NOT insulation or acoustical grade)
❖ Horticultural rockwool contains a surfactant to aid in water absorption
❖ Pore space is over 90 percent
❖ Available in slabs, cubes and granular form
❖ Not biodegradable

COIR

❖ Often called "coir dust" or "coir fiber pith"
❖ Consists of short fibers and dust from mesocarp of coconut fruits
❖ Variable pH, from 5.0–6.5
❖ Weighs 48–60 grams/liter (3–5 lb/ft^3)
❖ C:N ratio of 80:1
❖ Decomposes slowly
❖ May contain high amounts of salts, especially potassium; requires leaching
❖ Holds water well, up to 8 times its weight

Many recipes for soilless media are available, and a great number of commercially prepared mixes are listed for sale as well. The practical horticulturist must decide whether it is more economical to purchase a ready-made medium or to create a home-made mix. Table 8-2 lists ingredients for a peat–vermiculite medium that is suitable for growing a wide range of horticultural plants.

TABLE 8-2 Recipe for a peat-vermiculite medium.

INGREDIENT	AMOUNT REQUIRED FOR **0.9** CUBIC METER (**1** CUBIC YARD)
Horticultural vermiculite[a] (#2 grade)	0.39 cubic meter (11 bushels)
Shredded sphagnum moss peat	0.39 cubic meter (11 bushels)
Ground dolomitic limestone	2.27 kg (5 lb)
Single superphosphate (0-20-0), powdered	0.45 kg (1 lb)
5–10–5 fertilizer	0.91–5.44 kg (2–12 lb)[b]
Fritted trace elements	57–85 g (2–3 oz)
Wetting agent	28 g/22.7 liters (1 oz/gallon)

Source: Modified from the Cornell peat-lite mixes developed by J. W. Boodley and Raymond Sheldrake, Jr.

[a] No. 3 or no. 4 grades can be substituted, especially when a finer medium is desired for use as a seed germination medium. *Horticultural perlite* may be substituted for the vermiculite to produce a more freely draining medium. Such peat–perlite mixes are especially desirable as a medium for use in rooting cuttings.

[b] The lower level of 5–10–5 is preferable for a germination medium; higher levels can be used for bedding plants or transplants when longer term nutrition is required (Slow-release fertilizers can also be substituted for this purpose). Calcium nitrate at the rate of 0.45 Kg/m^3 (1 lb/ft^3) is often used instead of 5–10–5 where a constant or frequent fertilizer injection program is to be practiced.

In the United States it is uncommon to use a single medium rather than a mixture, but in Europe many growers utilize sphagnum moss peat as the only substrate component. When peat is used alone, great care must be taken with watering practices in order to avoid waterlogging of the medium. A waterlogged medium is undesirable because oxygen levels are reduced, and conditions may become favorable for the damping-off pathogens (see Chapter 16).

Another medium occasionally employed alone, usually on a small scale, is vermiculite. Vermiculite holds water well and is relatively well aerated because of the spaces inherent in its expanded plate-like layered structure. It is particularly effective in the rooting of leafy cuttings (Figure 8-18). However, vermiculite compresses readily with prolonged use, loses porosity, and is fragile when handled (see also Figure 8-19).

The choice of a medium and its constituents depends on a variety of factors, including intended use, availability, cost, and horticultural qualities. The agricultural advisory services and references listed at the end of this chapter provide further information on the subject.

FIGURE 8-18. A small-scale home propagation unit (often called a Forsythe Pot). The central clay pot has its drainage hole plugged and functions as a water reservoir. Because of the porous nature of the clay pot, water seeps into the surrounding medium, keeping it at a near-ideal water–air relationship conducive to the rooting of cuttings.

Containers

Containers are sold in a variety of shapes and sizes and component materials. As was mentioned in Chapter 7, containers for growing plants should have drainage holes to allow the water to drain from the medium, which will prevent pore spaces from filling with water and limiting oxygen

FIGURE 8-19. High volume soil mixing. Piles of medium constituents (*top left*); constituents being conveyed into a mixer (*top right*); steps in automated pot filling and planting (*bottom left* and *right*).

to the roots. Various sizes of flats, packs, trays, pots, and baskets are available, generally made of wood, metal, clay, plastic, or polystyrene. Flats, packs, multipacks, and trays are frequently used for propagation purposes or for producing plants that are sold when relatively small, such as bedding plants. Pots and baskets are used for a variety of purposes, ranging from propagation to finishing-off plants.

Flats are often wooden or plastic, however, metal flats are also available. Wooden flats are frequently made of redwood, cedar, or cypress because these woods are particularly resistant to decay. When other woods are used, they should be treated with a nonvolatile wood preservative such as a 2 percent solution of copper naphthenate or

copper azole ("Wolmanized"). Pressure treatment of wood with chromated copper arsenate (CCA) is considered to be unsafe. A new product, alkaline copper quaternary (ACQ) also offers promise as a wood preservative. Some common wood preservatives, such as pentachlorophenol and creosote should *always* be avoided for use around plants because they will volatilize extremely phytotoxic gases. Wooden flats can be steam-disinfested between uses, but they occupy a considerable amount of space during storage. However, plastic flats stack well and are much more efficient to store. Some plastic flats will melt or warp if they are steam-disinfested; they should be instead treated with a chemical such as a 1:9 liquid chlorine bleach: water solution.

Many flats and trays have preformed individual cells of various sizes and shapes. These cells eliminate the tangling of roots and thus reduce transplanting shock. Plug trays have small cells and are extremely popular in the commercial production of bedding plants. Because the volume of the medium is so small, careful attention must be paid to fertilizing and watering practices. Other flats resemble a series of tubes and are often used for the production of tree seedlings (Figure 8-21).

Packs are simply small flats. They are generally made of plastic and are often used for bedding plants. Many packs accommodate six bedding plants, which is often a convenient number for a consumer to purchase.

Pots are available in an assortment of shapes, sizes, and materials. Round pots are popular because they will allow for some air circulation around the plants even when they are placed on a bench pot to pot. Some growers prefer square pots however, because they feel they are providing more medium volume per plant and are using their bench space more efficiently.

Clay pots are popular because they are aesthetically pleasing when new. The walls of clay pots are porous and water moves through them readily. In contrast, the walls of plastic pots are nonporous to water. Dissolved salts will be carried along with water through the walls of the clay pots. As the water evaporates from the outer walls of the pots, however, the fertilizer salts will crystallize and form an unsightly white crust. There is no good way to remove these salt deposits once they build up on clay pots. Despite the salt deposits, many growers like clay pots, especially during the winter because they help the growing medium to dry out quickly so that the roots do not stay as wet. Clay pots are heavy, require considerable room for storage, and are breakable, but they can be easily disinfested by steaming. For novice growers, clay pots may be easier to use than plastic.

Plastic pots are light weight, are easily stored, and will often bounce rather than break when they are dropped, especially if they are new, but many will warp or melt if disinfested with steam. Plastic pots are available in a variety of colors, so the planter should be aware that, when placed in full sunlight, the medium in the darker colored pots will become warmer, even to the point of causing damage, than the medium in lighter-colored pots.

Disposal of plastic containers can become an environmental concern because they may require 200 to 300 years to break down when placed in a landfill. This topic is addressed in Chapter 10, where plastic mulch and other plastic disposal is discussed.

Many other container types and shapes are available to horticulturists. Growers may wish to experiment with or read about other types when making decisions on which types to use.

A common problem with container grown trees and shrubs is self-girdling of the roots. The problem is frequently discovered long after planting in the field or landscape (Figure 8-20). Self-girdling is the result of circling root growth in smooth round containers and these plants are said to be root-bound or pot bound. When a root encounters the side of the pot, it tends to grow around the interior of the pot. After planting out, as such roots grow and enlarge in girth, they often girdle the stem or main root. This may limit the uptake of water and nutrients, causing the plant to grow poorly or even to die. Anchorage of such plants is also poor, so trees may easily be uprooted by strong winds.

Pots are now made so that root circling is avoided (Figure 8-21). These containers have ribs, ridges or square corners on their insides that force the roots to grow downwards. In addition, many of these containers have large holes on the bottom. When placed on wire mesh benches, the roots that grow out through the holes are exposed to air, which dries their tips and thus effectively **air prunes** them. This air-pruning results in increased root branching and the resulting plant will have a fibrous root system without circling roots.

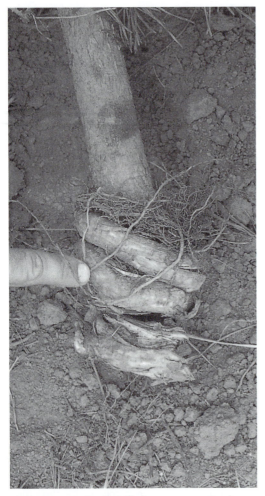

FIGURE 8-20. A self-girdled tree as a result of root circling in a round container.

Plants grown in containers without such modifications and without air pruning should be root pruned at the time of transplanting to avoid problems associated with root girdling. This can be accomplished by making two, three, or four vertical cuts through the external roots that are about 2 cm deep into the root system. Damage to the roots is minimal and root branching and outgrowth into the soil will be encouraged. An alternative is to split the root ball with a shovel vertically up from the bottom, and pull apart the two halves when transplanting. This "butterflying" of the root ball disrupts circling roots and encourages outgrowth resulting in a healthier plant with a much longer life expectancy.

The tremendous value of a fertile and productive soil resource cannot be overestimated. Since earliest recorded history production of food and other horticultural crops has been dependent on soil resources. In this chapter we have focused on physical and chemical characteristics of soil and on the horticultural applications and cultural practices required to make maximum effective use of soil resources. Horticulturists must gain a thorough understanding of soils and soil management in order to conserve our precious soils, maintain their fertility, and optimize their management to provide maximum benefits to humankind.

FIGURE 8-21. Special containers (*left* and *right*) used to channel roots downward to avoid root circling.

REFERENCES

Baker, K. F., P. A. Chandler, R. D. Durbin, et al. 1957. *The U.C. System for Producing Healthy Container-grown Plants.* Univ. of California Div. Of Agricultural Sciences, Davis, CA 332 pp.

Boodley, J. W. 1981. *The Commercial Greenhouse.* Delmar Publishers, Albany, NY 12205.

Brady, Nyle C. and R. R. Weil. 2002. *The Nature and Properties of Soils* (13th ed.). Prentice Hall, Upper Saddle River, NJ. 960 pp.

Bunt, A. C. 1976. *Modern Potting Composts.* The Pennsylvania State Univ. Press, University Park and London. 277 pp.

Buol, S. W., F. D. Hole and R. J. McCracken. 1973. *Soil Genesis and Classification.* Iowa State Univ. Press, Ames, IA. 360 pp.

Cook, J. G. 1960. *Our Living Soil.* Dial Press, New York. 190 pp.

Flegmann, A. W. and R. A. T. George. 1977. *Soils and Other Growth Media.* AVI Publishing, Westport, CT. 170 pp.

Fried, M. and H. Broeshart. 1967. *The Soil Plant System in Relation to Inorganic Nutrition.* Academic Press, New York. 358 pp.

Phillips, R. E. and S. H. Phillips. 1984. *No-Tillage Agriculture.* Van Nostrand Reinhold, New York. 306 pp.

Plaster, E. J. 1985. *Soil Science and Management.* Delmar Publishers, Albany, NY. 454 pp.

Robinson, D. W. and J. G. D. Lamb. 1975. *Peat in Horticulture.* Academic Press, London. 170 pp.

Sopher, C. D. and J. V. Baird. 1982. *Soils and Soil Management* (2nd ed.) Reston Publishing, Reston, VA. 238 pp.

United States Dept. of Agriculture. 1977. *Gardening for Food and Fun, Yearbook of Agriculture.* USDA, Washington, D.C. 392 pp.

United States Dept. of Agriculture. 1957. *Soil. Yearbook of Agriculture.* USDA, Washington, D.C. 784 pp.

CHAPTER 9

MINERAL
NUTRITION

Αll of the elements known to be essential for growth of higher plants, except carbon, are taken up from the soil. Some experts believe that in some cases even carbon may be provided by the soil. Although over 60 elements have been found in plants, only 17 have been demonstrated to be essential. They are: carbon (C), hydrogen (H), oxygen (O), nitrogen (N), phosphorus (P), potassium (K), calcium (Ca), magnesium (Mg), sulfur (S), iron (Fe), manganese (Mn), zinc (Zn), copper (Cu), boron (B), molybdenum (Mo), chlorine (Cl), and nickel (Ni). The first three are normally provided by water from the rhizosphere and carbon dioxide from the atmosphere.

....

EARLY KNOWLEDGE OF
ESSENTIAL ELEMENTS

Von Helmont's classic experiment reported in the 17th century demonstrated that nearly all of the mass of a plant comes from water and carbon dioxide (CO_2), although he thought that it came from water alone. Von Helmont had planted a willow cutting weighing 5 pounds (about 2.3 kg) in a tub of soil weighing 200 pounds (about 91 kg); then he watered it with rain water. The plant gained 164 pounds over five years, but only about two ounces of the soil's weight had been lost. Modern soil scientists now know that both CO_2 and water contributed to most of the increase in the bulk of von Helmont's plant, and that only a small part of a plant's weight comes from the mineral nutrients contributed by the soil.

Nitrogen, phosphorus, and potassium are referred to as macronutrients and are the mineral elements absorbed by plants in the greatest quantities. Other macronutrients are calcium, magnesium, and sulfur, although they are not taken up in such large quantities as N, P, and K. Iron, manganese, zinc, copper, molybdenum, boron, chlorine, and nickel are utilized in minute quantities and hence are referred to as micronutrients. In spite of the relatively tiny amounts required, the importance of these micronutrients should not be minimized because an absence or deficiency of even one micronutrient can cause extreme injury or even death of a plant.

MINERAL NUTRIENT UPTAKE

Mineral nutrients are taken up by plant roots in the ionic form, following a pathway essentially like that described earlier for water (see Chapter 7). The ionic form or forms usually absorbed for a given element are listed in Table 9-1. It has long been recognized that plant roots have the ability to accumulate elements within the plant to levels many times greater than their concentration in the soil solution. Furthermore, certain toxic elements can be excluded or reduced in the plant to levels significantly below those in the immediate root environment. Such phenomena occur because of the differential permeability of the plasmalemma and because the plasmalemma has an ability to affect active transport. In general, soil scientists agree that most of the anions and certain cations (e.g., K^+) can be accumulated in the plant root even in the presence of low ambient concentrations,

and they concur that certain other ions such as Ca^{++}, Na^+, and H^+ can be actively transported out of the tissue into the surrounding rhizosphere. Initial uptake is considered to be *passive*; it may occur by simple diffusion and mass flow or may be aided by mycorrhizal fungi. After uptake into the cortical cells, metabolic energy in the form of ATP provided by aerobic respiration is required for many nutrients to traverse the plasma membrane (*active* uptake). The nutrients then move through the plasmodesmatal connections from cell to cell and then to the vascular tissues.

Crop species vary in their nutrient requirements. Some, such as corn (maize) and potatoes require large amounts of fertilizer and are thus regarded as "heavy feeders." Many leafy vegetables possess a high requirement for nitrogen, whereas large amounts of potassium are requisite for good growth of certain crops for which the marketable part is a subterranean organ (e.g., the tuber of potato or the tuberous root of sweet potato and dahlia). The amounts of nitrogen, phosphorus, and potassium removed from the soil by several vegetable crops is presented in Table 9-2. Low levels of nutrients can also exert profound effects important to horticultural practice. Raising the nitrogen level of a medium deficient in nitrogen results in a stimulus of lateral bud growth. This response is valuable for certain ornamentals in which a bushy habit is desired, and it also has implications for propagation (i.e., more branches will yield more available cuttings to the propagator).

The nutrient status of plants also can materially affect susceptibility to disease-causing organisms (see Chapter 16). Excess nitrogen can result in young, succulent growth that is more susceptible to attack by certain pathogens. Examples include fire blight (*Erwinia amylovora*) of pears and other susceptible members of the Rosaceae, and tobacco mosaic virus of tomato. Low nitrogen levels, however, may increase susceptibility of tomato to *Fusarium* wilt, of beets to *Sclerotium rolfsii*, and of most seed-lings to damping-off caused by *Pythium*. Increased levels of calcium help tomato plants resist *Fusarium* attack and reduce damage to peas caused by *Rhizoctonia* root rot. Other nutrients also seem to be related to susceptibility of plants to diseases and pests, but much is yet to be learned in this important field. In general a balanced nutri-tional approach will produce plants with increased ability to resist attack by undesir-able organisms.

TABLE 9-1. Characteristics of Plant Nutrients

ELEMENT	FUNCTION	FORM(S) PLANTS ABSORB	DEFICIENCY SYMPTOMS
Nitrogen (N)	Constituent of: amino acids (which build proteins), chlorophyll, nucleic acids, and certain lipids.	NO_3^-, NH_4^+ NO_2^- (rarely)	Stunting-pale green color (general); firing of lower leaves, which may lead to death of lower leaves. Hastens maturity.
Phosphorous (P)	Important in energy release, (constituent of ATP and nucleic acids), necessary for normal development of flowers, fruits and seeds.	$H_2PO_4^-$ (predominant form in acid soils), $HPO_4^=$ and $PO_4^{\equiv}$ (alkaline soils)	Lower leaves dull or darker than normal green; purpling; flowers abort before forming; fruits and seeds may fail to develop.
Potassium (K)	Important in translocation; affects cell permeability, stomatal activity and plant water status; activator of many enzymes.	K^+	"Scorch" or necrotic spotting of lower leaf margins; although little stunting, dry weight is reduced. Chlorosis may precede scorch or necrosis.
Calcium (Ca)	Membrane and cell wall structure, neutralizing agent, constituent of chromosomes, affects cell elasticity.	Ca^{++}	Cupping and/or hooking of leaves; wilting; apical meristem breaks down.
Magnesium (Mg)	Constituent of chlorophyll, enzyme activator and constituent.	Mg^{++}	Interveinal chlorosis of older leaves.
Sulfur (S)	Constituent of the amino acids methionine, cysteine, and cystine thus part of certain proteins; affects chlorophyll synthesis.	$SO_4^=$	Looks like N deficiency (general pale green color); usually appears only on new leaves; nodulation of legumes may be reduced.
Iron (Fe)	Necessary for chlorophyll synthesis; important in respiration and photosynthesis (electron transport reactions).	Fe^{++}	Interveinal chlorosis of young leaves.

(Continued)

TABLE 9-1.   Characteristics of Plant Nutrients (*continued*)

ELEMENT	FUNCTION	FORM(S) PLANTS ABSORB	DEFICIENCY SYMPTOMS
Manganese (Mn)	Enzyme component and activator important in chlorophyll production and breakdown of H_2O in photosynthesis.	Mn^{++}	Chlorosis of young leaves of some species and of middle or lower leaves of others. Small necrotic spots frequently present in the chlorotic area.
Boron (B)	Exact role unclear, apparently important in cell division (or right after), may be involved in sugar translocation.	H_3BO_3	Corky tissue on or in roots of root crops; hollow and black internal portions of stems (especially cole crops); death of terminal mersistem and/or flowers of tomato and woody plants.
Copper (Cu)	Catalyzes certain enzyme systems, important in chlorophyll formation.	Cu^+, Cu^{++}	"Dieback" of citrus, chlorosis.
Zinc (Zn)	Functions in auxin synthesis, component of several enzymes.	Zn^+	"White bud" of corn, pecan rosette, etc. — shortened internodes small leaves, and dying back of branches.
Molybdenum (Mo)	Functions in N fixation; possibly involved in metabolism of P.	$MoO_4^=$	"Whiptail" of cauliflower; N-deficiency symptoms in legumes; cupping of leaves of broccoli (prior to "whiptail" appearance)
Chlorine (Cl)	Enhances electron transport reactions in photosynthesis, unknown role in root development.	Cl^-	Not well defined as yet; wilting and amino acid build-up may occur.
Nickel (Ni)	Precise role not clearly defined.	Ni^+	Not well understood.

Silicon, sodium, vanadium, and cobalt are elements which have been shown to be necessary for certain plant species, but the exact nature of their function is not known. Extremely minute amounts appear necessary. They have not been generally accepted as essential for all plants. Later experimental work may reveal the necessity of these and other elements for plant growth, in addition to demonstrating their functions.

Soil as a Source of Nutrients

Soils serve as a repository or reservoir for most of the essential plant nutrients. A soil's effectiveness in serving this function is governed in part by its texture and structure. The clay particles of a soil are made up of a series of flat laminated particles. Because of their fineness, they collectively possess a great external surface area and many have *internal* surfaces as well. The external surface area of one gram of clay particles may be 1000 times that of 1 gram of coarse

TABLE 9-2. Approximate Absorption of Nutrients by Some Vegetable Crops

VEGETABLE	YIELD (KG/HECTARE)	PLANT PART	NUTRIENT ABSORPTION		
			N	P	K
Broccoli	112	heads	22	2	50
		other	162	9	185
		total	184	11	235
Celery	1120	tops	190	39	426
		roots	28	17	62
		total	218	56	488
Muskmelons	252	fruits	106	19	134
		vines	67	9	39
		total	173	28	173
Potatoes	448	tubers	168	21	224
		vines	67	12	84
		total	235	33	308
Spinach	224	plants	112	13	112
Sweet corn	146	ears	62	9	34
		plants	112	13	84
		total	174	22	118
Sweet potatoes	336	roots	90	18	180
		vines	67	4.5	45
		total	157	22.5	225
Tomatoes	672	fruits	112	11	202
		vines	90	12	112
			202	23	314

Adapted from O. A. Lorenz, and D. N. Maynard, *Knott's Handbook for Vegetable Growers*, (2nd ed.)Wiley, New York, 1980.

sand. These external and internal surfaces have electronegative charges that enable them to hold positively charged cations such as H^+, NH_4^+, K^+, Mg^{++}, and Ca^{++}. These cations are exchanged onto and off of the surfaces of the clay particles and organic matter. This cation-exchange capacity (CEC)is expressed in milliequivalents per 100 g of dry soil. In addition to holding large numbers of cations, clay particles also hold large amounts of water.

The organic matter of a soil attracts and binds both water and ions. As mentioned earlier, organic matter aids in aggregation of soils, but organic colloids have both the ability to adsorb cations and to hold nitrogen, phosphorus, and sulfur in organic forms. The latter elements are released to the soil solution by mineralization (breakdown by microbial activity) of the soil organic matter, releasing such ions as NH_4^+, NO_3^-, $SO_3^=$, $SO_4^=$, $H_2PO_4^-$, and $HPO_4^=$. Because this process is dynamic and ongoing, soils rich in organic matter can provide a more or less continuous source of nutrients for plant growth. The soil microflora—fungi, actinomycetes and bacteria—accomplish this release by breaking down the organic matter to inorganic forms. This release is achieved for organic nitrogen, for example, by microbial release of NH_4^+, which is in turn reduced by *Nitrosomonas* bacteria to NO_2^- (nitrite). NO_2^- is subsequently oxidized to NO_3^- (nitrate) by the

Nitrobacter group of bacteria. Similarly, other nutrients such as sulfur and phosphorus become available to plants as a result of microbial activity.

In addition to serving as a source of nutrients, organic matter and the clay particles constitute most of the CEC of soils, as well as aiding in its **buffering capacity**. Buffering capacity is the ability of a soil to resist rapid changes in pH level that may be brought about by environmental influences and by effects of chemicals such as herbicides. Thus buffering capacity of a soil affects herbicide rates required for effective weed control and minimal crop phytotoxicity.

Specific Nutrients

Details of specific mineral nutrient functions, the ionic form(s) taken up by plants and common deficiency symptoms are delineated in Table 9-1. Deficiency symptoms appearing in the older leaves (e.g., N, P, K, Mg) are a result of the inherent mobility of these elements in the plant. If such elements are insufficient in the soil solution, the plant "mines" the element from the older leaves and it is translocated to the growing point and young leaves. Conversely, plants deficient in less mobile elements such as Fe, Cu, Mn, and Zn exhibit their deficiency symptoms in the younger leaves and new growth.

CARBON, HYDROGEN, AND OXYGEN

Plants manufacture their own food—sugars, proteins, fats, and so on—by photosynthesis and subsequent metabolic pathways. Although mineral elements or fertilizers are commonly called plant "food," they should instead be termed **nutrients**. Carbon, hydrogen, and oxygen are utilized by plants in the manufacture of their food, and as we have already seen, they obtain these elements from CO_2 in the atmosphere and water from the rhizosphere. Horticulturists can irrigate to prevent water from becoming a limiting factor (see Chapter 7).

In field plant production, CO_2 is seldom limiting, because the air contains about 350 parts per million (ppm) CO_2 and because the movement of air, along with diffusion, provides a continuous supply to the stomates. On very still days, however, CO_2 may become temporarily deficient in fields of closely spaced plants, such as closely planted potatoes. In closed containers or in tightly closed greenhouses, however, a large population of rapidly photosynthesizing plants may reduce the CO_2 concentration to a very low level. Producers of some greenhouse crops such as roses, carnations, and lettuce, can realize gains in production by adding supplemental CO_2 to the greenhouse atmosphere. One way to add CO_2 is to burn small amounts of natural gas, which will release CO_2 upon combustion (Figure 9-1). Burners must be kept properly adjusted, because if combustion is incomplete, phytotoxic gases such as ethylene (C_2H_4) and carbon monoxide (CO) may be released. Enrichment of the greenhouse atmosphere may also be achieved by the release of CO_2 from dry ice (solid CO_2) or from compressed CO_2 cylinders (liquid), but these methods are expensive.

Levels of 1000 to 1500 ppm CO_2 have been demonstrated to significantly improve production of many floricultural crops. An understanding of this response helps explain the beneficial effects seen from the historical practices of using manures and organic mulches in greenhouse production. The breakdown of these organic materials released CO_2, which enhances growth of the crops. It is also possible that a part of the growth enhancement typical of transplanted vegetables grown with plastic mulch may also be because of localized CO_2 enrichment. As organic matter decomposes under the plastic, the CO_2 produced moves up through the transplant hole, bathing the undersides of the leaves with higher-than-ambient CO_2 levels.

FIGURE 9-1. A CO_2 generator employed for enrichment of the greenhouse atmosphere with carbon dioxide, which is produced by combustion of fuel such as natural gas.

The use of "carbonized mist," i.e., carbonated water in a mist propagation system, has also been reputed to enhance rooting of leafy cuttings. Reports regarding the efficacy of this approach conflict, however, and the use of carbonized mist has not been adopted widely by commercial plant propagators. Several examples of nutrient deficiencies are presented in Figure 9-2.

MACRONUTRIENTS

Nitrogen

Nitrogen may be the most important mineral nutrient; it is used in greater amounts than any of the other macronutrients and generally constitutes the largest percentage of a plant's mineral weight. Nitrogen is a constituent of amino acids and thus is found in all proteins, including enzymes. Nearly all of the components of the plant's protoplasm contain nitrogen, so it is little wonder that a nitrogen deficiency may severely stunt a plant. A majority of the soil nitrogen available to plants is produced by the decomposition of organic matter, with lesser and varying amounts resulting from electric storms and nitrogen-fixing microbes and algae. Atmospheric nitrogen is also made available (fixed) by symbiotic organisms associated with the roots of a number of plant species. In leguminous plants (members of the *Fabaceae* or legume family), the symbiosis is accomplished in root swellings (*nodules*) formed by invasion of the root hairs and subsequent entrance into the cortex by bacteria (*Rhizobium* spp., *Bradyrhizobium* spp., or *Rhizobacter* spp.). These bacteria are capable of taking free nitrogen from the soil atmosphere and synthesizing more complex nitrogenous compounds. In fact, field trials

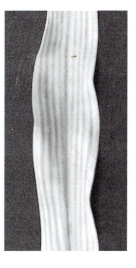

FIGURE 9-2. Examples of nutrient deficiencies. Nitrogen deficiency on birch (*left*), magnesium-deficient tomato (*middle*); and iron deficient corn (*right*).

have shown that a commercial crop of peas may fix over 130 pounds of nitrogen per acre in one growing season (about 150 kg/hectare). The mode of plant absorption and utilization of this nitrogen is not completely understood, but it is clear that the nitrogen nutrition of the host plant is influenced in a very positive way. Sloughing of the nodular cells and plowing under herbage of crops like alfalfa (green manuring) adds materially to the nitrogen content of the soil.

A large number of nonleguminous species including members of the genera *Casuarina*, *Eleagnus*, and *Alnus* also have the ability to fix atmospheric nitrogen. The symbiotic organism involved in *Alnus* nitrogen fixation is an actinomycete (*Frankia* spp.), rather than a bacterium. However, because these plants are not as commonly cultivated as the many legume species, less is known about their mechanisms of nitrogen fixation.

Nonsymbiotic fixation can also be accomplished by free-living bacteria and blue-green algae found in many soils. Examples of free-living bacteria include *Azotobacter* spp. and *Clostridium* spp. Soil populations of these species vary greatly, but there may be as many as 100,000 *Clostridium* cells per gram of arable soil. Blue-green algae may contribute appreciably to the nitrogen balance of water lilies and other aquatic ornamentals.

Phosphorus

Phosphorus is required in smaller quantities than is nitrogen. However, phosphorus nutrition is extremely important because it is a component of DNA, the genetic material of plants, and of ATP, the energy rich compound so crucial to most metabolic processes. Decomposition of organic matter provides a source of phosphorus for plant growth, but some soils may be phosphorus-deficient and require fertilizers containing phosphorus. Because soil phosphorus compounds are extremely insoluble and are often precipitated in the soil solution, phosphorus fertilization requires careful attention; timing and

pH control are especially important in this regard (see again Figure 8-7). The problem of unavailability is particularly acute in cold soils because of low microbial activity, so additions of phosphorus may be required in the spring, even though a soil test may show that phosphorus content is adequate.

Potassium

Often incorrectly referred to as *potash*, potassium is the third macronutrient required in relatively large quantities by higher plants. This element is derived in the soil by physical and chemical degradation of clay minerals such as biotite, muscovite, and illite. Leaching and weathering release the potassium ions from the clay lattices, making them available to the plant roots. Although it is not a major component of cellular molecules, potassium is required for survival of all living cells. Its roles in activating enzymes involved in protein synthesis and carbohydrate metabolism and in influencing the permeability of cell membranes makes it crucial for normal plant growth. It is also a pivotal compound for photosynthesis because it functions in opening and closing of stomates. As mentioned in the section on leaves, when the guard cells become turgid, the stomates will open. This turgidity occurs as a result of light-induced active movement of potassium ions into the guard cells.

Calcium

Calcium content of soils is highly variable. When calcium levels are low, soils tend to be acid. Hence, calcium additions ("liming") are commonly employed to correct excess acidity in soils because the calcium ion (Ca^{++}) will replace the hydrogen ion (H^+) in the soil.[1] Potassium, calcium and

....
[1] Agricultural lime applied to correct soil acidity usually consists of ground limestone, which is primarily composed of $CaCO_3$. *Dolomitic* limestone is a limestone that includes both $CaCO_3$ and significant quantities of $CaMg (CO_3)_2$, or dolomite.

magnesium are essential nutrients present as cations that influence each other's availability. The order of strength of adsorption is Ca, Mg, K. For example, when large amounts of magnesium are present, potassium may become less available. This factor must be taken into consideration when liming and fertilizing soils. Calcium's importance as a nutrient should not be overlooked, because it is a part of the pectin that helps cement the adjacent cell walls together. It is also important in the gelling of the cytoplasmic fluid; a calcium-deficient cell will lack sufficient viscosity for cell functions to proceed normally.

Magnesium

As pointed out earlier, magnesium is a constituent of the chlorophyll molecule and is thus essential for photosynthesis. It also has the aforementioned influence on availability of other cations such as potassium and calcium. It further influences cell metabolism by virtue of its role as an enzyme activator. Magnesium may be lacking in some agricultural soils, but this has not always been recognized because significant amounts are usually present in agricultural lime (especially dolomitic) applied to correct acidity of these soils.

Sulfur

As a component of the amino acids, methionine, cysteine, and cystine, sulfur is essential to protein synthesis. A deficiency of sulfur is rare, because it is commonly present in rain water and as a constituent of many low analysis fertilizers; unlike phosphorus, it tends to be present in relatively soluble forms that are not rendered insoluble by other soil compounds. Various forms of sulfur are used to lower pH of soils that are too alkaline for the crop to be grown. Addition of sulfur reduces the pH because the sulfur is gradually changed to sulfuric acid by soil biological oxidation:

$$2S + 3O_2 + 2H_2O \rightarrow 2H_2SO_4.$$

MICRONUTRIENTS

Iron

A deficiency of iron is an extremely critical problem because iron has so many important functions in plant growth. It is required for chlorophyll synthesis, is important in the electron transport system (respiration), and is essential for normal nitrogen metabolism. Although chemical analysis of nearly all agricultural soils will reveal a presence of large amounts of iron, iron deficiency is the most commonly observed micronutrient deficiency. Like many other micronutrients (e.g., Mn, Zn, Cu), iron becomes unavailable at high pH levels. It is also usually present in relatively small quantities in sandy and organic soils, or it may be complexed, or tightly bound, in organic soils. To overcome these problems, iron (and the other metallic micronutrients, Mn, Zn, and Cu) may be applied to the soil or plant in the form of a *chelate* (Figure 9-3).[2] A chelate (from the Greek *chele*, meaning claw) is a complex organic molecule that encloses the iron ion in such a way as to prevent it from being precipitated or complexed by other ions or substances in the soil solution. Although chelate chemistry and chelate-plant interactions are not thoroughly understood, some experts believe that the chelate molecule containing the micronutrient ion is absorbed in its entirety and then the substances in the plant cell "out-compete" the chelate for the enclosed ion. Micronutrients may also be supplied by other carriers.

Manganese

Manganese is important in activation of enzyme systems and is a component of many enzymes. It also is involved in the

----

[2] Examples of chelates include: EDTA (ethylenediaminetetra-acetic acid), DTPA (diethylenetriaminepentaacetic acid and EDDHA (ethylenediaminedi-*o*-hydroxyphenylacetic acid). Several commercial formulations of chelates are available for agricultural use.

FIGURE 9-3. Spraying chelated iron on nursery container stock. *Inset*: chelated iron source.

cleavage of water essential to the process of photosynthesis. Although it is not a constituent of the chlorophyll molecule, its deficiency symptoms usually involve chlorosis. This suggests an important role for manganese in chlorophyll synthesis or maintenance. Manganese behaves remarkably like iron in most soils, although it is seldom present in such high amounts as iron. However, manganese can reach toxic levels in potting media containing soils found along the Atlantic coasts (U.S. and Europe) when soils are steam treated at too high a temperature for too long. It becomes more readily available as pH is reduced and is relatively insoluble at high pH levels. In some cases of manganese deficiency, chelated manganese can be applied to correct the problem.

Zinc

Zinc is a component of many enzymes and is involved in synthesis of the amino acid, tryptophan. It is critical to plant growth because tryptophan is a precursor of the important plant hormone **indoleacetic acid** (auxin) (see Chapter 11). Deficiency symptoms often reflect the lack of auxin—for example, "white bud" of corn, "little-leaf" disease, and pecan rosette—all suggesting problems related to auxin-mediated growth processes. In most mineral soils zinc *availability* is more often a problem than actual lack of zinc. However, some sandy soils may be truly deficient in zinc content. Zinc availability follows a pattern similar to that of iron and manganese; that is, it is more available at low pH and becomes relatively insoluble at higher pH levels.

Copper

Copper is a component of many enzymes and of one of the electron carriers in photosynthesis. Copper deficiency is considerably more rare than a deficiency of iron, zinc, or manganese. Because it can be complexed by organic matter, copper deficiency is most commonly observed in plants grown on organic soils. Onions produced on organic soils may exhibit copper deficiency symptoms, usually manifested by reduced bulb size and abnormally light-colored exterior scales.

Boron

Boron is thought to be involved in DNA and RNA synthesis and is important in translocation of sugars in plants. Cell division, cell development, or both may be drastically reduced by a boron deficiency. Because boron is relatively immobile in the plant, a deficiency results in cessation of root and stem apical meristem growth, which often leads to death of the terminal meristem in boron-deficient plants. Boron content of soils varies considerably, but it is generally soluble at low pH and becomes less available at higher pH values. Borax is therefore added to many fertilizer formulations to be used in regions of high soil pH. Boron is often held in organic compounds in the topsoil where it is available for crop use. Because it is relatively soluble at acid pH levels, leaching losses may occur in sandy soils.

Molybdenum

Molybdenum is required in such minute amounts that there are more than 10 million hydrogen atoms in a healthy plant for each atom of molybdenum. The addition of 35 to 70 *grams* of molybdenum per hectare (0.5–1.0 ounce/acre) will suffice to correct a deficiency in a soil lacking adequate molybdenum. Deficiency may be severe in certain acid soils, especially in Australia and New Zealand. In such cases, liming may be sufficient to make existing molybdenum levels adequately available. Molybdenum's functions in plants are not well delineated, but it appears to be involved in phosphorus metabolism. It also has been demonstrated to be essential for nitrogen fixation by symbiotic and free-living N-fixing bacteria. Deficiency symptoms in legumes therefore often appear as nitrogen deficiency symptoms. Restriction of normal expansion of the leaf lamina in cole crops, especially cauliflower, is another example of molybdenum deficiency, but the reasons for this restriction are not clear.

Chlorine

Chlorine appears to have an as-yet-undefined role in root development and enhances the electron transport reactions of photosynthesis. Deficiencies of chlorine in the field are nearly impossible to demonstrate because chlorine is added naturally to the soil by precipitation and is commonly present as an impurity in many fertilizer salts. These facts probably account for the inability to prove chlorine's essentiality until relatively recently, since it was previously nearly impossible to create chlorine-free test solutions.

Nickel

Nickel is the element most recently discovered to be essential for plant growth. In legumes, the enzyme urease is now known to contain nickel. Nickel deficiency appears as a necrosis of leaflet tips in soybeans and certain other legumes, probably because of the buildup of toxic levels of urea due to an insufficiency of the urease necessary for urea metabolism.

Micronutrient Toxicity and Interactions

Micronutrients are by definition required in very small quantities. However, they are all harmful when large amounts of the available form are present in the soil solution. For example, 1.0 ppm of boron in irrigation water may be required for good crop growth, but slightly higher levels may cause toxicity for some crops. Sensitive crops such as beans and Jerusalem artichoke may be damaged by as little as 0.5–1.0 ppm boron, whereas beet, lettuce, and asparagus may tolerate up to 10 ppm. Similarly, the range for other micronutrients in which plants will grow is narrow.

One manifestation of excess amounts of a given nutrient, in addition to evidence of a direct toxic effect, is induction of a deficiency of another nutrient. Iron, manganese, zinc, and copper can interact competitively with one another; thus an excess of iron, copper, or zinc can cause an induced manganese deficiency even when the soil test may indicate the presence of adequate manganese levels. In the same manner, excess zinc, copper, or manganese can induce iron deficiency in soils containing apparently sufficient amounts of iron. This often occurs in Florida vegetable production areas where copper sprays have been applied to control bacterial diseases (e.g., bacterial spot of tomato). Because copper is relatively immobile in the soil, use of copper sprays over several years' time frequently leads to a build-up of copper in the soil and a copper-induced iron deficiency on crops grown in subsequent years.

FERTILIZER PRACTICES

How does the horticulturist determine a crop's fertilizer requirements—which fertilizer to use, in what quantity, how often, and how it should be applied? These questions can be answered by careful analysis of the nutrient status and requirements of the plant and of the nutritional characteristics of the soil or medium in which that plant is being grown.

Soil Analysis

Generally, the best information about a soil's nutrient status is obtained by use of a professional soil test. Most testing services will analyze a field soil for organic matter, phosphorus and potassium content and will determine the physical analysis and the pH. Services will provide more detailed analyses for container soils, assessing also nitrate and ammonium N, soluble salts, calcium, magnesium, iron, and other micronutrients. Recommendations based on the tests generally tell the grower a range and type of fertilizer to apply for a particular crop and suggest whether pH adjustment is necessary (Figures 9-4 and 9-5).

If soil pH is decidedly acid, or more acid than appropriate for a particular crop, application of agricultural lime may be required (Table 9-3). Agricultural lime is finely ground limestone composed primarily of calcium carbonate ($CaCO_3$) or combinations of limestone and **dolomitic** limestone [$CaMg(CO_3)_2$]. The application of agricultural lime is termed **liming**, and it is generally accomplished by spreading ground limestone onto the land prior to plowing.

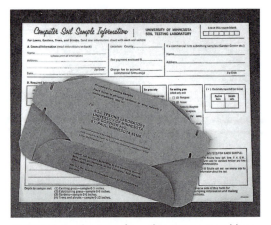

FIGURE 9-5. A copy of a soil test report and box for submitting sample. Many soil testing services are now computerized, which enables rapid reporting of results.

Liming with dolomitic limestone is preferable for soils that are low in magnesium. Most soil testing services will recommend an amount of agricultural lime to apply based on the desired pH, the crop requirements, and the soil type. When more rapid pH changes are desired, calcium oxide ("burned lime," CaO) or calcium hydroxide ("slaked lime," $Ca(OH)_2$) can be used. The materials, regardless of source, are added to cause a displacement of hydrogen ions from the soil colloids and to replace them with calcium or magnesium ions, thus rendering the soil less acid.

It is more difficult to change alkaline soils to a neutral or more acid condition. Acidification of soils by the addition of acidic organic matter such as acid peats is often practiced in order to make the soil favorable for the growth of species such as blueberries, rhododendrons, and azaleas. The effect of such a practice is short-lived because the peat is rapidly oxidized by soil microorganisms and is impractical on a field scale, so the addition of sulfur or sulfur-containing compounds may be recommended (Table 9-4). Ferrous sulfate works well for many of the acid-loving crops mentioned. It not only acidifies the soil, but also provides soluble iron, which is especially helpful in the nutrition of these

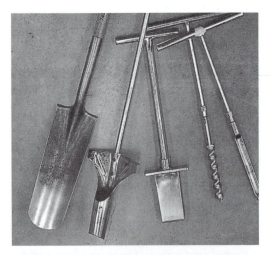

FIGURE 9-4. Equipment for taking a sample for a soil test. When an auger or probe is unavailable, the spade is used to take a 2.5 cm slice of soil 15 to 20 cm deep. The sides are cut away leaving a 2.5 cm-wide strip, which approximates the core taken by a soil probe.

TABLE 9-3. Limestone Needed to Change the Soil Reaction

CHANGE IN pH DESIRED IN PLOW-DEPTH LAYER	LIMESTONE (LB/ACRE)					
	SAND	SANDY LOAM	LOAM	SILT LOAM	CLAY LOAM	MUCK
4.0–6.5	2600	5000	7000	8400	10,000	19,000
4.5–6.5	2200	4200	5800	7000	8,400	16,200
5.0–6.5	1800	3400	4600	5600	6,600	12,600
5.5–6.5	1200	2600	3400	4000	4,600	8,600
6.0–6.5	600	1400	1800	2200	2,400	4,400

Source: Lorenz, O. A. and D. N. Maynard. *Knott's Handbook for Vegetable Growers* (2nd ed.). Wiley Interscience, New York, 1980.

TABLE 9-4. Approximate Quantity of Soil Sulfur Needed to Increase Soil Acidity to about pH 6.5

CHANGE IN pH DESIRED IN PLOW-DEPTH LAYER	SULFUR (LB/ACRE)		
	SANDS	LOAMS	CLAYS
8.5–6.5	2000	2500	3000
8.0–6.5	1200	1500	2000
7.5–6.5	500	800	1000
7.0–6.5	100	150	300

Source: Lorenz, O.A. and D. N. Maynard. *Knott's Handbook for Vegetable Growers* (2nd ed.). Wiley Interscience, New York, 1980.

species. **Flowers of sulfur** is a nearly pure form of sulfur in a powdered form. It is oxidized by soil microflora and is four to five times as effective as ferrous sulfate in reducing pH, depending on soil buffering capacity and original pH. Where only slight acidification is desired, or for maintenance of acidity level, use of acidifying fertilizers may be effective. The choice of fertilizer depends on the degree of acidification desired, because different fertilizers have different acidifying potentials. For example, ammonium sulfate [$(NH_4)_2SO_4$] has about three times the acidifying power of ammonium nitrate (NH_4NO_3) or urea, with equal amounts of nitrogen applied. However, all nitrogenous fertilizers are not acidifying; sodium nitrate ($NaNO_3$) and potassium nitrate (KNO_3) will actually make a soil slightly more alkaline.

Tissue Analysis

Plant tissue analysis is becoming increasingly important in making decisions about fertilization practices. Such analyses are often performed on leaves or parts of leaves; hence the terms **foliar analysis** and **tissue analysis**, which, in relation to fertilizer practice, are essentially synonymous (Figure 9-6). There are two major barriers to application of the results of tissue analyses to fertilizer practice. First, the chemical composition of plant tissues varies with the physiological age and the relative position and type of organ. Second, climatic conditions vary over the growing season and can exert a profound effect on the chemical composition of plant tissues. Furthermore, for annual crops, tissue analyses must be done quickly in order for the grower to make fertilizer applications in a timely and effective fashion. It is therefore necessary to standardize the sampling procedure in terms of timing and selection of plant parts for analysis. These sampling procedures vary considerably among plant species and types, so most critical values for a given plant have been determined empirically. Unfortunately, a systematic effort has been made for only a small number of crops. For this reason, it is difficult to generalize about required levels of specific nutrients.

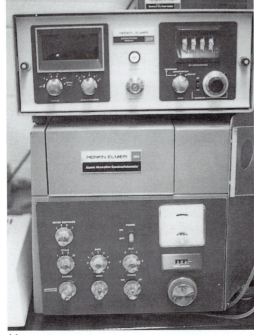
(a)

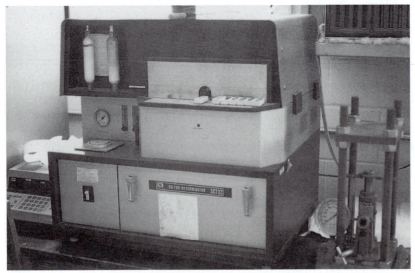
(b)

(c)

FIGURE 9-6. Equipment used for analysis of nutrients in plant tissue samples. (a) atomic absorption spectrophotometer; (b) nitrogen; and (c) sulfur analysis devices.

Both soil and tissue analyses may often be prone to error and are frequently difficult to interpret accurately. Many times this makes it difficult to make precise fertilizer recommendations based on either testing procedure alone. As a consequence, recommendations may be based on results of both soil tests and foliar analyses. This approach has reached a high level of sophistication in pineapple production and for many floricultural crops. The nutrient requirements peculiar to a specific crop and the crop production expectations, such as desired yield level and certain product quality characteristics, may further modify the way soil and foliar analysis information is applied to recommendations for that crop.

C:N Ratio

Another form of tissue evaluation involves an idea known as the **carbon:nitrogen (C:N) ratio**. This concept, sometimes referred to as the carbohydrate:nitrogen ratio, refers to the interrelation of the nitrogenous and carbonaceous components of a plant. In this regard, the work of E. J. Kraus and H. R. Kraybill is of historical, and to a certain degree practical, significance. Their work published in 1918, suggested that the C:N ratio governs the floral initiation process. They used tomato plants as a model and demonstrated that flowering was delayed in plants receiving large amounts of nitrogen while growing under environmental conditions conducive to rapid photosynthesis. If the nitrogen level was low, however, flowering was stimulated when plants were grown under the same environmental conditions. Both vegetative growth and flowering were reduced if the plants were grown under low nitrogen levels and an environment limiting photosynthetic activity.

In deciduous tree fruit production, growers have applied this knowledge to encourage flowering by employing various practices that tend to shift the C:N ratio toward a balance favoring carbohydrate storage in the tissue. Examples include reduced nitrogen fertilization, use of cover crops to help deplete soil nitrogen, root pruning, bending branches downward to slow growth, and ringing (cutting) the bark to create a temporary phloem blockage and thus an accumulation of carbohydrates above the "ring." Although it is apparent from this and other evidence, that flowering may be reduced by high nitrogen levels that lead to stimulation of vigorous vegetative growth, floral *initiation* is no longer thought to be controlled exclusively by nutritional factors. Rather, it is now thought that initiation of flowers in most species is also influenced by hormones (discussed in Chapter 12), although the flower initiation and developmental processes may be modified by C:N ratio and other nutritional factors. Also C:N ratios have little importance in floral initiation of species in which photoperiod or temperature is the trigger, as noted in Chapters 5 and 6.

An often overlooked but important facet of Kraus and Kraybill's work is their observation that the C:N ratio had a significant influence on adventitious root formation by cuttings. Stem cuttings taken from plants receiving abundant nitrogen and growing under conditions facilitating photosynthesis did not root as well as those taken from the less vigorous plants that had received less nitrogen. It has been suggested that the rapid shoot elongation induced by the higher levels of nitrogen prevents accumulation of carbohydrate reserves essential to adventitious root formation. In this way, the rapid shoot growth is competitive with the rooting process. Propagators take advantage of this knowledge by exercising control over nitrogen fertilization of stock plants; they must maintain a balance conducive to sufficient shoot growth to provide an adequate supply of cuttings, but the shoot growth must not be so vigorous that rooting will be impaired. The highly succulent shoots produced as a result of high nitrogen levels may also be more vulnerable to the attack of certain insects and disease organisms.

FERTILIZER SOURCES

Fertilizing With Organic Materials

Both plant and animal wastes may be termed **manures** when they are used to fertilize the soil. When used as fertilizers, they add nutrients to the soil in amounts that depend on the organic matter source. Even materials that contain low amounts of mineral nutrients may materially benefit the soil, because many breakdown products of the organic matter (often referred to as *humus*) serve to improve soil structure. We will consider such organic matter additions to the soil under the categories of plant residues and animal wastes.

Plant Residues

Plant waste products or residues include a wide range of materials, which will be decomposed by microbial action when added to the soil. When the bacteria, fungi, and other organisms involved in this process die, their remains will add nutrients to the soil that then become available for uptake by plant roots. In this fashion nutrients are *recycled*. If the plant residue added to the soil has a high C:N ratio, the soil microorganisms will use nitrogen from the surrounding soil for their growth as they metabolize the carbohydrate, thus creating a temporary nitrogen deficiency for plants growing in that soil. Therefore, if a plant residue with a high C:N ratio is added to the soil, application of nitrogenous fertilizer is recommended. Straw is a plant material that is composed primarily of the stems of cereals left after threshing to remove the grain; it has a relatively high C:N ratio, generally about 50:1. Consequently, when straw is used as a mulch or is incorporated into the soil to add organic matter, addition of a nitrogenous fertilizer is necessary to provide nutrition for the microorganisms to facilitate decomposition of the straw, and to avoid an induced nitrogen deficiency. This simple measure enables the energy in the straw to be made more rapidly available to the microorganisms without affecting crop growth.

Similarly, other plant residues may be incorporated into the soil to add nutrients and improve soil structure. We can readily see that their value for these purposes varies with their nutrient content and C:N ratio. Materials such as bark and sawdust have an extremely wide C:N ratio and a relatively low level of nutrients, making them less valuable as organic additions to a soil than materials such as alfalfa and clover. Because proteins contain large amounts of nitrogen, high-protein plants or plant residues added to the soil generally require little or no additional nitrogen fertilizer. By-products from the oil-seed industry such as cottonseed meal and other materials often used as protein supplements for livestock feeds generally fall into this category.

Green Manures

Fresh green plant material grown to be incorporated into the soil for soil improvement is termed **green manure**. Cover crops such as rye therefore serve a dual purpose: When the cover crop is plowed under they help with erosion control, provide additional nutrients, and improve soil structure. Leguminous species used as cover crops, or green manure crops, may add significant amounts of nitrogen to the soil because of their nitrogen-fixing ability, thus reducing the need for additional fertilizer nitrogen.

Composting

Plant residues can also be made into useful soil additives by the process known as *composting*. Almost any vegetable matter can be converted into a useful soil amendment by this procedure, but coarse materials and materials with high C:N ratio are slow to decompose. Animal by-products and wastes should be avoided for composting, since materials such as meat scraps may cause an objectionable odor upon decomposition and may also attract vermin. For composting to be successful, several essential elements must be included: (1) vegetable matter such as stems, leaves, crop wastes, and lawn clippings; (2) a small amount of field soil to serve as a source of inoculum with microorganisms required for decomposition of the vegetable matter; (3) a source of nitrogenous fertilizer to ensure rapid microbial activity (avoiding C:N imbalance); and (4) ground limestone to keep the pH at a suitable level for the composting process to proceed rapidly (Figures 9-7 and 9-8). A bin, pit, or enclosure may help to confine the material being composted and to facilitate composting operations. Ideally, the vegetable matter should be a mixture of both coarse and fine particles. Coarse materials aid in aeration, but when used alone will cause excessive drying out of the compost leading to poor composting; finer materials add water-holding capacity to the mixture, but if used alone may pack tightly and create anaerobic conditions that retard the composting process. For composting to progress rapidly, as is true for microbial decomposition of organic matter in a soil, moisture and air must be in balance. This is why compost should be turned about once a month to provide aeration and should be soaked periodically if rainfall is infrequent.

Most of the bacteria and fungi that are active in the composting process are aerobic and exothermic (i.e., they require oxygen and they give off heat). It is this latter characteristic that leads to a build-up of heat in the compost. The interior of a properly made compost pile will heat up to temperatures in the range of 50° to over 60°C (150° to 170°F). This temperature build-up is one of the desirable features of a compost pile, because it will kill most insects, weed seeds, and pathogenic fungi and bacteria that may be present. The product that results from the composting process is therefore a useful, relatively clean material that can be added to the field or garden for soil improvement. By composting, the horticulturist is effectively

recycling nutrients and creating valuable humic materials, thus enabling a more efficient use of available resources.

Animal Wastes

Dung, or fecal materials from animals readily come to mind when one thinks of animal wastes for use as fertilizers. Such materials are often referred to as farmyard manures, or simply manures, and their application to soils probably represents the

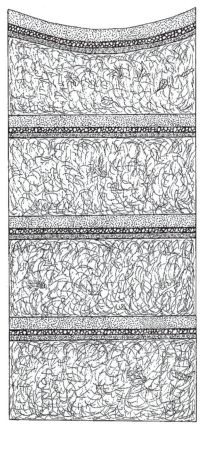

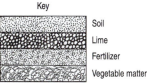

Key

Soil

Lime

Fertilizer

Vegetable matter

FIGURE 9-7. Cut-away view of a compost bin, showing alternating layers of vegetable matter, fertilizer, ground limestone, and soil.

FIGURE 9-8. View of compost pile illustrating alternating layers of coarse and fine textured plant matter (*left*) and a compost pile showing a depression near its center resulting from the breakdown of composted materials (*right*). Note that the exothermic reaction of the composting process has melted the snow.

most ancient of fertilizer uses. Animal manures may consist in various combinations of: (1) the material that was used as bedding or litter for the animals, such as straw, wood shavings, or other dry plant matter; (2) the food that has been passed through the animal and voided as a somewhat solid feces; and (3) the urine, which is composed of water and various soluble metabolic wastes. The mineral nutrient content of such manures will vary greatly with the relative proportions of bedding, feces, and urine; with the species of animal and its diet; and with the age and manner of storage of the manure. A significant amount of an animal's nitrogenous wastes are voided in the urine. Losses of such nitrogenous substances by volatilization, run-off, and leaching may be extremely high, depending especially on the bedding material and storage methods. Nitrogen and other soluble nutrients can also be leached from the solid portions of manures if manures are stored outdoors and are subject to rainfall, thus reducing the value of the manure as a fertilizer and potentially polluting ground water supplies.

Manures should generally be thought of as low analysis fertilizers (see Table 9-5), but they have value beyond their contribution of nitrogen, phosphorus, and potassium. Like most organic materials, they can provide small but significant amounts of

micronutrients. They also are particularly valuable for their influence on soil structure, producing break-down products that are effective in the aggregation of soil particles described earlier. The humic substances produced by their decomposition also aid significantly in improving the nutrient-and water-holding ability of a soil (especially coarse-textured soil types).

The potential of salt or ammonia burn from the urea in fresh animal manure is high, so, well-rotted (composted) manures should be applied to gardens and fields. Well-rotted manures contribute substantially to the soil structure but little to the fertilizer value. However, it is a common practice for farmers to spread relatively fresh manure directly onto field soil. Because the crop will be planted after weathering and leaching occur, no damage results.

Meat Processing and Other Animal Wastes

Numerous materials occur as by-products or wastes from slaughterhouse, meat-packing, fish processing, and similar enterprises. Some of the more commonly available of these materials include dried blood, steamed bone meal, hoof and horn meal, tankage, feather meal, and fish meal. These and other similar products vary greatly in their nutrient content and rate of availability.

TABLE 9-5. Organic Materials as Sources of Nutrients

	AMOUNT PROVIDED (PERCENT ON DRY WEIGHT BASIS)		
ORGANIC MATERIAL	N	P_2O_5[b]	K_2O[b]
Dried blood	13.0	2.0	1.0
Fish meal	10.0	6.0	—
Steamed bonemeal	1.0	15.0	—
Soybean meal	7.0	1.2	1.5
Wheat straw	0.6	0.15	1.0
Alfalfa hay	2.5	0.55	2.5
Cow manure[a]	1.0–2.1	1.0–3.1	1.9–3.0
Chicken manure[a]	1.5	1.8	2.0
Pig manure[a]	2.3	2.1	1.0

[a]Manure only, contains no bedding materials.

[b]Numbers given are P_2O_5 and K_2O equivalents.

Dried blood varies from about 10 to 14 percent nitrogen and is relatively rapidly decomposed. As such, it should be thought of as a relatively fast-release fertilizer; it is, in fact, nearly as fast-release as many of the highly soluble inorganic fertilizers. Therefore fertilizer burn, or soluble salts injury, can be as serious a problem with some organic fertilizers as it can be with inorganic fertilizers if the rate of application is too high. By contrast, materials such as hoof and horn meal and steamed bone meal are slowly decomposed and thus should be thought of as very slow release fertilizers. Steamed bone meal, in particular, is popular with home gardeners because over-fertilization is rare (i.e., it can be applied in large amounts without detrimental effects on the plants being fertilized).

Industrial and Municipal Wastes

Vast quantities of waste materials are generated by industries and municipalities every day. These wastes contain appreciable amounts of mineral nutrients that offer potential benefits as fertilizers. In recent years a great deal of attention has been given to research aimed at utilizing municipal sewage as a fertilizer. The majority of this research has focused on solid or semi-solid sewage waste materials that have received secondary treatments rendering them free of organisms that might constitute a health hazard to humans.[3] The mineral content of these materials tends to be slowly available and somewhat variable in content. However, most sewage materials generally contain appreciable but small amounts of nitrogen, phosphorus, and potassium as well as many other nutrients. One serious concern about these materials is the tendency to contain excesses of heavy metals such as lead and cadmium. The presence of certain industries in a municipality will contribute to the heavy metal content of the sludges, occasionally to such an extent that the sludge is not recommended for use in fertilization of food crops, especially those where the edible portion is the leaves, stems, or roots.

....

[3]Raw human wastes have been used as fertilizer in developing countries, but disease-causing organisms may be present, necessitating treatment to inactivate them.

INORGANIC FERTILIZERS

It is difficult to separate **organic** from **inorganic** fertilizers. Because sources and processing methods vary greatly, the form in which nutrients are absorbed by the plant is ionic, and therefore inorganic. Generally speaking, fertilizers manufactured from sources other than plant and animal remains are called **inorganic fertilizers.** The classic experiments conducted in England by John Lawes in the first half of the 19th century (ca. 1840) resulted in the first production and use of an "artificial" chemical as a fertilizer. He treated bones with sulfuric acid, producing a new material which he called **superphosphate.** He subsequently discovered that he could manufacture a similar superphosphate by treating rock phosphates with sulfuric acid. This material proved to be an effective fertilizer that provided a readily available source of phosphorus for crop growth. Lawes continued his experiments at his farm in Rothamsted, producing additional manufactured fertilizers and testing their effects on several crops. He laid out strips of crops treated with the various fertilizers in order to study the subject of soil fertility scientifically. Those experiments have been continued for well over a century and Rothamsted today represents one of the oldest of many similar agricultural experiment stations found throughout the world.

Fertilizer Carriers

Nitrogen

Organic fertilizer carriers such as cottonseed meal, dried manures, and fish meal constitute less than two per cent of the nitrogen sold as commercial fertilizers, probably because of high costs and limited availability. Because of this fact and because nitrogen is the mineral element required in the largest quantities by plants, inorganic nitrogen sources assume a great level of importance in the fertilizer industry. Several inorganic nitrogen sources are listed in Table 9-6. One of the most economical sources is ammonia, a gas produced from nitrogen gas taken from the atmosphere and combined with hydrogen gas from natural gas (CH_4), as follows:

$$N_2 + 3H_2 \longleftrightarrow 2NH_3$$

This simple reaction is extremely important, because it results in production of the least expensive and by far the most heavily used fertilizer nitrogen source. In fact, about 98 percent of the world's fertilizer nitrogen is ammonia or an ammonia product. Ammonia is the first step in formation of many other important fertilizer materials and precursors, such as anhydrous ammonia, ammonium nitrate (NH_4NO_3), urea [$(NH_4)_2CO_3$], ammonium hydroxide (NH_4OH), nitric acid (HNO_3), and ammonium phosphates.

Other important fertilizer nitrogen carriers include sodium nitrate ($NaNO_3$), ammonium sulfate [$(NH_4)_2SO_4$], and calcium cyanamid. Sodium nitrate is of particular historical significance as a nitrogen source because it occurs as a natural, relatively pure salt that is mined from saltpeter deposits such as those found in Chile (hence the term "Chilean nitrate"). A considerable portion of the sodium nitrate used in fertilizers is now produced by treating Na_2CO_3 with nitric acid. Ammonium nitrate usage is increasing, but it is important to remember that it is a potentially hazardous material because of its explosive properties. For this reason it normally is supplied in the "prilled" form—pelleted and coated with an inert material.

Phosphorus

Superphosphates are the primary form of phosphorus for fertilizer use. Ordinary superphosphate is essentially the same superphosphate demonstrated by Lawes' experiments and contains about 20 percent P_2O_5 equivalents, with appreciable amounts of calcium and sulfur present in

TABLE 9-6.   Common Inorganic Sources of Some Essential Nutrients[a]

	Source	Approximate Percent of Element	Other Nutrients Supplied
Nitrogen (N)	Ammonium sulfate	21%	S
	Ammonium nitrate	33.5[b]	—
	Calcium nitrate	15.5	Ca
	Potassium nitrate	13.0	K (44.0%)
	Urea	46.0	—
	Sodium nitrate	16.0	—
	Anhydrous ammonia	82.0	—
Phosphorous (P)	Diammonium phosphate	48.0	N (18%)
	Phosphoric acid solution	54.0	—
	Single superphosphate	20.0	Ca, S
	Concentrated superphosphate	46.0	—
Potassium (K)	Potassium nitrate	44.0	N (13.0%)
	Potassium sulfate	53.0	S
	Potassium chloride	62.0	C1
Iron (Fe)	Ferrous sulfate	19.0	S
	NaFe DTPA (chelate)	10.0	—
	NaFe EDDHA (chelate)	6.0	—
Manganese (Mn)	Manganese sulfate ($3H_2O$)	28.0	S
Copper (Cu)	Copper sulfate ($5H_2O$)	25.0	S
Zinc (Zn)	Zinc sulfate ($7H_2O$)	23.0	S
	NaZn NTA (chelate)	13.0	—
Boron (B)	Borax	11.0	—
	Boric acid	17.0	—
Molybdenum (Mo)	Sodium molybdate	39.0	—
	Ammonium molybdate	54.0	N[c]

[a]Calcium and magnesium fertilization is usually accomplished by application of agricultural lime to acid soils and is unnecessary for high pH soils.

[b]Theoretically 35.0 percent, but because of its explosive nature in the pure state, it is "prilled" with an inert clay additive.

[c]Of negligible consequence, since ammonium molybdate is normally applied at extremely low rates [rarely above 450 grams/hectare (1 pound/acre) and usually much lower].

the form of gypsum ($CaSO_4$). Concentrated superphosphates, often called triple superphosphates are produced by treating ordinary superphosphate with phosphoric acid (H_3PO_4) and they typically contain about 46 percent P_2O_5 equivalents. The ammonium phosphates are made by combining phosphoric acid with ammonia. Phosphoric acid itself may be used as a fertilizer, but because of its corrosive nature is more commonly used in the manufacture of the more easily handled materials already mentioned. Basic slag, a by-product of the steel industry, is more commonly used in Europe than in the United States, but is a desirable phosphorus source for acid soils because of its high content of calcium hydroxide.

Potassium

The principle source of potassium is salt beds or mines found in various parts of the world. The potassium may be present as crude salts, which must be refined, or as relatively pure salts. The two main potassium salts are muriate of potash (KCl) and potassium sulfate (K_2SO_4). Both are quite soluble. Some fertilizer use is also made of potassium-magnesium sulfate, especially where soils are deficient in magnesium.

Other Nutrients

Many of the other nutrients are supplied as components of materials employed to supply one or more of the three primary nutrients, nitrogen, phosphorus, and potassium (see Table 9-6). Calcium and sulfur are commonly provided when ordinary superphosphate is used, and, as mentioned earlier, both calcium and magnesium are supplied in nutritionally important amounts when dolomitic limestone is applied to soils to correct acidity. Many of the micronutrients are present in small quantities as impurities in other fertilizers, especially in low-grade salts and mixed fertilizers. When purer materials are chosen to provide the macronutrients, pure salts (often sulfates) or chelates may be employed to provide requisite micronutrients.

Fritted trace elements are nutrients contained in finely ground glasslike materials to which the desired nutrient or nutrients were added while the glass was still molten. The molten glass was selected for degree of solubility so that as the glass gradually dissolves, the nutrients are released over an extended period of time. Fritted nutrients of this type are primarily used in container media for production of bedding plants and various other greenhouse crops.

Mixed Fertilizers

Mixed fertilizers are commonly used for both field and greenhouse crop production. Generally speaking, a mixed fertilizer contains at least two, often all three, of the primary nutrients, nitrogen, phosphorus, and potassium. A mixed fertilizer is made by combining two or more fertilizer carriers and is said to be a **complete** fertilizer if it contains all three elements. The **analysis** of a mixed fertilizer is the percent of total nitrogen (N), available phosphorus, stated as P_2O_5 equivalents, and water-soluble potassium, expressed as K_2O equivalents, in that order (i.e., $N–P_2O_5–K_2O$) (Figure 9-9). Although modern analytical methods now make it possible to express fertilizers as pure elements—that is, N–P–K—nearly all commercially available fertilizers sold in the United States still bear the traditional $N–P_2O_5–K_2O$ analysis on the container label. However, it is a simple matter to convert P_2O_5 to P by multiplying by 0.437, and K_2O to K by multiplying by 0.83. For example, a fertilizer labelled 20–20–20 is 20–8.74–16.6 when expressed in the elemental form. It is preferable to express fertilizers as percentages of N–P–K, because most do not, per se, contain P_2O_5 or K_2O.

The fertilizer analysis must always be stated on the container, along with the weight, name, or brand and the name and address of the manufacturer. In the United States, the laws governing additional information vary from state to state but often may also require the specific amounts of the nitrogen from nitrate and ammonium sources and the acid-forming tendency, expressed as pounds of calcium carbonate required to neutralize a ton of the fertilizer. Fertilizer labels do not usually state exact ingredients, the presence or absence of filler material, or the amount of nutrients other than N, P, and K.

The fertilizer **ratio** expresses the fertilizer analysis as a sequence of the lowest whole numbers that can be obtained by dividing the analysis by the same number (e.g., the ratio of a 4–8–4, 5–10–5 or 10–20–10 fertilizer is 1–2–1). This knowledge is helpful when a particular analysis is unavailable and another analysis of the same ratio can be substituted. It is also

FIGURE 9-9. Fertilizer bag showing analysis statement (*top* and *bottom*).

FERTILIZER APPLICATION METHODS

The form of the fertilizer material, crop requirements, equipment available, and environmental conditions dictate the method to be used when fertilizing horticultural crops. Fertilizers may be applied as either solids or liquids, but applications of fertilizers in the solid (usually granular) form have been most common.

Broadcast

In the broadcast method, the fertilizer is spread uniformly over the field or plot. For row crops this is generally done before plowing, or at least prior to planting, and is incorporated into the soil by various implements. For turf, the fertilizer is broadcast directly onto the plants, often referred to as **top-dressing**. Indeed, most landscape plantings are fertilized by various broadcast methods. An example of a commonly used type of spreader for home use is depicted in Figure 9-10. Irrigation may be employed to dissolve the fertilizer, to wash it into the root zone, or to reduce the possibility of salt injury by diluting a concentrated fertilizer. Various implements are available for spreading (or broadcasting) fertilizers, including use of airplanes in special cases. Applications of limestone to adjust pH are also usually applied by broadcasting.

beneficial when comparing fertilizer prices, because fertilizers of the same ratio should have roughly the same fertilizer value if equal amounts of nitrogen are applied. For example, if a 10–20–10 fertilizer costs less than twice as much per pound of nitrogen as a 5–10–5, the 10–20–10 is more economical because it would require only one-half as much fertilizer to deliver the desired amount of nutrients.

FIGURE 9-10. Commonly used types of fertilizer spreaders for broadcast applications of fertilizer or materials for adjusting pH.

Banding

Fertilizer placement in a concentrated location, or **band**, at the time of planting is intended to stimulate early crop growth and is called **banding** (Figure 9-11). For large-seeded crops, banding is often done at a position slightly below and to the side of the seeds in the row, although recommendations of distance and position vary with the crop. When employing this method, growers must take care to avoid placing a too-concentrated, or high-nitrogen fertilizer in proximity to the seed, because high salt levels could cause severe damage to the germinating seed or young seedling.

Side-dressing

Application of fertilizer along the row or adjacent to the plant after the crop is growing is referred to as **side-dressing**. This method is particularly appropriate for plants such as sweet corn and potatoes that have a heavy demand for nitrogen during the growing season. It also is used to replace soluble nutrients leached by early season irrigation or rainfall, especially in sandy soils. Timing of side-dress applications requires good judgement and experience with the crop.

Soil Injection

Anhydrous ammonia is applied by special equipment that is designed to inject liquid ammonia into the soil, usually to a depth of

FIGURE 9-11. Band placement of fertilizer. In this case, the fertilizer was placed along the row of seeds, 5 cm below and 5 cm to the side.

15 cm (6 inches) where it combines with water to form NH_4^+. This approach is necessary because the liquid ammonia will volatilize at ambient temperatures and will be lost into the air if applied to the soil surface. Other liquid fertilizers can be applied in a similar fashion, but it is not a common practice.

Direct Liquid Applications

Some nonpressurized liquid fertilizers and liquid manures from animal production facilities with "lagoon" systems may be applied directly to the soil. Volatilization losses may occur with this approach, especially if conditions are conducive to liberation of ammonia gas (high pH, for example).

Fertilization Through the Irrigation

As purer and more concentrated fertilizer forms have become available, the application of fertilizer through irrigation systems has become feasible. This may be accomplished by including very dilute amounts applied with every irrigation or by use of more concentrated fertilizers included periodically. **Fertigation**, as it is sometimes called, can be a highly economical method for delivering soluble nutrients, especially nitrogen. Even phosphorus can be effectively applied to crops through irrigation if added as phosphoric acid. For sprinkler and drip irrigation systems, special devices for injecting concentrated solutions into the irrigation water are employed to meter the proper amount into the water. Similar systems utilized for fertilizing greenhouse and nursery crops employ equipment usually referred to as **proportioners** or **injectors** [not to be confused with the aforementioned injection equipment used for anhydrous ammonia application] (Figure 9-12). When using fertigation techniques, it is important to use on-line filters and pure fertilizer salts to avoid plugging of nozzles and damage to pumping equipment.

FIGURE 9-12. Injection equipment (*top* and *bottom*) used to add concentrated liquid fertilizers to water used for irrigating greenhouse crops. Similar equipment is used for large-scale fertigation of field-grown horticultural crops.

Foliar Fertilization

Spraying soluble fertilizers onto crop foliage is a means of delivering supplemental amounts of major nutrients but is more commonly employed to correct deficiencies of micronutrients. Because the plant leaf is an organ that cannot absorb nutrients efficiently, soils experts do not recommend attempting to supply all or even most of the plant's major element requirements by foliar applications. However, because only minute amounts of micronutrients are required to correct a deficiency, foliar applications are often used for this purpose, and they give a quicker response than soil applications.

Slow Release Fertilizers

Plants generally require nutrients over an extended period, usually for a large portion of their active growth. Because of this requirement, the use of fertilizers that release their nutrients gradually should theoretically enable the horticulturist to supply a plant's entire nutrient needs with a single fertilization at planting time (Figure 9-13). This would reduce leaching losses of soluble nutrients and for some crops it would eliminate the need to apply side-dressings, top-dressings, sprays, or other supplemental fertilizer applications during the growing season. Many organic fertilizers are naturally slow release, because their nutrients are gradually released by the mineralization activities of soil microorganisms. The rate of release is dependent on environmental conditions that influence rate of microbial activity, such as levels of soil moisture and temperature. Some inorganic fertilizers can be considered slow-release also, such as magnesium-ammonium-phosphate minerals and agricultural limestone. The slow release characteristics of such materials depend primarily on particle size and the natural rate of dissolution. However, manufactured urea-formaldehyde complexes are released by soil microorganisms feeding on these long-chain polymers, breaking the linkages and releasing urea that can be further broken down for subsequent use by the plant.

Many fertilizer producers have sought to manufacture relatively pure fertilizers that are constructed in such a way as to release their nutrients at a preordained gradual rate. The slow-release characteristics of most of these fertilizers are based on coating either solid or liquid fertilizer with substances that ensure a gradual release of the nutrients. Such coatings may consist of semipermeable membranes, impermeable membranes with precisely spaced pin-holes to facilitate gradual release, impermeable membranes subject to microbial decomposition, and coatings of

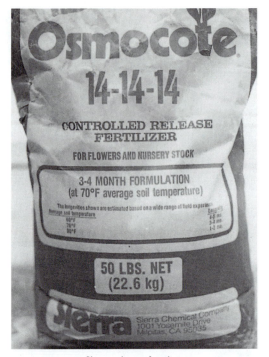

FIGURE 9-13. Slow-release fertilizer.

slowly soluble substances (e.g., sulfur-coated ureas). Another approach to slow release of nutrients has been the development of chemicals that delay and prevent nitrification. This keeps the nitrogen in the ammonium form for a longer period of time, thus slowing plant uptake and reducing leaching losses. Some researchers question the value of this technology and suggest that more research is required to determine the full extent of its usefulness for horticultural crops.

Timing Fertilizer Application

Timing of fertilizer application can be critical to achieving the desired stimulatory effect, without causing negative consequences. Generally, plants benefit most when fertilizer application is timed to coincide with the period of maximum need by the plants. Dry fertilizers should be applied when rainfall is expected or irrigation can take place. However, when soluble fertilizers such as nitrogenous fertilizers are followed by heavy rain or excessive irrigation, much of the nutrient value will be lost through leaching. Also, if fertilizers, especially those of high analysis, are applied onto leaves or turf without subsequent rain or irrigation, fertilizer "burn" (soluble salts injury) may occur.

It is recommended that landscape trees and shrubs be fertilized in the spring, taking care to avoid getting fertilizer on the foliage. An exception to this is that occasionally foliar application of dilute fertilizers will be applied to correct micronutrient deficiencies such as high pH-induced iron deficiency. It is essential that fertilization of most woody plants be terminated early enough in the summer to allow the plant to harden off. This is particularly important for most fruit crops and for many landscape plants. Late fertilization, especially when combined with adequate to excessive moisture, will often result in a new flush of vegetative growth or a continuation of soft vigorous shoot growth. Such new growth is highly vulnerable to early frost or freeze events and serious damage or even death of the plant may be the result.

In contrast to the above, if turfgrasses are only to be fertilized once, that single best time is in the early fall. Since vegetative growth and good green color are desired attributes of most turf installations, fall fertilization encourages much growth because fall is the most active period of growth for many turf species.

Fertilizer timing is also very important for houseplants and interiorscape plantings. Because such plants are growing very slowly in the winter months, fertilization is usually not recommended. When active growth resumes in the spring, application of fertilizer will result in healthy new growth by such plants. However, in some interiorscapes where growth is preferred to be limited, only sufficient fertilizer necessary to maintain healthy appearing plants will be applied.

Hydroponics

Sometimes referred to as "aquaculture" or "nutriculture," **hydroponics** is the cultivation of plants in water containing dissolved nutrients. Much of modern hydroponics is based on the solutions described by D. R. Hoagland and D. I. Arnon, and are derived from those proposed in 1860 by J. von Sachs (Table 9-7). It has been demonstrated that crops comparable to those grown in soil culture can be grown using hydroponic systems. However, hydroponics should not be regarded as a panacea for problems encountered in soil culture systems, since hydroponic systems may be more costly and require specialized management skills. For the most part, claims for the superiority of hydroponics are largely unfounded.

A commonly employed modified hydroponics approach has been to grow crops in beds of sand, gravel, or more recently, rockwool, and to periodically flush the beds with a complete nutrient solution such as Hoagland's. Between flushings, the solution is held in storage tanks containing submersible pumps that pump the solution back into the beds for the next flush. Rockwool is a substrate manufactured from molten rock that is spun into fibers in much the same fashion that cotton candy is made from sugar. The fibers are combined with a chemical binder and molded into shapes suitable for use as a root medium. Rockwool's porosity and water-holding ability make it a medium conducive to use as a substrate for hydroponic culture.

A hydroponics system that involves the use of no aggregate or substrate has received much recent attention. It involves bathing the roots of the plants in a thin *film* of nutrient solution, hence the term **nutrient film technique**, or NFT. NFT systems were popularized by A. J. Cooper in the 1970's and various modifications of his NFT system have been adopted by growers, especially in Europe, for production of greenhouse vegetables (Figure 9-14). Advantages of NFT versus gravel or sand bed systems include: lower volumes of nutrients and water are required, obtaining a substrate and disinfesting it for re-use are unnecessary, and more rapid crop and system replacement are possible. The reduced water requirement seems to be especially attractive to horticulturists in arid areas of the world.

TABLE 9-7. A Modified Nutrient Solution for Use in Hydroponic Plant Production

HOAGLAND'S NUTRIENT SOLUTION

SALT	STOCK SOLUTION (G TO MAKE 1 LITER)	FINAL SOLUTION (ML TO MAKE 1 LITER)
SOLUTION 1		
$Ca(NO_3)_2 \cdot 4H_2O$	236.2	5
KNO_3	101.1	5
KH_2PO_4	136.1	1
$MgSO_4 \cdot 7H_2O$	246.5	2
SOLUTION 2		
$Ca(NO_3)_2 \cdot 4H_2O$	236.2	4
KNO_3	101.1	6
$NH_4H_2PO_4$	115.0	1
$MgSO_4 \cdot 7H_2O$	246.5	2

MICRONUTRIENT SOLUTION

COMPOUND	AMOUNT (G) DISSOLVED IN 1 LITER OF WATER
H_3BO_3	2.86
$MnCl_2 \cdot 4H_2O$	1.81
$ZnSO_4 \cdot 7H_2O$	0.22
$CuSO_4 \cdot 5H_2O$	0.08
$H_2MoO_4 \cdot H_2O$	0.02

IRON SOLUTION

Iron chelate, such as Sequestrene 330, made to stock solution containing 1 g actual iron/liter. Sequestrene 330 is 10 percent iron, thus 10 g/liter are required. The amounts of other chelates will have to be adjusted on the basis of their iron content.

Procedure: To make 1 liter of Solution 1, add 5 ml $Ca(NO_3)_2 \cdot 4H_2O$ stock solution, 5 of KNO_3, 1 of KH_2PO_4, 2 of $MgSO_4 \cdot 7H_2O$, 1 of micronutrient solution, and 1 of iron solution to 800 ml distilled water. Make up to 1 liter. Some plants grow better on Solution 2, which is prepared in the same way.

Source: Adapted from D. R. Hoagland, and D. I. Arnon, *The Water-Culture Method for Growing Plants Without Soil*. California Agricultural Experiment Station Circular 347 (1950).

FIGURE 9-14. A nutrient film hydroponic system (*left* and *right*) for production of greenhouse vegetables.

Hydroponics as an alternative approach to soil culture will undoubtedly continue to receive attention. As new developments and modifications occur, hydroponics may become a more feasible and economically attractive option for production of many horticultural crops.

REFERENCES

Baker, K. F., P. A. Chandler, R. D. Durbin, et al. 1957. *The U.C. System for Producing Healthy Container-grown Plants*. Univ. of California Div. Of Agricultural Sciences, Davis, CA 332 pp.

Broughton, W. J. 1983. *Nitrogen Fixation*. Vol 3: *Legumes*. Oxford University Press, New York.

Devlin, R. M., and F. H. Witham. 1983. *Plant Physiology* (4th ed.). PWS Publishers, Boston. 577 pp.

Eskew, David, L. Ross, M. Welch, and W. A. Norvell. 1984. Nickel in higher plants. *Plant Physiology* 76:691–693.

Food and Agriculture Organization of the United Nations. 1981. *Crop Production Levels and Fertilizer Use*. FAO/UN, Rome. 69 pp.

Fried, M., and H. Broeshart. 1967. *The Soil Plant System in Relation to Inorganic Nutrition*. Academic Press, New York. 358 pp.

Ingels, J. E. 2004. *Landscaping Principles and Practices* (6th ed.). Delmar Learning, Clifton Park, New York. 494 pp.

Jones, U. S. 1982. *Fertilizers and Soil Fertility* (2nd ed.) Reston Pub., Reston, VA. 464 pp.

Lorenz, O. A., and D. N. Maynard. 1980. *Knott's Handbook for Vegetable Growers,* (2nd ed.) John Wiley & Sons, New York.

Maynard, D. N., and G. J. Hochmuth. 1997. *Knotts Handbook for Vegetable Growers* (4th ed.). John Wiley & Sons, New York. 456 pp.

Mortvedt, J. J., P. M. Giordano, and W. L. Lindsay (Eds.). 1972. *Micronutrients in Agriculture.* Soil Science Society, Madison, WI.

Nelson, L. B. 1965. Advances in fertilizer. *Advances in Agronomy* 17:1–84.

Nelson, P. V. 2003. *Greenhouse Operation and Management* (6th ed.), Prentice Hall, Englewood Cliffs, N.J.

Simpson, K. 1986. *Fertilizers and Manures.* Longman Group Ltd., London and New York. 245 pp.

Sutcliffe, J. F. 1962. *Mineral Salts Absorption in Plants.* Pergamon Press, London. 194 pp.

The National Fertilizer Association. 1949. *Hunger Signs in Crops: A Symposium.* The American Society of Agronomy and The National Fertilizer Association, Washington, D.C.

Tinker, P. B., and A. Lauchli. 1984. *Advances in Plant Nutrition.* Praeger Scientific, New York. 298 pp.

United States Dept. of Agriculture. 1977. *Gardening for Food and Fun, Yearbook of Agriculture,* USDA, Washington, D.C. 292 pp.

Wallace, A. 1965. Micronutrient deficiencies in plants and their correction with chelates. *Agricultural Science Review* 3:18–24.

Welch, R. M., and E. E. Carey. 1984. Plant scientists plug nickel into essential category. *Agricultural Research,* June, 1984:14–15.

MULCHES

....

OUR PURPOSE IS NOT TO

DO WELL, BUT TO DO

BETTER.

Adapted from a Quaker theologian.

C onservation of moisture clearly makes good sense, from the local arena of tending a home garden to the global scope of husbanding the earth's precious asset of useable water. One cultural practice that can greatly aid in conserving water, thus increasing water use efficiency, is the use of mulches. Mulches can be loosely defined as substances spread over the ground to protect soil and roots from undesirable environmental factors. They have several advantages for horticultural practices.

ADVANTAGES OF MULCHES

Among the advantages often attributed to mulches are the following:

1. **Water conservation** Mulches can greatly reduce evaporation from the soil surface and can reduce irrigation needs.
2. **Weed control** Mulches can prevent weed growth, thus reducing competition with the crop for light, water and nutrients.
3. **Reduced labor requirement** Much less labor is expended in weed removal and in applying water to properly mulched crops.
4. **Temperature modification** As a general rule, organic mulches tend to cool soil, whereas clear and black plastic mulches may be used to warm the soil. Mulches also generally reduce temperature fluctuations.
5. **Reduced costs** Economic considerations may vary with crop value, labor cost and size of enterprise, but mulches are often cost-effective for high-value crops.
6. **Protection of soil structure** Soil is protected because mulches reduce the impact of rain or irrigation water striking the soil (see Figure 7-45).
7. **Soil structure improvement** Soil structure is improved when the lower layer of an organic mulch decays or following incorporation of organic mulches into the soil.
8. **Reduced nutrient and water losses** Leaching of nutrients may be reduced, and run-off of water is retarded by most organic mulches.
9. **Keeps fruit clean** Soil particles will not be splashed onto fruit by rain drops or irrigation water.
10. **Reduced insect numbers** Reflective plastic mulches (white, aluminum-coated) have been reported to reduce aphid infestations in vegetables grown on such plastic mulches. In addition, certain organic mulches such as cedar chips are reputed to repel insects.
11. **Aesthetics** Many mulch types such as wood chips, shredded bark and some types of crushed rock help present a neat, tidy, and finished appearance in landscape plantings and designs.
12. **Turfgrass establishment** Following planting of turfgrass seedings, a light layer of mulch (usually straw) will aid germination and establishment by keeping the germinating seeds moist and protected from wind.
13. **Protecting trees and shrubs** Trees and shrubs surrounded by organic mulches are protected by keeping mowers and string trimming equipment away from trunks, thus preventing damage to the bark, or "mower blight."

USE OF ORGANIC MULCHES

Many organic materials can be effectively employed for mulching purposes. Straw, hay, sawdust, wood chips, sphagnum peat, lawn clippings[1], and shredded bark are frequently used, as are by-products of various industries, including ground corn cobs, peanut shells, rice hulls, sugar cane waste (*bagasse*), sunflower hulls, cottonseed hulls, and even the cocoa bean shells from the chocolate factory. Regardless of the organic material used, once such a mulch has been incorporated into the soil and has decomposed, it will improve soil structure—especially of "heavy" soils such as clays, clay loams, and silty soils—by aggregating fine soil particles. This aggregation improves drainage, and thus aeration of the root zone. Water-holding ability of more well drained soils also will be improved as the organic matter is broken down.

Because such breakdown is accomplished as a result of microbial activity, the astute horticulturist recognizes that additional

---

[1] We recommend that use of clippings from herbicide-treated turf be avoided.

nitrogen fertilizer must be applied when incorporating organic materials into the soil. This is necessary because the bacteria and fungi involved in the decomposition process will utilize large amounts of nitrogen as they multiply rapidly during the break-down of the organic matter. Because these microbes can compete more effectively than the "higher" (crop) plants for the nitrogen, an induced nitrogen deficiency of the horticultural commodity may result. This phenomenon and other nutritional aspects of organic additions to the soil are discussed further in Chapters 8 and 9.

One of the most important reasons for mulching is weed control. Organic mulches can be placed directly on top of weeds that are 2.5 cm tall or less. If a sufficient thickness of mulch is used (e.g., 15 cm of straw or hay or 6–8 cm of sawdust), the light will be blocked, photosynthesis will be prevented, and the weeds will die (Figure 10-1). Because larger weeds, especially perennial weeds such as curly dock (*Rumex crispus*), dandelion, and Canada thistle (*Circium arvense*), have greater food reserves (energy stored from earlier photosynthesis), they may push through an organic mulch. Therefore, it is recommended that weeds over 2.5 cm be removed by cultivation, hand-pulling, or herbicide treatment prior to application of the mulch. Once the mulch is in place, neither herbicide application nor mechanical cultivation will be required for weed control. Therefore, the diminished need for cultivation activities will reduce the damage to crop roots and will lessen the compaction of the soil caused by equipment, human traffic and water droplet impact. Organic mulches may improve water absorption by the soil and reduce leaching, erosion, and runoff losses. These benefits occur because the particles of the organic mulch reduce the impact of the water drops and diffuse or spread the water into finer droplets that slowly percolate through the mulch and into the soil. The slowed water movement will cause less leaching of nutrients through the soil and less loss of soil, water, and nutrients that may result from rapid runoff.

FIGURE 10-1. The organic mulch has completely suppressed weed growth in this planting.

The fibrous nature of most organic mulches enables such materials to exert a significant effect on soil temperature. Early in the growing season, organic mulches such as straw have been reported to cause soil temperature to be 5°C (9°F) or more lower than bare soil temperatures. In addition to the insulating properties of most organic mulch materials, cooling of the soil occurs because of the greatly increased surface area available for evaporation of the water. This response is somewhat analogous to that of the fibers used in conventional fan and pad cooling of greenhouses discussed earlier, which takes advantage of water's heat of vaporization. In warm climates and warm seasons of temperate climates, this cooling effect of organic mulches can help keep the crop root zone from becoming too hot. If the roots become overheated, excessive respiration occurs, which rapidly exhausts stored food and reduces growth. By contrast, in cool climates or during cool seasons (spring in temperature zones), organic mulches should probably be avoided because the soil may already be too cold.

In fact, warming of the soil may be preferable in the spring to encourage germination of early-planted seeds and root growth of perennials and transplants. An example of the latter approach is seen when warm air is blown through pipes buried in the root zone beneath mulched beds for culture of greenhouse vegetables, resulting in earlier and better yield and improved fruit quality. Rapid soil temperature fluctuations are also reduced by organic mulches as a result of their excellent insulating characteristics.

WINTER MULCHING

Mulching perennial plants, both woody and herbaceous, to provide protection from winter cold and frost heaving is a common horticultural practice. Organic mulches, such as straw, wood chips, and shredded bark are commonly employed and are usually applied in the fall. Thickness of the mulch can vary with the type of mulch and species being protected, but a depth of about four inches (10 cm) is common. In specialized cases, such as mulching strawberries with straw (Figure 10-2), the mulch is not applied until the soil is frozen. Winter mulches help protect plants by serving to insulate the soil, which prevents soil temperature fluctuations and reduces root damage because the soil is kept warmer than would be true for bare (nonmulched) soil. As a result of the warmer soil conditions, root growth continues later into the fall, thus enhancing the overall health of the plant. Mulches prevent reradiation of heat from the soil, so the above-ground parts of the plant will often be colder than is the case for nonmulched plants. However, it should be noted that research has shown that root tissues are much more vulnerable to cold temperature damage than stem and other aerial plant parts, so the net effect of mulching in most cases is significantly beneficial to plant survival in winter. Another feature of mulching that can be either positive or negative is the fact that organic mulches keep

the soil cold longer into the spring than is true for bare soil. This phenomenon is positive for species that tend to resume growth too early in the spring, because delaying growth may enable the plant to escape damage from late frost events. However, for plants that are targeted for early market, or that require the entire growing season to achieve proper maturity, later bud break or delayed growth may be detrimental.

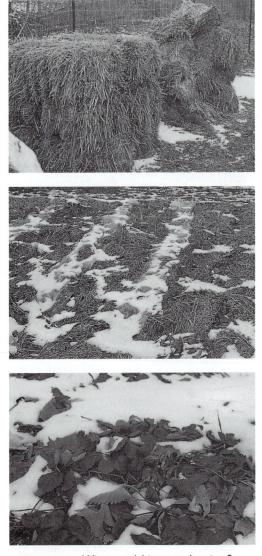

FIGURE 10-2. Winter mulching strawberries. Straw bales (*top*) are broken up and spread over beds or rows of strawberries (*middle*). Close-up of mulched strawberry plants (*bottom*).

Much research has been conducted with the goal of controlling weeds in container-grown nursery crops such as potted shrubs and trees. Specialized discs made of various materials (e.g., plastics, fabric, corrugated cardboard, pressed peat) are circular in shape with a slit to the center to facilitate fitting around the stem of the plant. Such discs fit snugly into the container, are permeable to water and suppress weed growth. Another approach combines the use of organic mulch with herbicides for weed control in container-grown crops. In this method, "nuggets" of douglas fir bark are impregnated with herbicide and placed on the surface of the container medium, thus suppressing weed growth that would otherwise compete with the containerized plant. Both of these methods dramatically reduce labor costs because the need for hand weeding is essentially eliminated.

Another specialized use of mulches is **hydroseeding** (also called **hydromulching**) which is used extensively in the turf industry. Specialized equipment is used to mix grass seeds, water, fertilizer, and wood pulp fibers (or chopped straw) together. The mixture is then sprayed on the area to be seeded. This procedure has been widely employed for seeding roadside slopes and other extensive turfgrass areas (Figure 10-3).

FIGURE 10-3. A hydroseeder being used to apply grass seed, fertilizer, water and a wood pulp fiber to a roadside slope.

Photo courtesy of Finn Corporation, Fairfield, OH.

Using a *living mulch* is another organic mulching method, which involves planting a cover crop, often a forage species such as bromegrass (*Bromus* spp). After sufficient growth has occurred, a herbicide is applied to kill or stunt the cover crop that is left in place. Then the grower plants seeds of the desired crop through slits in this "living mulch." A variation of this concept has recently received a great deal of attention: A nonphytotoxic growth retarding chemical is applied to a cover crop planted between rows of the economically important crop, holding the living mulch at a size or stage that will suppress weed growth but not compete with the crop. Living mulches also often reduce soil compaction and may harbor and encourage beneficial insects, but care must be taken to minimize competition with the crop plants for water, light, and nutrients.

NONORGANIC MULCHES

Plastic Mulches

The use of inert, nonorganic materials as mulches has been explored extensively over the years, but the most successful material of this type has been polyethylene (plastic) sheeting, which can be applied mechanically for field use (Figure 10-4). Opaque black polyethylene, usually about 0.04–0.06 mm in thickness, has been used most extensively. Its opaque nature prevents sunlight from reaching weeds, so it is effective in controlling them. Clear polyethylene, on the other hand, allows weed growth under the plastic, so growers commonly utilize soil fumigation at the time the plastic mulch is applied or employ herbicides prior to mulching with the clear plastic. Both clear and black polyethylene help prevent evaporative water loss from the soil and increase soil temperatures beneath them. Black plastic mulches have been reported to cause an increase in soil temperatures near the soil surface of 2° to 3°C when compared to bare soil, but clear

FIGURE 10-4. Field application of polyethylene mulch. Field view (*left*); close-up of equipment (*right*). Note that mulch is black for weed suppression (on roll), but has a white reflective surface for a cooling effect.

plastic causes increases of as much as 8°C(15°F). Clear plastic increases the soil temperature more and to a greater soil depth than black plastic does (Table 10-1). When the light penetrates the clear plastic and is absorbed by the soil, it is transformed to heat energy (long-wave radiation). This heat energy is prevented from reradiation by the film of moisture on the underside of the plastic, by the "dead" air space that reduces convective loss, and to some extent by the plastic itself. Although the temperature of black plastic can become very hot because black objects absorb large amounts of radiant energy, the soil is not warmed as much as with clear plastic because the contact with the soil is not continuous. The heat is conducted directly to the soil particles it touches, but moves more slowly across the air spaces (Figure 10-5).

A recent development in plastic mulches is the introduction of infrared transmitting (IRT) plastic films. Such films are more properly referred to as **wavelength-selective** mulches. They warm the soil while inhibiting weed growth because they selectively transmit specific wavelengths of incident energy, including infrared, while restricting photosynthetically active radiation (PAR). Wavelength selective mulches are often blue-green or brown in color and offer the advantages of clear plastic mulch (soil warming) while delivering weed control

approaching the level achieved with black plastic. It should also be noted that white and other reflective plastic may reduce soil temperatures by 2° to 3°C.

Plastic mulch use has dramatically increased in recent years and has become the norm for high-value horticultural crops such as tomatoes in Florida and strawberries in Florida and California. In the case of tomatoes, a "plug-mix" composed of a mixture of peat, vermiculite, and seeds may be mechanically planted directly through holes burned in the plastic as it is being laid in the field. For strawberries, the plants are

TABLE 10-1. A comparison of soil temperatures under mulches and for bare soil tomato plantings, Freeville, NY[a]

	SOIL TEMPERATURE	
MULCH	AT ONE INCH (2.5 CM) DEPTH	AT FOUR INCH (10 CM) DEPTH
None (bare soil)	26.7°C (80°F)	24.5°C (76°F)
Straw	21.7°C (71°F)	20.5°C (69°F)
Black polyethylene	27.2°C (81°F)	25°C (77°F)
Clear polyethylene	32.2°C (90°F)	24.5°C (76°F)

Source: Schales, F.D., and R. Sheldrake, Jr. 1963. Mulch effect on soil conditions and tomato plant response. Proceedings of the Fourth national Agricultural Plastics Conference: 78–90.
[a]Readings taken at 3 PM, July 11, clear sky, air temperature 27.5°C (82°F)

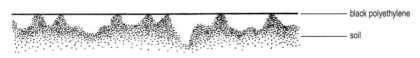

black polyethylene

soil

FIGURE 10-5. Schematic representation of black polyethylene laid over soil, illustrating the relatively small amount of direct contact.

planted through the plastic a few days after it is laid, because soil fumigants are normally used and time must be allowed for the fumigant to dissipate. Transplanted vine crops such as muskmelon can also be stimulated to mature earlier if planted through plastic mulch compared to when no mulch is used, thus enabling the early fruit to command a higher market price. Conversely, as mentioned earlier, organic mulches can delay maturity of many crops because of the cooling effect of most organic mulches.

Unusually high yields have been obtained for crops such as muskmelon grown with plastic mulch. This yield enhancement is difficult to explain completely on the basis of weed control, increased temperature, and moisture conservation. An additional factor suggested by Sheldrake is related to the increased production of carbon dioxide under the mulch. This CO_2 escapes primarily through the planting holes, thus bathing the undersides of the leaves of the young plants with a CO_2-enriched atmosphere. The elevated levels of CO_2 are thought to enhance photosynthesis, leading to more rapid plant growth and an increase in early fruit production.

Application of plastic mulches to control weeds and conserve moisture around the base of trees or shrubs in landscape plantings has been a frequently used practice. A circle of polyethylene or landscape fabric is placed around the stem after planting and a layer of organic mulch such as wood chips is spread on top of the plastic. The organic mulch is more aesthetically pleasing and prevents sunlight from reaching the plastic; this is an additional advantage contributing to the plastic's longevity, because ultraviolet light causes breakdown of polyethylene.

A modified type of plastic mulch employed for weed control is referred to as a **weed mat** or landscape fabric; it has largely supplanted plastic mulch for many landscape uses. Sold under various trade names, weed mats are plastic mulch that have fibers woven into them or are constructed in such a way that they look like a loosely woven cloth. They are most commonly made of woven or textured plastic, black or dark in color and are quite strong. Because of the perforations, or woven characteristic, water readily penetrates the weed mat, but weeds are still suppressed. Such mats or woven landscape fabrics are frequently employed in the landscaping industry at the time of installation of trees and shrubs.

Another practice employed with plastic mulches and weed mats for permanent or semi-permanent plantings is to install low-volume irrigation under the plastic or with the weed mat. Drip irrigation systems with emitters placed near the plant, bi-wall pipes and "ooze-tubes" or other types of low-volume irrigation may be utilized for this purpose.

Many other inert materials have been considered for use as mulches. Used tires are an example of a waste product that have posed a disposal problem. Employing shredded tires as a mulch is being studied, but further research is required to determine the utility and value of such a product. Additional examples include paper sheeting, aluminum foil, aluminum-coated paper, aluminum-coated polyethylene, newspapers, and old carpets. For various reasons, none have gained wide acceptance. Table 10-2 provides details of several mulching materials' characteristics.

TABLE 10-2. Characteristics of Common Mulching Materials

MATERIAL	SPECIAL FEATURES	ADVANTAGES	DISADVANTAGES
ORGANIC MATERIALS			
Grass clippings	Fine texture, moist	Readily available for home gardeners; inexpensive	Tend to mat down and reduce water and air movement; may contain herbicide residue; require frequent replacement
Straw, hay	High C:N ration	Inexpensive; readily available (in some areas)	Can cause nitrogen deficiency; may contain seeds that become weeds
Wood chips, shredded bark, shredded wood	Very high C:N ratio	Attractive; long lasting, slow to break down	May wash away on steep slopes; will tie up nitrogen, especially if incorporated into soil
Leaves	Should be chopped or shredded before use	Readily available for home gardeners, inexpensive, improve soil	Can mat down and restrict water and air movement; may blow away in high winds
Pine needles	Often fragrant	Decompose slowly; may improve soil; may acidify alkaline soils	Limited availability; may mat down restricting water and air movement
Sphagnum moss peat	Highly acidic (low pH)	May acidify soils over time or if incorporated; decomposes slowly	Repels water when dry; very difficult to re-wet once it is dry; expensive
Rice hulls	Regional availability; light weight	Attractive color; improves soil	Float or blow away unless composted
Cocoa bean hulls	Pleasant odor (chocolate); regional availability	Improve soil; attractive	Expensive; float or blow away easily
INORGANIC MATERIALS			
Crushed rock, gravel	Heavy, best if underlain with a plastic or woven weed barrier	May give a formal look	White rock near building may warm the building and plants too much; can be expensive; safety hazard when thrown by mower blades
Lava rock	Lightweight; best if underlain with a plastic or woven weed barrier	Fairly good insulator, gives a formal look	Not good on slopes, will roll away easily; expensive

(Continued)

TABLE 10-2. Characteristics of Common Mulching Materials (*continued*)

MATERIAL	SPECIAL FEATURES	ADVANTAGES	DISADVANTAGES
INORGANIC MATERIALS			
Landscape fabrics	Usually made of closely woven, black or dark colored fabric; best if used under aesthetically pleasing mulch	Good control of most weeds; allows water penetration; quite durable	May degrade in light; unattractive appearance; nutsedge and some other perennial weeds may penetrate; weed seeds may germinate on top of fabric
Plastic mulch	Made of polyethylene sheeting; breaks down slowly in ultraviolet light (sunlight); clear plastic warms soil significantly, black warms it slightly and white cools the soil. Silver and white on black plastic repels aphids; yellow attracts insects	Inexpensive; suitable for large acreage use (e.g. vegetable production); can be used in conjunction with soil fumigants; black and clear enhance early yields of several vegetables. Acceptable for organic crop production if removed at end of growing season.	Cracks or shatters when cold and from ultraviolet light; prevents water penetration (place water source under plastic); removal is labor intensive and expensive
Infrared transmitting plastic mulch (IRT)	IRT mulches allow infra-red light penetration, but absorb photosynthetically active radiation; blue-green or brown in color	Prevents weed growth; warms soil intermediate to that of clear and black plastic	More expensive than conventional plastic mulches; similar removal problems
Photodegradable and biodegradable plastic mulches	Break down under sunlight, or are broken down by soil organisms, respectively	Eliminate need for removal (less labor cost)	More expensive than conventional plastic mulches; not acceptable for organic crop production
Newspaper (intact or shredded)	Modern inks on conventional newsprint are biodegradable; best if covered with an organic mulch (intact newspapers); shredded must be at least 4 to 5 inches (10 to 12.5 cm) deep	Improves soil when incorporated; recycles a waste product; inexpensive; good moisture retention; good decomposition rate	Intact newspapers can limit water penetration; unattractive; must not use 'slick'/colored sheets; blows easily when dry

See Chapter 9 for a more detailed discussion of C:N ratio.

LIMITATIONS OF MULCHING

In additional to the aforementioned problems with mulches, such as induced nitrogen deficiency, cooling or heating of the soils at the wrong time of the year, and weed growth if not properly applied, some other potential problems of mulches should be considered. When organic mulches are used, problems with diseases and vermin may result if proper precautions are not observed. On poorly drained soils, mulches can keep the soil too moist, thus restricting oxygen in the root zone. On some crops, if the mulch is applied close to or touching the stem, moisture may be trapped, creating an environment conducive to development of disease organisms. Many organic mulches also encourage and provide refuge or breeding locations for snails, slugs, mice and other animal life that may attack the crop plants. Furthermore, certain types of mulches such as hay and straw may contain seeds that, upon germination, become weeds.

Other problems inherent to some organic mulches are related to events that may occur if the mulch dries out. Some mulches may tend to cake or crust, reducing oxygen levels, water penetration, or both. Winds may blow away all or part of the mulch, thus exposing the roots that have a tendency to form at the soil-mulch interface. Similarly, heavy rains may cause the mulch to be washed away from the plants, especially on sloping fields or gardens. When this happens, such exposed roots may desiccate, which of course will make them more vulnerable to winter injury. In the case of potato crops, another problem that may occur if the mulch is applied too thinly or if some or all is blown or washed away, is the greening of the tubers, which usually will form at or near the soil surface, when grown under an organic mulch. This greening is the result of the tubers (which are really stems) forming chlorophyll upon exposure to light. Unfortunately, a concommittent formation of glycoalkaloids, including **solanine** occurs, causing the potatoes to have an objectionable flavor and rendering them poisonous in extreme cases! Some mulches can become fire hazards when they dry out, so care must be taken to avoid using such mulches near buildings and to prevent them from drying out.

Nonorganic mulches also have problems. A major difficulty with polyethylene mulches is the disposal problem at the end of the crop season, raising a concern for potential damage to the environment. Even though some manufacturers have developed types of polyethylene and other plastics that break down quickly in response to ultraviolet light, there is still a problem of disposal since the polyethylene is secured by burying the edges in the soil. This means that the edges are not subjected to ultraviolet light and are therefore more or less intact. However, new recyclable high density polyethylene films are now being considered as an alternative to conventional plastic mulches. Such recyclable plastics are applied to the field, removed at the end of the growing season and reclaimed by the manufacturer to be recycled into useful products such as plastic lumber and pipe. Similarly, polyethylene used for greenhouse glazing and many kinds of plastic pots can also be recycled. Another approach to disposal of used plastics is to incinerate them. Polyethylene burns cleanly to produce carbon dioxide and water, with around 18,000 BTUs of heat produced per pound (454 grams) of plastic. However, problems with the logistics of incinerating plastics have not all been solved, so more research will be required for this approach to become a practical solution for waste plastic disposal.

Because they are impermeable to water, another difficulty sometimes encountered for plastic mulches is that the plant roots may exhaust soil moisture under the plastic mulch. Provision must therefore be made for adding water through holes in the mulch or by placement of drip irrigation emitters under the mulch. The latter approach is commonly practiced in Florida.

An important general consideration for all types of mulches is cost. Labor costs for application and perhaps maintenance of mulches can be a prohibitive expense for some crops. Herbicide use may be a more economically attractive alternative for large-area plantings or for low-value crops.

SUMMARY

Depending on crop requirements and value, material and application costs, and availability and the intended use, mulching may be an appropriate, environmentally friendly and highly beneficial horticultural practice.

REFERENCES

Chong, C. 2003. Experiences with weed discs and other non-chemical alternatives for container weed control. *HortTechnology* 13:23–27.

Decoteau, D. R., M. J. Kasperbauer, and P. G. Hunt. 1989. Mulch surface color affects yield of fresh market tomatoes.

Journal of the American Society for Horticultural Science 114: 216–219.

Emmert, E. M. 1957. Black polyethylene for mulching vegetables. *Proceedings of the American Society for Horticultural Science* 69:464–469.

Ingels, J. E. 2004. *Landscaping Principles and Practices* (6th ed.). Delmar Learning, Clifton Park, NY. 494 pp.

Mathers, H. 2003. Novel methods of weed control in containers. *HortTechnology* 13:28–34.

Plaster, E. J. 1985. *Soil Science and Management.* Delmar, Albany, NY. 454 pp.

Schales, F. D. and R. Sheldrake, Jr. 1963. Mulch effects on soil conditions and tomato plant response. Proceedings of the Fourth National Agricultural Plastics Conference: 78–90.

Sheldrake, R., Jr. 1963. Carbon dioxide levels in the microclimate as influenced by the permeability of mulches. Proceedings of the Fourth National Agricultural Plastics Conference: 93–96.

Waggoner, P. E., P. M. Miller, and H. C. DeRoo. 1960. Plastic mulching. Principles and benefits. Conn. Agr. Exp. Sta. Bulletin 634, University of Connecticut, Storrs, CT.

PLANT GROWTH
SUBSTANCES

PLANT HORMONES

Plant species grow in a myriad of shapes and sizes. This diversity of phenotypes results from an interaction between the genotype and environment. Plant genes, through transcription and translation, code for proteins, including enzymes that are involved in the biosynthesis of plant hormones. In turn, plant hormones stimulate the growth and differentiation of cells, tissues, and organs. Hormonal biosynthesis is also affected by environmental stimuli, including temperature, water, light, nutrition, pests, stresses, and human manipulations of plants such as pruning. Therefore, the phenotypes that we observe in nature or those that are produced by humans are a result, in large part, of the hormonal balance of plants.

The commercial application of plant growth regulators has important economic and horticultural significance in the production of plants. Plant growth regulators are used to propagate species, to increase yield, to improve plant quality, to alter plant growth habit or structure, or to aid in harvesting or postharvest storage.

Special words with specific meanings are associated with plant hormones and other plant growth regulators. Some of the words are frequently misused (e.g., plant hormone), so it is important for students to learn the correct use of the terminology of plant growth substances. These words have been given precise meanings for clear and accurate communication that is essential among horticulturists and other plant scientists. Important terms used by those who work with plant growth substances are given in the following list.

❖ TERMINOLOGY

ACROPETAL. Transport toward the apex. In relation to the soil line or plant axis, one should remember that acropetal transport is therefore up the stem and down the root, toward the growing point of each organ.

ASSAY. An analysis to determine the presence, absence, or quantity of one or more compounds.

BASIPETAL. In a direction away from the apex. As with acropetal transport, stems and roots have reversed polarity.

BIOASSAY. An assay that utilizes living material. Certain plants or plant parts will respond in a predictable fashion to applied compounds; if they give a response similar to that of a known compound, the chemical being tested is said to be "like" the known compound.

ENDOGENOUS. Refers to something that originated within the body of a plant. A compound manufactured by the plant is said to be of endogenous origin.

EXOGENOUS. Refers to something that originated from or is derived from external causes. Chemicals that are manufactured in a laboratory or factory are applied exogenously to a plant.

IN VITRO. This Latin term literally means "in glass." In usage, however, it means "in an artificial environment." Plant tissue culture is done in glass vessels, including culture tubes, flasks, or even baby food jars, or in plastic vessels such as Petri dishes. In all cases, it is called in vitro culture.

IN VIVO. Refers to something that occurs within a living organism. For example, photosynthesis occurs in vivo.

LOW CONCENTRATION. When considering plant growth substances, this generally means concentrations below 10^{-6} M, or below 1 ppm. Higher concentrations of exogenously applied substances are frequently utilized; however, a small fraction generally gets into the plant and to the active site. Within the plant, hormones frequently are expressed in fractions of moles or grams as follows:

1 mg = 10^{-3}g = milligram, or thousandth of a gram

1 µg = 10^{-6}g = microgram, or millionth of a gram

1 ng = 10^{-9}g = nanogram, or billionth of a gram

1 pg = 10^{-12}g = picogram, or trillionth of a gram

1 fg = 10^{-15}g = femtogram, or quadrillionth of a gram

In addition, plant growth regulators frequently are applied to plants in parts per million (ppm), and can be expressed as follows:

1 ppm = 10^{-6} = 1 mg · liter^{-1} (milligrams per liter of water) = 1 µg · ml^{-1}; 1% = 10,000 ppm.

MAJOR PLANT HORMONE. A hormone or class of hormones that is present in all higher plants. These include **auxins, gibberellins, cytokinins, brassinosteroids, ethylene, abscisic acid,** and **jasmonates.**

PHARMACOLOGICAL RESPONSE. A response not representative of that found in nature, such as a drug or herbicide effect resulting from an overdose.

PHYSIOLOGICAL RESPONSE. A response representative of that found in nature.

PLANT GROWTH SUBSTANCE. A natural or synthetic *organic* material (carbon compound) that in *low concentrations* regulates plant growth. Nutrients (such as sugars) and vitamins are not included as plant growth substances. Plant growth substances do not have to be translocated by plants. Hormones are a subcategory of plant growth substances. A synonym is *plant growth regulator* (PGR).

PLANT HORMONE. for a substance to be considered a plant hormone, it must meet all of the following criteria:

1. It must be an *endogenous* plant growth substance.

2. It must be an *organic* compound.

3. In **low concentrations**, it must regulate plant physiological processes; these often relate to plant growth and development.

4. It must be *transported* from the site of biosynthesis to the location at which it acts. This is the definition of a **correlative signal** within the plant. This criterion was originally derived from the animal definition of hormone, because transport (circulation) is generally more straightforward than in plants. This factor alone has prevented some PGRs from being classified as plant hormones.

5. It cannot be a nutrient or a vitamin.

The term plant hormone is frequently misused in agriculture. We suggest that when in doubt, you use the broader term *plant growth substance* or *plant growth regulator*.

TROPISM. The turning, bending, or curving of an organism or one of its parts either toward (positive tropism) or away from (negative tropism) a source of stimulation, such as light (phototropism) or gravity (gravitropism or geotropism).

MAJOR PLANT HORMONES

Plants contain many growth substances. Many of these are considered to be secondary plant growth substances because they either lack specificity and must be present in abundant quantities to be active, or have a limited distribution among higher plants. The most important classes of endogenous plant growth substances are the major plant hormones. These classes represent many compounds, but can be placed in two general groups: growth-promoting hormones and wounding and stress hormones.

The following list includes the classes of major plant hormones, with an important function and use:

GROWTH-PROMOTING HORMONES

❖ **Auxins.** These compounds stimulate cell elongation. They often are used to stimulate the initiation of adventitious roots on cuttings.

❖ **Gibberellins.** Gibberellins also stimulate cell elongation. They serve an important role in seed germination.

❖ **Cytokinins.** Cytokinins are compounds that stimulate cell division. They are important in shoot initiation and branching of plants.

❖ **Brassinosteroids.** Brassinosteroids promote cell elongation in stems and other organs.

WOUNDING AND STRESS HORMONES

❖ **Ethylene.** At physiological temperatures, in its pure form, ethylene is a gas that stimulates swelling of stems and roots. Ethylene is the "aging hormone;" it is frequently used to stimulate ripening of some fruits; it also promotes senescence.

❖ **Abscisic acid.** ABA is a general inhibitor of growth and is considered to be a "stress hormone," since various environmental stresses, such as drought, stimulate its biosynthesis. It functions, in part, to allow plants to adapt and survive environmental stresses.

❖ **Jasmonates.** Jasmonates are involved in plant defense responses in response to wounding and pathogens.

Each major plant hormone has several different functions in plants. Well-documented interactions among the groups of hormones help to account for the various plant pheno-types that occur. In addition, the student should be aware that new classes of major plant hormones may be discovered as researchers probe deeper into the physiology and biochemistry of cells and tissues.

An important criterion that must be met for a compound to be considered a plant growth substance is the requirement that it must be active in low concentrations. In such minute quantities, PGRs exert major and dramatic influences on plants. This requires that the signal generated by the PGR becomes amplified, similar to the amplification of a voice over a loudspeaker system at a rock concert. Because many chemical reactions and physiological pro-cesses within plants are controlled by enzymes, considerable evidence supports the notion that hormones are elicitors and thus signal or affect enzymes and enzyme systems. They can up-regulate and down-regulate genes that are ultimately respon-sible for protein (enzyme) biosynthesis. In some cases, hormonal activity can change cellular conditions, such as pH, which directly affects enzyme activity. Plant growth substances can also have great effects in low concentrations by changing the permeability of membranes. They may consequently allow more or less leakage of substances that can influence developmental and stress-response phenomena.

GROWTH-PROMOTING HORMONES: AUXINS

The word **auxin** does not imply a specific chemical composition of a class of com-pounds; rather, members of this group of growth substances all cause similar biolog-ical responses among plants. Early research that ultimately led to the discovery of auxins began with a publication written by Charles Darwin in 1880. Darwin investigated phototropism in seedlings of ornamental canarygrass (*Phalaris canariensis*). He was the first to discover that the tip of the

coleoptile (the first leaf in members of the grass family that surrounds the shoot apical meristem with its leaf primordia) perceives light coming from one direction and that the growth response occurs lower, away from the tip, and causes bending toward the light source. It was later shown by other researchers, that the auxin indole-3-acetic acid (IAA) is produced by the coleoptile tip. IAA moves basipetally (down) through the coleoptile and stimulates cell elongation. Research on phototropism supports the theory that IAA is redistributed from the irradiated side to the shaded side and that the basipetal transport system is greater on the shaded side, resulting in greater cell elon-gation on the shaded side. Because the cells elongate more rapidly on the shaded side, the leaf bends toward the light source (Figures 11-1 and 11-2). This is a result of exposure to blue light. As discussed in Chapter 5, blue light absorbing phototropins regulate pho-totropism, although their mechanism is not yet well understood. Knowledge of the pho-totropic response demonstrates how an envi-ronmental factor (light), can change plant phenotype via a plant hormone.

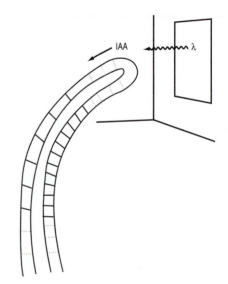

FIGURE 11-1. The bending of a coleoptile toward a unidirectional source of light. The IAA that moves downward on the side away from the light stimulates cell elongation here (thus the positive phototrophic response).

are: indole-3-butyric acid (IBA); naphtha-leneacetic acid (NAA); and the more active auxins that are commonly used as herbicides, 2,4,-dichlorophenoxyacetic acid (2,4-D); and 4-chloro-2-methylphenoxyacetic acid (MCPA) (Figure 11-4).

Auxin Biosynthesis

IAA and PAA are primarily produced from the amino acids tryptophan and phenylala-nine, respectively. However, tryptophan and IAA may have common precursors in some plants, such as corn, indicating an alternate route of synthesis in these plants. Sites with high auxin concentrations are shoot apices, especially the young, expanding leaves (<1/5 fully expanded), young expanding buds, root tips, developing seeds, and fruits. Near the shoot tips, the young expanding leaves are probably one of the most important sites of auxin biosynthesis. This knowledge explains why many cuttings root better if the growing point (apex) is left intact and leaves are retained on cuttings, rather than pruned off. Auxin produced by developing seeds is considered crucial for normal fruit development in several horticultural crops (see Chapter 3). When uneven fertilization occurs within the ovary of a plant such as apple or tomato and a locule contains no

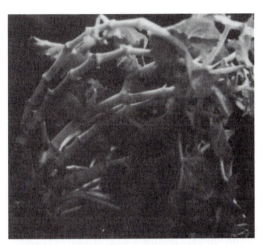

FIGURE 11-2. This house plant is showing a positive phototrophic response by bending toward a light source.

Examples of Auxins

Indole-3-acetic acid is considered to be the most important hormonal auxin (Figure 11-3). In some books, the term **auxin** and **IAA** are used synonymously. In 1982, phenylacetic acid (PAA) was identified as a natural auxin in plant shoots. It was shown to be present in greater quantities than IAA in seedling tomato, tobacco, sunflower, pea, barley, and corn; however, its activity is less than that of IAA.

IAA is broken down when exposed to blue light or ultraviolet-B radiation, so for this reason, IAA is not used commercially. In fact, if sprayed on plants, the photodestruction will occur within about 2 hours. Enzyme systems, including IAA oxidase and peroxidases will also breakdown IAA within the plant. Consequently, more stable synthetic auxins have been produced that have greater commercial utility. Examples of synthetic auxins

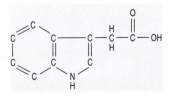

FIGURE 11-3. The chemical structure of the auxin, indole-3-acetic acid (IAA).

FIGURE 11-4. Damage caused by excess auxin. This tomato plant is showing typical symptoms of 2,4-D injury.

seeds, the fruit will not develop normally on that side and thus will be indented or flat because of the lack of auxin (Figure 11-5). It is interesting to note that in some species, exogenous auxin can replace seeds and cause parthenocarpic fruit. It is common to have poor fruit set on tomatoes in the spring or early summer because of low night temperatures. Fruit set can be induced if the flower clusters are dipped in an auxin solution or if the plants are sprayed with auxin. Garden centers often carry products containing synthetic auxin specifically for this purpose. Horticulturists can replace the natural auxin source (seeds) with an exogenously applied PGR in order to facilitate fruit development.

Auxin Activity

As mentioned previously, auxins stimulate **cell elongation**. It is through differential cell elongation that plants display tropic responses such as phototropism. (Stem elongation associated with etiolation was discussed in Chapter 5). When plants or plant parts are grown in the absence of light, photodestruction of IAA does not occur. Thus the endogenous levels of the hormone will become high, accounting for the marked internode elongation or stretching that is associated with etiolation.

Auxins also play an important role in **cell division**. Because of the important interactions between auxins and cytokinins, both classes are discussed relative to cell division in the cytokinin section of this chapter.

Auxins play a role in the **differentiation** and **dedifferentiation** (returning to the meristematic state) of cells and tissues. It has been demonstrated in callus grown in tissue culture and with stems of bean and birch that auxins influence dedifferentiation of vascular cambial tissue and differentiation of xylem. In fact, IAA is important in reactivating the vascular cambium of woody plants in the spring when growth resumes. Although plant cells all begin as meristematic cells, this is an example of the hormonal balance causing cells to become active and different from each other to allow the plant to function and express its phenotype.

The apical region of a plant generally exerts control over the development of buds found in the axils of leaves. This control can be nearly complete, such as in many young deciduous trees, or quite weak, as in tomato and some shrubs. The control of the apex over side shoot development is called **apical dominance** (Figure 11-6). Auxin is produced in the apical region, especially the young expanding leaves, and has been implicated clearly in apical dominance. Heading-back

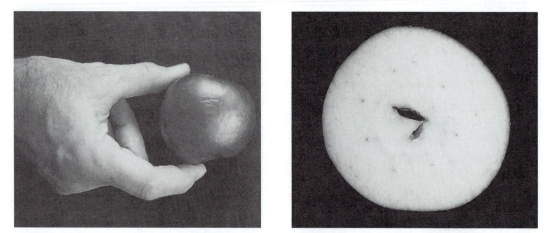

FIGURE 11-5. Misshapen apple fruit (*left*) and cross-section of the same fruit (*right*). Note that fruit enlargement is less on the side associated with locules lacking seeds because of less hormone production.

FIGURE 11-6. These bare-root trees have grown as single-stemmed whips because of strong apical dominance. Auxin produced by the apex has completely suppressed lateral bud outgrowth.

FIGURE 11-7. Many plants are pruned, thus removing stem apices and breaking apical dominance. Bushy plants with multiple branches result.

pruning or pinching out the apex results in free branching of many species. This practice removes a primary source of endogenous auxin. Auxins function in concert with other hormones such as cytokinins and possibly abscisic acid to control branching of plants. When plants are pruned, apical buds and young leaves are generally removed. As a consequence of removing these important sites of auxin biosynthesis, more heavily branched plants result (Figure 11-7). The hormonal balance of plants determines the different branching responses among diverse species and the influence of various environmental factors on branching, such as light quality, and levels of mineral nutrients.

Auxins function in the process (**abscission**) that results in the shedding of organs, such as leaves, flowers, and fruits from the plant. Abscission is generally associated with the formation of a layer of cells at the base of the organ; this is called an abscission layer (see Chapter 3). Auxin may either promote or inhibit abscission, depending on the timing of application, concentration, and the stage of development of the plant organ. The effect of exogenously applied auxin on fruit abscission has important horticultural and economic significance to the fruit industry. Most fruit crops will set more fruit than is desirable if conditions are optimal (Figure 11-8). Growers wish to thin the fruit to increase the size of individual fruits (less competition for available photosynthates), to lighten the

FIGURE 11-8. This spur has too many fruits set for adequate fruit size development. This problem may be prevented by promoting fruit drop, so growers often apply sprays of auxins such as NAA. The insecticide carbaryl is also applied to promote fruit abscission.

fruit load and thus reduce limb breakage, to stimulate flower bud initiation for the next year's crop, and to improve fruit quality and sometimes color. Although trees naturally shed some excess fruit early in the summer (e.g., "June drop" of apple and pear), it is usually not sufficient for commercial production. It is much more economical to thin chemically than manually. Growers therefore apply different chemicals such as the auxin NAA from 3 to 30 days after full bloom to chemically thin fruit crops such as apple and pear (stimulate fruit abscission). Depending on the timing or growing conditions, thinning may be insufficient or excess fruit may be lost, reducing yields. Growers therefore may spot-spray individual trees in an orchard. Their decision is based on the fruit load on the tree and prior experience with chemical thinning.

Later in the season, as the apple and pear fruits mature and ripen, the process of abscission begins naturally and can result in fruit drop, which reduces harvestable yield. A single application of a synthetic auxin such as NAA is therefore applied and will delay abscission from 7 to 10 days in apple. Trees usually are sprayed at the anticipated beginning of fruit drop. This demonstrates how growers utilize the basic knowledge of auxins to promote or delay abscission, which in turn increases production efficiency, crop quality, and yield.

Auxins stimulate the initiation of embryos and organs, especially roots (Figure 11-9). In some cell and tissue culture systems, such as carrot, the auxin 2,4-D is added to the medium to induce the formation of vegetative or **somatic embryos**. These induced cells and tissues are frequently transferred to an auxin-free medium to allow for the development of the embryos. Artificial seed coats made from sodium alginate and calcium chloride have been developed to protect these somatic embryos so that they can be planted with field-seeding equipment.

The influence of auxins on root initiation has had a profound impact on plant

FIGURE 11-9. Prolific adventitious root formation on this poinsettia cutting resulted from application of a rooting powder containing IBA.

propagation. Many plants are cloned by the **rooting of cuttings** if they do not come true from seeds (segregation), if they do not produce seeds, or if various economic reasons related to prolonged juvenility and other factors require it. Cuttings of some species root quite easily with no application of PGRs. These species appear to produce adequate quantities of endogenous auxin and other chemicals that naturally stimulate root initiation. Cuttings of many other species either require auxin applications for rooting to occur or need the growth regulators to improve rooting. It should be noted that even though the application of auxins in important to the rooting of cuttings of many species, other practices cannot be ignored, such as proper temperatures, light, and water relations, including humidity control. Many formulations that contain auxins (frequently IBA or NAA) are available on the market (Figure 11-10). The efficacy of these depends on the auxin or combination of auxins, their state (in solution or crystals suspended in an inert carrier such as talc), and the species and physiological state (relating to time of the year) of the cuttings (see Chapter 14). For specific details on timing, methods of preparation, handling, auxin formulation, concentration, and application methods, one should consult with references and follow published guidelines. If such information is not available, one should study information on related species and conduct a series of experiments to optimize rooting.

Auxins are also important in the formation of **compression wood** in conifers. This is a build-up of xylem on the undersides of branches that helps to support the weight of the limb.

Auxin Sensitivity

For many of the auxin activities listed on the previous pages, auxins can act as promoters or inhibitors, depending on the concentration of auxin and the plant tissue or organ. The growth of roots is extremely sensitive to the auxin concentration. Root elongation is stimulated only by extremely low auxin concentrations; at slightly higher concentrations, auxins will inhibit elongation and result in short, thick roots (Figure 11-11). This may seem contradictory with the use of exogenous auxins to promote rooting of cuttings. The student should understand that root initiation and root elongation are two different developmental phenomena. Generally, it takes more

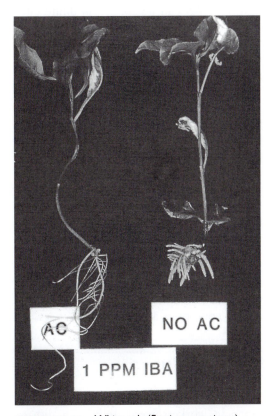

FIGURE 11-11. White ash (*Fraxinus americana*) microplants produced in vitro with excessive auxin in the medium. Note that the roots are short and thick, giving a stubby appearance when activated charcoal (AC) was not incorporated into the medium to adsorb the auxin.

FIGURE 11-10. Talc dilutions of auxin-type growth substances such as IBA are used commercially to stimulate rooting of cuttings. The ends of the cuttings in this propagator's hand have a white appearance from application of such a treatment.

auxin to stimulate root initiation than elongation. This may help explain part of the fallacy that if a little bit is good, a lot is going to be better. Too much auxin may indeed stimulate root initiation, but their subsequent development may be completely arrested. In addition, when potent auxins, such as 2,4-D, are applied in high concentrations they are excellent herbicides, therefore overdosing plants or cuttings with any auxin may damage them. Buds and stems are less sensitive to low levels of auxin than roots. It generally takes higher concentrations to stimulate bud growth than root growth and even higher doses to stimulate stem growth. However, if concentrations are too high, the growth of buds and stems will be effectively inhibited by auxins. In fact, high concentrations (one-half to one percent) of NAA can be applied in dilute interior household latex paint to pruned fruit trees to inhibit the outgrowth of latent buds into water sprouts. This inhibition can last for more than one year.

Plant species and their state of growth will also influence the response to endogenous or exogenous auxin. For example, 2,4-D is generally an effective herbicide for dicots but not for monocots. Woody plants show little response to auxin during the dormant season compared to the season of active growth. The differential responses of plant organs and species during different stages of growth is true for auxins, as well as for other plant growth substances. This knowledge should help horticulturists understand and expect different growth responses of plants under various environmental conditions and cultural practices.

GIBBERELLINS

During the early years of the 20th century, Japanese workers were studying a disease called "silly seedling" of rice. The symptoms included abnormally tall growth and a reduction in yield of approximately 40 percent. The disease is caused by a fungus with two names: *Fusarium moniliforme* (asexual form) and *Gibberella fujikuori* (sexual form). They found that extracts from the fungus could stimulate growth in a manner similar to the fungus. In 1935, the active fraction of the extracts was crystallized and named **gibberellin**. This work was published in Japanese journals in the Japanese language. Because of the language barrier and World War II, western scientists did not know until the early 1950s that the Japanese had discovered and had done scholarly work on gibberellins.

We now know that there are at least 126 different gibberellins, some unique to the fungus, some unique to higher plants, and some occurring in both. Unlike the auxins, gibberellins all have basically the same four-ring structure, as shown in Figure 11-12, but they differ from one another in that some have 20 carbons and some have 19 carbons, and they have different side chains. Naming of the gibberellins is related to their relative time of discovery, beginning with gibberellin A_1, or GA_1, then GA_2, GA_3, and so on. Only one gibberellin—GA_3—is known as gibberellic acid.

Many people have asked why there are so many gibberellins. Only approximately 12 gibberellins have high biological activity. It is estimated that within any one plant there are fewer than 15 gibberellins, with perhaps even fewer within a single organ. Some gibberellins are biosynthetic intermediates and others are breakdown products with little or no biological activity. It has been suggested that only a few (or even only one gibberellin) are active within a particular plant organ. GA_1, for example, may be the important gibberellin in corn, rice, and peas.

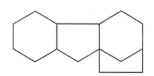

FIGURE 11-12. The basic four-ring structure of the gibberellins.

Gibberellin Biosynthesis and Commercial Availability

Gibberellins are diterpenoids, which means that they are composed of four isoprene units. An isoprene unit consists of five carbon atoms bonded to form a molecule shaped like the letter Y. The following is the structure of an isoprene unit without its double bonds:

$$C\text{-}C\text{-}\underset{\underset{C}{|}}{\overset{\overset{C}{|}}{C}} \text{ or } \textasciicaron$$

Gibberellin biosynthesis begins with a six-carbon compound, mevalonate, which loses one carbon as CO_2 to form an isoprene unit. Four isoprene units then string together and subsequently cyclize to form a compound, GA_{12}–aldehyde. Because this involves the linking of four five-carbon isoprene units, the resulting molecule has 20 carbons; for this reason all 19-carbon gibberellins are derived from the 20 carbon gibberellins. Many growth retardants function by blocking a step of GA biosynthesis; additionally, many dwarf mutant plants are missing enzymes that are involved in GA biosynthesis. A knowledge of GA biosynthesis is therefore essential to understanding more about the mechanism of action of growth retardants or understanding better why dwarf forms grow so slowly.

Gibberellins occur in high quantities in the following locations in the plant: the stem apex (especially young leaves), buds, roots, embryos, cotyledons, flowers, fruits, vascular cambium, and tubers. Most studies on GA biosynthesis have been done using developing seeds and fruits.

Gibberellins are fairly large and complicated compounds, compared to other plant hormones, so they are not synthesized chemically in laboratories for commercial use. To give an example of the difficult problems associated with gibberellin biosynthesis, 256 isomers of GA_3 are possible, but only one specific isomer has high activity. Consequently, companies that produce gib-

berellins for commercial sale grow the fungus *Gibberella fujikuori* in large vats and extract and purify the gibberellins. At the time of this writing, GA_3 is the only gibberellin that is offered for sale commercially without being mixed with other gibberellins or other growth regulators. It is fortunate that gibberellic acid is active in a wide variety of species. One can also purchase a mixture of GA_{4+7} together or in combination with a cytokinin (benzyladenine or BA). GA_4 and GA_7 are difficult to separate, so they are sold together.

Gibberellin Activities

Gibberellins promote **cell elongation**, and their effect on whole plants is much more dramatic than that of auxins. An application of gibberellic acid to bush beans or bush peas will cause them to grow as climbing pole beans or pole peas, respectively (Figure 11-13). When gibberellic acid is applied to cabbage, the internodes will elongate and the plant will grow two or more meters tall with a long stem rather than producing a head. Gibberellic acid is used commercially to promote internode elongation in sugarcane, which in turn raises the yield of sucrose obtained from this crop. It is also applied to grapes to elongate the peduncle and pedicels, resulting in looser clusters.

Gibberellins promote **cell division** in plants. With temperate woody species such as poplar (*Populus robusta*), the application of GA_3 causes an increase in the cell division of the vascular cambium and causes new phloem to differentiate into sieve tubes. However, when both IAA and GA_3 are applied, there is a great increase in cell division of the vascular cambium and both differentiated xylem and phloem are observed. This is an example of two growth substances working together to result in growth and developmental responses. Endogenous auxins and gibberellins are assumed to interact in a similar manner.

FIGURE 11-13. Dwarf peas are considered to be deficient in gibberellins. At left is 'Little Marvel' pea and at right, the same cultivar treated with 1 mM GA_3. Note the GA_3-stimulated internode elongation.

Seed germination seems to require endogenous gibberellins, as shown with barley seeds (Figure 11-14). During the process of germination of barley (and other monocot seeds) the following events occur:

1. The quiescent dry seed imbibes water that hydrates the cells.
2. Gibberellin is produced by the embryo. It has been demonstrated that a large number of gibberellins occur in barley seeds.
3. Gibberellin is translocated to the aleurone layer. This is a layer, three to four cells thick, that surrounds the endosperm.
4. In the aleurone layer, gibberellin causes the production of α-amylase and other hydrolytic enzymes.
5. The α-amylase (and other enzymes) moves to the endosperm, where it breaks down starch to sugar that can be used in respiration.

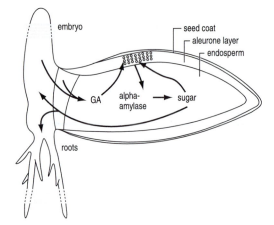

FIGURE 11-14. Schematic illustration of the involvement of gibberellins in seed germination. *USDA Agricultural Research Service photograph.*

6. The sugar is translocated to the embryo where it provides energy (through respiration) for growth. The sugar also lowers the cellular water potential leading to greater imbibition of water into the seed.
7. The germination process culminates with the emergence of the radicle through the seed coat. At this point seedling growth and development occur.

Naturally occurring gibberellins are considered to be required for seed germination. However, gibberellin soaks are almost never applied to nondormant seeds because little would be gained in germinative energy, and the resulting plants could be tall and spindly as a response to the gibberellin.

A knowledge of the function of gibberellins in seed germination has led to an interesting market for GA_3. Beer brewers, for example, use GA_3 because gibberellic acid can induce the production of hydrolytic enzyme during the malting of barley. During malting, barley seeds are germinated to convert the starch to sugar. When the the new sprout is two-thirds of the length of the grain, the starch is considered to be converted to sugar. To stop germination, the germinated seeds are then kiln dried and cut to flour (called grist) in a mill. Brewers found that by adding GA_3, the starch is hydrolyzed more rapidly to sugar, which ultimately shortens malting time. This can make beer brewing more economical.

Knowing the requirement for gibberellin in seed germination also has aided the understanding of seed (and bud) **dormancy**. For example, a seed that does not produce an adequate quantity of gibberellin cannot germinate and therefore it remains dormant, even if environmental conditions are favorable for germination. Application of gibberellic acid will often overcome dormancy. Other growth substances including cytokinins and abscisic acid also have important roles in seed dormancy. These interrelationships and methods to overcome seed and bud dormancy are discussed under dormancy in Chapter 12.

Gibberellins play an important role in the **sex expression** of flowers of several species. When applied correctly, gibberellins tend to promote maleness in flowers of many species, such as cucurbits.

Gibberellins also are involved in **flowering** in response to long days. Additionally, they cause **parthenocarpy** (development of seedless fruit) in many species, including grapes and apples. Gibberellins also can **delay senescence** of leaves and fruits. These are discussed in more detail in Chapter 12.

Gibberellin Sensitivity

Unlike their response to auxins, plants tend to continue to respond without inhibition to applied gibberellins over a wide concentration range. It is not uncommon to observe either a continued positive response or a leveling out rather than inhibition of growth with gibberellins. In some systems, however, gibberellins can be inhibitory, although this is more the exception than the rule.

CYTOKININS

Cytokinins were discovered in the 1950s. In early tissue culture work, researchers were looking for something to promote cell division for the production of callus. They learned that relatively undefined mixtures, such as liquid coconut endosperm and malt extract promoted cell division, but they did not know what fractional component was active. It was well known that when used alone, IAA caused cells to enlarge but cell division was lacking. When tobacco stem pieces (explants) were placed on the surface of a nutrient medium that was solidified with agar, and a purine base (adenine, a component of the nucleic acids DNA and RNA) was incorporated into the medium along with IAA, the cells began to divide and a callus mass was created (Figure 11-15). Researchers found that the adenine without auxin did not cause cell division and concluded that it was an interaction between adenine and auxin that resulted in the triggering of cell division.

It was shown further in early tissue culture studies, that autoclaved DNA from herring (fish) sperm and from calf thymus was active in stimulating cell division in tobacco wound callus tissue cultured in vitro. This activity was not present if the DNA was not autoclaved. The first cytokinin was then isolated from the autoclaved DNA and was given the name **kinetin**. It should be noted

FIGURE 11-15. Callus produced in vitro from tomato leaf explants on a medium containing both cytokinin and auxin.

that kinetin is not a natural component of plants; therefore, it is not considered a plant hormone. The first hormonal cytokinin was isolated from corn (*Zea mays*) seeds and was given the name **zeatin**. Zeatin seems to be present in most, if not all plants. Another important hormonal cytokinin is isopentenyladenine (*2iP*). A commonly used synthetic cytokinin is **benzyladenine (BA)**. Because adenine is an amino purine, BA also is known as benzylaminopurine (BAP); therefore BA and BAP are abbreviations for the same compound. A more recently discovered and very active cytokinin is **thidiazuron**.

In the early literature, this group of compounds was called *phytokinins* or simply *kinins*, but the latter term caused some confusion with a term that was used by animal physiologists. The term cytokinin was therefore adopted because **cytokinesis** means the process of the division of the cytoplasm of a cell.

Cytokinin Biosynthesis and Commercial Availability

Relatively large quantities of cytokinins are found in embryos, young fruits, young leaves, the cambial region, and roots, especially root tips. Roots and developing seeds are considered to be major sites of cytokinin biosynthesis. Therefore, when pruning roots, or taking stem or leaf cuttings for plant propagation purposes, horticulturists should be aware that these processes severely affect or even remove a major source of endogenous cytokinins, which in turn influence growth and development of the top.

Hormonal cytokinins generally consist of adenine with a five-carbon isoprene unit for a side chain. Adenine is a precursor for the majority of the molecule and the isoprene unit ultimately comes from mevalonate. Because gibberellins are derived from mevalonate, to a certain extent, gibberellins and cytokinins share a portion of the same biosynthetic pathway.

Although there are others, the four cytokinins, kinetin, BA, 2iP, and zeatin are all available for purchase and are listed here in order of cost. Kinetin is quite inexpensive; zeatin costs more than 100 times as much as kinetin. Although zeatin is a very active cytokinin, its cost prohibits its general use in horticulture.

Cytokinin Activities

Cytokinins promote cytokinesis. The requirement for auxin along with the cytokinins exists but is not always evident, because cytokinins can stimulate auxin biosynthesis. Knowledge of the strong stimulatory effect of cytokinins on **cell division** has a major application in plant tissue culture systems and may help explain meristematic activity.

Cytokinins contribute to **cell enlargement** (i.e., the growth of cells in all directions). Cell enlargement should not be confused with cell elongation phenomena that are promoted by auxins and gibberellins. In fact, cytokinins usually inhibit stem elongation. Cytokinins promote growth by swelling and stimulate leaf expansion. The influence of cytokinins on leaf expansion can be demonstrated easily in an indirect manner. Leaves of some species, such as coleus and rubber plant (*Ficus elastica*), can be induced to form adventitious roots fairly easily in a mist propagation bench. Neither species will produce adventitious shoots, so only rooted leaves will result. Roots are major sites of cytokinin biosynthesis; the rooted leaves will generally grow to a much larger size than leaves on intact plants or leaves on stem cuttings of the same species (Figure 11-16).

Cytokinins promote **differentiation**. In tissue culture systems, the relative levels and ratio of cytokinin and auxin in the medium will promote the differentiation of organs (organogenesis) and tissues. Cytokinins and auxins are produced by plants and can be applied to plants by people so there is therefore an interplay between the endogenous and exogenous plant growth substances.

FIGURE 11-16. Photograph of a rooted coleus leaf compared to a rooted leaf-bud cutting. Once rooted, the rooted leaf expanded in size considerably. This phenomenon was probably caused by the stimulation of cell growth by the cytokinins produced by the newly formed roots.

The relative ratios and expected responses are as follows:

1. **High cytokinin : low auxin** Promotes shoot (stem) initiation and development.
2. **High auxin : low cytokinin** Promotes root initiation and embryogenesis.
3. **Moderate to high levels of both cytokinin and auxin** Promotes callus growth.

The ratios and relationships listed here can be remembered easily with the phrase: "auxins root 'em and cytokinins shoot 'em." (Figure 11-17). Based on this knowledge propagators apply auxin-type rooting compounds to cuttings to stimulate adventitious root formation and cytokinins to leaf cutting or tissue culture systems to stimulate adventitious shoot formation. The understanding of this relationship also helps explain why some cells within a plant differentiate to become roots and others differentiate to become shoots.

Cytokinins have been shown to **delay or reduce senescence.** They function in two ways, relating to senescence: They help maintain the level of chlorophyll by slowing its breakdown and they increase chlorophyll biosynthesis. The promotion of biosynthesis, rather than just maintenance of chlorophyll synthesis, is a unique property of cytokinins. People have removed leaves from plants and applied a single drop of cytokinin to one spot on a leaf. They found that the rest of the leaf senesced while the spot where the cytokinin was applied remained green; in fact, nutrients migrated from the rest of the leaf to the green spot. Greener plants also have been observed in response to exogenous cytokinin application and in plants that were genetically engineered to overproduce cytokinin. When a gene for cytokinin biosynthesis (*ipt*) was inserted into tobacco plants, the genetically transformed plants had higher cytokinin levels and were greener than control plants. Yield increases might be achieved for some species if senescence could be delayed with a

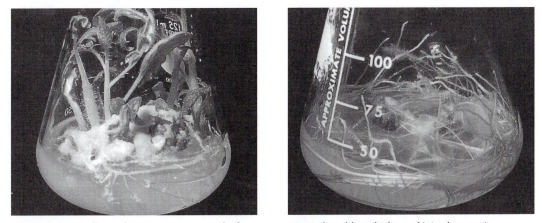

FIGURE 11-17. Shoot formation in tomato leaf segments was induced by a high cytokinin : low auxin ratio in the medium (*left*); roots promoted by a low cytokinin : high auxin ratio (*right*).

cytokinin spray or additional cytokinin biosynthesis genes.

Cytokinins play an important, indirect role in the **overcoming of bud and seed dormancy.** They interact with other growth substances, as discussed in Chapter 12.

Cytokinins promote the **opening of stomates.** Stomates tend to open in response to exogenous cytokinin applications and in genetically transformed plants that overproduce cytokinin. During times of stress, such as drought, the levels of cytokinins decrease in leaves (stress hormones, such as abscisic acid increase) and stomates close, thus allowing the plant to transpire less and thus conserve water.

Cell division is sometimes promoted in response to an infection. This is often associated with locally elevated levels of cytokinins. The following are examples of this phenomenon: (1) **Nodule formation** on legumes occurs in response to infection by *Rhizobium* or *Bradyrhizobium*; this cell division and infection results in an area where atmospheric nitrogen is fixed and utilized by the plant (see Chapter 9). (2) **Galls** form at the base of plants in response to infection by *Agrobacterium tumefaciens*, in a disease known as crown gall (Figure 11-18). This bacterium and knowledge of this disease are utilized in the genetic engineering of plants. The pathogen acts by inserting a gene for cytokinin biosynthesis (*ipt*) into the host plant cells, "engineering" them to overproduce cytokinin and thus form tumors that may have adventitious shoots. (3) **Witches' brooms** occur on plants in response to infections, insects, mites, parasitic plants, and genetic mutations (Figure 11-19). They are characterized by many branches with relatively short internodes growing from one location on a branch. Elevated levels of cytokinins have been associated with witches' brooms.

Cytokinins are also active in the release of lateral branches from **apical dominance,** or promotion of branching. They also play roles in **flowering** and **fruiting** of various horticultural species. These roles are discussed in Chapter 12.

FIGURE 11-18. Crown gall formation on geranium (*Pelargonium hortorum*). This massive cell proliferation, caused from the infection of the host plant by *Agrobacterium tumefaciens*, is induced by a high level of cytokinin production in response to the infection.

FIGURE 11-19. Witches broom on a branch of honeysuckle (*Lonicera* sp.). This is another phenomenon considered to be stimulated by an unusually high cytokinin level in the affected plant part.

BRASSINOSTEROIDS

The first brassinosteroid, brassinolide, was isolated from the pollen of the rapeseed plant (*Brassica napus*) in 1979. Since that time, they have been shown to be plant hormones that are widely distributed throughout the plant kingdom. They are steroids, as are the sex hormones in animals and the molding hormones in insects and crustaceans. There are more than 40 brassinosteroids that have been characterized structurally and functionally.

Brassinolide and its precursor castasterone are among the most important brassinosteroids because of their wide distribution among plant species. Brassinolide is about five times more active than castasterone. Brassinosteroids are present throughout the plant with the highest concentrations in pollen and immature seeds.

Biosynthesis

Steroids are ultimately derived from mevalonic acid, as are gibberellins, the five-carbon side chain of endogenous cytokinins, and abscisic acid. Therefore, they share parts of the same biosynthetic pathway as these other major hormones. Campesterol, which is derived from mevalonate, is the steroidal precursor of castasterone and brassinolide.

Brassinosteroid Activities

Brassinosteroids are considered to be essential for **stem elongation** and mutant plants that do not produce these compounds are dwarfs when grown in both light and dark. Stem elongation of these dwarfs cannot be overcome by applications of gibberellins or auxins. However, applications of brassinolide or castasterone will restore elongation growth, especially if the plants are grown in the light. Elongation of stems that is stimulated by brassinosteroids has been shown in bean plants to be a result of a stimulation of both **cell elongation** and **cell division**.

When the aerial portion of stems are treated with brassinolide, growth of their attached **roots** is promoted. However, there are some conflicting results because application of brassinolide to excised roots inhibits their growth. Therefore, it is possible that applications to intact seedlings may become important for stimulating root growth on young plants and transplants.

Applications of brassinosteroids can make plants more resistant to **environmental stresses**. Plants under water stress, salt stress, and herbicide stress are stimulated to grow more, or their root growth is stimulated following application of brassinosteroids compared to nontreated plants. This enhanced stress resistance is observable under field conditions and may have important implications for all of agriculture.

Brassinosteroids also stimulate **seed yields** by as much as 45 percent, even when applied at very low concentrations. Many of these studies have been on cereal grain crops, but they also have a stimulatory effect on yields of legume seeds and fruits (recall that there is a relationship between seeds in a fruit and fruit growth). A problem with application of brassinosteroids to field-grown crops is that timing in relation to when flowers open is critical. Generally not all flowers open on the same day, making results inconsistent.

A limitation of practical applications of brassinosteroids to field-grown plants is that these compounds are active for only about two or three days. There is a need for the development of a brassinosteroid with long-lasting activity. There are some candidate compounds being tested that may be persistent for longer and thus be of practical use.

WOUNDING AND STRESS HORMONES: ETHYLENE

Before the common use of electricity, city streets were illuminated with gas lamps that burned mainly carbon monoxide. In the mid-1800s, it was reported that trees in portions of German cities suddenly defoliated because broken mains emitted illuminating gas into the atmosphere. As early as 1901, ethylene was shown to be the active component of the illuminating gas causing the defoliation response.

Early in the 20th century, California lemon and tomato growers picked their crop green and put it into special buildings called **hot houses**. Because of the heat build-up (see Chapter 6), greenhouses are also sometimes called hot houses, but the hot houses in use in California were made of wood or other substances that were heated by kerosene stoves. Growers thought that the heat and humidity were responsible for ripening (tomatoes) or degreening (yellowing of lemons) their crop. When a farmer substituted another form of heating for the kerosene stove, no ripening or degreening occurred. It was then learned that fumes from the incomplete combustion of kerosene were necessary for ripening. The responsible agent in the fumes was ethylene. Today, fruit such as tomatoes are often harvested mature-green and ripened by exposing them to ethylene. Sometimes the flavor does not compare to vine-ripened tomatoes because many of these commercial cultivars are bred for enhanced shelf-life, rather than flavor and because the tomatoes may have been refrigerated. The low temperatures of refrigeration can cause chilling injury resulting in the flavor components not developing as well as tomatoes that ripen at higher temperatures (either on or off the vine). However, documented consumer tests show that people cannot tell any difference in flavor between vine-ripened tomatoes and tomatoes of the same cultivar harvested mature-green and ripened at room temperature. *Immature* green fruit will not properly ripen and will have an inferior flavor.

Ethylene (C_2H_4) in its pure form is a gas at the temperature range where plants grow, whereas most other growth substances are crystals in their pure forms. However, within plants most of these other hormones are in solution. Other compounds having ethylene-like effects (e.g., propylene, vinyl

chloride, and acetylene) require very high dosages to achieve the same response.

For a number of years, researchers seriously questioned whether ethylene is a plant hormone. Ethylene could be shown to meet all of the criteria to be considered a plant hormone, except the transport requirement. It has now been demonstrated to the satisfaction of most scientists that ethylene is transported within the plant, at least short distances. Therefore, most people consider ethylene to be a plant hormone.

Ethylene Biosynthesis

Ethylene biosynthesis occurs in many locations within the plant. However, various environmental factors and stages of maturity have profound influences on ethylene production. A knowledge of some of the basic steps of ethylene biosynthesis will help explain various phenomena relating to postharvest storage of fruits and vegetables and adaptive responses of plants to stressful conditions such as flooding. The major precursor for ethylene in the plant is the amino acid methionine. There are two intermediates between methionine and ethylene: S-adenosylmethionine (SAM) and 1-aminocyclopropane-1-carboxylic acid (ACC) (Figure 11-20). Oxygen is required for the conversion of ACC to ethylene. A concentration of less than two percent oxygen stops ethylene biosynthesis.

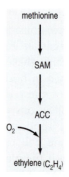

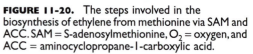

FIGURE 11-20. The steps involved in the biosynthesis of ethylene from methionine via SAM and ACC. SAM = S-adenosylmethionine, O_2 = oxygen, and ACC = aminocyclopropane-1-carboxylic acid.

Many things stimulate ethylene biosynthesis in plants, including the applications of exogenous chemicals, particularly auxin-type herbicides, such as 2,4-D. Wounding or otherwise disturbing plants also stimulates the production of elevated levels of ethylene.

Ethylene Activity

Historically, ethylene has been associated with the classic *"triple response" of etiolated pea seedlings*. Triple responses—stem stunting, swelling of the subapical portion of the stem, and diageotropism (or the tendency of the stem to grow at a right angle to the line of gravity)—have also been observed in other plants. At a concentration of one ppm and higher, ethylene inhibits both root and shoot elongation in many species. The triple response of ethylene may be what allows germinating seedlings to grow around rocks or other obstructions that restrict their growth in the soil and induce ethylene synthesis as they grow upward toward the soil surface. The horizontal growth allows the seedling to grow around the obstruction, and the swelling of the subapical portion of the stem makes the hook tighter, thus avoiding damage from the obstruction. Once the seedling is exposed to the light as it breaks through the soil surface, ethylene production is reduced as the hook opens.

Ethylene causes leaves to curl in a downward direction in a phenomenon called **epinasty** (Figure 11-21). Epinasty is caused by the cells on the upper side of the petiole growing more rapidly than the lower cells, which results in the downward orientation of the leaf. The angle of curvature is sometimes measured as an indication of the plant's response to ethylene (i.e., within limits, an increase in exposure to ethylene increases the curvature). If one is growing a crop in a greenhouse and observes epinasty on a crop such as tomatoes, the possibility of a gas leak should be investigated. Some herbicides, such as 2,4-D also cause epinasty; this is because they stimulate ethylene biosynthesis in plants.

FIGURE 11-21. Epinasty in tomato. Note the downward curling of the leaves, caused by more rapid growth of the cells in the upper part of the petiole.

Ethylene is also involved with the *induction of adventitious roots*. This is thought to be associated with localized accumulations of auxin. There appears to be a complementary role of both auxin and ethylene in adventitious rooting. If the endogenous sources of auxin are removed such as by pruning off the apical bud and cotyledons on a seedling, ethylene does not stimulate rooting. However, if ethylene biosynthesis is inhibited, rooting is likewise inhibited. Because of the stimulatory role of ethylene on rooting, stem cuttings of some species are treated with ethylene or ethylene-releasing compounds as root promoting agents.

Ethylene has dramatic effects on *latex production* (and other secretions) of different species including the rubber tree

(*Hevea brasiliensis*). Ethylene is a wound hormone and enhances the flow of latex because it delays the clogging of the latex vessels (laticifers). Latex is harvested by scraping (wounding) the bark or cutting a tapping groove in the bark of the rubber tree (Figure 11-22). A liquid ethylene-releasing compound (ethephon) is then applied to the bark or cuts, where it increases yield up to 100%. Early investigators found that the auxins 2,4-D, 2,4,5-T, and NAA stimulated latex flow, but this was probably because they stimulated ethylene biosynthesis.

Ethylene appears to play an important role in the adaptation of plants to flooding (anaerobic) conditions. The reasons relate to a knowledge of the ethylene biosynthesis pathway (see Figure 11-20). It should be recalled that oxygen is required for the reaction where ethylene is formed from ACC. When roots are deprived of oxygen because of flooding, the conversion of ACC to ethylene cannot proceed and ACC accumulates in the roots. The excess ACC moves up the stem into the leaves. In the leaves there is an adequate supply of oxygen for the reaction that converts ACC to ethylene. The result is that high levels of ethylene are found in the leaves. The ethylene, in part, accounts for characteristic flooding symptoms, including

leaf epinasty, stem swelling, formation of new roots on the stem, yellowing of leaves, abscission of older leaves, and formation of new roots and stems that are more tolerant of flooded conditions. The new roots and stems are more tolerant of flooded conditions because they contain more air spaces (aerenchyma) that function in the internal transport of oxygen. Greater aerenchyma formation occurs because ethylene stimulates production of enzymes such as cellulase that break down cellulose in the cell walls. Cells are destroyed, leaving more air spaces. Extensive aerenchyma are associated with aquatic plants and other species that grow in flooded areas, such as rice, and are thought to function in the internal transport of oxygen to organs growing under anaerobic conditions.

Ethylene promotes *abscission* of organs, including leaves and fruits. An example is the aforementioned tree defoliation resulting from the leaks of illuminating gas in German cities. Ethylene causes the sequential shedding of leaves, beginning with the oldest leaves, on up the plant with the youngest leaves abscising last. Why leaves of various ages abscise at different times is not fully understood. Early fruit abscission as harvest season approaches, which leads to a loss of harvestable yield, may be related to elevated levels of ethylene in ripening fruit. Exogenous application of auxin delays this fruit abscission, as mentioned earlier.

Ethylene also plays important roles in other plant processes: the promotion of senescence (ethylene promotes chlorophyll degradation); the release of some buds or seeds from dormancy; the stimulation of seed germination; the promotion of flowering of certain horticultural plants, including pineapple; sex expression in flowers (ethylene is usually associated with femaleness); the stimulation of lateral bud outgrowth; and fruit ripening. These topics are also discussed in the Chapter 12.

Because ethylene stimulates abscission, promotes chlorophyll degradation leading

FIGURE 11-22. Latex flow from tapping grooves of commercially grown rubber tree (*Heavea brasiliensis*).

Photograph courtesy Kimberly Ripley, Southern Illinois University, Carbondale, IL.

to senescence, and promotes fruit ripening, it is sometimes called the "aging hormone." To delay aging or to enable longer term storage of plant materials including fruits, vegetables, and cut flowers, treatments and conditions are selected that minimize the effects of ethylene, either by removing it, slowing its biosynthesis, or somehow interfering with the molecule(s) in the plant that recognize ethylene (receptor molecule). Postharvest storage has important horticultural significance and is critical to the nutritional and cultural needs of humans (See Chapter 15).

Fruit Ripening

Ripening refers to the complex changes that begin when a fruit is physiologically mature and end when it is in the most favorable state for consumption. These changes include color, firmness, texture, flavor, and aroma. After a fruit is fully ripe, it begins to senesce. In many fruits, the ripening process is stimulated by ethylene (see Chapter 15).

Ethephon

Because ethylene, in its pure form is a gas at normal temperatures, it dissipates too rapidly to be effective if it is applied to field-grown plants. A liquid form of ethylene is useful, however. The most commonly used liquid chemical form is ethephon, which has the chemical name of 2-chloroethylphosphonic acid and is sometimes written as CEPHA. Ethephon is taken up by plants and, once inside, breaks down to release only ethylene, chloride, and phosphate. Ethephon has been cleared for use on many food crops. It is used to promote ripening on many crops including apple, tomato, and coffee; to loosen fruit as a harvest aid for crops such as cherries and walnuts; to promote the flowering of pineapple; to increase latex flow in rubber trees and for rooting cuttings of various species.

ABSCISIC ACID

Abscisic acid (ABA) is an interesting compound because it can serve as an inhibitor of growth and can be a promoter of growth and is especially important as a plant stress hormone. It is in a class by itself and is a major plant hormone. ABA was specifically identified in 1965 and was found to be the principal component of what was previously known as dormin from sycamore maple (*Acer pseudoplatanus*) leaves and Abscisin II from cotton (*Gossypium hirsutum*) bolls. The specific identification of ABA showed people from different labs that they were all working with the same compound.

ABA Biosynthesis

ABA is present in all organs of higher plants and in phloem and xylem sap. A precursor for ABA is mevalonate, which is also the precursor of the gibberellins. Both ABA and gibberellins are terpenoids. Mevalonate is incorporated first into carotenoids that are in turn metabolized into ABA. ABA biosynthesis apparently occurs in roots, stems, leaves, fruits, and seeds. Many plant stresses stimulate ABA biosynthesis.

ABA Activities

At extremely low concentrations, between 10^{-10} and 10^{-8}M, ABA can act as a growth promoter and can stimulate growth phenomena such as elongation. This promotion seems to be especially true for roots growing under stressful conditions such as drought. However, ABA generally counteracts the effects of growth promoting hormones such as auxins and gibberellins. This apparent contradiction may be a dose response, but exemplifies our lack of knowledge of the complex nature of the underlying factors that control plant growth and development.

ABA was originally associated with abscission, the process by which plant organs such as leaves, flowers, and fruits separate from the plant. ABA seems to have only a slight effect in intact plants, however; its association with abscission was determined from studies on isolated plant parts. Ethylene has a much more pronounced effect than does ABA in the stimulation of abscission.

ABA appears to play a role in the promotion and maintenance of dormancy in seeds and buds. Although there are variations among species, other hormones, including gibberellins, cytokinins, and ethylene also seem to be involved in the onset, sustaining, and breaking of dormancy. Because of these apparently complex interactions, dormancy is covered in Chapter 12.

ABA is considered to be a stress hormone, since its levels change in plants that have been subjected to water stress, salinity, waterlogging, and temperature stress. Water stress has been studied the most extensively in relation to ABA. When ABA is applied exogenously, it causes fairly rapid closure of stomata, often within 3 to 10 minutes of application. Additionally, the size of the aperture is concentration dependent (i.e., the greater the amount of ABA, the smaller the stomatal pore). It appears that the water status of the soil is the initial signal for closure of stomata, rather than the water status of the leaf. As the soil dries, the amount of ABA in the roots rises up to 100-fold and this ABA is translocated in the xylem to the shoot and leaves where it inhibits shoot growth and causes the stomata to close resulting in the plant conserving precious water during times of drought. The newly synthesized ABA accumulates throughout the plant, especially in the leaves; in fact up to a 40-fold increase in ABA has been reported in wilted leaves. Because the stomates are closed, other processes, such as photosynthesis are also brought to a halt. Upon rewatering, the ABA level gradually declines and the stomates reopen.

This wonderful mechanism by which plants can control moisture loss and conserve whatever water they possess enables survival of many species during periods of drought. It would have tremendous economic importance if we could use a substance as an antitranspirant and thus reduce the water requirement of plants. However, ABA is rather expensive and this problem is further exacerbated by the fact that light causes the photodestruction of ABA. Hence, the search continues for an antitranspirant that will effectively reduce transpiration, cause no adverse reactions, and be relatively inexpensive.

ABA also plays a role in the maintenance of apical dominance, or inhibition of branching. The role of ABA in branching is discussed in more detail in the Chapter 12.

JASMONATES

Jasmonates are in the oxylipin class of compounds and are considered by many to be major plant hormones. They have wide distribution in the plant kingdom, regulate plant growth and development, and are important in plant resistance to insects and pathogens. Jasmonates are represented by *jasmonic acid* and *methyl jasmonate*. Of these two, methyl jasmonate is more active when applied exogenously.

Methyl jasmonate is a fragrant methyl ester of jasmonic acid and occurs as a gas at the temperatures at which plants grow. It gives flavor and scent to jasmine, flowers, and other herbs such as rosemary. Methyl jasmonate makes up about 2 to 3 percent of the essential oil of jasmine. Some studies have indicated that jasmonates are effective anticancer agents in humans.

Jasmonate Biosynthesis

Jasmonates are synthesized from linolenic acid that can be released from plant membranes or can occur freely. Biosynthesis can occur in most parts of plants, including leaves, flowers, roots, seedlings, and young immature fruits. Enzymes for jasmonate biosynthesis are in plastids, including chloroplasts.

Jasmonate Activity

Wounding of plants causes an almost immediate increase in the production of jasmonates. In turn, the jasmonates influence the production of various proteins (gene products) such as proteinase inhibitors and chitinases that have important functions in the defense of plants to microbial pathogens and insects. Additionally they stimulate the production of volatile aldehydes and oxoacids that also function in the defense of plants.

The coiling of tendrils in plants is an interesting and poorly understood phenomenon. Coiling can occur as a result of abrasion or the tendril touching a support. Tendrils that have no contact with other objects or the plant itself do not coil. It has been shown that methyl jasmonate and an intermediate in the biosynthesis of jasmonic acid, 12-oxy-phytodienoic acid, will cause tendrils to coil.

Jasmonates can act as inhibitors of stem and root growth, seed germination, and photosynthesis. This may be part of plant response to various environmental stresses. In a manner similar to ABA, jasmonate levels increase under stresses. The pathways are probably not the same for ABA and jasmonates, but growth inhibition helps plants survive environmental stresses.

Vegetative storage proteins are made in tubers, bulbs, and seeds, are used in subsequent growth and therefore are important for plant survival. The genes coding for vegetative storage proteins are induced by jasmonates. Not only are these storage proteins important when these storage organs and seeds begin growth, but they are important for human nutrition.

Tuberonic acid is a derivative of jasmonic acid. Tuberonic acid, jasmonic acid, and methyl jasmonate all induce the tuberization of potatoes in tissue culture systems. Their role in whole plants is not yet well-defined.

Ethylene and jasmonates share some responses. Similar to ethylene, jasmonates stimulate fruit ripening, leaf and flower senescence, and abscission. Therefore, jasmonates are wound induced and aging hormones, similar to ethylene.

OTHER PLANT GROWTH SUBSTANCES

Hundreds of natural plant products have plant growth regulating activity as promoters or inhibitors. These compounds may be required in large quantities to be effective, may be active in low concentrations or may be species specific.

As we learn more, it will not be surprising if other classes of major plant hormones are elucidated or if some secondary PGRs gain major economic importance. Horticulturist should read current literature to learn of new and important developments in plant science. An excellent reference to consult for updates in this area is *Horticultural Reviews*. This publication contains chapters by experts who cover horticultural topics in detail.

REFERENCES

Arteca, R. N. 1996. *Plant Growth Substances Principles and Applications*. Chapman & Hall. New York. 332 pp.

Basra, A. S. (Ed.). 2000. *Plant Growth Regulators in Agriculture and Horticulture Their Role and Commercial Uses*. Food Products Press. New York. 264 pp.

Davies, P. J. (Ed). 1995. *Plant Hormones Physiology, Biochemistry and Molecular Biology*. (2nd ed.) Kluwer, Dordrecht. 833 pp.

Hooykaas, P. J. J., M. A. Hall, and K. R. Libbenga (Eds.). 1999. *Biochemistry and Molecular Biology of Plant Hormones*. Elsevier, Amsterdam. 541 pp.

Horticultural Reviews (yearly series). AVI Pub. Co., Westport, CT.

Khripach, V. A., V. N. Zhabinskii, and A. E. de Groot. 1999. *Brassinosteroids A New Class of Plant Hormones.* Academic Press, San Diego. 456 pp.

Sakurai, A., T. Yokota, and S. D. Clouse (Eds.). 1999. *Brassinosteroids Steroidal Plant Hormones.* Springer, Tokyo. 253 pp.

Srivastava, L. M. 2002. *Plant Growth and Development Hormones and Environment.* Academic Press, Amsterdam, 772 pp.

CHAPTER 12

CHEMICAL CONTROL OF PLANT GROWTH

This chapter explains plant growth and developmental phenomena and how they are controlled by the endogenous hormonal balance. It discusses the use of exogenously applied chemicals to modify plants where appropriate. Some exogenously applied substances are simply the same chemical that the plant produces. In this case, the grower is making the application to alter the plant's hormonal balance, which will in turn affect plant phenotype. Other substances may substitute for, mimic, or affect the metabolism of plant hormones, (e.g., many growth retardants function by blocking various steps in gibberellin and/or brassinosteroid biosynthesis). In addition, some plant growth regulators (PGRs) appear to act not by directly affecting the hormonal balance within the plant, but through various indirect mechanisms.

A knowledge of exogenously applied plant growth regulators is essential for horticulturists. The kinds of growth regulators and their usage varies among the horticultural commodities. Whenever applying plant growth regulators, growers should follow the label instructions carefully, in order to get the desired results, to avoid serious consequences of improper rates, to obey the law, and to protect the environment and individuals, be they growers, or consumers.

Because of the major economic importance of PGRs in agriculture, chemical companies are continually developing and releasing new products that modify plant growth. Horticulturists should always strive to learn, because much information quickly becomes outdated. The information in this chapter can be used as a foundation, but it should be continually supplemented with the current literature in this rapidly changing field.

DORMANCY

Dormancy is a condition of buds or seeds where growth or germination will not occur, even under favorable growing conditions. Typically, dormancy allows plants to avoid growth during times when weather is generally unfavorable, such as during cold winter months or during periods of drought. There are three generally accepted terms for dormancy, **endodormancy, ecodormancy,** and **paradormancy.** In this textbook the term dormancy is used to describe endodormancy conditions, where growth is inhibited by physical characteristics of the seed or bud or by their hormonal balance that typically happens in winter in temperate climates. Ecodormancy (when environmental conditions are not favorable for growth, but growth will occur when these conditions improve) is referred to as quiescence in this chapter and paradormancy is referred to as apical dominance in this chapter.

Bud dormancy is brought on by photoperiod and temperature. Light and temperature can also play important roles in breaking of dormancy of seeds and buds. In many cases the onset, maintenance, and breaking of dormancy can be related directly to the hormonal balance within the seeds or buds. No one hormone is solely responsible for dormancy, but the presence or absence of substances such as ABA, gibberellins, and cytokinins, is associated with this no-growth state. Gibberellins are thought to have a primary role in overcoming dormancy. As

explained in Chapter 11, gibberellins are considered essential for seed germination because they stimulate the production of enzymes, such as α-amylase, that are necessary for the germination process. Therefore, seeds that lack adequate levels of gibberellins cannot germinate and thus remain dormant (see Table 12-1). The application of gibberellic acid will sometimes overcome dormancy in both seeds and buds if the dormancy was a result of inadequate levels of gibberellin. Exogenous applications change the growth substance balance within the seed or bud to a level favorable for growth.

Abscisic acid is thought to promote dormancy of seeds and buds. It has been shown that ABA specifically blocks the formation of enzymes such as α-amylase, which is stimulated by gibberellins. Therefore, even if gibberellin levels are adequate to allow germination or bud break, if ABA is present in sufficient quantity it will block the effect of the gibberellins and prolong dormancy (Table 12-1). Under these circumstances, the

TABLE 12-1. The Khan model, illustrating the possible relationships among gibberellins, cytokinins, and abscisic acid[a]

GIBBERELLINS	CYTOKININS	ABSCISIC ACID	RESULTING CONDITION
+	+	+	growth
+	−	+	dormant
+	+	−	growth
+	−	−	growth
−	+	+	dormant
−	−	+	dormant
−	+	−	dormant
−	−	−	dormant

Source: Khan, A. A. 1971. *Science* 171 (3974):853–859. Used with permission.

[a]The minus symbol (−) means that the level of the substance is below the minimum level (threshold) at which it has an effect; a plus (+) means that it is at or above the threshold level.

application of exogenous gibberellins will not overcome the dormancy.

Cytokinins are thought to play a permissive role in breaking dormancy. Although they do not influence germination or bud break directly, the effect of ABA can be overcome by cytokinins. Thus, when cytokinins overcome the ABA effect, they permit germination by allowing the gibberellins to serve their normal growth-promoting functions without the blockage by ABA (Table 12-1). Because they serve only a permissive function, cytokinins usually display low activity in dormancy compared to either ABA or gibberellins. However, cytokinin activities are prominent when combined with other promotive agents, such as gibberellins or ethylene.

Cytokinins and ethylene play important roles in overcoming **thermodormancy** of seeds, which is induced by high temperatures (see Chapter 6). Thermodormancy may be either secondary or induced dormancy and serves as an important survival mechanism by preventing germination when soil conditions are too warm for some cool season plants. It is a serious concern for horticulturists, because some cool season crops perform well when grown in the autumn. Seeds of a lettuce crop, for example, should be sown in the late summer or early fall for a fall or winter harvest, depending on the location. The temperatures are often still high at the time of planting, and thermodormancy may block germination. Applications of growth regulators, such as a combination of gibberellic acid, kinetin, and ethephon can overcome such thermodormancy and allow germination. It is possible to apply these growth regulators in an organic solvent such as acetone to the dry seeds, allow them to re-dry, and continue to store them dry until they are planted. The absorbed PGRs will remain effective. Thermodormancy can also be avoided by sowing germinated seeds, primed seeds, or planting in the late afternoon for sprouting during the cool night. This heat effect occurs only with unsprouted seeds.

There are important economic reasons for horticulturists to promote or maintain dormancy. PGRs are routinely used to delay sprouting of onion bulbs, potato tubers, and sometimes water sprouts on fruit trees, and on trees located under power lines. Maleic hydrazide, which is widely used to control sprouting of potato and onion (Figure 12-1). It is applied by spraying it on actively growing plants and is absorbed and translocated throughout the plants. It nonselectively inhibits cell division, thus it controls

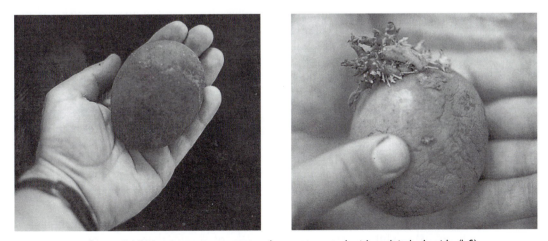

FIGURE 12-1. Sprout inhibition in storage potatoes. A potato treated with maleic hydrazide (*left*); untreated potato (*right*).

sprouting. This increases the storage life of these important vegetable crops.

The auxin, naphthaleneacetic acid (NAA), is used to control the regrowth of sprouts of fruit trees (apple, pear), or other trees, such as maples or elms. This approach is unique, in that the NAA is applied in either an asphalt base or interior household latex paint. It can be painted onto the bark of the trunk or limbs of the trees, or the asphalt can be applied as a wound dressing to fresh cuts and the surrounding bark. This treatment keeps buds that are present under the bark (latent buds) from growing into water sprouts; it often is effective for one year or longer. Use of NAA may result in considerable savings in pruning costs.

APICAL DOMINANCE

The shoot apex generally grows more rapidly than lateral buds, and it exerts control over the outgrowth of lateral buds and branches. The dominance of the apical bud over the axillary buds can be removed by pinching or pruning out the apex, resulting in a much more bushy plant. Horticulturally, heading-back pruning and pinching are common practices because bushy or well branched shrubs, flowering plants and other ornamental plants, and fruit crops often are considered desirable (see Chapter 13). Conversely, in some crops such as greenhouse tomatoes and standard chrysanthemums, branching is considered undesirable and growers spend much time and money in the removal of side shoots, a technique called **disbudding**.

The branching habits of plants vary considerably among species, cultivars and growing conditions. Therefore, branching is a good example of the interaction between the genotype and environment influencing the phenotype of the plant. Some species—such as some cultivars of sunflower (*Helianthus annuus*) and vines including pothos (*Epipremnum aureus*)—exhibit very strong apical dominance, which extends nearly the

complete length of the stem. Other species, including many shrubs, some cultivars of poinsettia, and tomato have relatively weak apical dominance and thus branch fairly freely. Other plants branch more or less depending on their age or stage of growth. For example, young trees have rather strong apical dominance and therefore grow as whips, whereas older trees usually branch freely and provide shade. Branching is also frequently associated with the conversion of the apical bud from vegetative to flowering. Environmental factors such as light quality (see Chapter 5) and the general nutritional status of the plant (see Chapter 9) also can have a profound influence on branching.

The branching of plants and its control has tremendous economic significance to horticulturists, and an understanding of the underlying mechanisms that control branching enable a grower to manipulate the phenotype of the plants which they grow. Several theories have been put forth to explain how the apical bud can control the outgrowth of lower lateral buds. Even with the great number of published papers on the subject, however, there is still no generally accepted theory to explain the inhibition of lateral buds; this is perhaps true because there may be different specific mechanisms for various species or because scientists have not taken the correct approaches to the problem (i.e., most studies deal with the application of specific chemicals to plants rather than measurements of endogenous substances). Additionally, there may simply be insufficient data to fully explain various branching phenomena.

A relationship is apparent between the hormonal *balance* and branching of plants. IAA is produced in the apical region, especially the young expanding leaves. Removal of the apex releases the axillary buds from their inhibited state and allows them to grow. The apical buds can be replaced by auxin and the axillary buds will be suppressed, but to achieve this, much higher levels of auxin than were present in the apical bud must be applied. The fact that a signal is produced in one site, the apical bud, and is perceived in another site, the axillary buds, has given rise to the term **correlative inhibition** as a synonym for apical dominance. Unfortunately, the exact function of auxin in the inhibitory response of lateral buds is unknown.

Cytokinins also appear to play a role in apical dominance. They are produced primarily in the roots and move upward in the plant. When applied exogenously, cytokinins stimulate the outgrowth of lateral buds for a short period of time; this effect is more pronounced if the cytokinin application is followed by an auxin application (gibberellins can sometimes be substituted for auxin). Endogenous cytokinins may also play a role in stimulating the outgrowth of axillary buds. Following decapitation of the apical portion of the shoot, there is an increase in the amount of cytokinin coming from the roots into the shoots and this seems to cause the axillary buds to grow. If the shoot apex is pruned from the plant and replaced by IAA or NAA, there will not be this increase in cytokinin from the roots and apical dominance will be maintained.

Abscisic acid inhibits the outgrowth of axillary buds when applied exogenously. It is not clear whether endogenous ABA is important in blocking the outgrowth of correlatively inhibited buds, similar to its role in bud dormancy. There is no convincing evidence of an important function for ABA in apical dominance.

The role of endogenous gibberellins in branching is also not clear. Sometimes exogenous applications of gibberellins promote branching and sometimes they inhibit branching. Generally, applications of ethylene or ethephon lead to a more heavily branched plant. Because ethylene serves to block the transport of endogenous auxin, it may be through this mechanism that ethylene plays a role in branching phenomena.

In addition to applications of members of these groups of major plant growth substances, there are other compounds that are applied experimentally or in practice, that can modify the branching habits of plants. Many of these compounds function by disrupting apical growth, resulting in shorter plants. Therefore, many PGRs that promote branching are also growth retardants.

APICAL DOMINANCE

Pinching Agents

Some plant growth regulators function by chemically destroying apical buds (chemically pinching) and thus promote branching. A spray often can be applied much more quickly and economically than manual pinching, but some chemical pinching agents can cause temporary phytotoxicity, evidenced by yellowing of the foliage. Permanent chlorosis or other damage restricts use of some of these chemicals on many crops and may eliminate use of certain chemicals altogether. Even temporary yellowing can worry a grower who has expended great effort and resources in growing a high-quality crop, only to see it apparently damaged. If temporary chlorosis is a side effect of a chemical, the grower must be educated about the transitory nature of the yellowing and be reassured that regreening will occur. The chemical dikegulac, combinations of methyl esters of fatty acids, mixtures of fatty alcohols, and single fatty alcohols have been used successfully on a wide variety of crops as chemical pinching agents (Figure 12-2). These are sold under various trade names. Growers should read the labels for specific spray recommendations and exercise the proper precautions.

Some PGRs function by blocking auxin transport downward in the stem from the apical buds. This serves essentially the same function as pinching, however no parts of the plant are destroyed. Examples of compounds of this type which have been used with some success are 2,3,5-triiodobenzoic acid (TIBA) and a group of substances called morphactins (a name derived from "*morph*ologically *acti*ve substa*nces*").

FIGURE 12-2. Top view of an eggplant treated with dikegulac. This chemical plant growth regulator moves throughout the plant and removes apical dominance, thus allowing lateral buds to develop into shoots.

Photograph courtesy of Kenneth Sanderson, Auburn University, Auburn, Alabama.

GROWTH RETARDATION

Growers generally strive to achieve the ideal conditions for growing a crop by controlling plant genotypes through the selection of cultivars or graft combinations (e.g., the use of dwarfing rootstocks) and through careful modifications of environmental conditions (including fertilization, irrigation, temperature, and light control). This can lead to healthy plants that grow too tall. Large plants may require additional pruning, or be less aesthetically pleasing than more compact plants. For example, potted flowering crops will be less top-heavy and may have more consumer appeal if they are kept short.

Plants can only be made shorter by pruning. Growth retardants function only by slowing or stunting growth, not by shrinking plants. Therefore, a grower can use a chemical growth retardant to slow or stop growth, but if the crop is already too large vegetatively, pruning is the only possible recourse.

As discussed in the previous section, killing terminal buds or otherwise reducing apical control are effective means of retarding plant growth. The discussion here considers compounds that function primarily to inhibit internode elongation without disrupting apical meristematic functions.

When such growth retardants are applied, the plants should initially be healthy and vigorous in order for the grower to see an effect. Growth retardants should generally be applied in the spring before vigorous growth occurs on woody perennial crops such as apple or while the plants are growing vegetatively for herbaceous crops such as chrysanthemum.

Many growth retardants keep plants short by blocking various steps in gibberellin biosynthesis. Because gibberellins are potent promoters of cell elongation, anything that lowers their endogenous levels will retard plant growth by inhibiting internode elongation. The specific mechanism of action of some growth retardants is not known, but scientists usually approach the problem by investigating the effect on gibberellin biosynthesis. Because they share some of the same pathway, some of the triazole growth retardants that block gibberellin biosynthesis also block brassinosteroid biosynthesis. Brassinosteroids can also cause stem elongation, so inhibiting their biosynthesis will also contribute to growth retardation.

Chemical growth retardants generally may be applied as a foliar spray or a soil drench. The foliar spray is considered more desirable because of greater ease and speed of application. Other means of delivering chemical growth retardants include soaking clay pots in the solution or dipping bulbs for various periods of time.

Some growth retardants that have been used on various crops include chlormequat, daminozide, ancymidol, mepiquat chloride, paclobutrazol, uniconazole, and flurprimidol. These PGRs retard growth by inhibiting internode elongation (Figure 12-3). Horticulturally, major areas of use of these plant growth regulators are on fruit crops, floriculture crops, woody ornamentals, and vegetable transplants. For information on their trade names availability, specific crop usage, timing, concentrations, and precautions, growers are directed to the internet and various trade and scientific journals, extension publications, and supply catalogs.

FIGURE 12-3. Retardation of growth of sweet corn with ancymidol (21 days after treatment) (*top*); and celosia with uniconazole (four weeks after treatment) (*bottom*). Note typical stairstep height response to increasing concentration of both chemicals.

Photographs courtesy of Terri Starman, Texas A&M University, College Station, TX.

Growth Retardants on Turf

Turfgrasses are distinctly different from most other horticulture crops. They grow in a closely-spaced community and are therefore in constant competition with each other, and they are walked upon and mowed regularly. With turfgrasses, leaf growth, not stem growth is of primary concern. Turfgrasses occupy tremendous areas of land, more than 10 million hectares in the United States. Major uses of turfgrass include lawns, roadsides, cemeteries, golf courses, sports fields, and parks. In most of these areas the turf is mowed either regularly or periodically, which requires a great deal of time and money. Therefore, there is much interest in controlling turfgrass growth and development chemically.

An ideal turfgrass growth retardant should meet several criteria:

1. It should have low phytotoxicity to the turfgrass, because the plants are grown for their aesthetic qualities.

2. It should not weaken the turfgrass plants. Turfgrass management is sometimes referred to as the management of competition, because the plants are in such close proximity and compete for light, nutrients and water. In addition a good competitive turfgrass should not have weed problems. If a growth retardant weakens the turfgrass plants, they will lose their competitive advantages and weeds and possibly diseases will become serious problems.

3. An ideal turfgrass growth retardant should specifically inhibit only vertical leaf growth and not tillers, rhizomes, stolons, or roots. Inhibition of these organs will impair the competitive ability of the turfgrass.

4. It should inhibit seedhead formation because seedheads are considered unsightly.

5. A growth retardant should not make the turfgrass more susceptible to environmental stresses such as water stress and temperature stress.

6. It should have relatively long-lasting effects, so that repeat applications will not become a major expense.

7. It should not be too expensive, which will limit its market to only very high value turfs or to use only by the wealthy.

8. Because people, pets, and wild animals walk and play on and otherwise contact turfgrasses, the retardant should not be toxic to animals or have any long-term effects.

Turfgrass growth retardants are divided into three categories, cell division inhibitors (also called type I inhibitors), herbicides, and gibberellin biosynthesis inhibitors (also called type II inhibitors). Cell division inhibitors include amidochlor, chlorflurenol, maleic hydrazide, and mefluidide. These are absorbed primarily by the foliage and will retard turfgrass leaf growth and inhibit the formation of seedheads. Because these can be somewhat phytotoxic and cause yellowing, they are used mainly on low to medium maintenance turfgrasses. They will control weeds as well as reduce mowing.

Some turfgrass growth retardants are herbicides that are used at low rates. Because of phytotoxicity they are used on low maintenance turfgrasses, such as along highways to reduce mowing and suppress seedheads. Example of herbicides that inhibit turfgrass growth are chlorsulfuron, glyphosate, imazameth, imazethapyr + imazapyr, metsulfuron, sethoxydim, and sulfometuron.

There are two basic types of gibberellin biosynthesis inhibitors used for turfgrasses, those absorbed through the foliage and those absorbed through the roots. Trinexapac-ethyl is absorbed through the foliage, is one of the least injurious to turfgrasses, and is thus applied to fine and high maintenance turfgrasses. Flurprimidol and paclobutrazol are absorbed by the roots. If turfgrasses are environmentally-stressed, they can cause some injury that is not evident with trinexapac-ethyl. However, gibberellin biosynthesis inhibitors cause less leaf burn than the cell division inhibitors or herbicides, but they do not effectively suppress seedhead formation.

VEGETATIVE GROWTH PROMOTION

Many different chemicals can shut down reactions in plants and retard growth; however, only a very few chemicals that are known to promote plant growth. Consequently, many of the PGRs that are commercially available are growth retardants. Gibberellins, including $GA_4 + GA_7$ and especially gibberellic acid (GA_3) can be used on several horticultural crops. GA_3 has been used to promote petiole elongation on celery and rhubarb and to stimulate larger and more open clusters in grapes. It can

increase crop yield in parsley, increase stem length in roses, accelerate growth of geraniums used to create unique tree geraniums and give an earlier stand for turfgrass sod production. A major agronomic use is on sugar cane, where GA_3 promotes internode elongation. Sugar cane plants treated with GA_3 develop higher sucrose amounts than the levels achieved by non-treated plants.

FLOWERING

Flowering is an extremely important process in virtually all fields of horticulture, because many plants are grown specifically for their flowers, fruits, or seeds. Flowering is undesirable in some species, including turfgrasses used in lawns, and on certain ornamental species in which the flower structures can be unsightly or putrid smelling. Precise control of flowering is relatively easy on some crops, such as chrysanthemum (see Chapter 5). Better control of flower bud formation will reduce problems associated with biennial bearing of some fruit trees. On these species, a uniform and predictable crop from year to year is an economically important advantage. A better understanding of the underlying mechanisms that control flowering in plants could change horticultural practices or simplify means of controlling flowering.

Plant genotype, age (especially as it relates to juvenility and phase change), plant size, and environmental factors such as light, temperature and nutritional status all affect flowering. Photoperiodic effects are discussed in the Chapter 5, vernalization in Chapter 6, and nutrition relating to the carbohydrate:nitrogen ratio in Chapter 9. Here the discussion is limited to the influence of plant growth substances on flowering.

Various grafting experiments and studies in which different parts of plants were exposed to floral induction stimuli have led to the theory that a chemical message related to flowering is produced in one site (the leaves in many plants) and moves to another site, the apical or meristematic regions. It is not known if flower promoters, flower inhibitors, or a balance between the two controls flowering, or if such substances truly exist. However, theoretical substances or groups of substances have been proposed: These are called **florigens** (promotors) and **antiflorigens** (inhibitors). Although many investigators have conducted extensive studies trying to isolate and identify florigens and antiflorigens, they have not been successful. Nevertheless these categories remain useful conceptually to explain flower induction. Possible reasons for the failure to isolate and identify these substances are as follows: (1) They may be present in extremely minute quantities, making their detection difficult with the current state of science. (2) The compounds may be unstable (labile) and may rapidly deteriorate upon extraction. (3) They may be complex balances of several chemicals rather than single substances. (4) They may be much more complex than known plant hormones such as proteins or nucleic acids and therefore require special techniques to be isolated. (5) They may not exist; flowering may result from some kind of nutritional or other change by the plants. As of this writing, there is no universally correct statement about a hormonal control of flowering, except that ethylene stimulates flower bud initiation in Bromeliads and that gibberellins stimulate flowering in some species and inhibit it in others.

PGR applications have been employed to both stimulate and prevent flowering. Although there are exceptions, some generalizations about exogenous applications of plant growth regulators can be made relating to the photoperiodic response groups of long-day plants (LDPs) and short-day plants (SDPs). When placed under nonflowering inductive periods (short days for LDPs and long days for SDPs), applications of gibberellins often promote flowering of LDPs and inhibit flowering in SDPs. Under long days, endogenous gibberellin levels are higher,

regardless of the photoperiodic response group. Generally, the LDPs that respond to gibberellin applications are those that grow as a rosette under short days and have rapid stem (internode) elongation (**bolting**) under long days; examples of these plants include lettuce, radish, and spinach. Because gibberellins also are well known for causing internode elongation, the relationship between flower formation and internode elongation has been studied. The two phenomena depend on separate processes, and some rosette plants, such as beets, can elongate without flowering. Gibberellins also will promote flowering in other crops that do not bolt, including coneflower (*Rudbeckia bicolor*), petunia, Douglas fir (*Pseudotsuga menziesii*), and stimulate flower bud elongation in cyclamen (Figure 12-4). Additionally, gibberellins will substitute for the cold treatment (see Chapter 6) in plants that require vernalization, such as celery, sugar beet, foxglove (*Digitalis purpurea*), and flowering stock (*Mathiola incana*).

Gibberellins are known to inhibit flowering in apples and other spring-flowering woody plants that initiate flowers in the previous summer. In fact, applications of GA3 in the autumn can delay flowering in plants such as grapes and stone fruits and

FIGURE 12-4. Acceleration of flowering in cyclamen. At left, no GA$_3$ treatment was applied; at right, the crown was sprayed with 10 mg liter^{-1} GA$_3$.

Photograph courtesy of R.E. Widmer, University of Minnesota, St. Paul, MN.

thus help avoid damage associated with spring frosts. However, this approach is not yet well enough understood for consistent commercial application. With some short-day plants such as fuschia and apple, the application of growth retardants that block gibberellin biosynthesis (e.g., chlormequat, Amo 1618, and phosphon D) will result in flowering. Although gibberellins are associated with flowering of some plants, they are considered to be neither florigen nor anti-florigen, but may cause a change in the balance of these theoretical compounds.

Auxins such as NAA and 2,4-D were used in the past to promote flowering in bromeliads, especially pineapple. They are thought to function by increasing ethylene biosynthesis, with the increased ethylene levels then inducing flowering. Pineapple growers routinely apply ethephon to vegetatively mature plants, generally 6 to 8 months before the desired harvest date. This results in synchronized and earlier harvests than would be possible without plant growth regulators. Ethephon offers important economic advantages because growers can produce more crops from a given area in a given time period. Ethylene also promotes flowering in sweet potato, litchee (*Litchi chinensis*), apple, and mango.

Other plant growth substances, notably cytokinins and ABA, have been used to stimulate flower initiation in a limited number of species. It should be noted, however, that at present, the flowering of most species cannot be accounted for in terms of PGRs. If in the future a flowering hormone or inhibitor is discovered, the economic importance will be tremendous.

SEX EXPRESSION

Plant growth substances are involved in the sex expression of some dioecious and monecious species. Sex expression can be related to the endogenous hormonal balance or can be controlled, in some species, by exogenous applications of PGRs.

Exogenous applications of plant growth regulators may be used by plant breeders of responsive species to ensure that they have an adequate number of pollen and seed parents to make their crosses. Applications of gibberellins will cause maleness in several crops, including papaya (*Carica papaya*), cucumber, and melon (*Cucumis melo*), but there are exceptions; they induce femaleness in castor bean (*Ricinus communis*), Chinese chestnut (*Castanea mollissima*), and begonia. Cytokinins, however, can cause femaleness in grape and cleome. Auxins will induce femaleness in monoecious cucurbits and begonia, and ethylene will also promote femaleness in many crops. Because of its effectiveness on a wide number of species, ethylene, applied as ethephon, is probably the most widely used growth regulator for induction of femaleness in horticultural crops.

During the growth and development of some monoecious species, such as acorn squash (*Cucurbita pepo*), the sex expression of the flowers changes frequently (Figures 12-5 and 12-6). Initially, underdeveloped male flowers form, then as the plant gets older, normal male flowers are produced. This is followed by a mixture of both normal male and female flowers, after which normal female flowers predominate. Finally female flowers may be produced that result in parthenocarpic fruit formation. In plants of this type, gibberellic acid promotes maleness, while both ethylene and auxin promote femaleness. Many plant physiologists believe that there may be corresponding changes in the endogenous hormone balance that governs the normal sex expression in such plants. It is not uncommon for home gardeners to become concerned because the first flowers that form on their cucurbits—for example, zucchini (*Cucurbita pepo*)—do not produce fruit, but simply fall off. As we can see from the foregoing example, this happens because the first flowers are male and cannot give rise to fruit.

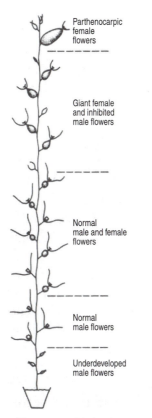

Parthenocarpic female flowers

Giant female and inhibited male flowers

Normal male and female flowers

Normal male flowers

Underdeveloped male flowers

FIGURE 12-5. The sequential development of the flowers during the ontogeny of an acorn squash plant.

From Nitsch, J. P., E. B. Kurtz, J. L. Liverman, and F. W. Went. 1952. "The development of sex expression in cucurbit flowers." American Journal of Botany 39:32-43. Used with permission.

FIGURE 12-6. Typical female flower, *left*, and male flower, *right*, of cucumber. Note inferior ovary in the female flower.

The timing of application of PGRs is important in their effectiveness on sex expression of flowers. Cucurbits should be sprayed when they have between one and three true leaves for maximum effectiveness. If applications are made too early, sufficient growth and development has not occurred, and if the plants are too old, flowers may already be irreversibly developed.

FRUITING

Fruit Set

The flower grows and develops following flower initiation. In many species the growth of the ovary stops either at the time of or shortly after anthesis, unless pollination occurs. Without pollination, the flowers will drop off the plant (there will be no fruit set). Fruit set is critical if a yield is to be realized in crops grown for their fruit such as tomato or apple. Many factors, such as unfavorable environmental conditions at the time of flowering, can result in the lack of fruit set and the subsequent loss of yield. For example, flowering of apricots can occur too early during some years, before there is sufficient bee activity for adequate pollination; poor fruit set and yield result.

Associated with fruit set is a burst of growth of the ovary; this generally occurs as a result of successful pollination (not fertilization). The rapid increase in ovary size occurs before there has been sufficient time for the pollen tubes to grow into the ovary. Faster fruit growth occurs in crops such as tomato when there is heavier pollination. The rapid increase in ovary size can be induced to occur in some species if pollen extracts are used instead of intact pollen. This knowledge has helped people understand that it is a hormonal (often auxin) signal that causes the fruit to set and begin rapid growth. Not only is the pollen a rich source of auxin, but pollination results in a stimulation of auxin formation in the ovary. It is now well established that on some species an application of auxin can induce fruit set. Some growers and homeowners apply auxin to their plants—for example, on tomatoes early in the growing season; this will prevent abscission of the flower buds and thus increase yield. Auxin applications are effective on several crops, including tomato, pepper, eggplants, holly, okra, figs, and cucurbits. They are ineffective for 80% of horticultural crops.

Exogenous applications of other PGRs can also induce fruit set in a limited number of species. Gibberellins have been shown to have an effect on fruit set of tomatoes, similar to the effect of auxin; however, gibberellin applications will also cause fruit set on some species where auxin is ineffective, including grapes, many of the stone fruits, apples, and pears. Cytokinin applications have been shown to induce fruit set in figs and improve fruit set of pollinated cucurbits. Ethylene may also be involved, because in some species there is a great increase in its biosynthesis following pollination. This probably accounts for the rapid fading, wilting, and abscission of flower petals following pollination. Fruit set is an extremely important process and, at least in some species, is governed by the changing hormonal balance associated with pollination.

Fruit Growth

Fruit growth generally follows a set pattern. Initially, because growth comes primarily from cell divisions, the fruit gets larger at a relatively slow rate (lag phase). This phase is followed by a rapid increase in size because of cell enlargement, with the fruit ultimately leveling off in size. Sometimes a second lag time occurs during the middle of the growth of an individual fruit, but this varies with the species. Generally, this can be depicted graphically in an S-shaped, or sigmoid, curve (Figure 12-7). Cherry has a second lag phase, so the fruit growth is represented by a double sigmoid curve.

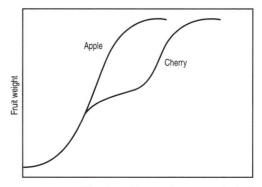

FIGURE 12-7. The sigmoid growth curve typical of apple and the double sigmoid growth curve of cherry fruit.

From Leopold, A. C., and R. E. Kriedmann. 1975. Plant Growth and Development, 2nd ed. McGraw-Hill Book Co., New York. Used with permission.

Fruit have the capability for strongly directing translocation or mobilization of foodstuffs, which permits fruit growth at the expense of materials in the leaves. Considerable evidence shows that a signal (probably hormonal) emanates from the fruits; the rapid growth and development of the fruit is the result of this hormonally directed transport. This is another example of hormonal importance in the internal communication system of a plant.

Fruit growth generally seems to be related to the mobilization of substrates into the various tissues associated with the ovules or seeds. Seeds, in fact, regulate many aspects of fruit growth. Removal of the fertilized ovules frequently terminates the growth of that portion of the fruit from which they were excised. Figure 12-8 shows that the strawberry receptacle only grew where the achenes remained. If all seeds are removed, receptacle growth generally stops completely.

Pollination efficiency and the pattern of fertilization (sexual union of gametes) often will have a profound influence on fruit size and shape. In many fruits the numbers of seeds and the overall fruit size at maturity are strongly correlated; that is, more seeds indicate larger fruit, if all other environmental conditions and the genotype are the

FIGURE 12-8. The relationship between developing seeds and fruit development. Receptacle tissue developing in association with a fertilized achene (*a* and *b* magnified). Removal of the achenes early in the fruits' development resulted in lack of receptacle development on the sides (*c*) and top (*d*) compared to normal fruit. Treatment of an acheneless fruit with auxin (*e right*) results in a fruit similar to the normal fruit (*e left*); an acheneless fruit not receiving auxin treatment is in the center.

From Nitsch, J. P. 1950. "Growth and morphogenesis of the strawberry as related to auxin." American Journal of Botany 37:211-215. Used with permission.

same (Figure 12-9). Uneven pollination will frequently cause a lack or low number of seeds in one or more locules, which will influence the geometry of the fruit. As shown in Figure 12-8, the fruit development is more complete and normal looking on the side where there are seeds and poor on the side with no seeds.

Seeds are potent sources of plant hormones. They often contain relatively high amounts of auxins, gibberellins, and abscisic acid. All have been implicated in playing a role in fruit growth as it relates to seed number and distribution. In some species one application of auxin will induce fruit growth, even when pollination and fertilization have not occurred; seedless or **parthenocarpic** fruit result. Parthenocarpic fruit also can result from other causes, including the stimulation from pollination without fertilization. Genetic reasons such as triploidy (having three sets of chromosomes)

FIGURE 12-9. Cross-section of apple fruits from the same harvest. The larger fruit (*left*) typically had a greater number of seeds than the smaller fruit.

may also cause fruit to be parthenocarpic. Some important horticultural fruits are parthenocarpic, such as navel oranges, some seedless grapes, and bananas. Many naturally occurring parthenocarpic fruits produce higher levels of auxins than seeded cultivars. These auxins may be responsible for the mobilization of assimilates from the leaves to the developing fruit similar to seeds in other fruits.

Other plant growth regulators are commonly used to increase fruit size. Gibberellic acid (GA$_3$) is effective for some fruit crops, including sweet cherries and grapes. GA$_3$ is used in 'Thompson Seedless' grapes to elongate and increase berry size and to encourage cluster loosening (Figure 12-10). Producers prefer loose clusters because they

suffer less rot than do tight clusters, and consumers prefer them because loose grapes are easy to remove from the cluster. Growers in certain areas often apply a mixture of the cytokinin benzyladenine and GA$_4$ + GA$_7$ to 'Red Delicious' apples to increase the size, by stimulating fruit elongation, and to modify the shape. Elongated fruit with pronounced lobes on the calyx end result (Figure 12-11). Consumers prefer these apples over more round, compact fruit and are often willing to pay the higher price commanded by such fruits.

ABSCISSION AND HARVEST AIDS

Abscission is the process of shedding or dropping nearly any plant part, including leaves, flowers, fruit, or branches. A weakened area or abscission zone forms at the point of detachment. Because of enzymes secreted by adjacent cells, the walls break down in a separation layer one to three cells wide, which creates a weak point at which the organ detaches from the plant (see Figure 3-36). The abscission process is keyed by many factors, including the environment (photoperiod, temperature, stress), age, pollination (which often causes petal abscission), or organ

FIGURE 12-10. Bunches of grapes: Untreated grapes (*left*); grapes treated with gibberellic acid (GA$_3$) (*right*). Note the larger berries and more open cluster as a result of the GA$_3$ treatment.
Courtesy of Abbott Laboratories (now Valent BioSciences Corporation), Long Grove, IL.

FIGURE 12-11. 'Red Delicious' apples untreated (*left*) and treated with a combination of benzyladenine, a cytokinin, and GA$_4$ + GA$_7$ (*right*). Note the elongated, almost pointed shape.
Courtesy of Abbott Laboratories (now Valent BioSciences Corporation), Long Grove, IL.

development. It should be noted that abscission is an active process confined to one specific area. Why abscission zones only form in certain areas and never in apparently similar neighboring cells remains a mystery.

Leaf abscission is a major concern with the major agronomic crop cotton, but in horticulture, fruit abscission is of major economic importance. Growers may thin fruit to reduce competition and thus produce a larger, more salable yield; to reduce biennial bearing of a fruit tree and ensure a more uniform year-to-year harvest; or to reduce the burden on a young plant (e.g., young coffee plants, which often flower profusely and produce so many fruits that the future performance of the plant is impaired unless the fruit load is reduced). Ethephon can be used for this latter purpose.

Abscission agents are used to facilitate the mechanical harvest of many fruit crops, including processing oranges and other citrus fruits, olives and cherries. Cherries, for example, are sprayed with ethephon 7 to 14 days before the anticipated harvest date. This treatment results in uniform ripening and loosening of the fruit. At the harvest date, a mechanical harvester (Figure 12-12) is attached to the trunk of the tree. It shakes the tree so vigorously that all of the loosened fruit drop onto a tarpaulin, from which they are easily loaded into cooling tanks. Many experimental and commercially available abscission agents are used, depending on the crop and its ultimate use. Some of these substances cause phytotoxicity; which is unacceptable on a crop that is sold based on its visual appeal, such as fresh

FIGURE 12-12. Tree shakers are often used for mechanical harvest of cherries or plums. Harvest by this method is made possible by treatment of the crop with a fruit abscission agent (ethephon). The shaker wraps around the tree trunk (*top left*) where it vigorously shakes the tree and catches the fruit as they drop (*bottom left*). The fruit are then carried by conveyer belt (*right*) to cooling tanks that can be loaded onto trucks.
Photographs courtesy M.J. Bukovac, Michigan State University, East Lansing, MI and B.H. Taylor, Southern Illinois University, Carbondale, IL.

oranges; peel injury would not be a major concern on oranges to be processed for juice. Additionally, some harvest aids may shorten the life of the tree. It this is the case, the grower must weigh the advantages of the efficiencies achieved through use of the growth regulator against the economics of having to replant more frequently.

SUMMARY

The genotype and environment alter plant phenotype largely via the hormonal balance in the plant. Horticulturists further alter the chemical makeup and ultimately the plant phenotype through physical manipulations, including pruning, and through application of plant growth regulators. Additional and perhaps more effective growth-regulating chemicals may become available to the grower in the future. The reader is advised to keep abreast of such developments. As more is learned about plant growth substances and as new chemicals become available, it is highly probable that such materials will have a profound impact on productivity and quality of horticultural crops.

REFERENCES

Arteca, R. N. 1996. *Plant Growth Substances Principles and Applications.* Chapman & Hall. New York. 332 pp.

Basra, A. S. (Ed). 2000. *Plant Growth Regulators in Agriculture and Horticulutre. Their Role and Commercial Uses.* Food Products Press. New York. 264 pp.

Davies, P. J. (Ed). 1995. *Plant Hormones Physiology, Biochemistry and Molecular Biology.* (2nd ed.) Kluwer, Dordrecht. 833 pp.

Horticultural Reviews (yearly series). AVI Pub. Co. Westport, CT.

Nickell, L. G. *Plant Growth Regulating Chemicals.* Vol. I and Vol. II. CRC Press, Boca Raton, FL.

Plant Growth Regulator Society of America. 1990. *Plant Growth Regulator Handbook.* 3rd ed. 146 pp.

Srivastava, L. M. 2002. *Plant Growth and Development Hormones and Environment.* Academic Press, Amsterdam, 772 pp. Pergamon Press. New York. 343 pp.

PART V

HORTICULTURAL
PRACTICES

CHAPTER 13

PRUNING

Plant size and shape can be controlled by changing the growing conditions, including water temperature, nutrition, light, or by exogenous application of PGRs. To achieve the desired results, growers often must also prune their crops. Pruning disrupts apical dominance and therefore stimulates branching resulting in a bushy plant. Additionally, plants can be pruned to maximize light interception, or to control size. There are many factors to consider when pruning plants, including how you want the plant to grow, diseases, and the overall response of the plants to the cuts. Pruning cuts must be considered carefully because they can have a lasting effect.

PRUNING

Pruning involves the removal of parts of the tops or the root systems of plants. Because plant parts are removed, pruning is a dwarfing process, but it also can revitalize or rejuvenate a plant, and thus stimulate growth. When pruning, one should not remove more than one-third of the growth (tops or roots) each year, because additional loss of plant tissue will potentially weaken the plant and make it more susceptible to diseases. There are some exceptions, however, such as hydrangeas grown in areas with cold winters that are pruned to the ground each year with no adverse effects.

Shearing is a specialized pruning technique in which tips of outer branches are trimmed off the plant. As a result, branching is stimulated near the cuts, and dense growth occurs toward the outside of the plant. Shearing is used frequently on hedges and topiary. The appearance of the outside of these plants is most important, and the fact that shearing and the subsequent dense outside branching shades the inside of the plant is of little concern. Shearing is less appropriate for plants where light penetration within the plant is important. Although some fruit crops are sheared, most are pruned to allow for light penetration within the plant to increase yield and fruit quality. Shearing is equally inappropriate for plants growing where their "natural" shape or form is important. Some people find shearing unattractive because cut edges of leaves usually turn brown following shearing; and this remains visible until new growth occurs. Most plants are therefore pruned using techniques other than shearing.

Plants are pruned for a number of reasons, depending on the purpose for which they are grown or used. Pruning can improve plant health by removing dead or diseased portions and it can open up a plant and thus allow better air circulation and light penetration. Improving air circulation helps in keeping plant parts dry and reducing infections. Increasing light penetration can alter the carbohydrate:nitrogen balance, and thus result in a stronger, healthier plant. Additionally, because of more photosynthesis from more surface area receiving direct sunlight, yields of fruit crops such as apple are often improved by proper pruning.

Plants also are pruned to correct structural defects such as weak branch crotch angles. If these defects are not corrected, extensive breakage will occur under conditions of heavy fruit loads or high winds and

open a tree for invasion by pathogens (Figure 13-1). Plants are pruned for size control—turfgrasses are mowed, for example, and mugho pines are often sheared. Plants are pruned to give desired shapes, which range from natural to topiary (Figure 13-2), espalier (Figure 13-3) and bonsai (Figure 13-4). As mentioned in Chapter 11, plants often are pruned to increase branching and thus to encourage them to become full. Other plants are pruned to save space, so that they will not outgrow their site. Such overgrowth is relatively common in home landscapes (Figure 13-5). Pruned plants often are easier to care for and maintain, and they are easier to cover completely with pesticide sprays. Plants are also pruned to improve their appearance and function in relation to humans. The mowing of turfgrasses is a good example of pruning for this purpose.

FIGURE 13-1. This tree had a narrow crotch angle and split under a heavy fruit load.

FIGURE 13-2. Topiary art is the term for plants pruned into geometric or animal shapes, such as this Taxus "Teddy Bear."

(a)

(b)

FIGURE 13-4. Bonsai is a special horticultural art form in which a plant is kept small, but well-proportioned, by restrictive pruning and cultural practices. (*a*) This *Juniperus procumbens nana* is 15 years old and 20 cm tall. (*b*) The *Juniperus chinensis* plants in this picture are 25 years old and 67.5 cm tall.

Photos courtesy of Bob Hampel, Minnesota Bonsai Society.

FIGURE 13-3. Fruit trees are frequently pruned into a flat form against a wall such as the apple tree in this illustration. This pruning technique is referred to as **espalier**, a French term taken from the Italian word for "shoulder".

Plant Health

Regardless of the reason for pruning, diseases can easily be spread on pruning tools. It is a wise practice to dip pruning tools in

FIGURE 13-5. These overgrown landscape plants have obstructed the entrance to this house. Judicious pruning, selection of smaller plant materials and proper placement could all help alleviate this problem.

FIGURE 13-6. Two common types of pruning shears. The scissor-cut type (*right*) makes a cleaner cut and is preferred over the anvil type (*left*).

alcohol, liquid chlorine bleach solution, or another chemical sterilant between cuts, or at least between plants. Otherwise, the tool may be contaminated during the first cut and thus introduce pathogens into fresh wounds with cuts made thereafter. Because this practice is uncommon, spread of many diseases can be traced to the pruning operation. Pruning shears and attachments are now available in which a sterilizing solution is pumped or metered out so that the cutting blade remains wet and sterile during use. The sterilant thus contacts all new wounds and may reduce infection.

The choice of pruning tools can also influence disease susceptibility. Anvil-type pruning shears should *not* be used on plants because they tend to crush tissue. Crushed tissue does not heal as well as the clean cuts made by scissor-type pruning shears and is thus more prone to disease. Examples of both types of pruning shears are illustrated in Figure 13-6.

An understanding of the pruning of large branches from trees serves as a sound basis

for the effect pruning cuts and their treatment can have on the health of a plant. One must first understand that although they are also living organisms, plants do not respond to wounds like human bodies do, and they should not be treated in a similar manner. When people get cuts, they usually apply a bandage; in the past when people cut a tree, they applied a wound dressing, but there is faulty logic in this practice.

When a tree is wounded, processes begin that could lead to infection. However, a chemical defense system is activated in response to wounding and chemical barriers form. Microorganisms will colonize freshly wounded wood, but trees have the ability to compartmentalize the injured wood and microorganisms and form secondary barriers. Because of this ability to compartmentalize, that new wood formed after a tree is wounded does not become infected unless it becomes further wounded. An understanding of chemical barriers and compartmentalization by trees makes the logic behind proper pruning practices more clear.

A large branch on a tree should not be removed with one cut (Figures 13-7, 13-8 and 13-9). Large branches have considerable weight, and upon cutting, a breaking and tearing process will begin that can rip a considerable portion of bark from the trunk and leave a sizable wound. Large wounds are slow to heal and can lead to much decay in the tree. A horticulturist should first undercut a large branch at a point beyond the branch collar and then make a second downward cut beyond the undercut. This way, when the branch begins to break and the bark tears, it will break and tear only to the undercut causing minimal wounding, and the branch will fall to the ground. A third cut to remove the branch stub should be made just beyond the branch collar, which will cause minimal wounding and will leave the branch collar intact. The tree produces the chemical barrier within the branch collar. The old recommendation was to make this third cut as flush with the tree trunk as possible, but this technique often resulted in the removal of the branch collar and thus the chemical barrier. As a result,

FIGURE 13-7. This severe bark tear resulted because of incorrect pruning technique.

considerable decay occurred in improperly pruned trees. On the other hand, long stubs should not be left because the tree will take too long to grow over the wound.

Another old recommendation of painting of wounds also has been shown to be faulty. Painted tree wounds have no less decay and heal no faster than nonpainted (or non-dressed) wounds. Because the natural chemical barriers and compartmentalization of decay serve the tree well (Figure 13-10), it is now considered to be a waste of time and money to paint tree wounds.

Timing of Pruning

The *timing* of pruning can be important to the health of the plant. As mentioned in Chapter 6, late summer or early fall pruning of woody plants will encourage new succulent growth that may not cold-acclimate properly. This can lead to winter kill of these new tissues. The pruning of some trees such as maples, green ash (*Fraxinus pennsylvanica*), and honeylocust during times when sap is flowing can lead to a bacterial infection known as bacterial wetwood, commonly evidenced by staining of the bark. Pruning of susceptible oaks and American elms during times of high insect activity will attract pests carrying the oak wilt fungus or the Dutch elm disease fungus, and the likelihood of infection will therefore be quite high. Different plants should be pruned at various times, depending on their growth habits, internal physiology, ornamental characteristics (e.g., spring flowering shrubs should be pruned following flowering) or susceptibility to various diseases. Appropriate references should be consulted to find pruning recommendations for specific horticultural plants.

Size Control

Pruning often is practiced to control plant size. Pruning the tips of branches, a practice called **heading**, reduces the size of the plant and is used to direct future growth. Heading

(a)   (b)

(c)   (d)

FIGURE 13-8. Proper pruning for large branches involves a series of cuts. (*a*) First the branch should be undercut. (*b*) A second pruning cut should then be made beyond the first cut. (*c*) Normally the branch will break as shown with no major bark tearing. (*d*) The final cut should be made just beyond the branch collar.

back cuts should be made just beyond a bud. This bud will grow and normally assume apical dominance over lower buds. The direction (up, down, right, or left) that the bud is pointing will be the direction that the new branch will grow. People usually will not want the new growth to be directed back into the center of the plant, so they will make their heading cuts immediately beyond buds that point away from the plant.

Although it will make a tree considerably smaller, growers should avoid cutting off large branches in their middles to create a tree shaped like a "hat rack" because of the large branch stubs. Hat racking leaves large wounds well beyond chemical barriers in branch collars that are vulnerable to rot. Such severe pruning will encourage the rapid growth of water sprouts, which will change the natural shape of the tree and compete with each other for available light. Such shoots on each branch arise close to each other and cause weak, narrow crotch angles. Therefore, such hat racking is not considered to be a desirable way to prune a tree. In Europe, many trees are **pollarded**, meaning that each year the shoots are cut back severely to the main branch. A small, thick tuft of small branches results, keeping trees very small. Prunings are often used for firewood.

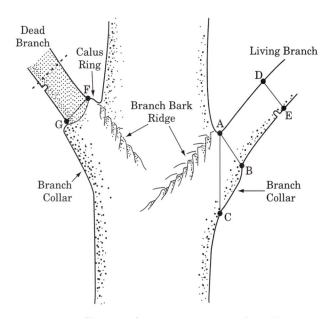

FIGURE 13-9. Diagram of a proper pruning cut (*a* to *b*), an improper flush cut (*a* to *c*), an improper stub cut (*d* to *e*), and proper removal of a dead branch (*f* to *g*).

(a)

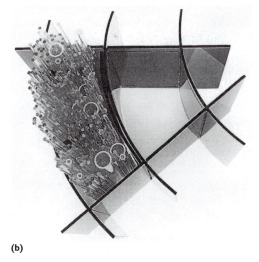

(b)

FIGURE 13-10. (*a*) This red oak was wounded with buckshot nine years prior to being cut. The decay-causing fungi have been compartmentalized. The walls, indicated by arrows, have resisted spread of the decay-causing fungi, thus confining it to the area in the photograph. (*b*) This model illustrates how trees are compartmentalized. The spread of infection is prevented by barriers to the inside, outside, and radially.

From A. L. Shigo, A New Tree Biology, Shigo and Trees, Associates, Durham, NH. Used with permission.

Shape Control

Plants are pruned to various shapes to improve their appearances, to increase yields, to encourage light interception, or to facilitate their care and possible harvesting. Plants may be pruned so that the central leader dominates, or is removed completely to create an open centered plant. Reasons for choosing to prune to a particular shape vary with the species and with the reason the plant is being grown.

Planning cuts carefully is critical to achieving the desired effect when pruning. Once a branch or shoot is removed it cannot be put back. Every pruning cut should have a reason and justification for which branch is to be removed and specifically where the cut is made. It is by carefully planning pruning cuts that growers achieve the desirable plant shape and effect. Often horticulturists stand back and survey the plant prior to making cuts. That way they can make planned cuts, each with a specific reason or objective. This reduces costly mistakes and can result in a more functional plant.

Pruning for Light Interception

Many fruit crops are pruned to maximize light interception throughout the tree. Better quality and often larger fruit will be produced on pruned trees than on non-pruned trees. Cuts must therefore be planned on some plants to allow for light to reach both outer and inner portions (Figures 5-9 and 5-10).

Hedges are pruned to allow maximum light interception on their exterior portions. A wider base and more narrow top will allow for best light interception (see Chapter 5 and Figure 5-12).

Pinching, Disbudding, and Disbranching

The soft new growth of many floriculture crops is removed (**pinched**) to stimulate branching and result in a fuller plant. Depending on the crop and situation, soft pinches (removal of only the apical buds) or hard pinches (cutting lower than just to soft, succulent growth) will be used. Plants may receive only one or several pinches depending on the species, time of propagation, and purpose for which they are grown. Pinches may be done chemically (see Chapter 12) or by hand.

Another commonly used pruning technique is **disbudding**. This is the process of removing axillary buds (side shoots) from crops such as chrysanthemum and carnation that are being grown as single stems. These species' weak apical dominance results in rapid outgrowth of undesirable side shoots; their removal allows the plant's energies to be channeled into the apical bud. Vegetables are usually not pruned. However, for greenhouse tomatoes and cucumbers, for field-grown staked tomatoes, or for other crops where a single stem growth pattern is preferred, disbudding or disbranching will be practiced. By removal of the axillary buds or branches of tomato, for example, a single straight stem can be achieved that can be tied to a stake or string for support.

Pinching and disbudding can be done by hand, mechanically, or chemically. The number of plants and potential side effects of the pruning technique must be considered when choosing the most efficient method. Chemical pinching is discussed in Chapter 12.

Renewal Pruning

Old shoots or branches are removed from many plants so they may be replaced by younger, more vigorous shoots. Renewal pruning cuts usually are severe. Although renewal pruning cuts frequently are made on fruit trees, they are also important for many landscape plants. Some shrubs, such as lilac, benefit from renewal pruning that removes old, large shoots, and makes way for younger stems. Such invigorated plants are often more attractive and may flower more than they did before being pruned.

Root Pruning

Roots are frequently pruned to stimulate root branching close to the base of the trees or shrubs. Root pruning results in the production of many new feeder (young) roots in the ball of soil that will be lifted when the plant is dug for transplanting. If root pruning had not occurred, many of the youngest root tips would be left in the soil when lifting, and transplanting success would be diminished. This is why nursery-grown trees and shrubs transplant more successfully than those dug from the wild. Nursery growers commonly undercut their field-grown trees and shrubs using an underground blade, although they also use other techniques. This operation is normally done in the autumn after active top growth has ceased.

To stimulate a well-branched root system in containers, growers often use open-bottom containers through which the roots can grow. These containers are placed on a wire mesh bench. When the roots encounter the air, the tips dry and die. Such air pruning stimulates root branching. Similarly, various planting containers cause root pruning. Some are made of a nylon mesh that girdles root tips, and others may contain copper to kill root tips. When these containers are removed and the plants transplanted, the well-branched root systems will support the plant and rapidly grow into the surrounding soil.

SUMMARY

Pruning techniques vary with the plant species and purpose for which it is grown. Even within a species or cultivar, growers may wish to use different pruning or training systems. Pruning is labor-intensive and thus expensive, but the cost must be weighed against the reasons the plant is grown and the ultimate result that is desired. For example, fruit growers prune to maximize fruit yield and quality and thus profit, whereas managers of topiary gardens do extensive pruning to maximize beauty or creat a specific effect. Growers of many field-grown vegetables do not prune at all because the benefits do not justify the expense of pruning.

Horticulture students study specific pruning techniques that relate to various crops. These techniques are specialized for the crop and are based on some of the general principles outlined in this chapter.

REFERENCES

Bailey, L. H. 1903. *The Pruning Book.* Macmillan, London. 545 pp.

Brown, G. E. 1972. *The Pruning of Trees, Shrubs and Conifers.* Faber and Faber, London. 351 pp.

Pirone, P. P., J. R. Hartman, M. A. Sall, and T. P. Pirone. 1988. *Tree Maintenance* (6th ed.). Oxford Univ. Press, New York. 514 pp.

Shigo, A. L. 1986. *A New Tree Biology.* Shigo and Trees Associates, Durham, NH. 595 pp.

Shigo, A. L. 1984. Compartmentalization: A conceptual framework for understanding how trees grow and defend themselves. *Annual Review of Phytopathology* 22: 189–214.

Staffek, E. F. 1982. *The Pruning Manual.* Van Nostrand Reinhold, New York. 152 pp.

PLANT

PROPAGATION

Millions of plants must be produced to meet the demands for horticultural commodities. Propagators use a variety of propagation units (*propagules*), such as cuttings or seeds (depending on the crop, economics, available technology, and efficiency). For example, it is reasonably easy to root cuttings of tomato, but much easier, more efficient, and cheaper to propagate tomatoes by seed. Some large commercial horticultural firms hire people specifically to propagate their plants; others may hire growers that propagate as only one of many steps in producing a crop.

Horticulturists strive to produce high quality plants in order to meet customer demands and to provide the consumer with a satisfactory product that will generate future sales. Many criteria are used to judge quality among horticultural commodities. One of the most important is crop uniformity. Uniformity is requisite for the grower to produce a product in which performance, quality, harvest or sale date, and profit margin all can be predicted. Consumers prefer to purchase the same high-quality horticultural products each time they shop and also prefer uniformity if several units of one item are to be chosen.

The growing of high quality, uniform horticultural crops is dependent on the choice of cultivar (genotype) and proper growing conditions (environment). These factors are discussed throughout this book. Once the cultivar is selected, the propagator begins the production of high-quality plants. If the propagator initially produced plants of poor quality, the grower may never be able to meet the consumer's expectations to produce a high-quality, timely, uniform crop. Therefore, the importance of the propagator in the overall success of a horticultural enterprise's effort to meet its demand for plant (or harvestable unit) numbers and quality must be strongly emphasized.

Propagators must consider many factors when producing high-quality plants for commercial production. They must select the cultivars best suited for the purpose for which the crop is being grown. Sanitation and pest control in the propagation area are of the utmost importance because conditions in most propagation areas favor rapid multiplication and spread of disease-causing organisms. Diseases can greatly decrease the number or performance of plants that a propagator can produce. Most propagation diseases can be avoided by practicing sanitation measures (see Chapter 16). Insects and mites are also unacceptable on plants that are being propagated.

The condition of the stock plants from which the propagules are collected is important. Diseases in the stock plants can carry over on or in the propagules. The general vigor of the stock plants will influence the production and performance of propagules. Stock plants, therefore, should be well cared for regarding their water, fertilizer, light, and temperature requirements, as well as pest and disease control.

Knowledge of the most efficient, and cost effective methods of propagating plants is important. This will be influenced by the availability of propagation facilities, by the particular crop, and by the level of knowledge and experience of the propagator. Some factors as simple as the timing of propagation can determine the degree of success in producing the crop and may

result in success or failure in propagating a plant. For example, stem cuttings of lilac (*Syringa*) can be rooted with reasonable success if taken during spring through an early portion of the summer. If lilac stem cuttings are taken during other times, they will not root, and time and labor will be wasted.

Propagators learn how to successfully propagate plants in several ways. They can take plant propagation courses, read books and journals, talk with other propagators or propagation experts, or visit other plant propagation facilities. There is no substitute for experience, however. By maintaining careful records and working with the plant materials, commercial propagators can become highly successful. Record keeping helps the propagator repeat successes of previous years and avoid past mistakes. Records can also be useful when determining how to schedule propagation space and calculate numbers of propagules to use to meet demands. For example, a propagator can look at previous years records to determine what percentage of rooting to expect from a cultivar that is propagated by stem cuttings. By knowing the percent rooting, the propagator can determine how many cuttings must be collected to meet the quota of plants needed to fulfill market requirements. By knowing the spacing in the rooting bench for the cuttings, the propagator can then determine how much space to allocate to that particular cultivar. Past records should also indicate the length of time the cuttings must remain in the rooting bench. This information is valuable in scheduling one crop of propagules after another.

Forecasts are important when choosing which plants to propagate and how many should be produced. Many plants can be propagated relatively easily, but commercial horticulturists must determine if they can be sold. Forecasts may be easier with plants that are sold shortly after they are propagated, such as annual bedding plants. Because the time between propagation and sale is short, it may be practical to line up sales even before the plants are propagated. However, many woody plants take from three to eight years from the time they are propagated to be large enough to be sold. Therefore, a woody plant that is needed in a landscape two years from now should have been propagated two to four years ago. Forecasts for these long-lived perennials can be based on past experiences and knowledge of other factors, such as the government's statistics on new housing starts. If a horticultural firm does not propagate sufficient numbers of plants to meet demands, not only will sales not be realized, but customers might be forced to shift to other firms that can provide them with sufficient numbers of a high-quality, uniform product.

Depending upon the crop and the purpose for which it is being propagated, the propagator has many ways to increase the numbers of the cultivar. Choices include propagation by sowing seeds, rooting stem cuttings, layering, or by using root cuttings, leaf cuttings, specialized plant structures, plant separation or division, or micropropagation.

Use of sexual propagules, such as seeds, can result in either a uniform or variable crop, depending on the genetics of the seeds (see Chapter 4). The various methods for handling and propagating by seeds are discussed in the last section of this chapter.

When seeds from heterozygous individuals are sown, the population will segregate and be variable. With some horticultural crops, such as oak and pine trees, this variability is acceptable because propagation techniques that result in uniform plants are not sufficiently efficient. Most heterozygous horticultural crops are asexually propagated and result in clones. A **clone** is a population of individuals that originated from (or can be traced back to) one individual; all members of a clone were asexually propagated and, barring any mutations, are genetically uniform. Growth differences among members of a clone therefore result from environmental factors.

VEGETATIVE PROPAGATION

Although achievement of clonal uniformity is a major reason for propagating plants asexually, there are often compelling reasons to avoid sexual (generally seed) propagation.

1. Vegetative propagation can be faster and easier than seed propagation with some crops, such as potato.
2. Some plants (such as bananas and navel oranges) do not produce seeds and therefore must be propagated vegetatively.
3. Various clonal propagation techniques, such as grafting may reduce or avoid juvenility and allow plants to flower and bear fruit faster than seedlings would.
4. Many plants are propagated asexually by grafting onto rootstocks that are resistant to or tolerant of soil-borne disease organisms, insects, or other stresses to which the top is susceptible if it is on its own roots.
5. Asexual propagation by grafting can be used to add a pollinizer to a different cultivar. For example, a grower could graft a compatible cultivar as a new branch on a self-unfruitful fruit tree.
6. Some asexual propagation techniques such as grafting can result in size control of the plant if the proper rootstock is selected.

CUTTING PROPAGATION

Humidity Control for Rooting Cuttings

Transpiration is an important factor in plant propagation by leafy cuttings. A stem or leaf cutting is a part of a plant utilized in vegetative propagation (cloning) wherein the plant part is removed from the parent plant (**stock plant**) and its natural source of water, the root system. Reducing transpiration in leafy cuttings is essential for their survival until they generate an adventitious root system.

Reducing Leaf Area

The propagator can deal with the transpiration problem in a variety of ways. One approach is simply to reduce the leaf area, thereby cutting down on the surface that is transpiring. This is accomplished by removing some of the leaves, usually lower leaves, or by cutting leaves in half (Figure 14-1). This latter method also allows the propagator to place the cuttings closer together, thereby making more efficient use of propagation space.

Shading Cuttings

Another way of reducing transpiration losses is to keep the leaves cooler by shading them. However, shaded cuttings receive less light, thus reducing photosynthesis. This may be undesirable, because less photosynthate is available for the metabolism and production of new cells necessary for adventitious root formation. Shading is often practiced in conjunction with other methods designed to increase relative humidity to nearly 100 percent. A high relative humidity in the vicinity of the leaf lamina will reduce the rate of transpiration; therefore, methods such as enclosing the propagation bed with a polyethylene tent will reduce transpiration and enhance the rooting process (Figure 14-2).

FIGURE 14-2. Methods employed to reduce transpiration in leafy cuttings: intermittent mist (*top foreground*); high humidity tent (*top background*). Interior of a high humidity tent during fog injection (*bottom*).

Humidity Tents

Placing cuttings within a white or clear polyethylene tent in a greenhouse can create a high humidity environment conducive to rooting. Shading materials often are placed over plastic propagation tents to reduce the buildup of heat at the level of the cuttings. Although higher temperatures may hasten the rooting of a few species, such as certain tropical plants, generally the excess heat is detrimental to the rooting process. High temperatures accelerate respiration rates that result in over-consumption of internal substrates that are required for the rapid cell division essential for development of new roots. Furthermore, by increasing evaporation, high temperatures

FIGURE 14-1. The leaves were cut in half on these rhododendron cuttings to minimize water loss through transpiration.

also result in causing the cuttings and the medium to dry out. Another problem inherent in many tent systems is that disease organisms may be encouraged by the stagnant, moist environment. In particular, the development of gray mold caused by *Botrytis* may be favored (see Chapter 16). These pathogens attack newly developed cells such as those formed by cuttings, thus causing a cutting rot and eventually serious damage or death. Reduction of heat and the stagnant air conditions is accomplished in some tent systems through the use of small fans for ventilation. Additional steps usually taken to prevent fungal attack include use of a pathogen-free rooting medium, good hygiene to prevent introduction of undesirable organisms into the propagation area, selection of cuttings from healthy stock plants, and judicious use of fungicides.

Sprinkling

Frequent sprinkling is another approach that many commercial propagators have implemented for relatively easy-to-root cuttings in field or greenhouse propagation beds. This can be accomplished by hand watering or through a conventional sprinkler irrigation system. Although inexpensive, this method has limited application, since it is often not practical to keep pace with the transpiration requirements for many species simply by using overhead irrigation.

Mist Propagation

Perhaps one of the most significant advances made in modern times in plant propagation has been the development of the intermittent mist system for rooting of cuttings (Figures. 14-3 and 14-4). Application of mist, achieved by forcing water under pressure through one of several kinds of nozzles, maintains a film of water on the leaf surface, thus reducing transpiration while keeping leaf temperatures from becoming excessive. By reducing leaf temperatures, the grower decreases respiration rate, minimizing excessive loss of carbohydrates. At the same time, cuttings can be exposed to full sun, which allows photosynthetic activity to continue unabated, thereby enhancing the rooting process.

A combination of two time clocks is the most common method of controlling intermittent mist systems (Figure 14-5). The first time clock turns the entire system on in the morning and off just before dark, and the second governs the frequency and length of each mist application. A common misting cycle is from four to six seconds every three to five minutes, although this varies greatly with species and conditions. As cuttings begin to root, they gradually become able to take up water more efficiently because roots are more effective organs for water absorption than the cut ends of the cuttings. At this point the propagator usually reduces the frequency of mist and shortens the duration of each cycle, effectively weaning the cuttings from the high humidity. This technique, often referred to as **hardening off**, enables the rooted cutting to survive in the environment where it will ultimately be placed. When this is impractical, growers often put the newly rooted cuttings in a shaded location to aid in the hardening-off process.

FIGURE 14-3. Cuttings being rooted under intermittent mist.

FIGURE 14-4. Another method for intermittently misting cuttings or irrigating established plants using a travelling spray boom (*top*). Cuttings being stuck (*bottom left*), and the water being sprayed with boom (*bottom right*).

Bailey Nurseries, Newport, MN.

Various devices have been developed to automatically correlate mist timing with evaporative losses as alternatives to the time clocks. One such system uses a mechanism that senses the change in weight of a simulated leaf, thus activating a switch that turns on the mist when water on the simulated leaf's surface has evaporated (Figure 14-6). Other control mechanisms include humidistats, solar radiation-sensing devices,

FIGURE 14-6. Mechanical leaf for controlling mist application. As the water evaporates from the screen "leaf" at right, the leaf becomes lighter and activates a switch that turns on the mist.

FIGURE 14-5. Time clocks used for intermittent mist systems. The first clock (*lower right*) turns the system on at the beginning of the day and off at night. The other clock regulates frequency of mist application.

and electronic systems. Regardless of the control mechanism, the solenoid (magnetic) valve selected should be normally open rather than normally closed. This provides a safety factor for the cuttings should the electrical current be interrupted. During an electrical power failure with the normally open solenoid valve, the intermittent mist would essentially become a constant mist. A normally closed system requires electrical current to open the valve, however and no water will flow if the electricity goes off. A power failure during a sunny day can result in drying out and even death of leafy cuttings if the system is normally closed and additional backup watering systems are unavailable. If the system is normally open, the constant mist will keep the cuttings wet and alive until the electricity is restored.

Propagators prefer to use intermittent mist rather than constant mist. Constant mist tends to reduce temperatures excessively of both the leaves and the medium, causing metabolic processes such as respiration and cell division to be slowed to less-than-optimal levels for good rooting of the cuttings. Constant mist also causes a greater amount of leaching of nutrients, especially nitrogen and potassium, than does intermittent mist. This potentially leads to growth-limiting levels of nutrients in the cutting tissues. Furthermore, because constant mist is running continuously, the cost for the water is greater.

Leaching of nutrients (especially nitrogen and potassium) from the leaves may occur as a result of the flow of water over the leaf surfaces with any mist system, although to a lesser degree with intermittent than constant mist. Research has shown it possible to replenish the leached nutrients by addition of a low concentration of soluble fertilizer to the mist water. Fertilizers containing nitrogen and potassium are usually used, but not many propagators employ this technique, possibly because many fast-rooting species will be well-rooted and removed from the mist bench before the leaching losses become a serious problem. Additionally, the nutrients favor algal growth, which can be unsightly and can render adjacent walkways dangerously slippery. (Various means are available to reduce the slippery algal growth, including the use of copper sulfate or bromide solutions.) Also, the nutrient mist can lead to accumulations of salt deposits in the mist nozzles, so the operator must examine the nozzles periodically for signs of clogging. Some propagators, therefore, choose to place slow-release fertilizers in the propagation medium, on the assumption that the slowly available fertilizer nutrients will be taken up by the cuttings at a rate equivalent to the losses caused by leaching from the mist. In both cases—that is, use of nutrient mist or use of slow release fertilizers—care must be taken to avoid excessive rates, because damage to the tender, newly developed roots could occur from high soluble salt levels.

Researchers have always questioned whether the high moisture levels inherent in a mist propagation system could encourage the development of serious disease problems. However, it has been demonstrated that this is not normally a problem when rooting leafy cuttings under mist for two reasons. First, the frequency of misting is such that any spores and bacteria that do land on the cuttings are washed off either prior to germination or prior to invasion of the plant tissue. Second, because the leaves are kept constantly moist, it is too wet for germination and/or growth of many species of pathogenic fungi and bacteria.

Excess moisture in the rooting medium can become a problem in mist propagation, but such difficulties are usually a consequence of inattention to requisites for a good propagation medium. A medium or container that drains poorly will easily become waterlogged and thus be poorly aerated. A poorly aerated medium limits the availability of oxygen in the root zone. Oxygen is required for new cell growth and maintenance of plant health, and it has an indirect influence on water uptake. Medium components often utilized to impart good drainage (aeration) include perlite, vermiculite, polystyrene beads, and coarse sand. Regardless of the composition, prior to its use, the medium should be treated with heat

or chemicals to kill undesirable organisms that might be present in the soil mix (see Chapter 16).

Another consequence of excessively wet soil is the fact that wet soils usually are cooler than dryer soils; lower temperatures depress metabolic rates and rooting. Propagators therefore often employ **bottom heat**, a source of warmth placed under the medium to enhance rooting of cuttings (see Chapter 6). Bottom heat is particularly effective with intermittent mist systems, because metabolic processes in the root-forming zone of the cutting will be stimulated by higher temperatures, whereas the cooling influence and high relative humidity of the mist will help prevent excessive losses from respiration and transpiration in the aerial portion.

Fog Propagation

Another approach for providing high relative humidity without excessively wetting the plants and medium is the use of fog. Fog in greenhouses, similar to the fog caused by clouds close to the earth's surface, keeps the relative humidity close to 100 percent, thus eliminating water loss from transpiration (Figure 14-7). The droplet size in fog systems is much smaller than that for mist systems, so the droplets remain suspended in the air, creating the fog effect. Droplets larger than 35 μm in diameter will settle out of the air. Indeed, fog droplets produced in many of the commercially available systems are less than 20 μm in diameter. Such small droplets are produced by forcing water at high pressure through narrow-diameter fog nozzles, or by using specially designed fans.

Fog is reported to offer several advantages over intermittent mist. Because the fog remains suspended in air, the rooting medium does not become too wet and remains warm, light, fluffy, and airy. Because evapotranspiration is negligible, it is not necessary to rewater the rooting medium after initially watering-in the cuttings. Researchers have reported fewer disease problems under fog than mist because the leaves remain dry.

FIGURE 14-7. A greenhouse in which a fog system is operating (*left*). Note the cloud-like effect, nearly obscuring the workers from view. Compare with natural fog outside a greenhouse (*right*).

Also, cuttings may be placed closer together under fog than they can under mist because water will not be applied from above that can run down and accumulate where leaves overlap. Some growers report higher percentages of rooting and shorter propagation times under fog than mist, possibly because the medium remains warmer and better aerated. Because the droplet sizes are much smaller, fog systems use about 25% less water than mist systems. This can amount to substantial savings, especially in areas where water costs are high.

Fog offers many advantages over other systems. However, when a fog system is in operation, it is difficult to see more than a few meters into the greenhouse. This is one of the few criticisms of fog systems: The fog may cut down on the sunlight that reaches the plants, reducing photosynthesis and growth. However, when used in geographic areas of high light levels such as Florida, fog has reduced rooting times by as much as 50 percent. It has also been effective for rooting cuttings of plant species that are traditionally propagated by air layering. That means less cost for labor, because air layering is a time-consuming hand operation.

Perhaps the biggest drawback of using fog is the fact that many people find it uncomfortable to work in an area with such high humidity. Workers have reported that they tire relatively quickly when working in fog houses and that their clothing becomes soaking-wet because their perspiration does not evaporate in the fog.

Proponents of fog point out that it can be an effective means of cooling greenhouses, and in some cases, may eliminate the need for fan and pad systems (see Chapter 6). If this is the case, costs for running fans may be reduced by 50 percent or more—a significant cost savings to the grower. In order for fog to function effectively as an evaporative cooling system, however, it is crucial that there be no leaks in the greenhouse; that is, no leaky glass panes, no vents that stick open and so forth. It is also important to note that use of bottom heat, a well-aerated pathogen-free medium, and other cultural practices important in any propagation system must be considered in order to employ fog systems successfully.

If fog systems offer so many potential advantages, why were they not more widely adopted in the past? The answer is relatively simple. For fog systems to work efficiently without high maintenance costs, the water quality must be high. This means that virtually no particulate matter can be present in the water. To achieve this level of purity, special filters must be used. In years past, such sophisticated equipment was unavailable or too costly; now modern filters and pumps are readily available and economically priced. Consequently, fog is becoming an increasingly popular system for propagation and greenhouse cooling.

CUTTING TYPES

Cuttings are used extensively to propagate many horticultural species clonally (Figures 14-8 and 14-9). Cuttings may consist of leaves or portions of leaves and are thus referred to as **leaf cuttings** or **sectional leaf cuttings**, respectively. **Leaf-bud cuttings** are single nodes and adjacent internode tissue; they have leaves attached. **Stem cuttings** are portions of shoots, either with or without apical buds. **Root cuttings** are root segments. The handling of these cutting types differs, but the overall goal of clonal reproduction remains the same.

To obtain plants from **leaf cuttings**, adventitious shoots with roots must form. The leaves of many species do not appear to have the capacity to form adventitious shoots, and although they may root, plants will not result (Figure 14-9). Leaves of some plants, like African violet and peperomia, are commonly used as propagules because they will form adventitious shoots with roots (Figure 14-10). Leaves that are used as cuttings offer an advantage over stem cuttings in that there are usually many leaves per plant, and thus relatively few stock plants are required.

FIGURE 14-9. Both leaf cuttings (*left*) and leaf-bud cuttings (*right*) of coleus will root; however, this leaf cutting cannot produce adventitious shoots, so a whole plant will not result. The axillary bud on the leaf bud cutting is growing to become the top of the new plant.

FIGURE 14-8. Two examples of propagation of blackberry by cuttings. Leafy two-node stem cutting (*left*), and root cutting (*right*). The axillary bud of the stem cutting will be the new plant, but the stem forming on the root cutting is adventitious (arrow points to the original root cutting on right).

Stem sections with one or two lateral buds and a leaf are called **leaf bud cuttings**. Leaf bud cuttings are frequently used when a propagator wishes to obtain many cuttings from plants whose leaves will not produce adventitious shoots (Figures 14-9 and 14-11). The axillary bud on the leaf-bud cutting simply elongates to become the top of the plant. Therefore, to obtain whole plants one must stimulate adventitious root formation on leaf-bud cuttings.

Stem cuttings must also be rooted to obtain clonal plants. Growth regulators, such as auxins are often applied for this purpose (see Chapter 11). Propagators use different types of stem cuttings, depending both on the season and plant species. These are outlined in Table 14-1.

To obtain plants from **root cuttings** one must stimulate adventitious shoot formation (Figure 14-12). Subsequently, these shoots often produce additional roots. Roots contain the most stored food as the growing season ends. These food reserves are depleted rapidly as new growth occurs in the spring. The stored food will provide energy to any new, adventitious shoots. Therefore, root cuttings should be taken during late winter to early spring before any new growth begins.

FIGURE 14-10. *Peperomia* leaf cuttings in a multicellular tray. Note the smaller leaf size on the plantlets (*top*). The plantlets can form either from the leaf blade on the left or the leaf petiole on the right of peperomia (*middle*). Sectional leaf cuttings of Rex begonia work well for producing adventitious shoots with roots (*bottom*).

FIGURE 14-11. Leaf-bud cuttings of potato respond in an interesting manner. They root and the axillary bud does not elongate very much but instead forms a tuber because it was located below the soil surface. In this picture the tubers appear to be black because they came from stock plants that produced purple skinned potatoes.

TABLE 14-1.   The timing and handling of different types of stem cuttings[a]

FACTORS TO CONSIDER	STEM CUTTING TYPE				
	HERBACEOUS	SOFTWOOD	SEMI-HARDWOOD	HARDWOOD (DECIDUOUS)	HARDWOOD (EVERGREEN)
Description	Succulent stems from nonwoody plants	New, soft, succulent growth on woody species	Partially mature wood on current season stems	Mature dormant or quiescent hardwood stems	Mature hardwood stems
Time of year	Year-round	Spring	Mid to late summer	Late fall to early spring	Late fall to late winter
Cutting length	7.5–12.5 cm (3–5 in)	7.5–12.5 cm (3–5 in)	7.5–12.5 cm (3–5 in)	10–76 cm (4–30 in)	10–20 cm (4–8 in)
Humidity	Intermittent mist, fog or humidity tent	Intermittent mist, fog or humidity tent	Intermittent mist, fog or humidity tent	Direct sticking into field, mist, fog, or tent	High humidity or light misting or fog
Auxin treatment	No auxin or IBA or NAA at up to 1000 ppm	IBA or NAA at 1000 ppm	IBA or NAA at 3000 ppm	IBA or NAA at 2500–20,000 ppm	IBA or NAA at 8000 ppm or higher
Plant examples	Poinsettia Chrysanthemum	Maple Lilac	Rhododendron Citrus	Willow Red-osier dogwood	Juniper Yew
Comments	Probably the fastest to root; can be susceptible to decay	Root relatively quickly (2–5 weeks); bottom heat helps 23–27°C (75–80°F)	Wounding may be beneficial; leaf area may be reduced to help control transpiration	One of the least expensive; easy to ship over long distances	Usually slow to root; bottom heat 23–27°C (75–80°F) helps; dipping in a fungicide will aid in preventing decay

FIGURE 14-12. Raspberry roots produce adventitious shoots readily and are therefore good propagules. This shoot is etiolated because it grew below the soil surface.

When handling cuttings, one must be aware of the following:

1. Initially, there are no roots (with the exception of root cuttings), so transpiration is a major concern. To reduce water loss, most cuttings are placed in a high humidity environment (Figure 14-13).
2. The stock plant and even the location on the stock plant from which cuttings are taken can have a profound effect on propagation success.
3. Sanitation is a major concern because disease organisms can greatly damage cuttings in a propagation bench.

FIGURE 14-13. Leafy cuttings should be kept cool and moist after they are harvested from the stock plants. Note the mist nozzles at the very top of the picture. These cuttings were being misted when this photograph was taken.

4. Plant growth regulator applications are often necessary to ensure that roots (or shoots on leaf cuttings) form faster and in greater numbers than would be obtained normally. There are several available choices of plant growth regulators, and the correct concentration and method of application is often critical to success.

5. Cutting preparation and handling is important. Clean cuts should be made when preparing cuttings. It is wise to first make holes in the rooting medium to accommodate the cuttings (Figure 14-14). Simply jamming a cutting into a medium can damage it considerably and cause poor rooting.

6. Attention must be paid to the orientation of cuttings. Cuttings of some species will root and grow into plants when inserted into a rooting medium upside down (Figure 14-15). However, plants will generally grow better if their correct polarity is maintained.

7. The choice of the medium, if used, can have a profound influence on propagation success. The propagation conditions, such as mist or fog, will influence this choice.

8. Many cuttings will respond positively to the application of bottom heat.

9. The time of year that cuttings are taken, especially with temperate zone woody plants, has a major influence on propagation success. Simply collecting cuttings at the proper time often means nearly 100% success, compared to total failure if the time of year is not correct.

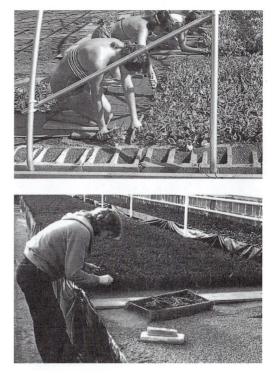

FIGURE 14-14. These workers (*top*) are properly making slits in the rooting medium (sand) before sticking their cuttings. One may wish to use a board as a straight edge to ensure that the slit will be straight and the cuttings will be in even rows (*bottom*).

LAYERING

The practice of layering involves the formation of adventitious roots while the cutting is still attached to the parent plant. There are several types of layering, but here we will discuss only two (a third type, **mound layering** is described in the Appendix, Apple). **Air layering** involves either cutting the stem or removing the bark from the stem of a plant and wrapping it with a

FIGURE 14-15. It is highly recommended that cuttings be stuck into a rooting medium rightside up. With some species like blackberry, it is easy to see that the cutting stuck upside down (*top, left*) rooted as well as the cutting stuck rightside up (*top, right*). The plant from the upside down cutting grew more or less normally, but was swollen and produced callus on the top, its physiological base (*bottom photograph*).

FIGURE 14-16. Because the white roots are evident on this air layered *Ficus*, the top is ready to be severed. It should then be planted and the bottom kept because it will resprout; two plants will result.

medium, often milled sphagnum moss, and usually clear polyethylene. After roots form, the rooted stem is cut from the parent plant and is planted, similar to a rooted cutting (Figure 14-16). Air layering is used for some foliage plants and several tropical fruit species, and it has been used successfully on some temperate trees.

Tip layering differs considerably from air layering in that the rooting occurs in the ground. Plants such as blackberry and black raspberry produce long primocanes. As the growing season progresses, the cane tips arch over and touch the soil. At this time, the shoot tips can be buried in the soil, roots will form, and a vertical shoot will emerge through the soil surface. A new plant is thus established that can be severed from the primocane and transplanted (Figure 14-17). Shoot tips may also form roots naturally (without the propagator's assistance) where they touch the ground. This also is referred to as tip layering.

SPECIALIZED PLANT STRUCTURES

Many plant species have special structures that are important in vegetative propagation, including corms, bulbs, tubers, stolons, and rhizomes (Figure 14-18; see also Chapter 3). They are handled in various ways, and references should be consulted for a particular species. Other plants grow in clumps or produce offshoots that can be divided or separated. The propagation term

FIGURE 14-17. In the autumn the tip of this black raspberry stem was buried in the soil. It rooted and the tip elongated. After this picture was taken, it was severed from the main stem and transplanted, much like a rooted cutting. This illustrates tip layering.

division implies that physical cutting of the plant is required in order to result in additional plants, whereas the term **separation** indicates that there is a natural place of detachment, so that cutting is not necessary.

GRAFTING AND BUDDING

Grafting and budding involve the joining together of plants, or plant parts, so that they grow as one (Figure 14-19). Budding is grafting only one bud onto a plant. Grafting is often used for species that are difficult to clone by other means, or because the grafted plant is somehow more desirable. The upper portion of a graft is called a **scion** (cion). A scion is grafted onto a **stock** (understock or rootstock) that becomes the lower portion of the plant. Sometimes another piece of stem may be grafted between the scion and the stock; this is referred to as the **interstem** (interstock, intermediate stock). The scion is generally chosen because it is the desirable

FIGURE 14-18. Plants with special structures require unique propagation techniques. With garden lily (*top left*) bulbils can be removed from the leaf axils and planted, similar to seeds. Scale pieces can be removed from a lily bulb (*center right*) or cut from a hyacinth bulb (*bottom left*) and placed into a greenhouse medium; within approximately two months, bulblets will be produced that can be removed and planted, similar to bulbils.

FIGURE 14-19. Grafting is a natural process. On the left, the roots of weeping fig grafted together where they crossed, and in the center where two branches on a tree touch, they are in the process of forming a graft union. People can make use of this phenomenon as is evidenced by the grafting together of the three braided *Ficus* stems on the right. In each case, no cutting was involved. The pressure exerted by growth in girth brought the vascular cambia in close contact and resulted in a grafted condition.

genotype or cultivar. The interstock is selected because it can influence the ultimate vegetative size and growth rate of the scion and understock. Interstocks will sometimes overcome an incompatibility between the scion and stock. Rootstocks are selected because they can control the size of the top of the plant; result in increased fruitfulness of a tree; influence the maturity of fruit; influence the winter hardiness of the plant; and offer disease, insect or other stress resistance. Plants are thus grafted for many different reasons.

To understand grafting techniques and the factors that influence success and failure of this form of propagation, one needs to be familiar with botany (see Chapter 3). A brief description of the formation of a graft union is helpful in explaining the factors that contribute to successful grafting. When two plant parts are grafted together, a healing process involving plant metabolism and cell division occurs. Therefore, when practical, temperature (usually 13°–32°C or 55°–90°F) and relative humidity (usually high) are controlled to promote metabolic activity and prevent the drying of the graft union. In

order for a graft to "take," the vascular cambia (see Chapter 3) of the scion and stock must be lined up and in close contact. This requires straight, smooth, and even cuts (Figure 14-20). Because of the necessity of vascular cambial contact, monocots, which

FIGURE 14-20. To ensure a successful graft union, straight, smooth cuts must be made on both the scion and understock so that the vascular cambia can be positioned in close contact. This type of cut can be seen in this bench grafting operation. Note the straightness of the cutting edge of the propagator's grafting knife; this allows him to make sharp cuts.

have scattered vascular bundles, rather than continuous vascular cambia cannot be grafted successfully. After grafting, the cut surfaces undergo a wound-healing response by forming a layer of necrotic (dead) material from the cut cell walls and contents. Callus cells form from the vascular cambia beneath the suberized regions of the stock and scion. These callus cells intermingle and a callus bridge forms. A new vascular cambium then forms across the callus bridge and gives rise to new xylem and phloem tissues, thus a functional vascular connection forms to serve the new plant.

Many different types of grafting are employed, both for propagation purposes or for damage repair. The names, such as **cleft grafting**, **wedge grafting** (Figure 14-21), **bridge grafting** (Figure 16-43), **saddle grafting**, **T-budding** (Figure 14-22), and **chip budding**, are descriptive. Plant propagation and grafting books should be consulted to learn how to properly make the cuts for these various types of graftage.

The taxonomic relationship (see Chapter 2) between the understock and the scion has a direct influence on whether a graft combination will be successful. Generally, the closer the relationship (such as within a clone or species), the better the chances that a graft will take. Although it is possible in some cases to successfully graft plants within a genus or family, such as tomato grafted onto potato, the chances of success are relatively small for most plants that are not closely related.

Graft unions that fail or grow poorly because of physical or chemical characteristics of the scion and understock are said to be **incompatible** (Figure 14-23). Grafts that are unsuccessful solely because of poor technique are *not* incompatible. Graft incompatibility may result in the failure of a graft combination to take, or may result in a grafted plant that grows as one for several years, then fails. This latter case is known as delayed graft incompatibility. There are several signs of graft incompatibility: (1) There may be a low percentage of bud or scion

FIGURE 14-21. Many species are asexually propagated by grafting. The avocado (*top*) illustrates wedge grafting, and the grafted cacti (*bottom*) have interesting ornamental value.

take. (2) The plants may exhibit premature autumn leaf coloration. (3) Young shoots may die back and the plant may defoliate, from the tip first. (4) The scion may grow more slowly or rapidly in girth than the understock. (5) The plant may die prematurely. (6) The graft union may be mechanically weak (the broken union is relatively smooth when this occurs). Incompatibility symptoms appear earlier and are more pronounced when the plant is grown under adverse environmental conditions.

FIGURE 14-22. There are several budding techniques. A T-shaped cut is made in the bark of the understock when T-budding. The bark flaps are peeled back and a bud is inserted. T-buds are generally wrapped with a budding rubber (*left*). When they grow out, the scion may initially create a wide crotch angle with the stem of the understock, as can be seen in the center, where a yellow-leaved honeylocust was budded onto a seedling rootstock with green leaves. After the graft takes, the top of the understock is cut off and the shoot from the bud is trained to grow vertically (*right*). Eventually evidence of the bud union will become less obvious as secondary growth occurs.

MICROPROPAGATION

Micropropagation is the multiplication of plants in vitro (literally "in glass") under sterile (aseptic) conditions, on or in a specific nutrient medium. The term **plant tissue culture** is often used more widely to include all applications of the aseptic culture of plants. Micropropagation is therefore only one aspect of plant tissue culture.

The general goal of micropropagation is to clone plants in vitro.

To multiply plants successfully using micropropagation techniques, one must first place small pieces of plant tissue or organs (**explants**), aseptically onto a specific medium. The medium usually consists of inorganic salts, organic substances, and sometimes a gelling agent. The inorganic salts contain macronutrients and micronutrients in appropriate ratios(see Chapter 9).

FIGURE 14-23. Graft incompatibility frequently shows up as the scion growing in girth faster than the understock on the apple tree (*left*). An interesting manifestation of graft incompatibility in a green/white ash combination appeared in the loss of cold hardiness at the graft union that led to the death of the tree (*right*).

The organic substances are mainly plant growth regulators, vitamins, cofactors, and a carbon energy source, usually sucrose. Agar, a carbohydrate extracted from seaweed, is often incorporated as the gelling agent. Agar comes in various grades or purity and is added primarily to provide support so that the plant tissue will have adequate oxygen for aerobic respiration. Other gelling agents also are available.

Because the ingredients in a tissue culture medium are ideal for cell growth, microorganisms such as bacteria and fungi grow well and often will kill the plant material being micropropagated. Consequently, much of the equipment and the practices for plant tissue culture are geared toward preventing the introduction of microorganisms into the culture vessel. Explants (pieces of plants placed in vitro) are therefore surface-disinfested, often in a liquid chlorine bleach solution with an added wetting agent. The latter is added to ensure better explant/bleach contact. Culture vessels and media are sterilized in autoclaves (Figure 14-24) at high temperature and pressure. Tools, such as scalpels and forceps are heat or chlorine bleach sterilized. All handling of surface sterilized tools, plant tissues, and opening of vessels is generally done in clean air benches that filter the air to partially or totally eliminate fungal spores and bacteria. In addition, personnel who are performing transfers must be properly trained to ensure that contamination by microorganisms does not occur.

Plants of either adventitious or axillary origin can be obtained with tissue culture techniques. Adventitious plants can be induced to form directly from the original explant or from callus (a mass of unorganized cells). Adventitious structures that form can be either unipolar (shoots, leaves, flowers, or roots) or bipolar (embryos or embryoids). Organs (e.g., shoots) form because **organogenesis** occurred; embryos form from vegetative (somatic) cells, this process is called **somatic embryogenesis** (Figure 14-25). When organogenesis or embryogenesis take place in callus, the

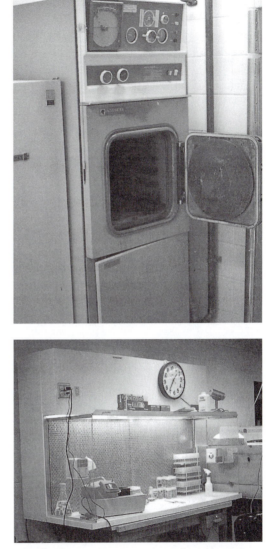

FIGURE 14-24. Asepsis is essential in plant tissue culture. An autoclave (*top*) is used to sterilize media and tools with heat (normally 121°C) and pressure (normally 1 kg cm^{-2}) for 20 or more minutes. A laminar flow clean air bench (*bottom*) filters the air to a point where fungi and bacteria are excluded to provide a sterile environment for placing explants in vitro or making transfers.

resulting plants are often quite variable (known as **somaclonal variation**), possibly owing to physiological, epigenetic, or genetic changes. Although this variability may be of value to plant geneticists and breeders, it is not beneficial in plant cloning.

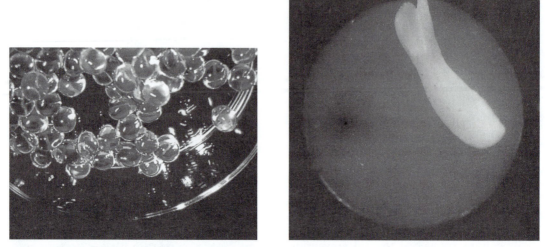

FIGURE 14-25. Plants can be regenerated, using in vitro techniques, via somatic (vegetative) embryogenesis. Somatic embryos can be encapsulated in artificial seed coats (*left*).
Photograph courtesy of Keith Redenbaugh, Plant Genetics, Inc.

An artificial alfalfa seed (*right*). The somatic embryo was embedded in a calcium alginate capsule that served as an artificial seed coat.
Photograph courtesy of Steven E. Ruzin, Plant Genetics, Inc.

When stem tip, nodal, or apical meristem explants are induced to branch, the resulting shoots tend to be quite uniform because they develop from axillary buds. Axillary shoots are less likely to produce off-type plants than plants that are produced adventitiously from callus. Therefore, obtaining axillary shoot proliferation by weakening apical dominance is generally the objective in commercial micropropagation (see Chapter 12). The choice of explant material and addition of plant growth regulators, often cytokinin, to the medium are critical for obtaining this goal.

Stages of Micropropagation

To understand the various objectives of micropropagation, many people divide it into four more or less distinct stages as follows.

Stage I: Explant Establishment

In this stage the grower places the explant in vitro so that it will remain alive and free from microbial contamination and begin to grow. The explant is often a shoot tip, a stem piece with one or more nodes, or an apical meristematic dome, because all are capable of producing axillary shoots. The proper handling of the explants varies considerably with the explant type, plant species, and condition of the stock plant.

Stage II: Axillary Shoot Proliferation

In this stage, the grower strives to obtain rapid and continuous axillary shoot production (Figure 14-26). The shoots also must elongate sufficiently so that they may be handled further. They may either be placed back onto a stage II medium so that their axillary buds will elongate and give rise to more proliferating cultures, or the shoots may be harvested and induced to produce adventitious roots.

Stage III: Pretransplanting

The shoots that form in stage II must be rooted so that clonal plants can be produced.

FIGURE 14-26. Micropropagation, via axillary shoot proliferation, is generally done under cool white fluorescent lamps (*top*). By using the proper medium and growing conditions, propagators can obtain good axillary shoot proliferation with many species, including this rhododendron culture (*bottom*).

Rooting may be done in vitro, often in an agar-gelled medium, or ex vitro in a typical greenhouse growing medium or plug tray under high relative humidity. Although with some species the microshoots will root well without exogenous plant growth regulators, it is sometimes helpful to apply an auxin to stimulate rooting.

Stage IV: Acclimatization to Ambient Conditions

The leaves that form in vitro on axillary shoots are usually photosynthetically inefficient, have poorly developed cuticles and unresponsive stomata (which, in fact, may remain open even at night), and contain much intercellular space. They are not well adapted for growth in the field or greenhouse. Direct movement from the in vitro environment to the low relative humidity of the ambient environment usually results in rapid wilting and death of the plantlets. They therefore must be gradually acclimatized to the ambient environment. This is often done under intermittent mist or fog (Figure 14-27) and can be one of the most difficult phases of micropropagation.

FIGURE 14-27. Microshoots are often rooted and micropropagated plantlets acclimatized to ambient conditions in areas where transpiration is minimal. A mist system such as the one on the track (*top*) or fog (*bottom*) both work well for this purpose.

SEED PROPAGATION

Homozygous or hybrid plants are generally propagated by seeds and produce uniform high-quality crops. Many heterozygous species are clonally propagated for uniformity (Figure 14-28). Clonal propagation requires more labor and technology and is therefore much more expensive than seed propagation. Also, some species produce abundant quantities of seeds, further reducing propagation expenses. Through breeding efforts, scientists have developed seed-propagated cultivars of some traditionally vegetatively propagated crops, such as potato. Seeds are smaller and lighter and are much less expensive to ship than most vegetative propagules. Although it would be desirable to propagate some heterozygous species such as oaks and beeches vegetatively, efforts have been largely unsuccessful. These plants therefore must be propagated by seeds.

Some clones have had problems with aging or "running-out" of the stock plants. For reasons such as buildup of viruses or microorganisms, or the plants' becoming more mature (less juvenile), vegetative propagules have responded progressively more poorly. For example, cuttings from juvenile plants (those not yet capable of flowering under normal flower induction conditions) generally root easier than cuttings from adult forms of the same species. Propagation by seeds can avoid this problem with declining propagule performance and can result in new, more juvenile stock plants.

Most vegetables, flowers, turfgrasses, and many herbaceous and woody perennials are propagated by seeds. A knowledge of the production, harvest, care, and handling of seeds is therefore essential for horticulturists.

Apomixis

Not all seeds form as a result of the fusion of a sperm and egg nucleus. In fact, it is possible to propagate clonally some species using seeds. The formation of an embryo (seed) without the sexual union of gametes is called **apomixis**.

Apomictic embryos can form for different reasons. The embryo can develop from an unfertilized egg nucleus that has not undergone complete meiosis and is still $2n$ (diploid) (see Chapter 4). Because it is the same ploidy level and genetic makeup as

FIGURE 14-28. Roses (*top*) that are grown for cut flowers are clonally propagated by grafting so they will produce a uniform, high-quality crop. Note the segregation and thus lack of uniformity in the seedling nursery bed of white pine (*bottom*). Both species are heterozygous.

the female parent, the parent's genotype is said to recur. This is known as **recurrent apomixis** and occurs in some seeds of apple and Kentucky bluegrass.

When the unfertilized egg nucleus has completed its meiotic divisions and develops into an embryo, it is haploid (1*n*). Because the female parent's genotype does not recur, it is known as *nonrecurrent apomixis*. This phenomenon is very rare.

The nucellus (maternal tissue) is the central portion of the ovule that contains the embryo sac. In some species (e.g., *Citrus* spp.) embryos form adventitiously from nucellar tissue or cells of the integuments. This is known as **nucellar or adventitious apomixis**. It is the reason why there are up to several embryos (polyembryony) in some citrus seeds. When each seed is sown, one to six or more seedlings may grow. Each seedling may be a member of the same clone, or one of the embryos may have developed sexually.

Although there may be nearly 100 percent apomixis in some species or cultivars, this phenomenon is not generally exploited to clone plants. This is because it is usually impossible to distinguish between clonal and sexually derived seedlings without growing the plants for some time or by destroying portions for chemical fingerprinting. Most seed propagation is therefore sexual propagation.

Seed Production

A large and important industry within the field of horticulture is devoted to the production of seeds. Growers produce crops specifically for their seeds. Although this practice can be highly profitable, very strict rules must be followed regarding seed quality control. General field sanitation is much stricter than it is when the same crops are grown for normal use.

Voluntary seed certification programs ensure high standards of genetic identity and purity of seeds. Seed certification agencies, whose function is to certify seeds

legally, may be governmental or international. Similar practices may be followed by commercial companies that produce vegetable or flower seeds.

There is a series of phases in the production of seeds of a new cultivar. Initially, breeders produce the new cultivar. They multiply the seeds to a sufficient quantity so the cultivar can be maintained. The maintenance is in a foundation planting that can be established only with **breeder's seeds** or other **foundation seeds**. Foundation plantings have a very high standard of genetic identity and purity that is maintained through inspections and testing. Although it is not quite as tightly controlled as foundation seed, **registered seed** can be the progeny of plants that grew from foundation seed or other registered seed. **Certified seed** are seeds for general distribution to growers. They may be the progeny of plants that grew from foundation seed, registered seed, or certified seed. They are inspected by a certification agency. Although certified seed may be more expensive than noncertified seed, many growers feel that the insurance of genetic purity and the resulting crop uniformity is worth the extra cost.

The regions where seeds are produced vary with the crop. Considerations are the natural means of pollination (wind blown versus insect vectors versus pollen that does not travel far) and the suitability of an area for a crop. Fields that are used for seed production typically are isolated so that uncontrolled cross fertilization will be minimized. Crops such as tomato, that are normally self fertilized and not wind pollinated need to be separated from production fields by five meters to one-half km (15 ft to 1/4 mile). If a crop is normally cross fertilized by wind-blown pollen or insects, however, fields should be more than two km (>1 mile) apart.

Within the United States a Federal Seed Act sets standards for seeds entering into interstate commerce. Additionally, state seed laws, which generally conform to the federal standards, impose further restrictions regarding labeling, sale, and transportation for sale of agricultural, vegetable, and other seeds. These laws have been implemented to protect the consumer. A familiarity with these laws is necessary if one is to produce or sell seeds. Restrictions may be sufficiently strict to disallow a specific harvest of a crop produced by one grower to be sold as seeds.

Many states identify specific weed species as being noxious. There may be two categories of **noxious weeds: primary** (or prohibited) noxious weed seeds and **secondary** (restricted) noxious weed seeds. Noxious weeds are considered to be especially troublesome or difficult to control. Tolerances of noxious weed seeds vary, but primary noxious weed seeds may not be allowed in any amount in seeds offered for sale. For this reason growers of seeds must pay particular attention to weed control within the fields being used for seed production.

Additional laws concern labeling and testing, and require that a seed container name the crop and cultivar on the label. Labels must include the particular crop and cultivar or hybrid. These names must be the recognized names of the plant materials and must have no misleading information. If seeds are blended, as is done frequently with turfgrass species, all kinds and cultivars must be listed if present in sufficient quantity (usually 5 percent). Germination information often must be provided as well as the calendar month and year that the germination test was conducted. All seed containers should be labelled with the name and address of the person or company who labeled the seeds or who offers the seeds for sale. It is reassuring that seeds must comply with certain standards; if a company's seeds do not comply with the law, the firm may be issued a "stop sale" order and be fined. Many states require that those who offer seeds for sale be issued an annual permit from the state Department of Agriculture.

Seed containers may contain the term **hard seed**. Hard seeds have impermeable seed coats and cannot absorb moisture. Treatments to enhance germination of

species with hard seed coats are covered later in this chapter.

Although most growers purchase vegetable, flower, and turfgrass seeds, seeds of woody plants may either be purchased or collected directly from nearby trees or shrubs. Growers of woody plants generally are interested in the **provenance** or location of geographic origin of the seeds, especially if seeds are collected from native plants. In cooler regions, plants may not survive harsh winters if they were propagated from seeds collected from a provenance with a longer growing season and milder weather. With species in some regions, however, plants from seeds collected from more remote provenances may outperform plants of local origin. Provenance test data are available in the literature, but many of these publications are directed towards a forestry use of the plant materials.

Variability can exist within a provenance, so many propagators will seek a uniform population, or a pure stand of the plant species. The uniformity may indicate that this group of plants is somewhat homozygous and that their progeny also may be uniform. Alternatively, some nurseries have their own seed orchards. They know the histories of these plants, know the performance of the progeny, and are confident that the planting will be cared for properly and not be destroyed. However, when first starting out, planters must germinate seeds and grow the resulting plants to determine their performance. If the plants are inferior or have undesirable characteristics, other seed sources should be tested.

Seed Harvesting

Embryos in seed must be sufficiently mature when seeds are harvested. Although embryos of some species are immature or rudimentary when they fall from the plant (e.g., ginkgo), most are relatively mature in ripe fruit. Indications of immature seeds include softness, watery or milky endosperm; gelatinous, light colored seed coat; and low specific gravity. Germination may be poor and storage life short if immature seeds are used. Immature seeds often shrivel when dried, and if they germinate, may grow into weak seedlings. Immature seeds also may be more wet than mature seeds and thus prone to rot during storage.

Fully ripe fruit collected immediately prior to seed dispersal generally yield sufficiently mature seeds. Indices of ripeness vary with species and range from color changes to chemical changes (e.g. sugar content) to changes in specific gravity (pine cones get lighter when ripe).

Specific gravity of seeds or fruits is tested by placing them in a liquid with a known specific gravity. With many seeds, sound or viable seeds are more dense than unsound seeds. Consequently, seed specific gravity often increases with maturity. Seeds of many species can be tested in water, fungicide solution (or suspension), or another suitable liquid. Sound (viable) seeds will sink; unsound seeds will float. This technique provides the propagator with an estimate of the percentage of sound seed in a lot and can be used to remove the nonviable seeds, resulting in a better stand of seedlings.

Species that produce seeds with hard coats are an exception. If seeds of these species are harvested when the embryo is sufficiently mature but the seed coat has not fully hardened, they can be sown directly without further treatment during the late summer or autumn. If the seed coats of these species are allowed to fully mature and harden, they will require a **scarification** treatment to penetrate the seed coat and allow germination to progress.

Cleaning and Extracting Seeds

It is more efficient to handle, transport, sell and sow seeds if they are free from debris and if various appendages are removed. Techniques for extracting seeds from fruit, removing debris, and removing appendages, if present, vary considerably with the plant species and type of fruit.

Seeds with appendages that facilitate dispersal by wind or air currents are often dewinged. The wings may be portions of the fruit or ovular tissue. Maples, pines and spruces are plants with winged seeds. The seeds can be dewinged by rubbing, placing them in a cloth sack and flailing, using mechanical dewingers that often work by tumbling the seed, or sometimes by wetting then drying and blowing off the chaff.

Seeds within fleshy fruits must be treated so they can be extracted efficiently. For example, tomato seeds are harvested and the fruit are cut by a machine in the field. The seeds and juice are separated from the pulp and skin. The remaining mucilaginous sheath and juices can be removed by acid treatment or fermentation. Hydrochloric acid treatment will quickly separate the seeds from the juice and pulp. Alternatively, the seed and juice mixture can be fermented for about two days at approximately 30°C. The pulp will float and can be removed. Fermentation has the advantage of killing bacterial pathogens that remain viable with the acid treatment. The seeds are then washed and dried, and clumps of seeds are broken up.

Some seeds are separated from fleshy fruits by macerating. For example, seeds are extracted from hawthorn (*Crataegus*) fruit by macerating fully ripe fruit in water. A blender can be used for this purpose, but care must be taken not to damage the seeds. The pulp is then removed by floating. Seeds may be washed and then dried for storage. Maceration and fermentation or acid treatment may be combined for some species.

Machines, such as hammer mills, threshers, tumbling pebble mills and debearding machines utilize various methods for threshing seeds, but these machines can break, crack, and otherwise damage the seeds if used carelessly. Various prairie grasses that are gaining in popularity in the landscape trade have their awns, glumes, or other appendages removed in hammer mills. Examples of species cleaned in this manner include the gramas (*Bouteloa* spp.) and bluestems (*Andropogon* spp.).

Pine and other conifer seeds are removed from cones by slowly drying the cones and then shaking the dried cones to remove seeds. Heat, which may be provided by a cone drying kiln, is necessary for cones of some species (e.g., Jack pine, *Pinus banksiana*) to open. Seeds are then shaken out of the cones.

After removal from the fruit and drying, seeds are cleaned to remove chaff, insect parts, soil, and other foreign debris. Fanning and screening are frequently used together for this purpose. The air screen combination is used to blow air up through the screens holding the seeds to remove light seeds and chaffy material. A series of screens with various sized holes is used. The uppermost screen has holes larger than the seeds to trap large litter and let the seeds pass through. A screen with holes smaller than the seeds allows very small seeds and other particles to drop through. Other screens with larger holes are used to catch other particles larger than the seeds and to grade the seeds for size.

Seed Storage

Most seeds should be stored in dry, cool conditions, however, seeds from some horticulturally important species rapidly lose viability when dry. Within limits and for most species, for each 10 percent decrease in seed moisture the life of the seed is doubled. Likewise, the life of the seeds doubles for each 10° drop in storage temperature. Many seeds can be stored at −18°C (0°F) for considerable lengths of time. Seeds of many species can be stored under ultracold conditions for long periods. Cryopreservation temperatures, such as that of liquid nitrogen (−196°C) have been successfully used for this purpose.

Seeds of some plants, including *Citrus*, oak, chestnut, and some walnut species should be stored moist under refrigerated conditions. Because the high moisture content encourages the growth of fungi, many propagators treat the seeds with a fungicide prior to storage. Moist chilling conditions are necessary to **stratify** seeds of many woody temperate species, which helps meet their dormancy requirements. Stratification is further discussed later in this chapter.

If stored in relatively small quantities, most seeds should be kept dry in a tightly sealed container such as a jar with a screw-on lid or a plastic container with a tight snap-on lid. Desiccants, such as silica gel treated with cobalt chloride ($CoCl_2$), can be added to the container to ensure that it remains sufficiently dry. The seeds should not touch the desiccant.

Seed Germination and Viability Tests

It is important for a propagator to know if seeds are alive (viable) and if they will germinate. Several tests can be used to determine seed viability as well as germination percentage and speed.

There are various methods to test for seed viability, including x-ray analysis, but one of the most common is the tetrazolium test. Seeds should be fully imbibed by first soaking in water in darkness. They are then cut in half longitudinally and soaked in a shallow 2,3,5-triphenyltetrazolium chloride (TTC) solution. Generally, within a half hour, living tissue will change the TTC into the insoluble red (usually appearing pink) pigment formazan. Seeds that are not viable will not become pink. The embryo, because of its high rate of respiration will turn color in viable seeds. The endosperm will remain white.

Seed laboratories and many propagators conduct germination tests. A person could sow known quantities of seeds in a greenhouse medium and measure emergence under greenhouse conditions or electric lamps. However, most germination tests are on filter paper, paper towels, other absorbent paper, or absorbent cloth. Known seed counts, often in multiples of 10 or 100, are placed on the blotter or other paper. These may be in petri dishes or other plastic containers or in rolled towels. A sufficient amount of water is added, then the seeds and container are placed in an environmentally controlled germination chamber. The relative humidity is maintained at a high level and the temperature is adjusted to be nearly optimal for germination. After a period of time, the seeds are removed from the germination chamber and counted. Propagators look for percent germination, normal and weak seedlings, hard seed, and nonviable seed.

The percent germination provides the grower or propagator with valuable information, but the speed of germination and the uniformity of germination also are important considerations. Batches of seeds that germinate rapidly should give earlier stands than those that germinate slowly. Earlier seedling emergence means that the seedlings will be exposed to light quicker and begin photosynthesizing and growing earlier. If all seeds germinate within a short time there will be a more uniform stand of seedlings than if there is a considerable delay between germination of the first and last seeds. Late-germinating seeds may be shaded by earlier germinating seeds and result in an uneven crop. Growers prefer uniformity in the time of germination of seeds from a given lot.

Seed Dormancy

Viable seeds of many species fail to germinate when given optimal conditions unless they receive appropriate pregermination treatments. These seeds are said to be dormant because their germination is inhibited by their own physiology, anatomy, or both.

In this discussion, one-seeded fruits, such as achenes, samaras, and caryopses are included with seeds, because they are handled similarly. An understanding of seed dormancy and knowledge of how it is overcome is necessary to propagate successfully many horticulturally important trees, shrubs, fruits, flowers, and some turfgrasses. Seed dormancy generally is associated with the ecology of the species. Often one can logically deduce treatments, which may overcome dormancy, by learning the characteristics of the native habitat of the species.

Many seeds may be dormant immediately at harvest, which is referred to as **primary**, or **innate, dormancy**. Seeds of other plants may germinate well if sown shortly following harvest; however, if they are exposed to adverse environmental conditions such as high temperatures, water stress, or low oxygen, they will become dormant. This is referred to as **secondary**, or **induced, dormancy** and is an important survival mechanism when environmental conditions become unfavorable for seedlings.

One type of primary dormancy is characterized by immature embryos. Although the seeds are shed by the plants, the embryo must continue to develop before germination will occur or normal seedlings will develop. Ginkgo embryos are very rudimentary, and American holly seeds contain essentially undifferentiated embryos, whereas some ash species contain complete, but very small embryos at harvest. Because they possess immature embryos, germination will not occur in freshly gathered seeds of species such as *Fraxinus excelsior* (European ash); however peach seeds and seeds of many members of the Ranunculaceae will germinate under the proper conditions, but will produce dwarfed, abnormal seedlings. The abnormal dwarf peach seedlings will grow normally if they are exposed to 0° to 5°C (32°–41°F) for 60 days.

Problems associated with immature embryos will be overcome if the seeds receive appropriate **after-ripening** treatments. After-ripening is a treatment applied to either dry or imbibed seeds (depending on the species) to stimulate further embryo development. Often high temperatures are required for after-ripening: Certain palm seeds require 38° to 40°C (100°–104°F) for 3 months, whereas many ash seeds require 20° to 30°C(68°–86°F) for 30 to 90 days.

Temperature is also used as a treatment to overcome other types of seed dormancy. Seeds of many plants require moist chilling conditions for a period of time to render them capable of germination. A moist chilling seed treatment is called **stratification**. It was so named because propagators used to place seeds between horizontal layers (strata) of moist sand in a flat or box in cold storage in order to facilitate germination. It is now known that moist chilling, not layering, is the essential treatment for overcoming dormancy of such seeds. However, the word **stratification** remains the term for this treatment. The moisture is necessary because the seeds must be imbibed to react to the cold stimulus. Chilling is generally between 0° and 10°C (32°–50°F) for 7 to 180 days, with 5°C (41°F) and 60 to 120 days being most common. For example, apple seeds require up to 60 days in a moist medium at 3° to 5°C (37°–41°F) and Japanese barberry (*Berberis thunbergii*) seeds require 90 days in a moist medium at –1° to 5°C (30°–41°F) to overcome dormancy. If left in cold moist storage for too long, seeds will often germinate, leading to weak etiolated seedlings (see Chapter 5), which may be difficult to manage.

Some seeds have two factors contributing to their dormancy and are therefore said to have **double dormancy**. Some ash species that exhibit double dormancy, have immature embryos and also require stratification—that is, the warm period must be followed by a cold period before germination can occur. Seeds of some species have such hard coverings (seed coats, fruit, or other structures) that they cannot germinate unless treated. Hard seed coverings can prevent imbibition of water, gaseous exchange, or may physically prevent the embryo from growing and emerging through the seed coat. It is thus necessary to make these seed coverings weaker or pervious to water and gases through the process of **scarification**.

There are several methods used to scarify seeds, including filing or nicking the seed by hand. With this technique, care must be

taken to avoid injury to the embryonic axis that will grow into the plant. Drums lined with an abrasive material can be used to efficiently scarify larger lots of seeds. Care must be taken not to scarify for too long because the seeds could be damaged and broken, which will lead to low germination rates.

Many people scarify seeds by soaking them in concentrated sulfuric acid for a period ranging from a few minutes to an hour or more. Concentrated sulfuric acid can be purchased from chemical supply companies; however, many commercial growers obtain it from automobile supply stores as battery acid. Concentrated sulfuric acid is extremely caustic and dangerous, so acid-proof containers must be used for scarification. Disposal also can be a problem. Because sulfuric acid can burn through clothing and skin, plenty of water must be available to dilute spills rapidly. The fumes should not be breathed because they can cause lung damage. An especially dangerous time is when the acid is rinsed from the seeds.

Scarifying seeds with sulfuric acid requires that the seeds be cautiously and slowly stirred, especially when the acid is first added. The acid causes chemical reactions that cause the seeds to stick together. This bonding can become so strong that a hammer must be used to free individual seeds. Heat is given off as a result of the oxidation reactions during scarification. Consequently, the container and its contents will become quite warm. Quantities of seeds should be limited to under 10 kg to avoid excessive heating. Nevertheless, acid scarification can be an efficient method for scarifying seeds of many hard-seeded species.

After the acid scarification treatment, the acid must be poured off and the seeds thoroughly rinsed with water. The seeds may then be dried for storage, stratified if necessary, or sown.

A hot water soak is another method to scarify seeds. Care must be taken not to cook the seeds and thus kill them. The grower should first heat water to boiling, remove it from the heat source, add the seeds, and then remove them when the water is cool, about 12 to 24 hours later.

Dry heat can be used to rupture seed coats of some species. This is effective on some species that release their seed during fires.

Light is essential for the germination of many species such as lettuce seeds (see Chapter 5). This is a phytochrome mediated response; therefore, red light is necessary for species that are **positively photoblastic** (i.e., require light for germination). If light is required, seeds can be sown shallowly; alternatively some species, including zoysia-grass (*Zoysia japonica*) can be given a red light treatment prior to sale and may be planted more deeply. Seeds must be imbibed to perceive the light treatment. Some species such as tomato and some lilies, are **negatively photoblastic** (their germination is inhibited by light) and should be sown more deeply for good germination.

Seed Sowing and Germination

The process of germination is described in Chapter 11. For germination to begin, cells must become hydrated, and for it to progress, metabolism, especially respiration, is important. When sowing seeds and providing for their germination, growers must consider several environmental factors. Seeds must be given adequate moisture, oxygen, and a proper temperature for germination to occur. These normally are provided by the germination medium or soil. An ideal germination medium must therefore have good water holding capacity, but be well drained. Temperature can be controlled by choosing the time to sow seeds in field soils or by applying bottom heat in a greenhouse, growth chamber, or hotbed.

Seeds vary considerably in size and therefore in their food reserves. Because smaller seeds have less stored food (usually carbohydrates, oils, and proteins) they must be sown at a more shallow depth than larger seeds. A general rule of thumb for sowing seeds is to cover them with soil at a depth of three times the seeds' average diameter. Except for extremely small seeds, this should be enough soil to keep the seeds moist until they germinate and the roots grow deeper into the soil. Often it is necessary to irrigate newly sown seed beds frequently to prevent the soil surface from drying. Irrigation can cause serious crusting of some soils and anticrusting agents may be necessary to enable tender seedlings to emerge. Organic mulches also can be used to keep the soil surface moist and prevent soil crusting.

Seeds of horticultural crops can be sown by hand or mechanically. Mechanical sowing of seeds can be more efficient and give a more uniform seed distribution than hand sowing. Seeds are sown mechanically or by seed spreaders for a variety of species.

When seeds germinate and seedlings emerge, they often must be given special attention compared to established plants. Frequent irrigation is necessary until root systems develop. Nutrients in the soil or medium are not necessary for germination or emergence, but they must be provided to produce vigorous, high-quality plants. With some horticultural crops, growth must be retarded when plants are seedlings because small, full, stocky plants are desirable. Therefore growers often apply chemical growth retardants or will control growth by manipulating temperature and moisture.

Various seed treatments are applied to aid mechanical sowing or to produce early, uniform stands. Seeds can be pelleted with various materials to make the shape of the propagule become round and larger and thus easier to plant using machines. Pelleting materials include fine vermiculite, clay, or a sand/sugar coating.

Embryos within seeds can be brought to an even level of maturity by soaking the seeds in an osmoticum. The solution allows the seed to take up sufficient water for the embryos to develop, but not enough water for them to germinate. The seeds can then be dried and sown at a convenient time. Such seed **priming** results in early uniform stands.

Alternatively, seeds can be germinated before they are sown. Germinated seeds with radicles one to two millimeters long must be sown in a viscous liquid to protect the tender tissues of the radicle. The process often is called **fluid drilling**. The seeds frequently are suspended in a gel made of a special hydrophilic (water attracting) polymer and sown using special planters.

Following germination, it is a common practice to transplant crowded seedlings into larger containers in order to provide greater soil volume and larger aerial space for growth. This transplanting, often referred to as "spotting out" or "pricking off," usually is done when the seedling has reached the first true leaf stage (Figure 14-29).

HARDENING-OFF

Hardening-off is necessary with many types of plants, including seedlings before transplanting from the germination flat, bedding plants before sale and transplanting into the garden, rooted cuttings prior to removal from high humidity conditions, and micro-propagated plants prior to and during the removal from in vitro conditions. During or prior to transplanting, roots are generally pruned, either purposely or accidently. Root pruning can lead to water stress because of the loss of surface area for absorbing water. In addition, temperature and light conditions are often different following transplanting, compared to the environment in which the plants were grown. Thus, transplanted plants are subjected to multiple stresses.

Hardening-off helps prepare transplants for the new stresses and thus reduces transplant shock. It is accomplished by manipulating watering, temperatures, and fertilization. Water should be gradually reduced and the plants should be slightly but not severely stressed because severe water stress will lower plant quality (see Chapter 7). For rooted cuttings, the mist, fog, or relative humidity should be gradually reduced. If appropriate for the species and the situation, temperatures should also be gradually reduced for hardening-off. For bedding plants, seeds often are germinated at 25°C (77°F). After germination, the temperature should be lowered gradually to 18°C (65°F) for a period of time and then

FIGURE 14-29. Transplanting seedlings. Holding the seedling correctly (*left*) by the cotyledon; and holding the seedling incorrectly (*right*) by the hypocotyl, which can easily crush the tender tissues.

may be further dropped to 13°C (45°F), depending on the species. Fertilization also should also be gradually reduced, but not eliminated. Nonfertilization can lead to nutrient deficiencies and poor root growth. The goal, therefore, is to keep the nutrients low enough to prevent excessive new growth but high enough to maintain quality plants.

Hardening-off is often done in cold frames or hotbeds. Cold frames are low structures covered with polyethylene or with glass, often window sashes, and with no supplemental heating other than solar heat. Hotbeds resemble cold frames, but they also include a heat source. These structures offer protection for young plants, especially at night, and they are also relatively inexpensive to construct and maintain. They are also used for overwintering plants, rooting cuttings, and growing transplants.

SUMMARY

Propagation is the beginning step for most horticultural operations. Plants are propagated by a variety of techniques, which must be mastered for each crop that is grown. Because newly propagated plants are young and tender, their care is critical. The production of high-quality plants begins with the propagator. It is easy to kill or damage newly propagated plants, so they must be given special attention. Mistakes made by the propagator may show up immediately or several years later. For example, if a propagator grafts a scion onto a rootstock and there is delayed incompatibility, the tree may not die until several years later when economic losses may be serious and replacement costs high.

Horticulturists must be aware of propagation, even if they work with established plants. Choices are available for purchasing plants of a given species that have been propagated differently (e.g., grafted plants versus rooted cuttings versus micropropagated plants versus seedlings) and which may grow quite differently. For example, landscapers must know what a graft union looks like in order to know how deeply to transplant and whether the propagation method was appropriate.

REFERENCES

Debergh, P. C., and R. H. Zimmerman (Eds.). 1991. *Micropropagation*. Kluwer Academic Publishers, The Netherlands.

Dirr, M. A., and C. W. Heuser, Jr. 1987. *The Reference Manual of Woody Plant Propagation*. Varsity Press, Athens, GA.

Hartmann, H. T., D. E. Kester, F. T. Davies, Jr., and R. L. Geneve. 2002. *Hartmann and Kester's Plant Propagation Principles and Practices* (7th ed.) Prentice Hall, Englewood Cliffs, NJ.

Kyte, L. 1987. *Plants From Test Tubes. An Introduction to Micropropagatio*n. (rev. ed.) Timber Press, Portland, OR.

Macdonald, B. 1986. *Practical Woody Plant Propagation for Nursery Growers*. Timber Press, Portland, OR.

Schopmeyer, C. S. (tech. coordinator). 1974. "Seeds of Woody Plants in the United States." *USDA Agriculture Handbook* 450. Forest Service, USDA, Washington, DC.

Stefferud, A. (Ed.). 1961. *Seeds. The Yearbook of Agriculture* 1961. USDA, Washington, DC.

Wells, J. S. 1985. *Plant Propagation Practices*. American Nurseryman Publishing, Chicago, IL.

CHAPTER 15

POSTHARVEST
HANDLING[1]

L osses in quantity and quality affect horticultural crops between harvest and consumption. The magnitude of postharvest losses in fresh fruits and vegetables is an estimated 5 to 25 percent in developed countries and 20 to 50 percent in developing countries, depending upon the commodity. To reduce these losses, producers and handlers must understand the biological and environmental factors involved in deterioration and use postharvest techniques that delay senescence and maintain the best possible quality. This chapter presents an overview of the biological and technological aspects of postharvest handling of fresh horticultural commodities.

Fresh fruits, vegetables, and ornamentals are living tissues subject to continuous change after harvest. Although some changes are desirable, most-from the consumer's standpoint—are not. Postharvest changes in fresh produce cannot be stopped, but they can be slowed within certain limits. Senescence is the final stage in the development of plant organs during which a series of irreversible events leads to breakdown and death of the plant cells.

....
[1]This chapter was written by Adel A. Kader, Professor of Postharvest Physiology, Department of Pomology, University of California, Davis, CA 95616.

379

Fresh horticultural crops are diverse in morphological structure (roots, stems, leaves, flowers, fruits, and so on), in composition, and in general physiology. Thus, commodity requirements and recommendations for maximum postharvest life vary among the commodities. All fresh horticultural crops are high in water content and thus are subject to desiccation (wilting, shriveling) and to mechanical injury. They are also susceptible to attack by bacteria and fungi, with pathological breakdown the result.

QUALITY AND MATURITY

Components of Quality

Quality is defined as "any of the features that make something what it is" or "the degree of excellence or superiority." The word quality is used in various ways in reference to fresh fruits and vegetables such as *market quality, edible quality, dessert quality, shipping quality, nutritional quality, internal quality,* and *visual quality.*

Quality of fresh horticultural commodities is a combination of characteristics, attributes, or properties that give the commodity value in terms of human food (fruits and vegetables) and enjoyment (ornamentals). Producers want their commodities to have good appearance and few visual defects. But for them a useful cultivar of a given commodity must score high on yield, disease resistance, ease of harvest, and shipping quality. To receivers and market distributors, quality of appearance is most important; they are also keenly interested in firmness and long storage life. Consumers judge fruits and vegetables as good quality if they look good, are firm, and offer good flavor and nutritive value. Although consumers buy on the basis of appearance and feel, their satisfaction and likelihood to buy that fruit or vegetable again depends on their perception of good eating quality.

The various components of quality, listed in Table 15-1, are used in commodity evaluation in relation to specifications for grades and standards, selection in breeding programs, and evaluation of responses to various environmental factors and/or postharvest treatment, including storage conditions. The relative importance of each of these quality factors depends on the commodity and its intended use (fresh or processed). Appearance factors are the most important quality attributes of ornamental crops. Fragrance may also be another factor in some flowers.

Numerous defects can influence appearance quality of horticultural crops. Morphological defects include sprouting of potatoes, onions, and garlic, rooting of onions, elongation of asparagus, curvature of asparagus and cut flowers, and seed germination inside fruits. Physical defects include shriveling and wilting of all commodities; internal drying of some fruits; and mechanical damage such as punctures, cuts and deep scratches, splits and crushing, skin abrasions and scuffing, deformation (compression), and bruising. Temperature-related disorders (freezing, chilling, sunburn, sunscald), puffiness of tomatoes, blossom-end rot of tomatoes, tipburn of lettuce, internal breakdown of stone fruits, water core of apples, and black heart of potatoes are examples of physiological defects.

Textural quality of horticultural crops is not only important for their eating and cooking quality but also for their shipping ability. Soft fruits cannot be shipped long distances without extensive losses owing to physical injuries. This has necessitated harvesting fruits at less than ideal maturity from the flavor quality standpoint in many cases.

Flavor quality involves perception of the tastes and aromas of many compounds. Objective analytical determination of critical components must be coupled with subjective evaluations by a taste panel to yield useful and meaningful information about flavor quality of fresh fruits and vegetables. This approach can be used to define a minimum level of acceptability. To find out

TABLE 15-1. Components of quality of fresh horticultural crops

Main Factors	Components

Appearance (visual)	Size: dimensions, weight, volume
	Shape and form: diameter/depth ratio, smoothness, compactness
	Color: uniformity, intensity
	Gloss: wax
	Defects: external, internal
	    Morphological (such as sprouting, rooting, and floret opening)
	    Physical and mechanical (such as shriveling and bruising)
	    Physiological (such as blossom end rot of tomatoes)
	    Pathological (caused by fungi, bacteria, or viruses)
	    Entomological (caused by insects)

Texture	Firmness, hardness, softness Crispness Succulence, juiciness Mealiness, grittiness Toughness, fibrousness
Flavor (taste and smell)	Sweetness Sourness (acidity) Astringency Bitterness Aroma (volatile compounds) Off-flavors and off-odors
Nutritive value	Carbohydrates (including dietary fiber) Proteins Lipids Vitamins Minerals
Safety	Naturally occurring toxicants Contaminants (chemical residues, heavy metals, etc.) Mycotoxins Microbial contamination

consumer preferences for flavor of a given commodity, large-scale testing by a representative sample of consumers is required.

Postharvest-life of fruits and vegetables based on flavor is generally shorter than postharvest-life based on appearance and textural quality. To increase consumer satisfaction, fruits and vegetables should be sold before the end of their flavor-life.

Fresh fruits and vegetables play a very significant role in human nutrition, especially as sources of vitamins (vitamin C, vitamin A, vitamin B_6, thiamin, niacin), minerals, and dietary fiber. They also contain many phytochemicals (such as antioxidant phenolic compounds and carotenoids) that have been associated with reduced risk of some forms of cancer, heart disease, stroke, and other chronic diseases. Postharvest losses in nutritional quality, particularly vitamin C content, can be substantial and are enhanced by physical damage, extended storage, higher temperature, low relative humidity, and chilling injury of chilling-sensitive commodities.

Safety factors include levels of naturally occurring toxicants in certain crops (such as glycoalkaloids in potatoes) that vary according to genotypes and are routinely monitored by plant breeders to ensure that they do not exceed their safe levels in new cultivars. Contaminants, such as chemical residues and heavy metals, on fresh fruits and vegetables are also monitored by various agencies to ensure compliance with established maximum tolerance levels. Sanitation procedures throughout the harvesting and postharvest handling operations are essential to minimizing microbial contamination. Proper preharvest and postharvest handling procedures must be enforced to reduce the potential for growth and development of mycotoxin-producing fungi.

Storage conditions can greatly influence the composition and quality attributes of fresh horticultural crops. In most commodities, the rate of deterioration in nutritional quality (especially vitamin C content) is faster than that in flavor quality, which is lost at a faster rate than textural quality and appearance quality. Thus, storage life based on appearance is often longer than that based on maintenance of good flavor (taste life).

Quality Standards

Grade standards are developed to identify the degrees of quality in a given commodity which aid in establishing its usability and value. Such standards are important tools in the marketing of fresh fruits and vegetables because they (1) provide a common language for trading among growers, handlers, processors, and receivers at terminal markets; (2) assist producers and handlers in preparing fresh horticultural commodities for market and labeling goods appropriately; (3) provide a basis for making incentive payment for better quality; (4) serve as the basis for market reporting (prices and supplies quoted by the Federal/State Market News service are meaningful only if they are based on products of comparable quality); and (5) help settle damage claims and disputes between buyers and sellers.

The U.S. Standards for grades of fresh fruits and vegetables are voluntary except when they are required under certain state and local regulations, by industry marketing orders (federal or state), or for export marketing. They are also used by many private and government procurement agencies in purchasing fresh fruits and vegetables. The U.S. Department of Agriculture Food Safety and Quality Service (FSQS) is responsible for developing, amending, and implementing grade standards. The first U.S. Grade Standards were developed for potatoes in 1917. Currently there are more than 150 standards for 80 different commodities.

The quality criteria used in these standards (see examples in Table 15-2) emphasize appearance factors in most commodities. Some of these factors (e.g., insect damage, growth cracks, mechanical injuries) affect the rate of deterioration during postharvest handling, but other factors (such as color uniformity and healed scars) are mostly cosmetic. In many cases, good appearance does not necessarily mean good flavor and nutritional quality. A fruit or vegetable that is misshapen or has some external blemishes may be as tasty and nutritious as one that is perfect in appearance. Thus, it is important to include quality criteria other than appearance that will more accurately reflect consumer preferences. Such quality indices must be relatively simple to evaluate, and objective methods for their evaluation should be developed to facilitate their use.

Maturity in Relation to Quality

For decades, horticulturists have directed substantial effort toward the evaluation of maturity indices. Extensive data are available on morphological, physiological, and biochemical changes in fruits and vegetables during development, maturation, and ripening. However, only a small portion of these data has been used in the establishment of maturity standards. In the U.S. standards for grades, maturity is considered

TABLE 15-2. Quality factors for selected fresh fruits and vegetables in the U.S. standards for grades

COMMODITY	QUALITY FACTORS
Apple	Maturity, color (color charts), firmness, shape, size; freedom from decay, internal browning, internal breakdown, scald, scab, bitter pit, Jonathan spot, freezing injury, water core, bruises, russeting, scars, insect damage, and other defects.
Grape	Maturity (as determined by % soluble solids), color, uniformity, firmness, berry size; freedom from shriveling, shattering, sunburn, waterberry, shot berries, dried berries, other defects, and decay. Bunches: fairly well filled but not excessively tight. Stems: not dry and brittle, at least yellowish-green in color.
Lettuce, crisphead	Turgidity, color, maturity (firmness), trimming (number of wrapper leaves); freedom from tip burn and other physiological disorders; freedom from mechanical damage, seedstems, other defects, and decay.
Muskmelon	Soluble solids (>9 percent), uniformity of size, shape, ground color and netting, maturity, turgidity; freedom from "wet slip", sunscald, and other defects.
Potato	Uniformity, maturity, firmness, cleanness, shape, size; freedom from sprouts, blackheart, greening, and other defects.
Strawberry	Maturity and ripeness stage ($>\frac{1}{2}$ or $>\frac{3}{4}$ of surface showing red or pink color depending on grade), firmness, attached calyx, size, freedom from defects and decay.
Tomato	Maturity and ripeness stage (color chart), firmness, shape, size; freedom from defects (puffiness, freezing injury, sunscald, scars, catfaces, growth cracks, insect injury, and other defects) and decay.

one parameter of quality for many fruits and vegetables. It is defined as "that stage which will ensure proper completion of the ripening process."

Table 15-3 includes a listing of maturity indices used for selected fruits and vegetables. It is necessary for some commodities to define maturity indices for specific cultivars, production areas, and seasons. Although numerous objective indices for maturity are available, only a few are actually used in practice because they are in most cases destructive and difficult to do in the field or orchard. Emphasis is placed on appearance factors; harvesting stage, for example, is determined by experience and judged largely by the visual appearance of the commodity.

Timing of harvest (based on maturity indices) is complicated by the great differences in the rate of development and maturation of individual plants or even of organs on the same plant, bush, or tree. This variability in maturation and ripening is especially important when once-over mechanical harvesting is used. Variability is related to preharvest cultural practices and environmental factors.

For many vegetables, the optimum eating quality is reached before full maturity (true for leafy vegetables and immature fruits including cucumbers, sweet corn, green beans, and peas). With these crops delayed harvest results in lower quality at harvest and faster deterioration after harvest.

TABLE 15-3. Maturity indices for selected fruits and vegetables

INDEX	EXAMPLES
Elapsed days from full bloom to harvest	Apples, pears
Mean heat units during fruit development	Peas, sweet corn
Development of abscission layer	Muskmelon
Surface morphology and structure	Cuticle formation on grapes and tomatoes Netting of muskmelons Gloss of some fruits (development of wax)
Size	All fruits and may vegetables
Specific gravity	Cherries, watermelons, potatoes
Shape	Angularity of banana fingers Full cheeks of mangoes Compactness of broccoli and cauliflower
Solidity	Lettuce, cabbage, Brussels sprouts

Textural properties
   Firmness
   Tenderness
   Toughness
	
	Apples, pears, stone fruits
	Peas
	Asparagus

Color, external	All fruits and most vegetables
Internal color and structure	Formation of jelly-like material in tomato fruits Flesh color of some fruits

Compositional factors
   Total solids
   Starch content
   Sugar content, acid content, and/or sugar/acid
     ratio
   Juice content
   Oil content
   Astringency (tannin content)
	
	Avocado, durian, kiwifruit
	Apples, pears
	Apples, pears, stone fruits, grapes, pomegranates,
	    citrus, papaya, melons
	Lemons, limes
	Avocados
	Persimmons, dates

Internal ethylene concentration	Apples

Maturity at harvest is the most important factor that determines storage life and final fruit quality. Immature fruits are more subject to shriveling and mechanical damage and are of inferior quality when ripe. Overripe fruits are likely to become soft and mealy with insipid flavor soon after harvest. Fruits picked either too early or too late in the season are more susceptible to physiological disorders and have a shorter storage life than those picked at the proper maturity.

All fruits and many fruit vegetables, with a few exceptions such as pears, avocados, and bananas, reach their best eating quality when allowed to ripen on the tree or plant. However, some fruits are picked mature but unripe so that they can withstand the postharvest handling system when shipped long distance. Most currently used maturity indices are based on a compromise between those indices that would ensure the best eating quality to consumer and those that provide the needed flexibility in marketing.

Fruits can be divided into two groups: fruits that are incapable of continuing their ripening process once removed from the plant and fruits that can be harvested mature and ripened off the plant. Berries (e.g., blackberry, raspberry, strawberry), grape, cherry, citrus (grapefruit, lemon, lime, orange, mandarin, and tangerine), pineapple, and pomegranate cannot continue to ripen once picked. Example of fruits that can ripen off plant, however, include apple, pear, quince, persimmon, apricot, nectarine, peach, plum, kiwifruit, avocado, banana, mango, papaya, and cherimoya. Fruits of the first group produce very small quantities of ethylene and do not respond to ethylene treatment except in terms of degreening (removal of chlorophyll) in citrus fruits and pineapples. Fruits capable of off-plant ripening produce much larger quantities of ethylene in association with their ripening, and exposure to ethylene treatment will result in faster and more uniform ripening.

BIOLOGICAL FACTORS INVOLVED IN DETERIORATION

Respiration

Respiration is the process by which stored organic materials (carbohydrates, proteins, fats) are broken down into simple end products with a release of energy. Oxygen (O_2) is used in this process, and carbon dioxide (CO_2) is produced (see Chapter 6). The loss of stored food reserves in the commodity during respiration hastens senescence as the reserves that provide energy to maintain the commodity's living status are exhausted; reduces food value (energy value) for the consumer; causes loss of flavor quality, especially sweetness; and causes loss of

salable dry weight (especially important for commodities destined for dehydration). The energy released as heat, known as vital heat, affects postharvest technology considerations such as estimations of refrigeration and ventilation requirement.

Respiration rate is related to deterioration rate of horticultural perishables; the higher the respiration rate, the faster the deterioration rate and shorter the postharvest-life of a given commodity. Respiration rate increases with temperature, exposure to ethylene, and physical and physiological stresses.

Ethylene Production

Ethylene, the simplest of the organic compounds affecting the physiological processes of plants, is a natural product of plant metabolism and is produced by all tissues of higher plants and by some microorganisms (see Chapter 11). As a plant hormone, ethylene regulates many aspects of growth, development, and senescence and is physiologically active in trace amounts (less than 0.1 ppm). It also plays a major role in the abscission of plant organs (see Chapter 12).

Generally, ethylene production rates increase with maturity at harvest, physical injures, disease incidence, increased temperatures up to 30°C, and water stress. On the other hand, ethylene production rates by fresh horticultural crops are reduced by storage at low temperature, and by reduced O_2 (less than 8 percent) or ethylene is competitively inhibited by elevated CO_2 (above 1 percent) levels around the commodity.

Compositional Changes

Many changes in pigments take place during development and maturation of the commodity on the plant. Some may continue after harvest and can be desirable or undesirable. Loss of chlorophyll (green color) is desirable in fruits but not in vegetables. Development of carotenoids (yellow and orange colors) is desirable in fruits such as apricots, peaches, and citrus; the desired red color development in tomatoes, watermelons, and pink grapefruit is due to a specific carotenoid (lycopene); beta-carotene is provitamin A and is important in nutritional quality. Development of anthocyanins (red and blue colors) is desirable in fruits such as apples (red cultivars), pomegranates, cherries, strawberries, cane berries, and red-flesh oranges; these water-soluble pigments are much less stable than carotenoids. Changes in anthocyanins and other phenolic compounds, however, are undesirable because they may result in tissue browning.

Changes in carbohydrates include starch-to-sugar conversion (undesirable in potatoes but desirable in apple, banana, kiwifruit, mango, and other fruits), sugar-to-starch conversion (undesirable in peas and sweet corn but desirable in potatoes), and conversion of starch and sugars to CO_2 and water through respiration (see Chapter 6). Breakdown of pectins and other polysaccharides results in softening of fruits and a consequent increase in susceptibility to mechanical injures. Increased lignin content is responsible for toughening of asparagus spears and root vegetables.

Changes in organic acids, proteins, amino acids, and lipids can influence flavor quality of the commodity. Loss in vitamin content, especially ascorbic acid (vitamin C), is detrimental to nutritional quality. Production of flavor volatiles associated with ripening of fruits is very important to their eating quality.

Growth and Development

Sprouting of potatoes, onions, garlic, and root crops greatly reduces their utilization value and accelerates deterioration. Rooting of onions and root crops is also undesirable. Asparagus spears continue to grow after harvest; elongation and curvature (if the spears are held horizontally) are accompanied by increased toughness and decreased palatability. Similar geotropic responses

occur in cut gladiolus and snapdragon flowers stored horizontally. Seed germination inside fruits such as tomatoes, peppers, and lemons is an undesirable change.

Transpiration or Water Loss

Water loss is a main cause of deterioration because it results not only in direct quantitative losses (loss of salable weight) but also in losses in appearance (wilting and shriveling), textural quality (softening, flaccidity, limpness, loss of crispness and juiciness), and nutritional quality.

The dermal system (outer protective coverings) governs the regulation of water loss by the commodity (see Chapters 3 and 7). It includes the cuticle, epidermal cells, stomata, lenticels, and trichomes (hairs). The cuticle is composed of surface waxes, cutin embedded in wax, and a layer of mixtures of cutin, wax, and carbohydrate polymers. The thickness, structure, and chemical composition of the cuticle vary greatly among commodities and among developmental stages of a given commodity.

Transpiration rate is influenced by internal or commodity factors (morphological and anatomical characteristics, surface-to-volume ratio, surface injuries, and maturity stage) and external or environmental factors (temperature, relative humidity, air movement, and atmospheric pressure). Transpiration (evaporation of water from the plant tissues) is a physical process that can be controlled by applying treatments to the commodity (e.g., waxes and other surface coatings and wrapping with plastic films) or by manipulating the environment (e.g., maintenance of high relative humidity and control of air circulation (see Chapter 7).

Physiological Breakdown

Exposure of the commodity to undesirable temperatures can result in physiological disorders. Freezing injury results when commodities are held below their freezing temperatures. The disruption caused by freezing usually results in immediate collapse of the tissues and total loss (see Chapter 6). Chilling injury occurs in some commodities (mainly those of tropical and subtropical origin) held at temperatures above their freezing point and below 5° to 15°C (41°–59°F), depending on the commodity. Chilling injury symptoms become more noticeable upon transfer to higher (nonchilling) temperatures. The most common symptoms are surface and internal discoloration (browning), pitting, water soaked areas, uneven ripening or failure to ripen, off-flavor development, and accelerated incidence of surface molds and decay (especially organisms not usually found growing on healthy tissue). Heat injury is induced by exposure to direct sunlight or to excessively high temperatures. Its symptoms include bleaching, surface burning or scalding, uneven ripening, excessive softening, and desiccation.

Certain types of physiological disorders originate from preharvest nutritional imbalances. For example, blossom-end rot of tomatoes and bitter pit of apples result from calcium deficiency. Increasing calcium content via preharvest or postharvest treatments can reduce the susceptibility to physiological disorders. Calcium content also influences the textural quality and senescence rate of fruits and vegetables; increased calcium content has been associated with improved firmness retention, reduced CO_2 and ethylene production rates, and decreased decay incidence.

Very low oxygen (less than 1 percent) and high carbon dioxide (greater than 20 percent) atmospheres can cause physiological breakdown (fermentative metabolism) of most fresh horticultural commodities. Ethylene can induce physiological disorders in certain commodities. The interactions among O_2, CO_2, and ethylene concentrations, temperature, and duration of storage influence the incidence and severity of physiological disorders related to atmospheric composition.

Physical Damage

Various types of physical damage (surface injuries, impact bruising, vibration bruising, and so on) are major contributors to deterioration. Browning of damaged tissues results from membrane disruption, which exposes phenolic compounds to the polyphenol oxidase enzyme. Mechanical injuries not only are unsightly but also accelerate water loss, provide sites for fungal infection, and stimulate CO_2 and ethylene production by the commodity.

Pathological Breakdown

One of the most common and obvious symptoms of deterioration results from the activity of bacteria and fungi (see Chapter 16). Attack by most organisms follows physical injury or physiological breakdown of the commodity. In a few cases, pathogens can infect apparently healthy tissues and become the primary cause of deterioration. In general, fruits and vegetables exhibit considerable resistance to potential pathogens during most of their postharvest life. The onset of ripening in fruits, and senescence in all commodities, renders them susceptible to infection by pathogens. Stresses such as mechanical injuries, chilling, and sunscald lower the resistance to pathogens. The relative importance of these causes of deterioration varies by commodity groups, as shown in Table 15-4.

ENVIRONMENTAL FACTORS INFLUENCING DETERIORATION

Temperature

Temperature is the environmental factor that most influences the deterioration rate of harvested commodities. For each increase of 10°C (18°F) above optimum, the rate of deterioration increases twofold to fourfold. Exposure to undesirable temperatures

TABLE 15-4. Principal causes of postharvest losses and poor quality for various groups of fruits and vegetables

GROUP	EXAMPLES	PRINCIPAL CAUSES OF POSTHARVEST LOSSES AND POOR QUALITY (IN ORDER OF IMPORTANCE)
Root vegetables	Carrots Beets Onions Garlic Potato Sweet Potato	Mechanical injuries Improper curing Sprouting and rooting Water loss (shriveling) Decay Chilling injury (subtropical and tropical root crops)
Leafy vegetables	Lettuce Chard Spinach Cabbage Green onions	Water loss (wilting) Loss of green color (yellowing) Mechanical injuries Relatively high respiration rates Decay
Flower vegetables	Artichokes Broccoli Cauliflower	Mechanical injuries Yellowing and other discolorations Abscission of florets Decay
Immature-fruit vegetables	Cucumbers Squash Eggplant Peppers Okra Snap beans	Overmaturity at harvest Water loss (shriveling) Bruising and other mechanical injuries Chilling injury Decay
Mature-fruit vegetables and fruits	Tomato Melons Citrus Bananas mangoes Apples Grapes Stone fruits	Bruising Over-ripeness and excessive softening at harvest Water loss Chilling injury (chilling sensitive fruits) Compositional changes Decay

results in many physiological disorders, as previously mentioned. Temperature also influences the effects of ethylene and controlled atmospheres (reduced oxygen and elevated carbon dioxide concentrations). Spore germination and growth rate of pathogens are greatly influenced by temperature; for instance, cooling commodities below 5°C (41°F) immediately after harvest can greatly reduce the incidence of *Rhizopus* rot.

Relative Humidity

The rate of water loss from fruits and vegetables depends on the vapor pressure deficit between the commodity and the surrounding ambient air, which is influenced by temperature and relative humidity. At a given temperature and rate of air movement, the rate of water loss from the commodity depends on the relative humidity. At a given relative humidity, water loss increases with the increase in temperature.

Atmospheric Composition

Reduction of oxygen and elevation of carbon dioxide, whether intentional (modified or controlled atmosphere storage) or unintentional (restricted ventilation within a shipping container, a transport vehicle, or both), can either delay or accelerate deterioration of fresh horticultural crops. The magnitude of these effects depends on commodity, cultivar, physiological age, O_2 and CO_2 levels, temperature, and duration of holding.

Ethylene

The effects of ethylene on harvested horticultural commodities can be desirable or undesirable, so it is a major concern to all produce handlers. Ethylene can be used to promote faster and more uniform ripening of fruits picked at the mature-green stage. On the other hand, exposure to ethylene can be detrimental to the quality of most nonfruit vegetables and ornamentals.

Light

Exposure of potatoes to light should be avoided because it results in formation of chlorophyll (greening) and solanine (toxic to humans). Light-induced greening of Belgian endive, garlic, and onion is also undesirable.

FOOD SAFETY

Over the past few years, food safety has become and continues to be the number one concern of the fresh produce industry. The U.S. Food and Drug Administration published in October 1998 a "Guide to Minimize Microbial Food Safety Hazards for Fresh Fruits and Vegetables." This guide is based on the following principles: (1) Prevention of microbial contamination of fresh produce is favored over reliance on corrective actions once contamination has occurred; (2) In order to minimize microbial food safety hazards in fresh produce, growers, packers, or shippers should use good agricultural and management practices in those areas over which they have control; (3) Fresh produce can become microbiologically contaminated at any point along the farm-to-table food chain. The major source of microbial contamination with fresh produce is associated with human or animal feces; (4) Whenever water comes in contact with produce, its quality dictates the potential for contamination. The potential of microbial contamination from water used with fresh fruits and vegetables must be minimized; (5) The use of animal manure or municipal biosolid wastes as fertilizers should be closely managed in order to minimize the potential for microbial contamination of fresh produce; and (6) Worker hygiene and sanitation practices during production, harvest, sorting, packing, and transport play a critical role in minimizing the potential for microbial contamination of fresh produce."

A training manual for trainers, entitled "Improving the safety and quality of fresh fruits and vegetables," was published by the USFDA in November, 2002 to provide uniform, broad-based scientific and practical information on the safe production, handling, storage, and transport of fresh produce. It is available electronically in English and Spanish at the following internet site: www.jifsan.umd.edu/gaps.html.

Clean, disinfested water is required in order to minimize the potential transmission of pathogens from water to produce, from infected to healthy produce within a single lot, and from one lot to another over time. Waterborne microorganisms, including postharvest plant pathogens and agents of human illness, can be rapidly acquired and taken up on plant surfaces. Natural plant surface contours, natural openings, harvest and trimming wounds, and scuffing can be a point of entry and provide safe harbor for microbes. In these protected sites, microbes are largely unaffected by common or permitted doses of postharvest water sanitizing treatments (such as chlorine compounds, ozone, peroxyacetic acid, and hydrogen peroxide). It is essential therefore, that an adequate concentration of sanitizer is maintained in water in order to kill microbes before they attach or become internalized in produce. This is important in some preharvest water uses (such as spraying pesticides or growth regulators) and in all postharvest procedures involving water, including washing, cooling, water-meditated transport (flumes), and postharvest drenching with calcium chloride or other chemicals.

HARVESTING

Harvest Methods

Fresh vegetables, fruits, and flowers are still harvested by hand (Figure 15-1). Only humans have the unique combination of eyes, brain, and hands that permits the rapid harvest of delicate and perishable crops with minimal loss and bruising. Harvesters can also be trained to select only those fruits or vegetables of the correct maturity, thus greatly reducing the amount of material that must be removed on the grading line in the packing shed. In fact, some crops can be harvested directly into shipping containers without further sizing or grading.

Although many fruits can be detached without the use of any implements, vegetables and certain fruits (for example many citrus fruits, bananas, and strawberries) are removed by applying a shear force (snapping) or by cutting with some sort of knife or shears. Hand harvesting is hard work, and many devices have been developed to ease the hand harvesting of crops. The efficiency of hand harvesting is increased by the use of chemical, genetic, cultural, and mechanical aids.

Many devices are used to assist the action of harvesting. These include ladders, picking

FIGURE 15-1. Hand harvesting is the predominant method used for fresh fruits.

baskets, knives, and specially designed shears. A plumber's helper on a pole can be used to harvest papaya and mango. Mechanical aids are often used to place pickers in the best position to carry out the harvest operation (e.g., motorized harvesting platforms and person movers).

Mechanical separation of the commodity from the plant is widely used in harvesting dried products (nuts and grains) or perishables destined for immediate processing. A wide variety of mechanical harvesters has been developed to harvest horticultural crops mechanically (see Chapter 12). Mechanical harvesters are usually sophisticated and have a very high unit cost. They may require a smaller but more skilled labor force. Savings may be realized because the harvest can be accomplished in less time.

Removal of the commodities from the plant is usually accomplished by application of shear or tensile forces by such devices as pinch rollers, combs or rakes, shakers, beaters, cutters, or augers. The inertia of the product itself may often be used as part of the separation mechanism. The amount of force required to separate the commodity from the plant can cause damage to the commodity. Fruit can be damaged by falling through the tree: Small twigs can puncture fruit, large limbs can bruise the fruit, and fruit can hit each other or the surfaces of the harvester. Crops are often damaged, or poorer grade, and more susceptible to decay when mechanically harvested. Mechanically harvested commodities often are fit only for processing.

Management of Harvesting Operations

Management of harvesting and delivery to the packinghouse operations to maintain the commodity's quality may involve some or all of the following aspects:

1. Organization, training, supervision, and motivation of the harvesting crew to achieve maximum efficiency while maintaining effective quality control (selection of proper maturity, discarding unmarketable units, and avoiding physical damage).

2. Scheduling harvesting during the cooler part of the day, protecting the harvested commodity from the sun, and expediting transport to the packinghouse to reduce losses in quality due to high temperatures.

3. Reducing physical injuries during harvesting and transport to the packinghouse by ensuring that buckets, field boxes, bins, and gondolas used are clean and have smooth surfaces; by using air-suspension- equipped transport vehicles (which reduce damage by 50 percent relative to spring suspension); by grading roads to eliminate potholes and bumps; and by restricting speed and/or reducing

tire air pressure of transport vehicle to a level that will avoid free movement of the commodity.

PREPARATION FOR MARKET

Preparation for market eliminates unwanted material and selects items of similar grade in order to improve the value of the marketed portion of the crop. If the value of the crop is not improved by packing, it is a waste of money to run the crop through a packinghouse.

The preparation of fruits and vegetables for marketing can be a very simple operation, as in the harvesting and field packing of lettuce (Figure 15-2), or it may involve many separate handling steps. Field packing has the following advantages over a packinghouse operation: less material to transport and dispose, fewer handling steps, less damage to the commodity, better quality, and less initial cost. It has its disadvantages, however. It offers less control over quality, requires the presence of a large machine in the field, and calls for a skilled labor force. Operations also depend on good weather more than packinghouse operations do, and it may be difficult to cool some commodities in containers.

FIGURE 15-2. Field harvesting and packing of head lettuce.

Packinghouse Facilities

Selecting the best location for a packinghouse depends on may factors, including the availability of a dependable, clean water supply, an electric power source, and a waste disposal system; proximity to a good road that connects with the highway system; and proximity to a community that can provide the needed work force.

The packinghouse facilities must be designed with maximum flexibility to permit handling of several commodities produced in the area and to allow for future expansion. All equipment should be well padded and with minimal drops to minimize physical damage.

A flow diagram of the packinghouse operations for fresh fruits and vegetables is shown in Figure 15-3. The primary components of a packinghouse include the following:

❖ Shaded produce assembly area, unloading dock, and scale for weighting received products.

❖ Several packing lines that can be adjusted to accommodate various commodities. The number and size of packing lines will depend on the kinds and quantities of commodities that need to be prepared for market each day. In general, about 1.5 m^2 of packinghouse floor space will be needed for each ton of product to be handled per day.

❖ Area for storage of shipping containers and other packaging materials and for box making, and an automated system for delivery of these boxes to the packaging area.

❖ Forced-air cooling facility adjacent to a cold storage facility. About 0.7 m^2 of cold storage space will be needed per ton to be handled per day. Two cooling and storage facilities may be needed to accommodate chilling-sensitive and chilling-insensitive commodities (optimum temperatures of about 10°C and 0°C, respectively).

❖ A shaded loading dock connected to the cooling and cold storage facilities for loading the commodities into refrigerated transport vehicles.

❖ A ripening facility if mature-green tomatoes and other commodities that need ethylene treatment will be handled.

❖ An enclosed area for partial or complete processing of commodities that are not usable for the fresh market.

❖ A cull accumulation area.

❖ Offices for management personnel.

❖ Laboratory and office space for quality and safety assurance personnel.

❖ Designated area for workers including assigned lockers, tables and chairs, restrooms, and vending machines for snacks and drinks.

A stand-by power generator may be needed, if frequent power outages are expected, to maintain operation of the packinghouse and cooling/storage facilities.

Packinghouse Operations

Dumping may be accomplished either in water (Figure 15-4) or dry. The dump tank should be designed for rapid emptying and filling and for easy cleaning. Water must be chlorinated (100-150 ppm chlorine level maintained) and should be changed frequently to ensure cleanliness. A system for heating the water when needed should be included. Dry dump systems should be adequately padded to reduce physical injury. A padded lid can be used for covering bins and slowly inverting the bin and delivering the commodity through a controlled opening in the lid.

Cleaning vegetables and fruits may be done by dry brushing. In most cases, however, washing with water containing a detergent and/or 100 to 150 ppm chlorine will be required. A final rinse with clean water usually follows washing. Removal of excess moisture from the surface of washed commodities is essential and can be accomplished by sponge rollers plus air draft.

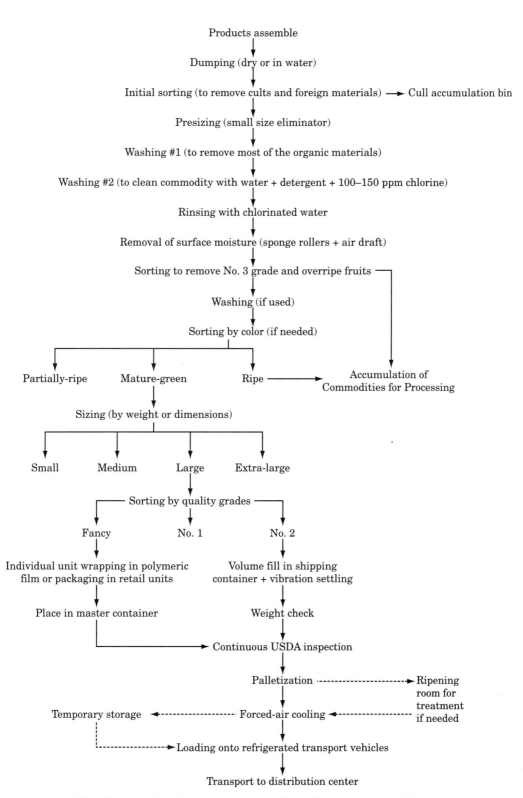

Products assemble

Dumping (dry or in water)

Initial sorting (to remove cults and foreign materials) ⟶ Cull accumulation bin

Presizing (small size eliminator)

Washing #1 (to remove most of the organic materials)

Washing #2 (to clean commodity with water + detergent + 100–150 ppm chlorine)

Rinsing with chlorinated water

Removal of surface moisture (sponge rollers + air draft)

Sorting to remove No. 3 grade and overripe fruits

Washing (if used)

Sorting by color (if needed)

Partially-ripe   Mature-green   Ripe ⟶ Accumulation of Commodities for Processing

Sizing (by weight or dimensions)

Small   Medium   Large   Extra-large

Sorting by quality grades

Fancy   No. 1   No. 2

Individual unit wrapping in polymeric film or packaging in retail units   Volume fill in shipping container + vibration settling

Place in master container   Weight check

Continuous USDA inspection

Palletization ⤏ Ripening room for treatment if needed

Temporary storage ⟵ Forced-air cooling ⟵

Loading onto refrigerated transport vehicles

Transport to distribution center

FIGURE 15-3. Flow diagram of packinghouse operations for fresh fruits and vegetables.

FIGURE 15-4. Unloading potatoes into washing tank using water pressure.

FIGURE 15-5. Waxing lemons.

Sorting for removal of defects and for quality grading is done manually. Effective sorting requires adequate belt space, ability to adjust product flow, assignment of responsibility among workers, adequate lighting, and worker training and supervision. The sizer must also be of adequate capacity in relation to the product volume. Other important factors include accuracy, minimum physical injuries, ease of adjustment to accommodate different commodities, ability to change commodity diversion pattern as peak sizes change, and ease of equipment cleaning and maintenance. Weight or machine vision sizers are more versatile and should be used for a multipurpose packing line.

Waxing of certain fruits and fruit-vegetables (Figure 15-5) is used mostly for enhancing appearance and reducing water loss by about 30 to 40 percent, especially if the commodity is exposed to less-than-optimal temperature and relative humidity conditions. An increasing number of consumers however, prefer unwaxed fruits and vegetables. A more effective (and possibly more expensive) alternative to waxing is wrapping individual or multiple units of the commodity with thin polymeric films that restrict water vapor movement without significantly altering diffusion of O_2, CO_2, and ethylene. Sorting and packaging apples in plastic bags for retail distribution (Figures 15-6 and 15-7) can also reduce water loss.

FIGURE 15-6. Sorting apples to eliminate unmarketable fruits.

FIGURE 15-7. Packaging apples in plastic bags for retail distribution.

The packing method should immobilize the commodity within the shipping container as much as possible to reduce vibration bruising. It should avoid overfilling, which causes compression bruising, and should be cushioned against impacts to reduce impact bruising. The shipping containers used should be selected for strength and stacking ability and should be adequately ventilated (at least 5 percent of the surface) for both vertical and horizontal air flow during cooling, storage, and transport. Shipping containers should have attractive labeling that includes the commodity's optimal holding temperature (color coded in accordance with the color coding proposed for the storage rooms at the distribution center). A code stamped on each box indicating production area, grower, cultivar, and shipping date would help trace possible causes of problems.

A continuous quality grade inspection by a USDA inspector is important: It will eliminate the need for subsequent inspection and avoid related delays. Use of an effective maintenance and sanitation program for all facilities and equipment is essential.

Cooling and Cold Storage Facilities

Forced-air cooling (Figure 15-8) is the most versatile cooling method provided that the relative humidity of the air is kept above 95 percent. Hydrocooling is a faster cooling method, but not all commodities tolerate wetting. Waxed cartons must be used if the commodity is hydrocooled after packing.

Cold storage rooms (Figure 15-9) should be adjacent to the cooling facility in order to facilitate transfer of the cooled com-

FIGURE 15-9. A cold storage facility for fresh produce.

modity into the storage area and to prevent rewarming of the commodity before it is loaded into the transport vehicle.

Ripening Facility

A ripening facility may be needed for initiating ripening of some commodities such as mature-green tomatoes, bananas, avocados, mangos, and some muskmelons. Such a facility may include several rooms equipped with systems for temperature and relative humidity control as well as for for ethylene introduction and uniform distribution within the room. The optimum temperature range for ripening is 15° to 25°C; the higher the temperature, the faster the ripening. The optimum relative humidity range is 90 to 95 percent.

Ethylene at 100 ppm is more than adequate and can be supplied in a continuous flow system using a compressed cylinder of ethylene diluted with an inert gas such as nitrogen. An alternative is to use ethylene generators where ethylene is produced by heating ethanol plus a catalyst. To ensure uniform distribution of ethylene within the room, the air containing ethylene can be forced through the pallets of the commodity. Carbon dioxide accumulation should be avoided (via introduction of fresh air) or corrected (by use of CO_2 absorbers such as hydrated lime or a molecular sieve) if needed.

FIGURE 15-8. Forced-air cooling of cut flowers

Transporation

The comparative advantages and disadvantages of truck versus rail transport (piggyback or refrigerated rail cars) should be examined in relation to cost, speed, reduction in handling steps, and extent of physical injuries due to transport. Many improvements in insulation, air flow systems, and thermostats have been recently introduced in some of the refrigerated transport vehicles. Whenever possible, such improved transport vehicles should be used because they provide better temperature maintenance. Also, trucks with air suspension systems significantly reduce vibration bruising of the commodity during transport.

As part of the quality control efforts, all transport vehicles should be inspected before loading for the following:

❖ Good condition of walls, doors, floor drains, and air delivery chute

❖ Cleanliness

❖ Proper operation of the refrigeration unit

Transport vehicles should be cooled to the desired temperature before loading the commodity. The pallets should be center-loaded, leaving air channels between the load and walls of the transport vehicle. The load should be secured to prevent shifting of the load during transport. A system for monitoring temperature during transport should be followed using recording thermometers or time-temperature indicators. Transportation during the night can provide better temperature management.

Quality Control

An effective quality control system throughout the handling steps between harvest and retail display is essential to provide a consistently good-quality supply of fresh fruits and vegetables to the consumers and to protect the reputation of a given marketing label. Quality control starts in the field with the selection of the proper time to harvest for maximum quality. Careful harvesting is essential to maintain quality. Each subsequent step after harvest has the potential to either maintain or reduce quality. Few postharvest procedures can increase the quality of individual units of the commodity. Quality control procedures include the following steps:

OPERATION	PROCEDURES
Harvesting	Check proper maturity and quality.
Preparation for market	Monitor effectiveness of the various steps (washing, sorting, waxing, sizing, fungicide treatment, and so on); check culls to determine causes of cullage and sorting accuracy; check shipping containers and other packing materials against specifications; check packed containers for compliance with grade, size, and weight regulations.
Cooling	Monitor product temperatures at key points in the handling system, especially before and after cooling.
Transportation	Check transit vehicles for cleanliness and cooling before loading, loading pattern, load immobilization, thermostat setting, and placement of recording thermometer.
Destination Market	Check quality and condition of the product and shipping container.

Quality control personnel should devote full time and attention to their function and should have the authority to make needed changes in the harvesting and handling operations when required to obtain and maintain the desired quality. They should also be involved in training workers within their organizations on the importance of quality attributes of each commodity and on procedures for postharvest quality maintenance.

Many attempts are currently being made to automate the separation of a given commodity into various grades and the elimination of defective units. The availability of low-cost microcomputers and solid-state imaging systems have made computer-aided video inspection on the packing line a practical reality. Solid-state video camera or light reflectance systems can be used for detection of external defects, and x-ray or light transmittance systems can be used for detecting internal defects. Further development of these systems to provide greater reliability and efficiency will be very helpful in quality control efforts.

STORAGE

Cold storage facilities should be well engineered and adequately equipped. They should have good construction and insulation including a complete vapor barrier on the warm side (usually outside) of the insulation; strong floors; adequate and well-positioned doors for loading and unloading; effective distribution of refrigerated air; sensitive and properly located controls; enough refrigerated coil surface to minimize the difference between the coil and air temperatures; and adequate capacity for expected needs. Commodities should be stacked in the cold room, leaving air spaces between pallets and room walls as well as among pallets to ensure good air circulation. Storage rooms should not be overloaded beyond limit for proper cooling. In monitoring temperatures, commodity temperature rather than air temperature should be used.

Under certain climatic conditions, ambient outside air may be used to maintain a desirable temperature range in ventilated storage facilities, which are less expensive to operate than mechanically refrigerated cold storage facilities. Recent advances in automated controls of ventilation fans in response to sensing of ambient temperatures have helped improve temperature maintenance in ventilated storage facilities, which are less expensive to operate than mechanically refrigerated cold storage facilities. Mechanical refrigeration may be used as backup system in such facilities to achieve a more consistent temperature control.

Storage Temperature and Relative Humidity

Temperature is the most important environmental factor influencing the deterioration of harvested commodities. Most perishable horticultural commodities last longer at temperatures near 0°C. At temperatures above the optimum, the rate of deterioration increases twofold to fourfold for every 10°C rise in the temperature. Temperature also influences how other internal and external factors influence the commodity and has a dramatic effect on the germination and growth of pathogens. Temperatures outside the physiological norm can cause rapid deterioration due to the following disorders:

1. **Freezing injury** (See Chapter 6). In general, perishable commodities have high water content and large, highly vacuolated cells. The freezing point of the tissue is high, and the disruption caused by freezing usually results in immediate collapse of the tissues and total loss. Freezing is normally the result of inadequate refrigerator design or poor setting or failure of thermostats. In winter conditions, freezing can occur if produce is allowed to remain for even short periods of time on unprotected transportation docks.

2. **Heat injury** High temperatures are also very injurious to perishable products. In growing plants, transpiration maintains temperatures in the optimal range (see Chapter 7). Organs removed from the plant lack the protective effects of transpiration, and direct sources of heat—for example, full sunlight—can rapidly heat tissues to above the thermal death point of their cells, leading to localized bleaching, necrosis (sunburn or sunscald), or general collapse.
3. **Chilling injury** Some commodities (chiefly those native to the tropics and subtropics) respond unfavorably to storage at low temperatures, even temperatures well above the freezing point but below a critical temperature termed the chilling threshold temperature. Chilling injury is manifested in a variety of symptoms including surface and internal discoloration, pitting, water soaking, failure to ripen, uneven ripening, development of

off flavors, and heightened susceptibility to pathogen attack.

The rate of water loss from fruits and vegetables depends on the vapor pressure deficit (VPD) between the interior of the commodity and the air around it (see Chapter 7). This value is determined by the temperature and the relative humidity of the external air. The optimum relative humidity during storage of fresh leafy and other nonfruit vegetables ranges between 95 and 98 percent, but it is best kept at 90 to 95 percent for fruit vegetables and fruits. Lower relative humidities are recommended for dry onions (65–70 percent), for pumpkins and winter squash (50–70 percent), and for tree nuts and dried fruits and vegetables (60–70 percent).

Fresh horticultural crops vary greatly in their storage potential, which is related to their degree of perishability (Table 15-5). Genotypic differences have been found in storage potential of many commodities.

TABLE 15-5. Classification of fresh horticultural crops according to their relative perishability and potential storage life in air within their optimal at near optimum temperature and relative humidity ranges

Relative Perishability	Potential Storage Life (weeks)	Commodities
Very high	<2	Apricot, blackberry, blueberry, cherry, fig, raspberry, strawberry; asparagus, bean sprouts, broccoli, cauliflower, green onion, leaf lettuce, mushroom, pea, spinach, sweet corn, tomato (ripe); most cut flowers and foliage; minimally processed (fresh-cut) fruits and vegetables
High	2–4	Avocado, banana, grape (without SO_2-treatment), guava, loquat, mandarin, mango, melons (honeydew, Crenshaw, Persian), nectarine, papaya, peach, plum; artichoke, green beans, Brussels sprouts, cabbage, celery, eggplant, head lettuce, okra, pepper, summer squash, tomato (partially ripe)
Moderate	4–8	Apple and pear (some cultivars), grape (SO_2 treated), orange, grapefruit, lime, kiwifruit, persimmon, pomegranate; table beet, carrot, radish, potato (immature)
Low	8–16	Apple and pear (some cultivars), lemon; potato (mature), dry onion, garlic, pumpkin, winter squash, sweet potato, taro, yam; bulbs and other propagules of ornamental plants
Very low	>16	Tree nuts, dried fruits and vegetables

These differences may be related to variation among cultivars in morphological structure, turgidity, composition (such as content of nitrogen, calcium, phenolic compounds, polyphenol oxidase activity, and organic acids), and softening rate. The relative susceptibility to pathogens can also be an important factor in determining storage life.

Mixing two or more commodities in the same storage facility and/or transport vehicle depends on their compatibility in terms of optimal temperature, ethylene production rate, and susceptibility to ethylene action. A classification of fruits and vegetables according to their sensitivity to chilling injury and ethylene production rates at optimal handling temperatures is show in Table 15-6.

The effects of light conditions in storage facilities on produce quality should also be considered. Radiant heat from high-intensity light sources can raise commodity temperatures. Light induces undesirable greening (chlorophyll synthesis) in some commodities such as potato, onion, and Belgian endive. In the case of potatoes, light may also induce synthesis of solanine, which to toxic to humans.

Ethylene Exclusion or Removal

Undesired accelerated softening and ripening of fruits during transport and storage result in shorter postharvest life and faster deterioration. For example, the presence of ethylene in cold storage facilities of apples, kiwifruits, and avocados kept in air or in controlled atmospheres can significantly hasten softening and reduce storage life of these fruits.

TABLE 15-6. Classification of fruits and vegetables according to their sensitivity to chilling injury and ethylene production rates* at optimum handling temperatures

RELATIVE ETHYLENE PRODUCTION RATE (µL/KG · HR) AT 20°C	NONCHILLING SENSITIVE	CHILLING SENSITIVE
Very low (<0.1)	Artichoke, asparagus, beets, cabbage, carrot, cauliflower, celery, cherry, garlic, grape, leeks, lettuce, onion, parsley, parsnip, peas, radish, spinach, strawberry, sweet corn, turnip	Ginger, grapefruit, lemon, lime, melons (casaba, Juan canary), orange, pomegranate, potato, snap beans, sweet potato, tangerine, taro (dasheen)
Low (0.1–1.0)	Blackberry, blueberry, broccoli, Brussels sprouts, endive, escarole, green onion, kiwifruit (unripe), mushrooms, persimmon (Hachiya), raspberry, tamarillo	Cranberry, cucumber, eggplant, okra, olive, peppers (sweet and chili), persimmon (Fuyu), pineapple, pumpkins, summer squash, watermelon
Moderate (1.0–10)	Figs	Banana, guava, lychee, mango, melons (cantaloupe, crenshaw, honey dew, Persian), plantain, tomato
High (10–100)	Apple, apricot, kiwifruit (ripe), nectarine, peach, pear, plum	Avocado, feijoa, papaya
Very High (>100)		Cherimoya, mammee apple, passion fruit, sapote

*Ethylene production rate by fruits and fruit-vegetables is greatest as they approach the eating-ripe stage.

There are many examples of the detrimental effects of ethylene on vegetables and ornamentals. The incidence and severity of ethylene damage depend on its concentration, duration of exposure, and storage temperature. Following are a few examples of ethylene injury: russet spotting of lettuce; increased toughness of asparagus; development of bitter flavor in carrots; leaf and flower abscission and yellowing of broccoli, cabbage, cauliflower, and ornamental plants; failure to open some cut flowers (such as carnations); and calyx abscission in eggplant. Ethylene damage can be greatly reduced by holding the commodity at its lowest safe temperature and by keeping it under modified or controlled atmospheres (reduced oxygen and/or elevated carbon dioxide). Under such conditions, both ethylene production by the commodity and ethylene action on the commodity are significantly reduced.

Ethylene may be excluded from storage rooms and transport vehicles by using electric fork-lifts and by avoiding mixing ethylene-producing commodities (those listed in classes moderate to very high in Table 15-6) with those sensitive to ethylene. Ethylene may be removed from storage rooms and transport vehicles by using adequate air exchange (ventilation) (e.g., one air change per hour provided that outside air is not polluted with ethylene) and using ethylene absorbers such as potassium permanganate. The air within the room or transit vehicle must be circulated past the absorber for effective ethylene removal. It is also very important to replace the used absorbing material with a fresh supply as needed. Catalytic combustion of ethylene on a catalyst at high temperatures (greater than 200°C) and use of ultraviolet radiation can also be used to remove ethylene.

Treatments to Reduce Ethylene Damage

Treating ornamental crops with 1-methylcyclopropene (1-MCP), which is an ethylene action inhibitor, provides protection against ethylene damage and has been used commercially since 1999. In July 2002, 1-MCP at concentrations up to 1 ppm was approved by the U.S. Environmental Protection Agency for use on apples, apricots, avocados, kiwifruit, mangoes, nectarines, papayas, peaches, pears, persimmons, plums, and tomatoes. The first commercial application is its use on apples to retard their softening and scald development and to extend their postharvest-life. As more research is completed, the use of 1-MCP will no doubt be extended to several other fruits and vegetables.

Controlled and Modified Atmospheres

The terms controlled atmosphere (CA) and modified atmosphere (MA) refer to atmospheres in which the composition surrounding the commodity is different from normal air, which contains 78.08 percent N_2, 20.95 percent O_2, and 0.03 percent CO_2. Usually this involves reduction of O_2 levels, elevation of CO_2 levels, or both. Low-pressure (hypobaric) storage is one method to establish a CA atmosphere in which the commodity is held under partial vacuum, which results in reduced O_2 levels and increased diffusivity of ethylene and other gases. MA differs from CA only in the degree of precision in controlling partial pressures of O_2 and CO_2; CA is more exact than MA.

During the past 50 years, uses of CAs and MAs to supplement temperature management have increased steadily and have contributed significantly to extending the postharvest life and maintaining quality of several fruits and vegetables. This trend is expected to continue as technological advances are made in attaining and maintaining CA and MA during transport, storage, and marketing of fresh produce. Several refinements in CA storage have been made in recent years to improve quality maintenance; these include low O_2 (1.0–1.5

percent) storage, low ethylene CA storage, rapid CA (rapid establishment of the optimum levels of O_2 and CO_2), and programmed (or sequential) CA storage (e.g., storage in 1 percent O_2 for two to six weeks followed by storage in two to three percent O_2 for the remainder of the storage period). Other developments, which may expand use of MA during transport and distribution, include using edible coatings or polymeric films to create a desired MA within the commodity.

Fresh fruits and vegetables vary greatly in their relative tolerance to low O_2 concentration and elevated CO_2 concentrations. These are the levels below which (for O_2) or above which (for CO_2) physiological damage would be expected. These limits of tolerance can be different at temperatures above or below recommended temperatures for each commodity. Also, a given commodity may tolerate higher levels of CO_2 or lower levels of O_2 than those indicated if the durations is short. The limit of tolerance to low O_2 would be shorter as storage temperature and/or duration increases, because O_2 requirement for aerobic respiration of the tissue increases with higher temperatures. Depending on the commodity, damage associated with CO_2 may either increase or decrease with an increase in temperature. CO_2 production in the tissue increases with temperature, but its solubility decreases. Further, the physiological effect of CO_2 could be temperature dependent. Tolerance limits to elevated CO_2 decrease with a reduction in O_2 level, and similarly, the tolerance limits to reduced O_2 increase with the increase in CO_2 level.

The primary objectives of CA storage of fresh produce are extension of postharvest life; maintenance of appearance, textural, flavor, and nutritional quality; and control of postharvest pathogens and insects. The extent to which CA is beneficial depends on the commodity; cultivar; maturity stage; initial quality; concentrations of O_2, CO_2, and ethylene; temperature; and duration of exposure to these conditions.

The extent of commercial use of CA and MA is still limited. Current CA use for long-term storage of fresh fruits and vegetables is summarized in Table 15-7.

TABLE 15-7. A summary of controlled atmosphere (CA) use for long-term storage of fresh fruits and vegetables

RANGE OF STORAGE DURATION (MONTHS)	COMMODITIES
> 12	Almond, brazil nut, cashew, filbert (hazelnut), macadamia, pecan, pistachio, walnut, dried fruits and vegetables
6–12	Some cultivars of apples and European pears
3–6	Cabbage, Chinese cabbage, kiwifruit, persimmon, pomegranate, some cultivars of Asian pears
1–3	Avocado, banana, cherry, grape (no SO_2), mango, olive, onion (sweet cultivars), some cultivars of nectarine, peach and plum, tomato (mature-green)
< 1	Asparagus, broccoli, cane berries, fig, lettuce, muskmelons, papaya, pineapple, strawberry, sweet corn; fresh-cut fruits and vegetables; some cut flowers

The use on nuts and dried fruits and vegetables (for insect control and quality maintenance including prevention of rancidity) is increasing and will likely continue to increase because it provides an excellent substitute for chemical fumigants (such as methyl bromide) used for insect control. Also, the use of CA on commodities listed in Table 15-7 other than apples and pears is expected to increase as international market demands for year-round availability of various commodities expand.

CA and MA use of short-term storage or transport of fresh horticultural crops (Table 15-8) has increased during the past few years and will continue to increase, supported by technological developments in transport containers, MA packaging, and edible coatings. Figure 15-10 shows an MA system used for strawberries, cane berries, cherries, and other commodities.

CA and MA conditions, including MA packaging (MAP), can replace certain postharvest chemicals used for control of some physiological disorders such as scald on apples. Furthermore, use of some postharvest fungicides and insecticides can be reduced or eliminated in cases where CA or MA conditions provide adequate control of postharvest pathogens or insects.

Use of CA or MA may facilitate the picking and marketing of more mature (better-flavored) fruits by slowing down their postharvest deterioration rate to permit transport and distribution. Another potential use for CA and MA is in maintaining quality and safety of minimally processed (fresh-cut) fruits and vegetables, which are increasingly being marketed as value-added, convenience product.

The residual effects of CA and MA on fresh commodities after transfer to air (during poststorage marketing operations)

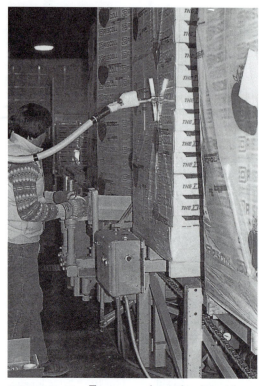

FIGURE 15-10. Treatment of strawberries with 15 percent CO_2-enriched atmosphere before transport to market.

may include the reduction of respiration and ethylene production rates, maintenance of color and firmness, and delayed decay incidence. Generally, as the concentration

TABLE 15-8. A summary of CA and MA use for short-term storage or transport of fresh horticultural crops

PRIMARY BENEFIT OF CA/MA	COMMODITIES
Delay of ripening and avoiding chilling temperatures	Avocado, banana, mango, melons, nectarine, papaya, peach, plum, tomato (picked mature-green or partially ripe)
Control of decay	Blackberry, blueberry, cherry, fig, grape, raspberry, strawberry
Delay of senescence and undesirable compositional changes (including tissue brown discoloration)	Asparagus, broccoli, lettuce, sweet corn, fresh herbs, minimally processed (fresh-cut) fruits and vegetables

of O_2 is decreased, the concentration of CO_2 is increased, and the duration of exposure to CA or MA conditions is increased, the residual effects become more prominent.

The use of polymeric films for packaging produce and their application in modified atmosphere packaging (MAP) systems at the pallet, shipping container (plastic liner), and consumer package levels continue to increase. MAP usually maintains 2 to 4 percent O_2 and 8 to 12 percent CO_2 and is widely used in extending the shelf-life of fresh-cut vegetable and fruit products. Use of absorbers of ethylene, carbon dioxide, oxygen, and/or water vapor as part of MAP is increasing. Although much research has been done on use of surface coatings to modify the internal atmosphere within the commodity, commercial applications are still very limited due to the inherent biological variability of the commodity.

··

KEYS TO SUCCESSFUL HANDLING OF HORTICULTURAL CROPS

Maturity and Quality

To ensure maturity of fruits and vegetables and to maintain high-quality products, harvest at the proper maturity stage relative to intended use and marketing practices and periods. Eliminate produce with serious defects, and inspect produce quality and condition when it is received. Separate out produce that must be sold immediately, and place it on display first. Trim produce before display, when needed, and date incoming produce containers to ensure rotation on a first-in, first-out basis. Rotate produce when replenishing displays.

Temperature Management Procedures

Temperature of produce is one of the most important factors; therefore, harvest during the coolest part of the day possible, and keep produce in the shade while accumulating fruits or vegetables in the orchard or field. Transport produce to the direct-marketing outlet as soon as possible after harvest. Protect produce on display from exposure to direct sunlight. Ship packed produce to the market in refrigerated transit vehicles unless shipping distances and durations are short (few hours), and maintain proper temperature and relative humidity in display cases and cold storage rooms. Avoid letting produce stand near a hot radiator, on a wet floor, or on a receiving platform in extreme cold, heat, or wind. Figure 15-11 includes a more complete list of actions needed to maintain the cold chain throughout the postharvest handling system for perishable horticultural crops.

Mechanical Injuries

Handle produce with care during harvesting and hauling to the market or produce stand. Use suitable materials-handling equipment (hand tracks, clamp-trucks, pallet jack, and so on). Avoid drops, impacts, vibrations, and surface injuries of produce throughout the handling system. Do not drop containers of produce or set them down roughly. Use shipping containers that will provide adequate protection for the commodity from physical injuries. Do not stack containers of produce very high, because the contents in the lower containers may bruise. Stack containers so that the pressure comes on the structure of the package, not on the produce.

Sanitation Procedures

Wash produce, then remove excess moisture if needed. Sort out and properly discard decaying produce. Clean harvest containers daily, and clean reusable shipping containers, display and storage facilities, and so on periodically with water, soap, and disinfectants.

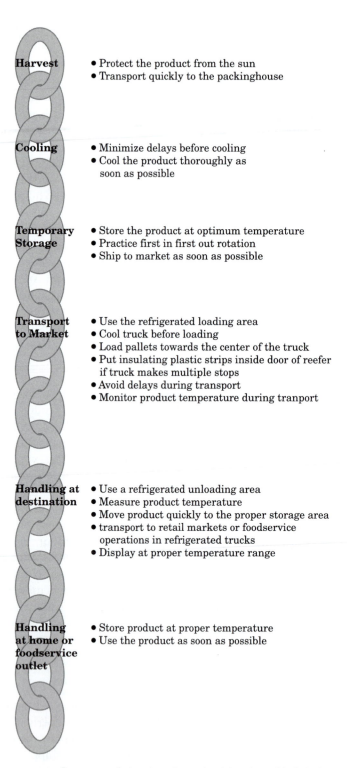

Harvest
- Protect the product from the sun
- Transport quickly to the packinghouse

Cooling
- Minimize delays before cooling
- Cool the product thoroughly as soon as possible

Temporary Storage
- Store the product at optimum temperature
- Practice first in first out rotation
- Ship to market as soon as possible

Transport to Market
- Use the refrigerated loading area
- Cool truck before loading
- Load pallets towards the center of the truck
- Put insulating plastic strips inside door of reefer if truck makes multiple stops
- Avoid delays during transport
- Monitor product temperature during tranport

Handling at destination
- Use a refrigerated unloading area
- Measure product temperature
- Move product quickly to the proper storage area
- transport to retail markets or foodservice operations in refrigerated trucks
- Display at proper temperature range

Handling at home or foodservice outlet
- Store product at proper temperature
- Use the product as soon as possible

FIGURE 15-11. Recommended actions for maintaining the cold chain for perishables.

....

REFERENCES

Abeles, F. B., P. W. Morgan, and M. E. Saltveit. 1992. *Ethylene in Plant Biology*. 2nd ed. San Diego: Academic Press, 414pp.

Bartsch, J. A., and G. D. Blanpied. 1984. *Refrigeration and Controlled Atmosphere Storage for Horticultural Crops*. Northeast Regional Agricultural Engineering Service Publication 22. 42pp.

Bartz, J. A., and J. K. Brecht (Eds.). 2002. *Postharvest Physiology and Pathology of Vegetables*. 2nd ed. New York: Marcel Dekker, 744 pp.

Beaudry, R. M. 2000. "Responses of horticultural commodities to low oxygen: limits to the expanded use of modified atmosphere packaging." *HortTechnology* 10:491–500.

Blankenship, S. M., and J. M. Dole. 2003. "1-Methylcycolopropene: A review." *Postharvest Biology and Technology*. 28:1–25.

Calderon, M. and R. Barkai-Golan, (Eds.). 1990. *Food Preservation by Modified Atmospheres*. Boca Raton, FL: CRC Press, 402pp.

Couey, H. M. 1989. "Heat treatment for control of postharvest diseases and insect pests of fruits." *HortScience* 24:198–202.

Gorny, J. R., Ed. 2001. *Food Safety Guidelines for the Fresh-Cut Produce Industry*. 4th ed. Alexandria, VA: International Fresh-cut Produce Association, 218pp.

Gross, K., C. Y. Wang, and M. E. Saltveit (Eds.). 2002. "The commercial storage of fruit, vegetables, and florist and nursery stocks." USDA Agricultural Handbook. 66 (http://www.ba.ars.usda.gov/hb66/index.html).

Harvey, J. M. 1978. "Reduction of losses in fresh market fruits and vegetables." *Annual Review of Phytopathology* 16:321-341.

Hyson, D. 2002. *The Health Benefits of Fruits and Vegetables. A Scientific Overview for Health Professionals*. Produce for Better Health Foundation, Wilmington, DE, 20p.

International Institute of Refrigeration. 2000. *Recommendations for Chilled Storage of Perishable Produce*. Paris, France: IIR, 219pp.

Kader, A. A. 1983. "Postharvest quality maintenance of fruits and vegetables in developing countries." In Lieberman, M. (Ed). *Postharvest Physiology and Crop Preservation*. New York: Plenum, 455-470.

Kader, A. A. 1985. "Ethylene-induced senescence and physiological disorders in harvested horticultural crops." *HortScience* 20:54–57.

Kader, A. A. 1986. "Biochemical and physiological basis for effects of controlled and modified atmospheres on fruits and vegetables." *Food Technology* 40(5): 99–100, 102–104.

Kader, A. A. 1986. "Potential applications of ionizing radiation in postharvest handling of fresh fruits and vegetables." *Food Technology* 40(6):117–121.

Kader, A. A., (Ed.). 2001. CA bibliography (1981–200) and CA Recommendations (2001), CD. Davis: University of California, Postharvest Technology Center, Postharvest Horticulture Series No. 22 (The CA Recommendations, 2001 portion is also available in printed format as Postharvest Horticulture Series No. 22A).

Kader, A. A., (Ed.). 2002a. *Postharvest Technology of Horticultural Crops*. 3rd ed. Oakland: University of California, Division of Agriculture and Natural Resources Publication 3311, 535pp.

Kader, A. A. 2002b. "Opportunities in using biotechnology to maintain postharvest quality and safety of fresh produce." *HortScience* 37:467–468.

Kader, A. A., D. Zagory, and E. L. Kerbel. 1989. "Modified atmosphere packaging of fruits and vegetables." *Critical Review of Food Science and Nutrition*. 28:1–30.

Kay, S. J. 1991. *Postharvest Physiology of Perishable Plant Products*. New York: Van Nostrand Reinhold, 532p.

Kitinoja, L., and J. R. Gorny. 1999. *Postharvest Technology for Small-Scale Produce Marketers: Economic Opportunities, Quality and Food Safety*. Davis: University of California, Postharvest Horticulutre Series 21.

Kitinoja, L., and A. A. Kader. 2002. *Small-Scale Postharvest Handling Practices: A Manual for Horticultural Crops*. 4th ed. Davis: University of California, Postharvest Horticulture Series 8E, 260pp.

Knee, M., Ed. 2002. *Fruit Quality and Its Biological Basis*. Sheffield: Sheffield Academic Press, 320 pp.

Lamikanra, O., Ed. 2002. *Fresh-Cut Fruits and Vegetables: Science, Technology, and Market*. Boca Raton, FL: CRC Press, 456 pp.

Lurie, S. 1998. "Postharvest heat treatments of horticultural crops." *Horticultural Reviews*. 22:91–121.

Mackay, S. 1979. *Home Storage of Fruits and Vegetables*. Northeast Regional Agricultural Engineering Service Publication 7, Cornell Univ., Ithaca, N.Y. 30 pp.

McGregor, B. M. 1989. Tropical products transport handbook. *USDA Agricultural Handbook*. No. 668, 148 pp.

Mitcham, E. J. 2003. "Controlled atmospheres for insect and mite control in perishable commodities." *Acta Horticulture*. 600:137–142.

Mitra, S., (Ed.). 1997. *Postharvest Physiology and Storage of Tropical and Subtropical Fruits*. Wallingford, UK: CAB International, 423 pp.

Nell, T. A., and M. S. Reid. 2000. "Flower and plant care." *Society of American Florists*, Alexandria, VA.

Paull, R. E., and J. W. Armstrong, (Eds.). 1994. *Insect Pests and Fresh Horticultural Products: Treatments and Responses*. Wallingford, UK: CAB International, 360 pp.

Seymour, G. B., J. E. Taylor, and G. A. Tucker, (Eds.). 1993. *Biochemistry of Fruit Ripening*. London: Chapman and Hall, 454 pp.

Shewfelt, R. L., and S. E. Prussia, (Eds.). 1993. *Postharvest Handling: A Systems Approach*. San Diego, CA: Academic Press, 358 pp.

Snowden, A. L. 1990. *A Color Atlas of Postharvest Diseases and Disorders of Fruits and Vegetables. Vol. 1. General Introduction and Fruits*. Boca Raton, FL: CRC Press, 302 pp.

Snowden, A. L. 1992. *A Color Atlas of Postharvest Diseases and Disorders of Fruits and Vegetables. Vol. 2. Vegetables*. Boca Raton, FL: CRC Press, 416 pp.

Thompson, A. K. 1996. *Postharvest Technology of Fruits and Vegetables*. Oxford: Blackwell Science, 410 pp.

Thompson, A. K. 1998. *Controlled Atmosphere Storage of Fruits and Vegetables*. Wallingford, UK: CAB International, 288 pp.

Thompson, J. F., A. A. Kader, and K. Sylva. 1996. "Compatibility chart for fruits and vegetables in short-term transport or storage." Oakland: University of California Division of Agriculture and Natural Resources Publication 21560 (poster).

Thompson, J. F., F. G. Mitchell, T. R. Rumsey, R. E. Kasmire, and C. H. Crisosto. 1998. "Commercial cooling of fruits, vegetables, and flowers." Oakland: University of California Division of Agriculture and Natural Resources Publication 21567, 61 pp.

Thompson, J. F., P. E. Brecht, R. T. Hinsch, and A. A. Kader. 2000. "Marine container transport of chilled perishable produce." Oakland: University of California Division of Agriculture and Natural Resources Publication 21595, 32 pp.

United States Department of Agriculture. 2000. "Nutrition and your health: dietary guidelines for Americans." Home and Garden Bull. 232, U.S. Dept. Agr., Washington, D.C. (www.usda.gov/cnpp).

Wang, C. Y. (Ed.). 1990. *Chilling Injury of Horticultural Crops*. Boca Raton, FL: CRC Press, 313 pp.

Watkins, C.B. 2000. "Responses of horticultural commodities to high carbon dioxide as related to modified atmosphere packaging." *HortTechnology* 10:501–506.

Wills, R., B. McGlasson, D. Graham, and D. Joyce. 1998. *Postharvest: An Introduction to the Physiology and Handling of Fruit, Vegetables and Ornamentals*. Wallingford: CAB International, 262 pp.

....

INTERNET RESOURCES

http://postharvest.ucdavis.edu: University of California Postharvst Research and Information Center.

http://www.ba.ars.usda.gov/hb66/index.html: A draft version of the forthcoming revision to USDA Agricultural Handbook 66 (November 8, 2002).

http://www.fao.org/inpho/: Postharvest information site of the Food and Agriculture Organization of the United Nations.

http://www.postharvest.org/: Training in postharvest technology.

http://www.uckac.edu/postharv/: University of California Kearney Agricultural Center.

http://postharvest.ifas.ufl.edu: University of Florida Postharvest Group.

http://www.fdocitrus.com: Florida Department of Citrus postharvest information.

http://www.postharvest.tfrec.wsu.edu: Washington State University postharvest information.

http://www.bae.ncsu.edu/programs/extension/publicat/postharv/: North Carolina State University postharvest information.

http://www.postharvest.com.au/: Sydney Postharvest Laboratory information.

http://www.chainoflifenetwork.org: A comprehensive assembly of information about postharvest handling of floral crops.

http://www.ams.usda.gov: U.S. Department of Agriculture, Agricultural Marketing Service information on quality standards, transportation, and marketing.

http://www.ams.usda.gov/nop/: National organic program standards.

http://www.aphis.usda.gov: U.S. Department of Agriculture, Animal and Plant Health Inspection Service information on phytosanitary and quarantine requirements.

http://www.nutiton.gov: Gateway to U.S. government information on human nutrition and nutritive value of foods.

http://www.nal.usda.gov/fnic/foodcomp: Composition of foods.

http://www.5aday.org: Produce for Better Health Foundation's promotion of produce consumption.

http://www.5aday.gov: National Cancer Institute's promotion of produce consumption.

http://www.aboutproduce.com: Information about produce for consumers.

http://www.foodsafety.gov: Gateway to U.S. government information on food safety.

http://www.jifsan.umd.edu/gaps.html: U.S. Food and Drug Administration's Manual of "Improving the Safety and Quality of Fresh fruits and Vegetables."

PLANT PROBLEMS

CHAPTER 16

PLANT PESTS AND OTHER PROBLEMS

Not only must horticulturists provide the proper growing conditions, they have to battle pests to produce quality plants. Pest problems, such as grasshopper or locust plagues, or diseases, such as late blight of potatoes (caused by *Phytophthora infestans*, which resulted in the Irish potato famine) have led to starvation and death of great numbers of people. Even if a pest or other problem does not completely wipe out a crop, it can have a serious economic impact on the grower.

Plants will not grow well and have an acceptable appearance if the environmental conditions are unfavorable or if they are attacked by disease-causing organisms, insects, spider mites, or other pests. Pests also can adversely affect plants or plant parts after they are harvested and even during storage. It is well known that improper storage, rather than low production, is a major factor in food shortages in many countries of the world. Much of the time, the best strategy is to *avoid* problems rather than to wait until they occur and then try to solve them. Therefore, horticultural practices often are designed to prevent such problems. In this chapter we discuss sanitation practices that help to limit or prevent the spread of diseases. Often these practices require initial investment in time and expense, but they will provide an excellent insurance policy against problems and will be very cost-effective in the long run.

Horticulturists should strive to produce high quality plants. If buyers are not satisfied with crop quality, they will look elsewhere in the future. A professional horticulturist will generally choose the plant genotype and will strive to provide optimum environmental conditions for growing plants. This care includes attention to proper watering practices, temperature control, fertilization, soil conditions, light, plant growth regulator applications, and management of pests. Improper growing conditions will result in a lowering of productivity and quality; attacks by pests will produce a similar result.

It is important for the horticulturist to remember that correct diagnosis is essential when dealing with plant problems. To diagnose a given problem, one generally needs a history of the crop or plant. Unless the horticulturist is familiar with the complete history of the crop (e.g., if he or she is the grower), it is imperative that he or she asks pertinent questions. One often cannot look at a plant and tell what is wrong with it at a glance; in fact, some important clues may be overlooked without careful examination of the plant and its environment. Many symptoms have more that one unrelated cause; additionally, two different agents may cause identical symptoms. Asking the correct questions is sometimes difficult, because questions will vary with the symptoms, but determining how and when the plant was watered, its location in terms of site and soil type, the use of pesticides on or around the plant, fertility, and any other pertinent information may allow a more accurate diagnosis. Many problems that can be traced to poor plant culture cannot be corrected properly with pesticides alone. In fact some problems such as citrus canker are so serious that total eradication of infected and neighboring plants is the only acceptable control measure. Correct diagnosis is essential. The use of references, consultation with experts, and sound judgement based on horticultural principles and knowledge can enable the horticulturist to

recommend cures or remedies for many plant problems.

Plant losses are caused by abiotic (noninfectious) and biotic agents. **Noninfectious agents** are nonliving and include unfavorable environmental conditions and factors such as air pollution, acid rain, hail, blowing sand, and poor cultural practices. **Biotic agents** include the plant pathogens: fungi, bacteria, parasitic higher plants, mollicutes, and viruses; nematodes; weeds; insects; mollusks; and vertebrates such as deer, rabbits, birds, pets, and even people. This chapter introduces these subjects; students should consult the references listed at the end of the chapter and consider taking courses in plant pathology (study of plant diseases), entomology (study of insects), and weed science for further knowledge in this broad field.

When assessing plant quality, one must have minimum standards for levels of damage, blemishes, abnormalities, and pests for a crop. Standards vary with the crop, its use, the geographic location of the consumers, and other intangible factors. For example, consumers may accept peel blemishes on a fruit crop such as oranges because they know that a particular cultivar is especially flavorful, or people may accept lower quality in "organically grown" produce than other fresh fruits and vegetables because of their feeling about nutrition and chemicals. In addition, it may be acceptable to have a low level population of insects, such as aphids on the leaves of a field of tomatoes for processing, whereas if the tomato plants are to be sold as transplants, there is no acceptable level of aphids.

To produce high quality plants, the grower must control pests. Many pest control approaches are available to the grower; some horticulturists choose to use chemical pesticides, whereas others avoid chemicals entirely. A sound approach that is receiving much attention is called **integrated pest management** (IPM) or **integrated control**. Some people prefer the term **integrated crop management**. IPM systems are designed to combine a wide range of strategies, which when used together in a proper manner, will result in a sound approach for controlling pests and growing the crop while considering effects on the yield and quality of the crop, economics, and protection of the environment. IPM employs a knowledge of the ecology of the pest and the crop and of the genotype of crop and pest as it relates to resistance, the environment, cultural practices, natural biological predators, microbiological agents, chemical messengers (e.g., sex attractants), and pesticides. This program requires a sound background in different areas of science, including horticulture, ecology, chemistry, meteorology, economics, mathematics, entomology, plant pathology, and other areas of biology. IPM is a dynamic approach for controlling plant pests because of shifts in pest and predator populations, different plant growth phases throughout the growing season, and changing environmental conditions.

PLANT DISEASES

Plant pathology is the study and prevention of plant diseases. A disease can be defined as an interaction between a host (the horticultural plant) and the environment that results in an abnormal condition. Diseases can be caused by abiotic agents—such as temperature extremes, excesses or deficiencies of water and nutrients, pollution, and other unfavorable environmental factors—or biotic agents, **plant pathogens**, such as fungi, bacteria, viruses and viroids, mollicutes, and nematodes (Figure 16-1). Any organism that obtains nourishment at the expense of another living organism is known as a **parasite**.

Noninfectious Plant Diseases

Many plant disorders are caused by unfavorable meteorological conditions, soil conditions, air pollutants, acid rain, toxic substances, nutritional disorders, and other factors. Through experience, coursework, workshops, and reading, horticulturists gain a knowledge of symptoms that are associated with various abiotic factors (Table 16-1). It is not uncommon for fungi and bacteria to enter and grow in damaged plant tissue. Therefore, it can sometimes be confusing to see obvious signs of a fungus or other organism, even though they are not the cause of the disease. In addition, when environmental conditions are unfavorable for plant growth, a combination of factors may act together to cause damage. For example, drought often occurs during hot weather; the symptoms manifested on a plant will be the result of a combination of stressful events. Sometimes this combination can make diagnosis difficult for the horticulturist. By learning about the history of the plant, one can often learn which combination of factors, or which single event caused a noninfectious disease. In an examination of plants and plant damage, characteristic patterns, or sections of a field or greenhouse growing area may exhibit the problems more than other areas and may provide a clue that will aid in diagnosis.

Infectious Diseases

An **infection** occurs when a pathogenic organism lives within and takes nourishment from, or parasitizes, a host. When the infection can be spread from plant to plant, it is referred to as an infectious disease. Conversely, an **infestation** occurs when a great number of organisms are present on an inanimate object or a plant surface. A contaminated soil is said to be infested with fungi, bacteria, insects, and weed seeds.

Infectious plant diseases are caused by different pathogens (infectious agents), including fungi (singular, **fungus**), bacteria (singular, **bacterium**), viruses and viroids, mollicutes, nematodes, and parasitic higher plants. Specific pathogens will attack only certain plants; this varies with the genotype of the plant and the pathogen. Pathogens

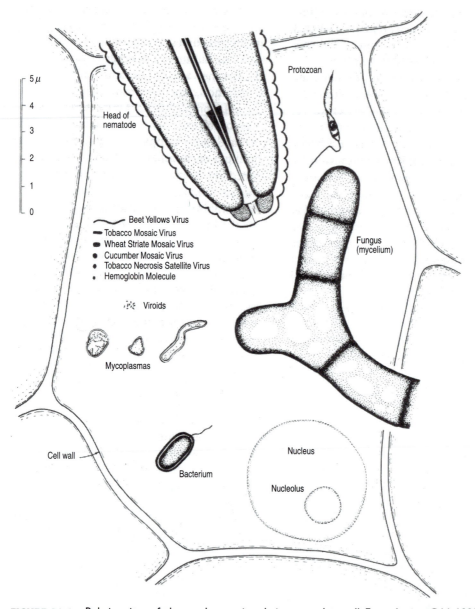

5 μ
4
3
2
1
0

Protozoan

Head of
nematode

Fungus
(mycelium)

⌐ Beet Yellows Virus
━ Tobacco Mosaic Virus
◖ Wheat Striate Mosaic Virus
● Cucumber Mosaic Virus
• Tobacco Necrosis Satellite Virus
· Hemoglobin Molecule

Viroids

Mycoplasmas

Cell wall

Bacterium

Nucleus

Nucleolus

FIGURE 16-1. Relative sizes of plant pathogens in relation to a plant cell. From Agrios, G.N. 1988.
Plant Pathology, Academic Press, New York. Used with permission.

have the ability to inherit their infection type, the degree of **virulence** or **avirulence**. A virulent pathogen has the ability to successfully infect the host plant and the relative degree of virulence is genetically controlled (see again Chapter 4). Additionally, pathogens often attack only specific organs or tissues, which varies with the age or stage of development of that plant part. Environmental conditions, cultural practices, time of the year, stage of growth of the pathogen relating to its life cycle, and the genotype of the pathogen also have a profound influence on the process of infection, spread of the disease, the degree of severity, and the practices required to control a specific disease.

TABLE 16-1. Abiotic plant diseases

Abiotic Factor	Source	Major Symptoms
Sulfur Dioxide (SO_2)	Fossil fuel combustion, especially power-generating	Interveinal chlorosis or necrosis of leaves
Ozone (O_3)	Auto emissions (forms from a photochemical reaction of pollutants and sunlight)	Stipple (at a glance, looks similar to spidermite injury)
Hydrogen fluoride (HF)	Phosphate fertilizer factories, steel and aluminum factories, superphosphate fertilizer	Marginal and tip necrosis of leaves
PAN (Peroxyacetyl Nitrate)	Auto emissions (forms from a photochemical reaction)	Silvery appearance of undersides of foliage
Ethylene	Gas leaks; incomplete combustion	Defoliation, leaf curling, abscission
Herbicide damage	Weed and feed fertilizer, spraying on a windy day, improper application rate	Damaged leaves, leaf cupping, thick, leathery leaves, plant death
Other chemical toxicity	Applying excessive concentrations, unfavorable environmental conditions at the time of application, drift, plant sensitivity	Burn, spotting, russeting, leaf chlorosis, leaf necrosis, plant death
Lack of light	Seasonal effects, distance of plants from electric lamps, or lamp age	Dropping of lower leaves, failure to flower, poor yield of fruit, spindly growth
Lack of water	Drought or failure to water and care for plants properly	Wilting, stunting, yellowing of leaves, dropping of lower leaves, plant death
Excess water	Excessive rain, improper watering practices, poor drainage	Wilting, stunting, yellowing of leaves, dropping of lower leaves, plant death
Nutrient deficiency	Insufficient fertilizer application, pH problems, nutrient interactions, specific soil problems	Discolored foliage, various other disorders, depending on the nutrient
Nutrient excess (salt injury)	Overfertilization, problems with native soils, irrigation water high in salts. Insufficient water to flush the root zone	Downward cupping of leaves, marginal and tip necrosis on leaves, excessively lush growth, brown or dead roots, salt burn around the stem at the soil line, plant death

(Continued)

TABLE 16-1. Abiotic plant diseases (continued)

ABIOTIC FACTOR	SOURCE	MAJOR SYMPTOMS
Wind damage	Severe storms, lack of windbreaks or other protection from wind	Torn leaves, sand blasting, broken or bent-over plants or plant parts
Sunscald	Sun striking an unwrapped trunk of a young tree when air temperatures are below freezing	Death of cells on surfaces exposed to prevailing sunlight source (SW side in northern hemisphere), bark splitting
Low temperatures (chilling or freezing)	Weather conditions, improper fertilization and/or pruning practices, heater failure	Die-back, lack of flowering, plant death
High temperatures	Weather conditions, cooling system failure, dark colored soils and mulches, lack of proper shading	Wilting, marginal and tip necrosis, plant death
Overpruning	Improper training, overzealousness	Excessive growth, many water sprouts, possible loss of winter hardiness, plant death
Mower blight (mechanical damage)	Careless driver of mower or trailer, mowing too close to plant material, using a "weed-whacker" around the base of plant material	Bark girdled or torn away from trunk, broken limbs
Hail damage	Hail storms	Torn leaves, defoliation, broken stems, pitted or damaged fruits

It should be noted that the pathogen is not the disease *per se*, it is the *cause* of the disease. This difference sometimes is confusing to the horticulturist, and can cause communication problems.

Because of the wide range of horticultural crops, different growing conditions, and myriad plant diseases, diseases of all crops cannot be described in this textbook. We therefore introduce some specific concepts and cite a limited number of examples in the following sections to provide a basic foundation of knowledge about the agents that cause infectious diseases of plants. These are not necessarily the most serious pests of the host plants, but they are examples that can be applied to other related pests and crops.

FUNGI AND FUNGAL-LIKE ORGANISMS

Fungi are organisms that produce no chlorophyll; their vegetative body consists of threadlike strands called **hyphae** (singular, **hypha**), which, when present in masses are called a **mycelium** (plural, **mycelia**) and their cell walls contain chitin and glucans. They produce fruiting bodies, with various morphologies, in which **spores** are produced. Fungi are spread vegetatively or by spores that are analogous to seeds of higher plants. Most of the more than 100,000 species of fungi live on dead organic matter and are called **saprophytes**. These are important in breaking down organic matter and thus play an important role in nutrient recycling (see Chapter 9). A much smaller number, more than 8000 species of fungi, cause diseases in plants. When considering control measures, one must think of the fate of beneficial microorganisms, since their loss can sometimes have adverse effects on crop plants.

Fungi are classified taxonomically according to their fruiting bodies and the production of spores. Some spores are produced as a result of the sexual union of gametes, and others are generated asexually. Within different phases of its life cycle, a single fungal species may produce either asexual spores (**conidia**) or sexual spores. Some fungi have such complicated life cycles that asexual spores may infect one plant species and sexual spores may infect a completely different and unrelated plant species. A knowledge of the life cycle is therefore important in understanding alternate hosts, if they exist, and in understanding the specific conditions that favor release and dispersal of the spores.

There are five general groups or phyla (singular: phylum) of plant pathogenic fungi (Figure 16-2), and two phyla of fungal-like plant pathogens as follows:

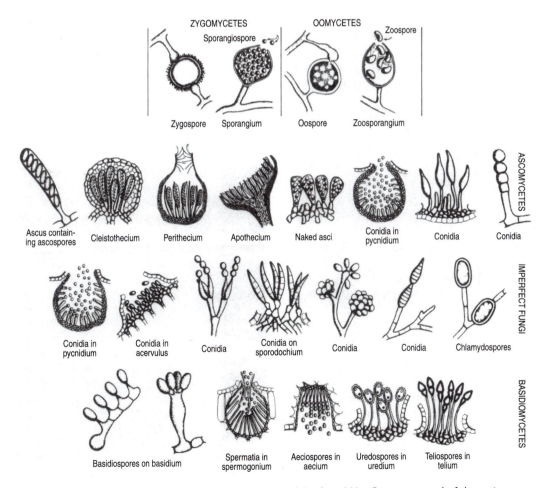

FIGURE 16-2. Representative spores and fruiting bodies of the fungal-like Oomycetes and of the main groups of fungi.
From Agrios, G.N. 1997. Plant Pathology, Academic Press, New York. Used with permission.

Fungal-like Organisms

In historical classifications of fungi, several types were considered "lower fungi." These groups are no longer classified as being in the Kingdom Fungi, but are now considered to be in the Kingdoms *Protozoa* (e.g., Myxomycetes and Plasmodiophoromycetes) and Chromista (Oomycetes). Within the Plasmodiophoromycetes, *Plasmodiophora* causes clubroot of crucifers and *Spongospora* causes powdery scab of potatoes. Within the Oomycetes *Synchytrium* causes black wart of potato and *Olpidium* infects plant roots and transmits plant viruses, including lettuce big vein "virus."

1. **Chytridiomycota** (Class: Chytridiomycetes) is a group that is characterized by hyphae with no cross-walls (septa).

2. **Zygomycota** (Class: Zygomycetes) produce asexual spores in sporangia and the spores are not motile. Within the Zygomycetes, *Rhizopus* is an example that causes soft rot of fruits and vegetables.

3. **Ascomycota** (Class: Ascomycetes) are a large group of fungi that includes many plant pathogens. They are characterized by bearing their sexual spores (**ascospores**), usually in groups of eight, in an enclosed sac, called an **ascus**.

4. **Basidiomycota** (Basidiomycetes) bear their sexual spores (**basidiospores**) externally on a **basidium**. Many species in this group have complicated life cycles. The hyphae of both the Ascomycetes and Basidiomycetes have septa.

5. The **Deuteromycetes** or **Fungi Imperfecti** are a group in which the sexual stage is unknown. They therefore spread asexually. In some species, the perfect (sexual) stage has been discovered, but because of historical reasons these are listed frequently with two different scientific names.

Within the Oomycetes, *Phytophthora* causes damping-off, late blight of potatoes, and root rots, and several other genera in this class cause downy mildews. Members of this group cause enormous economic losses of horticultural crops. Their control is essential for successful production. Many of the sanitation practices listed for the various crops in the Appendix are implemented specifically to avoid introduction of pathogens from this group.

Pythium sp. is another genus in the Oomycetes that causes root rot and stem rot diseases, but especially important horticulturally is the seedling disease called **damping-off**. *Pythium* is used here to illustrate a disease caused by a member of this group of fungi. Other fungi, including *Phytophthora*, *Rhizoctonia*, and *Fusarium* also can cause damping-off.

Damping-off is a disease that affects seeds or seedlings; an arbitrary age of six-week-old seedlings has been set as an upper limit in plant age for the damping-off disease. Plants older than this that are infected in a similar manner are said to have root rot or stem rot.

Pythium can attack seeds or recently germinated seedlings before they emerge through the soil surface. In this phase, the disease is called **pre-emergence damping-off**. The result is generally an exceedingly poor seedling stand. The disease begins as a small, dark or water-soaked spot on the seed or seedling. As it progresses, the cells collapse and the seed or seedling disintegrates (Figure 16-3). As a result, if a poor stand is noticed, and one digs in the soil, it usually is difficult to find the seeds or their remnants.

When *Pythium* causes damage after the seedlings have emerged through the soil surface, the disease is called **postemergence damping-off**. This fungal-like organism attacks the young, tender seedling stem at the soil line, the stem collapses in this area and the seedling falls over and dies (Figures 16-3 and 16-4). The fungus continues up the stem and the remaining plant parts also eventually dry up and die.

Pythium and most fungi that cause damping-off are not spread by airborne spores, but are carried on soil and plant fragments, and in water. Wet soil conditions favor the growth of *Pythium*; it is therefore often called a "water-mold." A knowledge of its spread and the conditions favoring its growth help a horticulturist to employ control measures effectively.

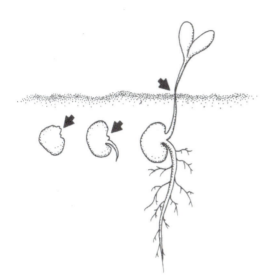

FIGURE 16-3. Illustration of pre-emergence (*left* and *middle*) and postemergence (*right*) damping-off. Arrows indicate points of injury.

FIGURE 16-4. Postemergence damping-off in marigold seedling. The fungal attack caused collapse of the cells near the soil line.

Organisms that cause damping-off are best controlled by eliminating them from the soil or medium prior to planting and by preventing their spread. If they are present or introduced into the medium or onto the seeds or seedlings, fungicides can be used; however, prevention is the best approach for dealing with these fungi. *Pythium*, other pathogens, weed seeds, insects, and other organisms can be eliminated from a soil or medium by applying a heat treatment, usually steam (Figure 16-5), or by using various chemicals (Figure 16-6), including formaldehyde, chloropicrin (tear gas), methyl bromide, a combination of chloropicrin and methyl bromide, or metham-sodium. It should be a common practice to wash tractor tires when moving from field to field to prevent the transportation of pests from one field to another.

Specific rules for both heat and chemical treatments of growing media have been developed for maximum effectiveness and safety. With steam, major considerations are to avoid killing all microorganisms (sterilization) or totally destroying soil structure. Raising temperatures in the medium to 60° to 71°C (140°–160°F) for 30 minutes will kill harmful organisms but preserve many beneficial organisms (partial disinfestation). This practice will provide competition if the medium becomes inoculated with *Pythium*,

FIGURE 16-5. Commercial unit for steam treatment of greenhouse media. A steam line will be attached to the coupling at the left end of the cart and steam will be forced upward through holes in the floor of the cart and thus into the moist medium. The cover confines the steam to the soil mass.

FIGURE 16-6. Soil is first prepared (*top*), then covered with an airtight seal (*middle*, here polyethylene). Methyl bromide (*bottom*), or an alternative fumigant, can then be injected into the soil to kill insects, disease causing microorganisms, and weed seeds. Alternatives are now being sought for this soon to be banned fumigant.

and *Thielaviopsis*. Examples of soil-less media can be found in Chapter 8.

The spread and introduction of *Pythium* and other damping-off-causing organisms is best controlled by employing some basic sanitation principles. Many of the basic practices and essential equipment at successful nurseries, greenhouses, and other controlled environment growing facilities are based on sanitation practices and the prevention of the spread of diseases. Tools (including trowels, clippers, knives, and shovels) and containers (including pots, flats, and benches) should be *disinfested* regularly. Many different products are available, but an effective treatment is liquid chlorine bleach that contains 5 percent sodium hypochlorite (NaClO). The bleach should be diluted to a 10 percent solution by mixing 1 part bleach with 9 parts water. This can be used to dip tools or to wash containers. Some containers can also be disinfested by using steam.

Introduction and spread of *Pythium* also can be prevented by not resting one's feet on benches and not stepping over flats or other plant containers. When walking outdoors, soil may be picked up on the soles of shoes and can easily fall off and inoculate a previously clean medium. The nozzles of hoses should always be kept off the floor, because the fungus can be introduced into the nozzle and spread upon watering (Figure 16-7). Placing of disinfested flats onto an infested floor can introduce *Pythium*, and allow damping-off to occur. Likewise greenhouse workers can spread fungi on their hands or even covers of flats containing seeds or seedlings. A logical approach to sanitation will prevent many problems and will help the grower produce a uniform, high quality crop. This integrated approach, along with use of fungicides[1] as an additional weapon, can help to avoid problems and effectively manage pests, such as those that cause damping-off (Figure 16-8).

will not interfere with nitrification activities (see Chapters 8 and 9), and is more economical than complete sterilization. In addition, the increased use of soil-less mixes as growing or propagation media has reduced the incidence of root-rot types of diseases such as those caused by *Pythium*

····················
[1] *Fungicides* are chemicals that kill fungal organisms. *Fungistats* prevent fungus growth without killing the fungus. Sphagnum moss and some sphagnum moss peats possess fungistatic activity.

FIGURE 16-7. Hoses should be hung up to prevent them from picking up pathogens from the greenhouse floor.

FIGURE 16-8. Marigold seedlings planted by the broadcast method (*top*) and planted in rows (*bottom*). Damping-off can spread more rapidly in broadcast seedlings.

True Fungi

Ascomycetes

Many important pathogenic fungi are members of this class. In their asexual stage, these fungi are similar to the *Fungi Imperfecti*; therefore, they are often mentioned together. Examples of pathogenic Ascomycetes include *Taphrina*, the causal agent of peach leaf curl, and the powdery mildews, and *Venturia inequalis* that causes apple scab (Figure 16-9). Many preventive spray programs are directed at *Ascomycetes* fungi.

Ceratocystis ulmi (imperfect or asexual stage is *Graphium ulmi*) is an Ascomycete that caused Dutch elm disease. This example illustrates a disease caused by an Ascomycete fungus. Elm species differ in their degrees of susceptibility, but one of the most susceptible, the American elm (*Ulmus americana*), with its graceful vase-shaped beauty, has been devastated by this disease.

Dutch elm disease is so named because it was first discovered in Holland in 1921; it was researched extensively by Dutch scientists. The fungus was first introduced into North America on elm logs imported from The Netherlands and was first found in Ohio and some east coast states in the 1930s. It has spread rapidly from there.

FIGURE 16-9. Apple scab, caused by the ascomycete *Venturia inequalis*, is a serious problem in commercial apple production.

Dutch elm disease is a vascular infection that causes wilting. Other fungi, including *Fusarium* and *Verticillium* also cause wilts. As a defense mechanism, plants seal off infections to prevent their spread. Plants are largely successful in confining vascular infections, and because plants have such great water transport capacities, they will continue to grow normally and the infection will not spread. However, in response to some vascular infections such as Dutch elm disease, a series of steps takes place that results in the plugging of areas of the xylem. Neighboring parenchyma cells produce protuberances (**tyloses**) that grow through the pits into xylem vessels. The tyloses become dense and effectively block the vessels. After the formation of tyloses, phenolic compounds are released and become oxidized and will polymerize. Various colors may be observed; thus vascular discoloration is a symptom that generally accompanies wilt diseases. When diagnosing probable wilt diseases, a horticulturist should scrape away some bark, or cut through a stem to look for discoloration or streaks in the xylem tissue. When plants are susceptible to a disease, vascular clogging can limit the amounts of water moving through the plant enough to cause death. This is indeed the case when American elms become infected by the fungus that causes Dutch elm disease. It is important to understand that the fungus itself does not block the vascular system. The blockage is a result of the plant's natural defense system, in which it blocks its own vascular system in an attempt to prevent the spread of the fungus that is present in its vascular system.

Depending on how it was spread and introduced, Dutch elm disease can cause the elm tree to wilt and die in one season, or it can cause leaves on individual branches to turn yellow, wilt, then turn brown. The branch will eventually die; however, the infected tree can remain alive for several years. When the tip of one or more branches turns yellow on an elm tree, a symptom called flagging, the horticulturist should suspect Dutch elm disease. Such a symptom should be further investigated because if it is Dutch elm disease, it should be attended to (generally through tree removal) in order to prevent its further spread (Figure 16-10).

Ceratocystis ulmi is generally spread in two ways, by elm bark beetles and by root grafts. The elm bark beetles, *Scolytus multistriatus* and *Hylurgopinus rufipes* develop on dead wood under the bark of older, established elm trees. They leave characteristic tunneling patterns on the dead wood (Figure 16-11). The asexual spores or conidia of the fungus are sticky and adhere to the beetles, which, upon emergence, fly to other trees and spread the infection. This is an example of insects as **vectors** of plant disease. After the bark beetles spread the disease, other trees may show the characteristic flagging symptoms. The disease also is spread in the vascular systems of trees, consequently the fungus will move from tree to tree via root grafts. When roots of trees of the same species cross or touch in the soil they can graft together naturally. This grafting is a result of the pressure created by secondary growth (diameter) of the touching roots causing a break in the bark and resulting in vascular contact. Callus forms and a vascular connection is created between the two roots. The result is a

FIGURE 16-10. This American elm was cut down because it had Dutch elm disease. The wood was subsequently destroyed.

FIGURE 16-11. Typical elm bark beetle larval tunnels on elm wood (bark peeled away for photographic purposes).

FIGURE 16-12. Diseases such as Dutch elm disease and oak wilt can spread between two trees that have developed root grafts such as these.

common root system for two or more adjacent trees (Figure 16-12). The fungus can thus move from tree to tree through this root system. A tree that is infected in such a manner generally responds by blocking the vascular system in the main trunk; rapid wilting and death will follow.

Although much research has been done on Dutch elm disease, no effective control has been found. Sanitation, accomplished by the removal and destruction of diseased trees (Figure 16-10), will prevent inoculation of healthy trees. Trees can be sprayed with an insecticide such as methoxychlor to control the beetles. Common root systems (Figure 16-12) can be disrupted by trenching between trees or by treating the soil with metham-sodium to kill the roots in a localized area. Success has been reported with fungicide injections into healthy elm trees for protection (not cure); however this treatment tends to be expensive. Alternatively, people are planting other tree species to replace susceptible elms, and some new elm hybrids that are resistant to Dutch elm disease have been developed. In the future, "Elm Street" may again be planted with stately elm trees.

Basidiomycetes

The common mushrooms, shelf fungi, and puffballs are all Basidiomycetes. Plant smuts such as corn smut and onion smut are caused by Basidiomycete fungi. Additionally, root and stem rots can be caused by members of this class, including *Armillaria*, *Poria*, and *Polyporus* that attack many woody plants and *Rhizoctonia* that attacks a wide range of horticultural crops. **Mycorrhiza** are symbiotic relationships between plant roots and certain Basidiomycete fungi. This association results in better plant growth because the plants absorb phosphorus better and are more resistant to attack by some soil-borne fungi, including *Pythium*. Basidiomycetes are economically important because of the damage that the pathogens cause, because of the positive attributes of mycorrhizae, and because of the culinary value of the edible, cultivated mushrooms.

Rusts are a group of diseases that are generally caused by Basidiomycetes; they are presented here as an example of important diseases caused by pathogens in this class of fungi. Rusts are generally characterized by reddish-brown or orange lesions. They have rather complicated life cycles because they can produce up to five different types of spores: pycniospores, aeciospores, uredospores, teliospores, and basidiospores. Some rusts produce all their spores on one plant, whereas others produce some spore forms on one plant species and other spore forms on another species that is not related to the first host. The alternate hosts can come from different plant subclasses—in stem rust of wheat, wheat (*Triticum aestivum*), a monocot, is one host and common barberry (*Berberis vulgaris*), a dicot, is the alternate host. Hosts can even come from different plant classes—in white pine blister rust, for example, white pine (*Pinus strobus*), a gymnosperm, is one host and currants and gooseberries (*Ribes* spp.), angiosperms, are the alternate hosts. Rusts are obligate parasites, so they cannot live on or grow in nonliving materials or dead organic matter. However, some special media have been developed so that now a few species can be grown in the laboratory.

An example of a rust disease, cedar-apple rust (Figure 16-13), is caused by different species in the genus *Gymnosporangium*, especially *G. juniperi-virginianae*. The alternate hosts are eastern red cedar (*Juniperus virginiana*) and apple, but pears, quince (*Cydonia oblonga*) and hawthorns (*Crataegus* spp.) also are susceptible. The symptoms on cedars are hard, brown, pockmarked leaf galls. In the spring, after rains, the galls produce columns of orange gelatinous ooze called **telial horns**, the site of teliospore production. To the untrained eye, these resemble orange flowers, and have been the focus of many telephone calls made to horticulture extension agents. Cedar-apple rust does minimal damage to eastern red cedars, but on apple, yellowish spots show up primarily on leaves and may also appear on young stems and fruit. These spots soon turn orange and generally will have a red band around each spot. Such spots reduce the photosynthetic area of apple leaves, thereby lowering fruit production, and when present on fruit the spots also lower the market quality of the apple. Similarly, the aesthetic value of ornamental crabapples is diminished.

Gymnosporangium juniperi-virginianae produces pycniospores and aeciospores on apple. The aeciospores blow in the wind, probably during late summer to early autumn, to cedars, where they cause infection. The fungus overwinters as mycelia in the galls on the cedars and after rain in the spring, teliospores are carried by the wind to apple trees, often up to distances of approximately two kilometers.

FIGURE 16-13. Cedar apple rust. *Left:* dormant gall on *Juniperus* (*top*) and rust spots on apple (*bottom*). *Right:* gelatinous telial horns have emerged on *Juniperus* following a rain.

Cedar-apple rust can be controlled, by locating apple orchards more than two kilometers from any cedar trees. Alternatively, fungicidal sprays applied to apple trees early in the spring will provide excellent protection. Once the symptoms are visible on apple leaves, however, nothing can be done except to include fungicides for cedar-apple rust in the spray program the next year.

DEUTEROMYCETES OR FUNGI IMPERFECTI

Many of the species originally classified in the *Fungi imperfecti* subsequently have been shown to be Ascomycetes. The asexual stages of both are quite similar. Because they have been classified historically as imperfect fungi (the perfect stage is the sexual stage), many common plant pathogens are known by their imperfect name. Important pathogenic Deuteromycetes include *Alternaria* that causes leaf spots and blights, *Fusarium* and *Verticillium* that cause wilt diseases, and *Thielaviopsis* that causes root rots. Control of these pathogens, or the use of resistant plants, is essential for successful production of many horticultural crops.

Botrytis cinerea is an imperfect fungus that has a wide host range and serves as an example of a pathogen from this group of fungi. It infects many different ornamental plants; can be a serious problem on fruit crops, including strawberries; and can lead to the rapid deterioration of various fruits, vegetables, and flowers, after they have been harvested and during storage. *Botrytis* causes diseases with names ranging from *Botrytis* blight to watery or soft rot and grey mold, and has sometimes been implicated in damping-off.

Botrytis is spread by air-borne spores or by water. The spores require moisture for germination, so any condition that results in water remaining on plant parts for a prolonged period of time—such as poor air circulation, crowding of plants, dew or other condensation, improper irrigation practices, or even improper misting of plants—will create ideal conditions for this ubiquitous fungus.

Botrytis can attack dead or living plant material, and can spread from dead areas of leaves to healthy areas. It produces a fuzzy, grayish mass of spores and mycelium and is usually identified by this gray-mold appearance (Figure 16-14).

This fungus can be controlled with fungicides or proper cultural practices. In greenhouse and other controlled environment facilities, practices that keep plants dry and minimize the time that their leaves, stems, flowers, and fruits are wet will minimize *Botrytis* problems. These practices include avoiding water splash onto the foliage when watering, ensuring good air circulation in the facility, and avoiding crowding of plants. Removal of all organic debris and infected material will aid in control of this disease. There are also fungicides available that provide excellent control of diseases caused by this fungus.

BACTERIA

Bacteria are microscopic single-celled organisms that contain no chlorophyll. Like the fungi, the majority of bacterial species are beneficial and function in various capacities

FIGURE 16-14. *Botrytis* on coleus. *Botrytis* is often characterized by a fuzzy gray mold (*arrow*).

including the breakdown of organic matter, nitrogen fixation, and mineralization (see Chapters 8 and 9). There are three groups classified according to their shape. The **cocci** are spherical, the **spirilla** are spirally curved, and the **bacilli** are rod or oval shaped. Bacterial plant diseases are caused primarily by bacilli, and are characterized by leaf spots, cankers, tumorous growth or galls, blights, soft rots, or vascular wilts. Four genera of bacteria are responsible for causing most bacterial plant diseases: *Pseudomonas, Xanthomonas, Erwinia,* and *Agrobacterium.*

Bacterial diseases decimate many horticultural crops. Some examples of these diseases are the following: (1) Bacterial leaf spots characteristically first appear as water-soaked spots that can become necrotic, or the necrotic areas can open and leave a shot-hole appearance. Many of these are caused by *Pseudomonas*. (2) Fire blight causes damage to important fruit crops, such as apple and pear; in addition, it can affect some ornamental plants, such as mountain ash (*Sorbus* spp.) and other members of the Rosaceae. The bacterium *Erwinia amylovora*, causes cankers, branch death, and fruit infections on these species (Figure 16-15). (3) Crown gall (Figure 16-16) is characterized by a tumorous mass, generally located near the soil line. It is caused by *Agrobacterium tumefaciens*, a species that genetically alters the host plant's cells by incorporating some of its DNA into the plant's chromosomes. Biotechnologists are using this natural system to incorporate new genes into some plant species (see again Chapter 4). (4) Soft rots of vegetables are generally caused by *Erwinia*. These can greatly reduce the numbers of vegetables that a grower can offer for sale. (5) Citrus canker is a serious disease caused by *Xanthomonas citri*. The best control measures are destruction by burning infected trees (Figure 16-17), and quarantine measures. By following these steps, horticulturists have successfully eradicated the disease from Florida in the past. Many other bacterial diseases

FIGURE 16-15. Fire blight of apple. Some apples and most pears are highly susceptible to this disease; the early symptoms appear on new growth and inflorescences as shown here.

affect plants and some are so severe that they limit the areas where some crops or cultivars can be grown.

Potato scab can be a serious problem on this important food crop. It is caused by *Streptomyces scabies*, an organism classified as an actinomycete. Actinomycetes are filamentous and often are considered to be intermediate between bacteria and fungi. They do not grow well under acidic conditions. Growers therefore select soils of low pH for potato-growing areas in order to control this serious disease (see Chapter 8).

FIGURE 16-16. Crown gall disease generally exhibits the symptom of a tumorous gall that forms at the base of the plant. This infected plant is a geranium (*Pelargonium*).

FIGURE 16-17. Destroying orange trees and other citrus by burning is one way to control the spread of several diseases.

Bacterial diseases are generally more difficult to control than those caused by fungi. Sanitation measures, as discussed under fungi (*Pythium*), are important in controlling the introduction and spread of bacterial diseases. Cultural practices including fertilizing and watering are also important. Water splash can spread bacteria, and water drops on leaves can promote bacterial growth and infection. Controlling the carbohydrate:nitrogen ratio (see Chapters 8 and 9) and thus the degree of succulence of the growth is important during periods that favor infection. Soft and succulent growth tends to be more susceptible to bacterial infections. Furthermore, because soft rot organisms frequently enter through wounds, care in harvesting and postharvest handling to avoid injuries to fruits and vegetables is of utmost importance. Keeping plant foliage and stems dry also creates an unfavorable environment for many bacterial infections. Some chemicals also are available to help control bacterial diseases. Antibiotics, especially *streptomycin*, are expensive but effective against some bacterial diseases. In addition, copper compounds also have activity against bacteria. Use of careful cultural practices, resistant cultivars (if they exist), and sprays help horticulturists to maintain quality and minimize losses caused by bacterial diseases.

MOLLICUTES (PHYTOPLASMAS AND SPIROPLASMAS)

Certain diseases that in the past were thought to be caused by viruses have been shown to be caused by mollicutes. Phytoplasmas resemble the mycoplasmas that cause diseases in humans and other animals in that they are microscopic and lack cell walls but instead are surrounded by a triple-layered membrane. Some similar organisms have a characteristic spiral or helical form and are thus called spiroplasmas.

Phytoplasmas cause a number of diseases in horticultural plants, including aster yellows, tomato big bud, peach x-disease, elm phloem necrosis, and coconut lethal yellowing. Spiroplasmas cause citrus stubborn disease and corn stunt. These organisms are spread mainly by leafhoppers (Figure 16-18), but other insects, including plant hoppers and tree hoppers also can contribute to their spread.

Phytoplasmas are susceptible to tetracycline and other antibiotics but not to penicillin. Antibiotics will not suppress diseases caused by phytoplasmas if applied to the foliage or as a soil drench, but are effective when injected into tree trunks. Control of the insect (leafhopper) vector on the crop and, for aster yellows, on nearby weeds that harbor the phytoplasmas, can also help to avoid the disease. In addition, control of the weeds that can harbor the organism will also help prevent introduction of the organisms into a susceptible horticultural crop such as carrots. Sanitation measures such as removal and destruction of infected plants will effectively reduce the inoculum. The selection of disease-free scionwood for grafted crops, including peach, pear, and citrus, will also help to control these diseases.

VIRUSES AND VIROIDS

Viruses are infectious agents that are much smaller than bacteria or mollicutes. In fact, they cannot be seen with a light microscope.

FIGURE 16-19. Healthy potato tuber (*left*) and one showing symptoms of potato spindle tuber disease (*right*), caused by a viroid.
(From: Fundamentals of Plant Pathology, 2nd ed. by Daniel A. Roberts and Carl N Boothroyd. Copyright© 1972, 1975, 1984. W.H. Freeman and Company. Reprinted with permission.

FIGURE 16-18. Flor's leafhopper (*top*) and a mountain leafhopper (*bottom*).
Photographs by R.E. Coville (top) and Jack Kelly Clark (bottom), California Agriculture, 41(3,4):27. Used with permission.

They consist of a nucleic acid (RNA or DNA) surrounded by a protein coat. Most plant viruses contain RNA, but only a few contain DNA, such as dahlia mosaic virus. Some infectious particles have been identified that consist of naked RNA (no protein coat); these are much smaller than viruses and are called **viroids**. The disease known as potato spindle tuber is caused by a viroid (Figure 16-19).

Viruses cause a wide number of symptoms on plants; in fact most are named after a major host and symptom that they cause (e.g., tobacco mosaic virus). The symptoms can be specific to the host plant but can vary with the plant's age and stage of development, the growing conditions, and the strain (genetic variant) of the virus. Some virus diseases are considered to be nearly

symptomless: The plant appears to grow normally, but the yield decreases. Many viruses cause characteristic symptoms to appear on the leaves. They can cause chlorotic flecks or patches to appear in the normal green tissue of leaves (Figure 16-20) or fruit; which are generally referred to as mosaic diseases. In addition, viruses can cause other characteristic symptoms including vein clearing; abnormal leaf morphologies, especially wrinkling and straplike leaf blades; leaf rolling; chlorotic streaks; necrotic spotting; and ring spots. Viruses can also cause stunting of plant growth, formation of tumors and galls, pitting of stems and fruits, sterility, flattened stems, cankers, and stem necrosis. Unless they are controlled, viruses and viroids can cause severe economic losses in horticultural crops.

Viruses reproduce only inside living cells. They enter plant cells through wounds and once inside, the RNA is freed from the protein coat. The virus then replicates itself, in part by inducing the host cell to produce certain molecules. It is thought that the symptoms that are produced in plants are a consequence of the viral replication process. Once inside the plants, viruses that infect

FIGURE 16-20. Typical virus symptoms in blackberry (*Rubus* sp.). Healthy leaf at left for comparison.

plants spread from cell to cell via the plasmodesmata. In addition, the particles can be spread much more rapidly in the phloem. Because of their systemic nature, viruses may spread throughout the plant. The distribution patterns can vary, however, from the virus being present in all parts of all organs to sectors of the plant being free from the virus.

No pesticides are effective against plant viruses. The most effective control measures for diseases caused by plant viruses are prevention of their spread to noninfected plants or use of resistant plants. Many viruses are spread by vegetative propagation, including grafting and budding, cuttings, bulbs, corms, rhizomes, stolons, and tubers. Not only will the resulting daughter plants contain the same virus as the mother plants, but the propagator's hands or tools (e.g., knife) can spread the disease to noninfected propagules. This spread is best prevented by selecting stock plants that have been indexed or tested to assure that they do not contain specific viruses. These plants are referred to as **virus indexed** or **specific pathogen tested** plants. They are said to be free from certain pathogens but have not been tested for all pathogens.

Some viruses are spread from plant to plant via pollen and can thus infect the ovules. Therefore, it is also possible to spread some viral diseases on seeds.

Insects are major vectors by which viruses are transmitted from plant to plant. Therefore, good insect control practices can significantly limit the spread of plant viruses. The most important insect vectors are aphids and leafhoppers (Figure 16-21). Other insects that transmit viruses include white flies, mealy bugs, scale insects, thrips, leaf miners, beetles and grasshoppers. Insects with piercing-sucking mouthparts can carry viruses on their stylets (style-borne) or internally, where the viruses may multiply. Most viruses transmitted by aphids are style-borne and most by leafhoppers are carried internally.

Spider mites, nematodes, fungi, and parasitic plants (dodder) are also known to transmit viruses. Viruses can move from plant to plant by leaves rubbing against each other or by human hands touching the plants. It is because of this last reason that people who smoke or otherwise handle tobacco are advised to wash their hands before handling tomatoes or other Solanaceous crops. The serious virus disease tobacco mosaic virus (TMV) may be present in the tobacco and can be spread in this fashion. In addition, viruses can be spread on knives and pruning tools.

FIGURE 16-21. Aphids such as these on ash leaves are among the most common vectors of virus diseases.

Plant viruses may be inactivated with a heat treatment. Depending on the tissue, species, and virus, hot water dips or soaks at 35° to 54°C or warm (35°–40°C) growth chambers have been effective. The apical meristem of a rapidly growing stem may also be excised and grown aseptically in tissue culture to escape viruses (Figure 16-22). The resulting plants (assuming success) will have escaped the virus. Heat treatment of the stock (donor) plant may enhance the success of this meristem culture approach to virus elimination. These plants may then be indexed for specific pathogens, and if free, can be grown for use as stock plants, or for production.

It is by recognizing viral symptoms, employing good cultural and propagation techniques, and strictly controlling pests, that one can avoid diseases caused by

FIGURE 16-22. Meristem culture of dahlia can be used to obtain plants free of dahlia mosaic virus. This culture was initiated from an in vitro heat treated dahlia meristem culture.

Photograph courtesy of Judith Kersten, University of Minnesota.

viruses. Many propagators produce virus indexed plants. In addition, some cultivars have been bred to be resistant to certain viruses. Such practices will allow the grower to produce high quality plants and will prevent problems associated with other pests, as well as viruses and viroids.

PARASITIC HIGHER PLANTS

Many species of seed-bearing plants gain all or a portion of their nourishment from other plants. Some of these parasites cause considerable damage to their hosts; others do not seem to hurt the hosts and may be beneficial, although this has yet to be verified. Parasitic higher plants may have leaves, stems, and roots, such as members of the family Santalaceae; some have stems and green photosynthetic leaves, such as the mistletoes; and some consist of vines and lack chlorophyll, such as the dodders (*Cuscuta* spp.).

The control measures vary for different parasitic plants. Burning of a field can be effective, as can the use of herbicides. The best way to prevent dodder infestation is to plant crop seed lots that contain no dodder seeds. Other parasites can be eliminated by pruning out infected branches on host trees.

A serious problem on many conifers in some areas is dwarf mistletoe (*Arceuthobium* spp.). This dioecious parasite causes swellings, cankers, and witches' brooms to form on branches. The trees can thus become deformed and may die. Dwarf mistletoe forcibly ejects seeds that are covered with a sticky substance. The seeds stick to the bark of trees, where they germinate and a rootlike body, called a **haustorium** penetrates the bark and enters the vascular system of the tree. This process allows the dwarf mistletoe to gain nourishment from the host but starves the portion of the branch located beyond the point of infection. Dwarf mistletoe can be controlled only by pruning out infected branches or removing infected trees. Because the seeds can be expelled distances up to 15 meters, one also can create a protective zone with no parasite present around healthy conifers.

··

NEMATODES

Nematodes are small, generally round wormlike animals sometimes called eelworms. Their size is so small that they are invisible to the naked human eye. Only about 10 percent of all nematode species are parasitic on higher plants, while the majority feed on microscopic plants and animals; some parasitize higher animals, including humans. Nematode anatomy and morphology have been well described (Figure 16-23), and specific techniques exist for separating them from soil and plant debris for identification.

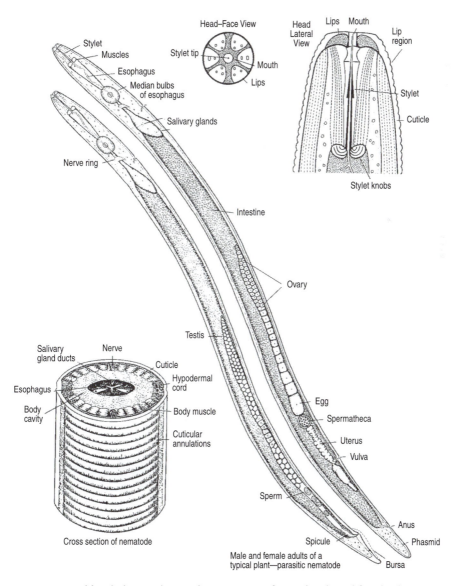

FIGURE 16-23. Morphology and main characteristics of typical male and female plant parasitic nematodes.
From Plant Pathology, 2nd ed. by G.N. Agrios, Academic Press, New York.

Most plant parasitic nematodes are soilborne and feed on underground plant parts, including roots, bulbs, corms, tubers, and rhizomes; however, some feed on leaves and flowers (e.g., chrysanthemum foliar nematode). **Endoparasitic** nematode species enter plant parts, such as roots, feed, and complete most of their life cycles there. **Ectoparasitic** species feed only on the surface cells of roots. These feeding habits of nematodes can cause different symptoms on plants.

Nematodes cause many plant diseases including root knots, cysts or galls, root lesions that result in rot, stubby root, and root branching. Root damage may cause the tops of the plants to show mineral nutrient deficiencies, chlorosis, wilting, poor stands, and reduced yields. When nematodes attack above-ground portions of a plant, they can cause necrotic spots, rots can set in, distorted growth can appear, or galls can form. Thus, the control of nematodes is essential for growing high quality horticultural crops. Problems with nematodes are generally greater in warmer climates because soil freezing tends to reduce populations.

Nematodes can be managed by crop rotation—switching to a species that is not a host for that particular pest. This technique is not always practical, however, because to be relatively certain of reducing the nematode populations, a 10-year nonhost rotation is often necessary. With perennial crops, such as tree fruits, growers should consider species that are not hosts for nematodes when choosing a cover crop. Specific chemicals, nematicides, can be applied to control nematodes. In addition, soil fumigation will also effectively control these pests. Quarantines can slow or prevent nematode spread and introduction into new areas.

A knowledge of the movement of nematodes is also important for understanding control practices. Nematodes are motile and can move relatively short distances, depending on the species, soil, and environmental conditions. They are dispersed over wider areas in soil, in water, or on plant parts. Floods, dust storms, tractors and other equipment tires, tools, and feet can all spread nematodes. These parasites can also disperse when people ship plants, and when rain or irrigation water splashes from leaf to leaf. Sanitation practices such as washing equipment and tools between fields can help reduce the spread of nematodes. Therefore, an integrated approach to proper cultural practices—such as selection of plant species and cultivars, sanitation, soil fumigation (or heat treatment), and the use of pesticides—is necessary for nematodes to be controlled effectively.

HORTICULTURAL PRACTICES AND PLANT DISEASE

As mentioned previously, common horticultural practices, such as vegetative propagation and handling of plants can contribute greatly to the spread of infectious plant diseases. Many other aspects of horticulture are geared toward the avoidance or prevention of infectious diseases, including greenhouse sanitation, partial disinfestation of media, not dipping cuttings directly into the can of rooting powder, washing equipment between fields, and disinfesting tools in liquid sterilants, such as diluted (nine parts water to one part bleach) household liquid chlorine bleach. Knowledge of infectious plant diseases, their spread, and the reaction of plants to pathogens is important in understanding the logic behind horticultural practices related to prevention and control of plant diseases.

INSECTS

An estimated 20 million species of insects exist, which is more than all other groups of animals and plants combined. At some stage during their life cycles (generally immature stages), as many as 50 percent of insect species utilize living plant material as a food. This creates serious economic losses in horticultural, agronomic, and forest crops each year. Some insects feed on specific plant species, whereas others have wide host ranges. Additionally, one host may be attacked by many different insects; for example, it has been estimated that approximately 400 insect species feed on apple trees.

Not all insects are harmful, in fact, many are beneficial. Insects play important roles as pollinators of various plants especially fruit and vegetable crops. Important insect pollinators include bees, flies, butterflies, and moths (Figure 16-24). Other insects, such as silkworm caterpillars, and honey bees, produce economically important commodities. Many insects are important foods of various animals, including birds and fish. In some societies, insects contribute significantly to the human diet. Insects also serve as biological control agents for other more harmful insects, or even plant weed species.

FIGURE 16-24. Honey bees help fruit and seed production by transporting pollen between flowers.

The fact that insects attack plants has been known since ancient times, yet because of various defense mechanisms, plants continue to survive. Some plant species will regrow after insect attacks; others complete their life cycles when insect activity is low or when weather conditions are unfavorable for the pests. Other plants have structures such as trichomes that make insect feeding difficult, and some plants produce secondary products that are toxic or distasteful to insects. Some plant species are simply dispersed over wide areas, which increases their chances of survival. Specialists in horticulture and other areas of agriculture tend to concentrate plant species and even specific genotypes in great numbers in fields, greenhouses, or other areas (monoculture), but this practice favors the build-up of large insect populations and can have disastrous consequences for a crop. To understand and effectively control insect pests, a horticulturist should have a knowledge of their biology and behavior.

Insect Biology

Insects are cold-blooded invertebrate (lacking backbones) animals that belong to the phylum Arthropoda. The following points are important distinguishing features of insects (Figure 16-25):

1. They have external skeletons called exoskeletons that contain a specific nitrogenous polysaccharide known as chitin.
2. Their bodies are divided or segmented into a relatively distinct head, thorax, and abdomen.
3. They have three pairs of jointed legs that function in movement.
4. They have one pair of antennae that serve as sensing organs.
5. They are bilaterally symmetrical.
6. They have specialized appendages called mouthparts for feeding.
7. They usually have two pairs of wings, especially in adults.

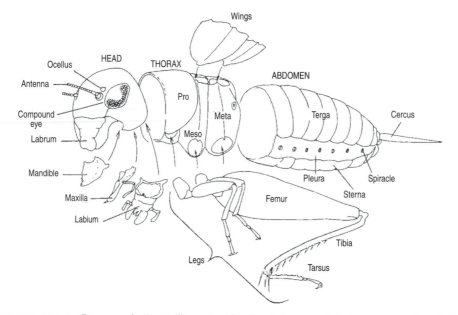

FIGURE 16-25. Diagram of a "typical" insect with a head, thorax and abdomen; two pairs of wings; one pair of antennae; and three pairs of legs. This common cricket has the typical mouthparts of a chewing insect.
From Pfadt, R.E. 1985. Fundamentals of Applied Entomology. 4th ed. Reprinted with permission of Macmillan Publishing Company.

Most insects reproduce through eggs, but some species (e.g., aphids) give birth to live young that develop inside the mother. Although in most species the eggs are fertilized by the sperm produced by the male, some insects reproduce without fertilization. Development of an unfertilized egg is called **parthenogenesis** and occurs in aphids and some weevils. These species therefore do not require males for reproduction; however, males do form under certain conditions and allow for more genetic diversity because of sexual union of gametes. In honeybees, the female, or queen bee, is able to control fertilization and lay unfertilized eggs that develop into drones and fertilized eggs that develop into workers or queens.

Insects generally lay their eggs in protected locations that favor their survival. Some insects lay their eggs in plant tissue to gain protection, moisture, and a food source for their young. Penetration and oviposition (the act of laying eggs), the plant's response to the eggs or larvae, or tunneling and feeding by insect larvae can lower the

quality of plants or cause complete losses of horticultural commodities. If this is likely to be the case, control measures must be geared towards prevention of oviposition.

After hatching, and during their growth and development, most insects go through changes in shape and form. This process is called **metamorphosis**. In some species few changes other than increase in size are observed, whereas in other species great differences are evident between the young forms and adults (Figure 16-26).

The growth process of an insect is limited by the size of its exoskeleton. Therefore, to increase in size, the insect must first shed its old exoskeleton and replace it with a new exoskeleton that is at first soft and is straightened and stretched by the insect; the exoskeleton then hardens. This process, called **molting**, generally occurs in a protected site, because the insect is particularly vulnerable with its new, initially soft exoskeleton. Most insects go through four to six molts. The insect that exists between molts is referred to as an **instar**. For example,

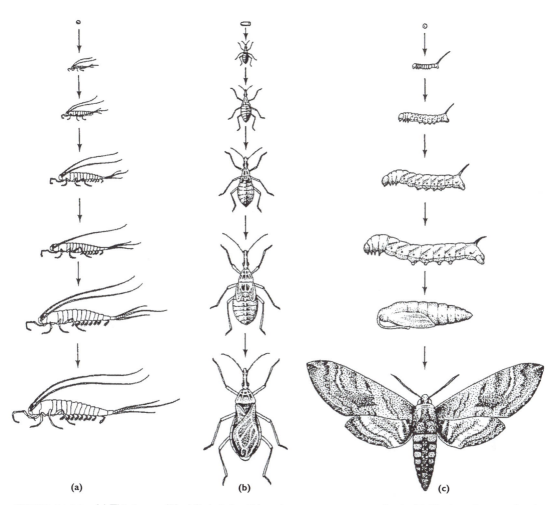

FIGURE 16-26. (a) This insect (*Machilis*, bristletails) undergoes no metamorphosis. (b) *Rhodnius* (assassin bugs) have some changes during development (simple metamorphosis). (c) *Manduca* (tomato hornworm) undergoes a complete metamorphosis.

From Evans, H.E. 1984. Insect Biology. A Textbook of Entomology. Addison-Wesley, New York. Used with permission from The Benjamin/Cummings Publishing Company.

upon hatching from an egg and before the first molt, the insect is in the first **instar**; after the first molt it is the second instar, and so on.

Instars are referred to by various names, depending on their stage in the entire process of metamorphosis (if it occurs), and the development of that particular species. After hatching, the young insects that have incomplete or simple metamorphosis are called **nymphs**. They are called **larvae** if they have complete or complex metamorphosis. Nymphs generally resemble adults, whereas

larvae look dramatically different from the adults. Many larvae such as fly maggots, caterpillars, and grubs are wormlike in appearance. Generally the larval state is occupied predominately by feeding. After several larval molts, the generally inactive, nonfeeding pupa is formed. The pupa is often surrounded by a cocoon structure and is a transition period before adulthood. The adults of most insect species are specialized for dispersal and reproduction.

Taxonomically, insects are in the class **Insecta**. Insecta is further divided into two subclasses, the wingless insects (Apterygota) and the winged insects (Pterygota). Most insects that are serious plant pests are in the subclass Pterygota. Within this subclass are 23 orders, 8 of which contain serious plant pests (Table 16-2).

Insect Mouthparts

Most of the damage to plants caused by insects results from feeding activities. Pests may ingest whole portions of plants, or suck in plant juices; this varies with the type of mouthparts possessed by a particular species. An understanding of insect mouthparts is important for assessing damage, and determines which type of insecticide (contact or stomach poison) will be effective on a specific pest.

Chewing Mouthparts

Chewing mouthparts are considered to be the least specialized in Insecta. The chewing action is accomplished by two **mandibles** (jaws) that are hardened and have cutting edges on the more distal portion and grinding edges near the hypopharynx

(tonguelike organ). Rather than moving up and down, like human jaws, the mandibles move across each other in a scissoring manner. Because of their chewing mouthparts, many insects chew through or cut holes in plant parts, or eat entire organs, such as leaves. Some insects, such as members of the Lepidoptera (butterflies and moths) possess chewing mouthparts only during the rapidly feeding or larval stage; their mouthparts are modified or lost when the insects become adults. Other insects, such as Coleoptera (beetles), may have chewing mouthparts throughout their lives.

Piercing-Sucking Mouthparts

Some insects, especially members of the Hemiptera and Homoptera, have mouthparts that are specialized to puncture and penetrate plant tissue and suck out the sap. These mouthparts consist of three to six (depending on the species) sharp, pointed, needle-like structures called **stylets** that are enclosed in a sheath or **labium** that holds the stylets in position. It is by moving the stylets up and down that they pierce tissues and enter into the plant. The labium holds

TABLE 16-2. Examples of Insecta orders

Order	Mouthpart Type	Examples
Coleoptera*	Chewing	Beetles, weevils, borers
Diptera	Chewing, piercing-sucking, or sponging	Flies, leaf miners, fungus gnats
Hemiptera	Piercing-sucking	True bugs, plant bugs, lace bugs
Homoptera*	Piercing-sucking	Treehoppers, leaf hoppers, aphids, whiteflies, scale insects
Hymenoptera	Chewing or chewing-lapping	Sawflies, wasps, bees
Lepidoptera*	Chewing in larvae, siphoning in adults	Butterflies and moths
Orthoptera	Chewing	Grasshoppers, crickets, mantids
Thysanoptera	Rasping-sucking	Thrips

*Indicates the orders with the greatest number of pests on horticultural crops.

the stylets together during this action but does not enter the wound. Some insects have a strong pumping mechanism to remove fluids. Aphids, however utilize the sap pressure generated by the plant itself (Figure 16-27).

The damage caused by insects with piercing-sucking mouthparts is obviously different from that caused by insects with chewing mouthparts in that portions of plant tissue are not removed and ingested. The presence of insects with piercing-sucking mouthparts is apparent from a sticky, shiny "honeydew" on the leaves. The honeydew is excreted by these insects. Ants, in fact, "farm" aphids so that they can collect the honeydew and use it as a food source. In early stages the honeydew can be washed off with water. One may also notice a black, sooty mold on plant leaves, stems, and fruit. This mold is sometimes misdiagnosed as a disease, when, in fact, it is a fungus that is growing on the honeydew. Therefore, sooty mold can be controlled by ridding the plants of insects that are excreting the honeydew. However, aphids and other insects with piercing-sucking mouthparts are major vectors of plant viruses. Viruses are frequently controlled best by protecting plants from these insect pests.

Rasping-Sucking Mouthparts

Thrips are the only insects that possess rasping-sucking mouthparts, an intermediate form between chewing and piercing-sucking mouthparts. A thrips (note that thrips is both the singular and plural form) has mouthparts that are modified to function as short stylets that abrade plant tissue and allow the insect to suck the juices. Elongated, scraped areas are typical of thrips damage on plants (Figure 16-28).

TYPES OF INSECT DAMAGE TO PLANTS

Root Damage

Roots can be attacked by species with piercing-sucking mouthparts, such as the root aphid that attacks ornamentals such as dahlia, sweet pea, and cosmos, and by species with chewing mouthparts such as grubs. Insects that feed on roots can cause many symptoms, including wilting, yellowing, thinning out, or plant death.

White grubs, the larvae of members of the Coleoptera including June beetles, oriental beetles, and Japanese beetles, cause serious damage to turfgrasses. The adults lay their eggs in the grass and the larvae (white grubs) hatch out and burrow around

FIGURE 16-27. Aphids, although small, multiply rapidly and can cause severe damage in some crops. They also may spread virus diseases.

FIGURE 16-28. Thrips damage on bean leaflet.

the roots of the turfgrass plants at a depth of two to three centimeters and feed on the roots. The damage to turfgrasses will be severe if the population of white grubs is two to five per 900 cm^2. This causes the turfgrass to become soft and spongy so that it can be lifted and rolled back by hand, like a rug, to reveal the insects. Subsequently, light brown or dead areas will appear in a lawn. People frequently observe increased mole and bird activity on lawns infested with grubs. Skunks can also be seen digging in lawns for grubs. Proper application of pesticides will prevent grub damage to lawns and will help to reduce the foraging activities of animals that eat white grubs.

Stem Damage

Insects with piercing-sucking or chewing mouthparts feed on stems of plants. Stems also can be damaged by females inserting their hard ovipositors to lay eggs, (e.g., female cicadas and tree hoppers). Plant damage from feeding can include wilting, outgrowths, galls, witches' brooms, holes, tunnels, dieback, and death of the stem or entire plant. Honeysuckle (*Lonicera* spp.) plants can respond to aphid feeding by producing unsightly witches' brooms, even though the aphids have piercing-sucking mouthparts (Figure 16-29). This reaction is a major concern to people who have planted honeysuckle in their landscapes.

Borers cause major damage to woody ornamentals, garden flowers, and fruit and vegetable crops. Insects that feed and tunnel within plant roots or stems are considered to be borers. They are generally in the larval stages of their life cycle, although some adult beetles can also be borers. As a result of their tunneling and feeding, borers can disrupt the vascular tissue or girdle the plant. Symptoms associated with borer infestations include wilting, yellowing, dieback, localized stem swelling, stunting of growth, weakened stems, exudation, and death.

FIGURE 16-29. Aphid feeding may cause a plant to react by forming a "witches broom" growth, as in the honeysuckle (*top*). *Center:* several witches brooms are visable on this tree (not caused by aphid feeding in this tree); *bottom:* a closer view shows the multiplicity of twigs produced in a witches broom.

Borers that attack woody plants have a preference for weakened individuals that are in poor health. Therefore, providing proper fertilization, watering, and growing conditions, and generally keeping the plant pest-free will make it less attractive to borers. Some borers, however, will attack healthy herbaceous plants; for example, the squash vine borer will cause the sudden wilting and death of an otherwise healthy squash or pumpkin plant. In addition to the sudden wilting, a "sawdust" material indicates the presence of an insect in the stem. Borers often cause further problems for the plant, because they leave wounds through which plant pathogens can enter. Severe damage or death of the plant may result. Bacterial soft rot, caused by *Erwinia carotovora*, often enters iris rhizomes through wounds made by the iris borer. This foul-smelling soft rot can wipe out an iris planting. To prevent iris soft rot, it is essential to prevent borer damage by removing old leaves that protect the eggs laid by the iris moth and then by spraying the plants with an appropriate insecticide.

Galls or abnormal tumor-like outgrowths of plant tissues form on leaves, flowers, and stems. This abnormal growth or localized cell division is the way that some plants or plant parts respond to insect pests. Galls provide a protective environment for the insect, reduce the pest's water loss, and shelter them from predators. Some galls, such as bladder galls that form on leaves, are of little consequence, but other stem galls can stunt the growth or lead to death of the terminal portions of the shoot.

Leaf Damage

Leaves may be deformed, eaten, or tunneled into and they may have galls, spots, or otherwise be damaged by insect pests. Many species of insects attack leaves and cause various kinds of damage, so we have chosen two types of insects as examples: an insect that ingests leaves and a leaf miner.

The imported cabbage worm (*Pieris rapae*) causes serious damage to crucifers, including cabbage, cauliflower, broccoli, radish, kale, and mustard (Figure 16-30). It also can feed on other crops including lettuce, horseradish, sweet alyssum, and nasturtium. The white adult butterflies lay eggs on a host such as cabbage. The young larvae primarily eat the cell layers off the undersides of leaves, and when they are older, the green larvae eat all cell layers and make holes in the leaves. The larvae also are capable of boring into cabbage heads, where they continue to feed on leaf tissue. These holes, and the presence of the pests in the heads renders the cabbages unmarketable. Another sign of imported cabbage worms is the greenish-brown excrement. The cabbage looper (*Autographa brassicae*), which causes similar damage has a looping movement. Because of the holes in the leaves that it creates with its feeding activities (chewing mouthparts), the cabbage looper is also considered to be one of the most serious pests of crucifers.

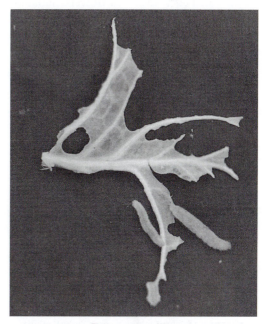

FIGURE 16-30. Damage to cabbage caused by the imported cabbage worm, a chewing insect.

Leaf miners are larval forms of sawflies, moths, or flies that bore or tunnel within leaves (Figure 16-31). They have chewing mouthparts and ingest the internal leaf cells as they make their tunnels. These miners can seriously damage a plant by greatly reducing the photosynthetic area of the leaves. In severe cases, the leaves will abscise prematurely as a result of leafminer feeding activities. Adult leaf miners lay their eggs on the leaves and the young hatch and enter the leaves. It is easier to control the adults or young before they enter the protected environment within the leaf. Once inside a leaf, leafminers are much more difficult to control. Floriculturists are especially concerned about leaf miners because the miners reduce the aesthetic appeal of crops such as chrysanthemum.

Flower Damage

Flowers can be damaged by insects at any stage of development. Injury can range from deformities to complete loss of flowers or flower buds. Although insects cause great amounts of damage to flowers, they usually do not feed exclusively on flowers. Japanese beetle grubs, for example feed on roots and cause damage to turfgrasses and the adults

feed on leaves, flowers, and fruits of a wide number of horticultural species. They will consume entire flowers, but their feeding is by no means limited to these organs. Thrips also damage flowers, in addition to other plant parts, although there is one species known as *flower thrips*. Flower thrips, because of their rasping-sucking mouthparts produce tiny streaks or flecks on the petals and can cause the petal tips to turn brown (Figure 16-32). They also can cause flecking or a silvering of leaves. Flower thrips spread from plant to plant on wind currents, and their continual arrival through greenhouse vents makes their control difficult. They can cause serious losses by reducing the marketability of various floriculture crops.

Fruit Damage

Fruits can be damaged by ovipositors and by feeding. In addition, it is not uncommon for insects to be present within fruits. Many of us have had the unpleasant experience of biting into an apple and discovering a worm, or even worse, half a worm.

Insects can cause total destruction of fruits or cause blemishes, scars, or russeting on the fruit surface. Pests that attack fruits are of major concern to orchardists and pomologists, who employ extensive spray

FIGURE 16-31. Leaf miner tunnels in columbine (*Aquilegia*) leaves.

FIGURE 16-32. Thrips damage (white specks) on a streptocarpus flower.

programs to be able to produce high quality, pest-free marketable fruit.

The *codling moth* can cause serious damage to the fruit of many horticultural crops, including apple, stone fruits, English walnut, and pear. The adult moths lay their eggs on leaves or fruit; when the larvae hatch they enter the fruit through the calyx end or bore into the fruit where a leaf touches a fruit or two fruits touch. The white to pinkish larvae with dark heads then feed on the core (apple) and seeds (thus the fruit becomes wormy), then tunnel out and leave the fruit with ugly worm holes. Codling moth larvae also excrete brown to black colored masses and pellets from the core to exit hole, which greatly reduces the fruit quality (Figure 16-33).

The *corn earworm* is perhaps the most serious pest on sweet corn. It also attacks tomatoes and cotton; it is then called the *tomato fruitworm* or *cotton bollworm*, respectively. The adult female moth lays her eggs on corn silk; they hatch and follow the silk down to the ear, where they feed on the young leaves, silks, and kernels (Figure 16-34). Corn earworms usually will be seen as green, pink, or brown longitudinally striped worms in the tips of ears, and their presence is unacceptable to the consumer. Because of the damage that they cause to ears of sweet corn, they result in millions of dollars in losses to growers in the United States every year. These pests can

FIGURE 16-34. Corn earworms cause serious losses to sweet corn crops.

be controlled by regular sprays. The damaged portion is usually trimmed off the end of the ears by home gardeners. This also can be an effective commercial practice to reduce pesticide applications and is a good example of integrated crop management.

Insect Pest Management

Insects may be specialized to feed on specific plant organs or may attack plants indiscriminately. Because they cause severe reduction in the yield and quality of all horticultural crops, measures to control these pests are an important aspect of horticulture.

Effective management of insect pests requires a combination of many factors. This approach may include using plants with resistance to pests when practical and available; designing cultural practices to

FIGURE 16-33. Codling moth larval damage in apple.

reduce pest populations, such as disrupting overwintering habitats; using sterile insects; using insect sex attractants; and employing biological controls, chemical pesticides, and legal sanctions, such as plant quarantines. The most effective approach to insect pest control depends on several factors:

1. Identifying an acceptable population level of a specific pest on a crop. This level depends on the use of the crop and the seriousness of the damage caused by that pest. Sometimes eliminating a minor pest kills predators, resulting in much more serious infestations by a major pest.

2. Establishing threshold levels of economic injury on a crop.

3. Learning the biology of a specific pest. If a pest multiplies very rapidly, it may be best to apply control measures before the population becomes so large that it results in major economic losses.

4. The economics of control must be considered. The best method may be so costly that profits to the grower are seriously reduced.

Cultural Techniques

Providing proper growing conditions—including watering, proper light levels, and fertilization—may reduce insect problems on some plant species. Many insects overwinter on dead or decaying plant parts or within living plants. The removal and proper disposal of dead plant material or pruning or elimination of infested plants can reduce pest problems during the next growing season. Other factors, including disinfesting a soil or medium with chemicals or heat, can reduce insect populations and infestations the next growing season. The rotation of crops can also greatly reduce pest problems. This approach is most effective on insects that have a limited host range, are relatively immobile, and have long life cycles. One must be careful when choosing the crops to be rotated, so that conditions that favor new pests will not be created. Timing planting, growing, and har-

vesting crops so that these occur when pest populations are low can also be effective in avoiding pests and problems. An alternative is to choose plants that are resistant to an important pest (Figure 16-35).

Reproductive Controls

These controls include the use of sterile insects and sex attractants and have been used when outbreaks of the Mediterranean fruit fly (medfly) have threatened the fruit and vegetable industries in California. Insects can be sterilized chemically or by radiation, or can be genetically manipulated. Sexual attractants (pheromones) can be used to attract the wild-type or fertile insects to an area with a high population of released sterile insects. Sexual attractants can also be used to lure insects into traps, or

FIGURE 16-35. Because this potato plant contains the gene for leptine, a chemical that repels insects, the Colorado potato beetles will only take one or two bites before they are repelled. Selecting such resistant plants can be part of an integrated approach to pest management.
Photograph courtesy of the Agricultural Research Service, USDA.

confuse males so that they cannot locate female insects. It will be interesting to see if reproductive controls play a major role in insect control in the future.

Biological Controls

The term **biological control** is applied to the use of living organisms (parasites or predators) to control pests. For biological control to be successful, the population of the parasite must be sufficient to reduce the number of insect pests to acceptable levels. Many parasites will not completely eliminate a pest. If the insect pest population in wiped out, the parasite will not have a food source and will die out also. Therefore, an equilibrium must be established between the parasite and the pest. Biological controls will also be less effective if the parasites or predators are too mobile, or if the grower uses pesticides that are toxic to the biological control agent.

A wide variety of biological control agents have been utilized, including: birds, mites, spiders, toads, insects, and microbial agents (Figs. 16-36, 16-37, 16-38). One of the most successful of the biological control agents, the bacterium *Bacillus thuringiensis* (BT), is marketed as a commercial insecticide and will kill several kinds of caterpillars. It is used on crucifers because of its effectiveness against imported cabbage worms and cabbage loopers. Because BT does not maintain itself in the environment, it must be applied with each infestation. Scientists have also **genetically engineered** the gene for the BT toxin into several plants (e.g., corn, potato, walnut) to render them resistant to lepidopterous insects (see again Chapter 4).

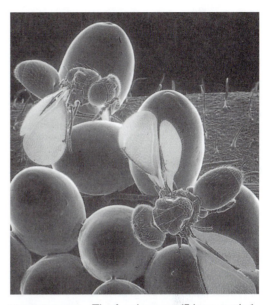

FIGURE 16-37. The female wasps (*Edorum puttleri*) in this photograph are so small that they can stand on a pinhead. They are preparing to oviposit into the eggs (larger oval structures) of the Colorado potato beetle. The wasps have been released as a form of biological control of this serious potato pest.
Photograph courtesy of the Agricultural Research Service, USDA.

FIGURE 16-36. The seven-spotted lady beetle (*Coccinella septempunctata*) (known as C-7) is larger than the American lady beetle. C-7 has been introduced to control aphids. It is pictured here eating a pea aphid (*arrow*).
Photograph courtesy of the Agricultural Research Service, USDA.

FIGURE 16-38. This reddish-brown parasitic wasp (*Microplitis croceipes*) is laying her eggs in the tobacco budworm (*Heliothis virescens*). The eggs will hatch and the new developing wasps will eat and live on the insides of the caterpillar. The caterpillar becomes sluggish, then will eventually die as a result of the parasites.

Photograph courtesy of the Agricultural Research Service, USDA.

Chemical Pesticides

Insecticides historically have been classified by the manner in which they enter the insect's body. **Stomach poisons** enter through the digestive system and thus require that the pest ingest the plant materials. Therefore, stomach poisons are only effective on insects with chewing mouthparts, because they eat plant parts. Insects with piercing-sucking mouthparts insert their stylets past any insecticide that is present on the plant surface and thus avoid consuming surface-applied stomach poisons. Insects with piercing-sucking mouthparts must therefore be controlled with **contact poisons** that are sprayed directly onto the pest or by **systemic poisons** that are ingested with the plant sap. **Fumigants** are another type of poison, because they enter the insect as a gas.

Because modern science has given us so many more and different types of pesticides, it is not surprising that many do not fit neatly into the categories mentioned in the preceding paragraph. Insecticides are therefore generally classified as inorganic and organic. Most synthetic insecticides are organic compounds and can be further classified as oils, biological derivatives, and synthetics. Some insecticides can be taken up by and translocated throughout the plant and are therefore called *systemic* insecticides. Other insecticides are nonsystemic. Insecticides are continuously introduced to and removed from the market because of new mammalian toxicity data and pest resistance. We therefore, will not list any specific insecticides in this textbook. Extension bulletins, trade journals, other publications and specialists should be consulted for current information on pesticides. *When using any pesticide, one should first read the label and carefully follow all instructions. Rules regarding protective clothing, proper application methods, disposal, reentry time, and timing before eating food crops must be adhered to for safe handling and usage. It is illegal to use any pesticide for purposes other than those specifically listed on the label.*

Physical Methods

Physical methods may have merit for reducing insect populations in certain specialized circumstances. Applying a strong spray of water to house plants or those used in interiorscaping is a simple method for removal of white flies, aphids, insect eggs, and spider mites. Such a spray must be forceful enough to remove the pests while gentle enough to avoid damage to the plants. Because this technique involves use of large water volumes, provision must be made for ease of water removal or drainage. Hand-picking of large insects such as tomato hornworms may be practical for small plantings or where labor is plentiful. Another physical method for insect removal has been explored for reducing lygus bug (tarnished plant bug, *Lygus lineolaris*) populations in strawberry plantings; it employs a large machine that literally vacuums the insects from the crops (Figure 16-39). The efficacy of this method is still

FIGURE 16-39. This machine employs an over-the-bed vacuum to suck up and remove a serious pest of strawberries, tarnished plant bugs (*Lygus lineolaris*), from this strawberry field.
Courtesy Plant Sciences, Inc. Watsonville, CA.

under investigation, but it exemplifies the potential for nonpesticide approaches to insect management.

Additional physical methods are often employed for greenhouse insect management. For example, screens are placed over cooling pads, vents, and other openings to prevent insects from entering the greenhouse. A commonly used screen is made of polyethylene and has 400 mesh size holes, but it may be necessary to wash off the screen to eliminate accumulation of dust that could otherwise reduce necessary airflow. Another physical technique employed in greenhouses is the use of yellow sticky cards. These cards are placed just above the height of the plants and are inspected frequently for the presence of insect pests such as aphids, fungus gnats, and white flies that are attracted to the yellow color. Although such sticky cards may help to directly reduce insect populations, their primary use is to enable the grower to monitor presence and levels of pests and thus determine necessity and timing of insecticide application.

Legal Control

States and the federal government have specific rules regarding the transportation of plant materials. These rules have been developed to prevent the importation of pests. Sometimes, plants are quarantined for a period of time to allow officials to determine that plants or plant parts are free from specific insect pests. United States federal quarantines are administered by APHIS, the Animal and Plant Health Inspection Service. APHIS inspects plant material that enters the country to ensure that it is free from certain pests. Quarantines protect growers of horticultural crops because they reduce potentially serious pest problems.

It is through an integrated approach that horticulturists can produce quality crops with acceptable low levels of insects and insect damage. Many basic horticultural practices are geared directly to insect pest problems. These practices can provide either protection from or control of various insect pests.

MITES

Like insects, spider mites are members of the phylum Arthropoda. These pests are in the class Arachnida, which is characterized by animals possessing four pairs of legs, an exoskeleton, an abdomen, and a cephalothorax, which is a body region with characteristics of both a head and a thorax. Mites are very small individuals of the order Acari; they are smaller than ticks and some are even microscopic in size. The most important mite pests on plants are the spider mites, especially the twospotted spider mite. However, other mites, including the cyclamen mite and the bulb mite are also serious horticultural crop pests.

The twospotted spider mite is a serious pest on a wide number of floricultural crops, woody landscape plants, vegetable crops, and fruit crops. These tiny animals are less than 0.5 mm long. They reproduce sexually and lay eggs that are clear to light green in color. After hatching, the mites pass through four motile nymph stages. They cause a yellow stipple or speckled appearance on leaves and flowers (Figure 16-40).

FIGURE 16-40 Tiny discolorations or stipples result from feeding by spider mites. This injury could be confused with ozone damage.

When infestations are heavy, webbing will be visible, and leaf drop and a lowering of plant vigor will be evident. Spider mites usually feed on the undersides of leaves. One can see twospotted spider mites on the undersides of infested leaves by looking at the side directly under the stipple. If the mites move, they are alive, which indicates that the previous pesticide spray was ineffective. Another way to confirm the presence of spider mites is to shake the infested leaf or branch over a clean piece of paper. Specks will fall onto the paper and if the tiny specks move, it is a sure sign of spider mites.

Spider mites are sensitive to temperature and humidity and reproduce faster under relatively warm and dry conditions. Conditions that favor rapid reproduction result in serious damage to horticultural crops. Because of the speed of reproduction during warmer weather, infestations can "sneak up" on a grower before he or she is aware of the problem.

Control of twospotted spider mites can be difficult because they develop resistance to miticides fairly rapidly. Numerous miticides are available on the market however. Some success also has been reported with predatory mites such as *Phytoseiulus persimilis*. In fact, a strain of a predatory mite has been bred for resistance to organophosphate and carbaryl insecticides, which makes it possible to use an integrated approach of a biological control agent in combination with certain pesticides to control the twospotted spider mite.

SNAILS AND SLUGS

Snails and slugs are members of the animal phylum Mollusca, which is characterized by invertebrates with soft bodies that are usually protected by a hard outer shell. Slugs are mollusks that lack such a shell (Figure 16-41). Bodies of both snails and slugs have four distinct parts: (1) a head with sensory tentacles, eyes, and mouth parts; (2) a foot for locomotion; (3) a mantle, which is a fold of skin over the back; and (4) the visceral mass that contains most of the internal organs. Some mollusks also have a shell.

Within the mouth is a structure called the radula that is covered with rows of teeth. Therefore, snails and slugs are capable of ingesting plant tissue.

Damage to horticultural crops by snails and slugs depends on the population density of the pests, the species, and the degree of movement. Movement and feeding are influenced by environmental factors

FIGURE 16-41. Slugs such as this one often cause severe damage to garden vegetables and flowers.

including temperature, relative humidity, light level, soil conditions, and the plant population that is present. To learn the behavioral patterns of snails and slugs, one must understand how they respond to light and moisture. Wetness, in particular, attracts slugs and they will come out to feed during the daylight hours during or after a rain or prolonged irrigation. Snails and slugs generally hide in dark, moist, protected locations, such as under living or dead plant material, under mulches, in the soil, or under boards or logs. They emerge during dusk and feed during the night and return to their hiding places at dawn. One often can trace portions of their paths by following the mucus-covered trails that they leave.

Snails and slugs can be controlled chemically, culturally, or biologically. Molluscicides generally are applied in baits that attract and then poison snails and slugs. Some molluscicides, however, act as repellents at the proper concentrations; thus the pests will not reach the plants to feed and cause damage. Cultural practices, such as plowing can help control snails and slugs, especially if large clods are broken up. This gives fewer hiding places for the pests. Barriers that dehydrate the mollusks, such as table salt, or completely dry quick lime can be effective barriers to slug and snail movement. It also has been demonstrated that shallow pans filled with beer, or yeast and water, placed at soil level, will attract slugs that will crawl in and drown in the liquid. Various parasites of snails and slugs have been shown to provide good control. Therefore, a variety of approaches are available to control these pests.

VERTEBRATE PESTS

Higher animals can seriously damage horticultural crops. Examples of vertebrate animals that cause damage to plants include moles, voles, mice, gophers, rabbits, deer, beavers, birds, and pets (Figure 16-42). The damage that they cause is variable and potentially devastating. Control measures vary, including traps and poisons, methods that exclude the pests (fences), techniques for scaring or repelling them (e.g., scarecrows) and various cultural and biological controls.

For illustrative purposes, we will discuss damage to fruit trees by voles. Voles are members of the genus *Microfus* and are small mouselike rodents with short tails. They either run along the soil surface (meadow voles) or along tunnels below the soil surface (pine and prairie voles). Voles damage fruit trees by gnawing through and removing bark, including the phloem and vascular cambial tissues of the main trunk or large lateral roots. This damage can weaken or kill fruit trees by facilitating the entry of pathogens through the wounds or from the roots starving for photosynthate. Because the xylem is intact, the top of the tree will continue to be provided with water and nutrients for a time, but death will eventually occur.

It is possible to repair damage done by voles to tree trunks by bridge grafting (Figure 16-43). A variety of control measures are available to a fruit grower, ranging from reducing vole populations to protecting the trees. Tree guards and hardware cloth around the base of a tree can offer some protection, especially from meadow voles (Figure 16-44). Placing sand or pea gravel around the base of the trunk also can discourage the pests. Trapping can help reduce vole populations, as can elimination of vegetation under the trees by tillage or herbicides. Various biological controls can be used, including grazing by hoofed animals to crush tunnels; however, these larger animals may cause more damage to the trees then they prevent. Snakes can be introduced into an orchard; however, pickers and other workers tend to rebel and may not enter orchards because of this practice. In addition, a number of chemicals are available for the control of voles.

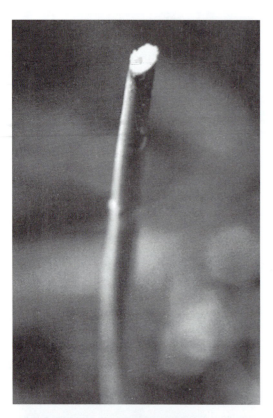

FIGURE 16-42. Vertebrate pests can cause serious damage to plants. The young maple tree at the top was damaged by a rabbit. Rabbits typically make a clean cut with their teeth. Beavers can devastate landscape trees near ponds and lakes (*bottom left*). Male deer (bucks) rub the velvet off their new antlers each year. They prefer to rub on young trees that will bend. They will strip off bark (*bottom right*) or even break trees as a result of their rubbing.

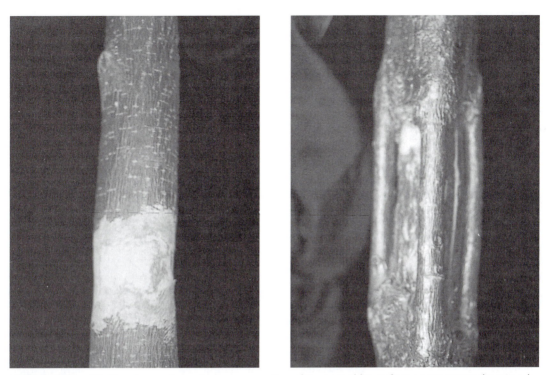

FIGURE 16-43. Girdling damage by rodents can sometimes be repaired by grafting scions across the injured area. This *bridge graft* reunites the cambial and phloem regions above and below the damaged area.

A knowledge of the types of damage and the biology of vertebrate pests will greatly add to one's understanding of their control. In many cases, constructing barriers to prevent the entry of the animals will reduce or eliminate losses. For example, netting over fruit trees, bushes, or canes on a small scale will prevent birds from eating valuable fruit. Problems can arise, however, when one or two birds somehow manage to get under the net (Figure 16-45).

FIGURE 16-44. Hardware cloth is often placed around the base of fruit trees to prevent rodent damage.

FIGURE 16-45. Bird damage to garden peas. In this case, the damage was caused by a cardinal, a protected species, so exclusion is the best control measure to employ.

WEEDS

A weed can be defined simply as a plant out of place. This plant may or may not be of economic value when grown for its own attributes; for example, a tomato plant growing in a field of onions would probably be considered just as much of a weed as if it were a redroot pigweed (*Amaranthus retroflexus*).

Weeds greatly reduce yields and plant quality because of competition, allelopathy, and the fact that they can serve as hosts for pests. Weeds compete with economically important plants for water, nutrients, and light. Many weeds are aggressive and grow rapidly, quickly establishing deep root systems, and crowding out the horticultural plants. Weeds, such as quackgrass, not only compete with crop plants, but also exude inhibitory substances (allelochemicals) from their living or dead parts that can adversely affect the growth of crops. Weeds can also serve as hosts for insect pests and diseases that affect horticultural plants. Weed control is therefore critical for the production of high-quality plants.

Typical weed species are extremely successful in part because of their propagation strategies. Weeds may spread vegetatively and/or by seeds. Additionally, they may have dormancy mechanisms controlling their spread and multiplication, making weed control difficult.

Individual weeds may produce seeds in the hundreds, thousands, or, as in the case of tumbling pigweed, several million. Some weeds produce abundant seed crops that have special features that enable wide dispersal. The tumbling pigweed, as its name suggests, spreads its seeds as the wind blows the top of the plants across the landscape. Seeds of other weeds, such as the dandelion, are adapted to be spread in wind currents (Figure 16-46). Weed seeds, such as those of cocklebur and sand bur, cling to animal hair or to clothing and may be carried over great distances. Birds often ingest fleshy fruits that contain seeds and carry them for a considerable distance; as a result it is not uncommon to see plants such as blackberries and asparagus growing as weeds along fence lines. Weed seeds also may be spread in water and on soil. Washing tractor tires between fields will reduce the dispersal of weed seeds as well as preventing the spread of disease-causing organisms and nematodes.

Weed seeds also are spread over wide areas when present as contaminants in seed lots of horticultural and other crops. Within the United States, information pertaining to weeds and the control of weed seeds present in seed lots offered for sale varies considerably among the individual states. States also differ in their lists and tolerances (minimum numbers present) of noxious weeds (weeds arbitrarily determined by law to be troublesome) that are allowed to be present in seeds that are purchased. People producing or selling seeds should be aware of their state laws that pertain to tolerances allowed for seeds of weed species (see Chapter 14).

In addition to spreading by seeds, many weeds reproduce vegetatively. Some serious weeds, including quackgrass (*Agropyron repens*) (Figure 16-47) and Canada thistle (*Cirsium arvense*) have underground stems or rhizomes from which new shoots emerge some distance from the parent plant. Mechanical cultivation practices can chop up and disperse the rhizomes of these weeds

FIGURE 16-46. Dandelion seeds (achenes) are dispersed by wind.

and have the negative effect of propagation and spreading them over the entire field or garden. Other underground stem structures, such as the corms that form on nutgrass (*Cyperus rotundus*), also serve as vegetative propagules. Another example of a weed that spreads vegetatively is wild garlic (Figure 16-48), which forms bulbils in its inflorescence (vegetative apomixis). When they get heavy enough to cause the stem to bend over, they root and form a new clump where they contact the ground.

A knowledge of the dormancy of weed dispersal units (seeds and vegetative propagules) is important to the understanding of weed biology and control. The seeds of many weed species are positively photoblastic; therefore when the soil is tilled or otherwise turned over, weed seeds are brought to the surface, the light overcomes their dormancy, and the seeds germinate. In part, because of dormancy mechanisms like this, weed seeds can remain dormant in the soil for many years, and germinate as soon as they are moved to the soil surface. Therefore, although tillage can be an excellent weed control practice, it can aid in the propagation of these unwanted plants.

Integrated approaches are aimed toward the prevention, control, or eradication of weeds. These include individual or combined use of physical, chemical, and biological control measures.

FIGURE 16-48. Wild garlic (*Allium* sp.) produces bulbils in its inflorescence. This is an example of adventitious apomixis.

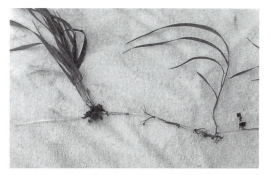

FIGURE 16-47. Quackgrass is a species that can spread over distances of several meters by its slender, penetrating rhizomes. A new plant can be produced at each node.

Physical approaches to weed control include pulling weeds by hand, hoeing, cultivating, mowing, burning, flooding, and mulching; they do not include chemicals or biological measures. Physical methods can be labor-intensive and may or may not be adaptable for large plantings; however, they may be combined efficiently with chemical control. Weed control is probably the primary reason for cultivation. Because of concerns about soil erosion, however, many growers are reducing tillage operations in efforts to save soil, reduce tractor fuel costs, and save time. Simply scraping weeds off at the soil surface with a hoe has been shown to be equal to and sometimes better than cultivation. Burning can efficiently reduce weed populations under greenhouse benches or in fields. Fire offers the additional advantage of destroying viable weed seeds, and the ashes that remain can be a source of nutrients. Mulches can effectively control weeds if they are applied correctly. Weeds can grow in or through organic or gravel mulches, but this problem can be prevented by applying the mulch thickly, putting chemicals or other barriers under the mulch, or by pulling weeds that grow through the mulch. Mulching is covered in more detail in Chapter 10.

Many herbicides (chemicals that kill plants) are available for the chemical control of weeds. The use of herbicides has increased rapidly since the early 1940s, following the introduction of 2,4-D. Like other pesticides, commercially available herbicides are changing constantly as new chemicals are introduced, and others are removed from the market.

Herbicides may kill plants selectively, or nonselectively and may act systemically (be translocated) or by contact. Many of the selective herbicides are more effective on either monocot or dicot weeds, but not both. This can be advantageous, if for example, the crop is a dicot, a grass herbicide should take out monocot weeds and not damage the crop. In other cases, it is an advantage to be able to kill anything that is green in an area for complete weed control. Systemic herbicides can offer the advantage of killing below-ground plant parts and thus provide better weed control than contact poisons on some weed species because new shoots may sprout from roots if the chemical only kills the top of the plant.

Herbicides also are classified according to the timing of application, which may be determined with respect to the growth pattern of the weed or the crop. Many chemicals are applied prior to planting and are therefore referred to as preplant herbicides. Preplant herbicides are sometimes incorporated into the soil prior to planting the crop to avoid waiting for a rain or irrigating to wash the herbicide into the soil and are referred to as preplant soil-incorporated herbicides. Other herbicides are applied prior to the emergence of either the crop or weed seedlings, or both (pre-emergence

herbicides). Many herbicides are postemergence herbicides and are applied after the crop or weed seedlings have emerged. These herbicides must be very selective, or be applied directly to the weeds to avoid contact with the crop. Utilizing genetic engineering techniques, scientists have incorporated genes for herbicide resistance into crop plants to allow for the spraying of broad-spectrum herbicides over the crop to kill the weeds and not the economically important plants. For specific application methods, students should consult the reference books listed at the end of this chapter and consider taking courses in weed science.

Weeds also can be controlled by utilizing living organisms, such as insects and other animals (Figure 16-49). A biological control

FIGURE 16-49. This small orange-and-brown flea beetle (*Aphthona flava*) was imported from Italy to feed on leafy spurge. The larval stage of the beetle eats roots and root hairs, thus helping control this costly weed of the rangelands and pasturelands in the Great Plains.

Photograph courtesy of the Agricultural Research Service, USDA.

agent must selectively control the weed(s) and not damage economically important plants, should reduce the weed population to acceptable levels, should be adapted to the specific environment, should seek out the weeds, and should not be severely limited by predators or parasites. The cactus moth meets these criteria and has been used effectively against the prickly pear cactus (*Opuntia*). Another interesting biological weed control is the use of geese. Because they prefer young grassy weeds, geese have been used frequently for weed control in strawberries and other crops, but the geese normally require supplemental feeding and care. Although some biological weed controls are available to growers, they may require special practices such as reductions or limitations in pesticide usage.

ORGANIC APPROACHES TO PEST MANAGEMENT

In recent years there has been an increased interest in reducing dependence on the use of pesticides and an increased focus on environmentally responsible approaches to pest management. Cultural and biological techniques discussed earlier in this chapter are important components of organic horticulture, as well as integrated pest management (IPM). National standards have been developed that must be met in order for a grower's produce to be termed "organic." Three general categories of practices and treatments must be considered in order to qualify:

1. **Approved**

   This category includes such examples as crop rotation, providing habitat for natural predators and parasites, hand or mechanical control, botanical pesticides of low toxicity and species specific effect, botanical repellents such as garlic and hot peppers, dormant or summer oils of animal or plant origin, biological control agents, insecticidal soaps, composting to reach 160°F, use of resistant varieties, hot water seed treatments, certified disease free seeds and natural nonliving or noncompetitive mulches.

2. **Regulated**

   Plastic mulch (must be removed at end of growing season); electrical and flame weeding, salt on asparagus (for weed control); botanical insecticides such as rotenone, pyrethrum, neem oil, sabadilla, citrus oil, and hellebore; biological insecticides and biological control agents to obtain bacterial, viral and hormonal extracts for insect control; insect extracts; diatomaceous earth; petroleum based oils (*only* for woody plant summer and dormant pest control); Bordeaux mixes (copper sulfate, lime and oil); certain copper compounds; elemental sulfur; hydrogen peroxide; and chlorine for seed treatment and as a disinfestant for pots, flats, greenhouse benches, and tools.

3. **Prohibited**

   All synthetic herbicides, insecticides, antibiotic sprays, fungicides, bactericides, rodenticides, acaricides, molluscacides, ovicides, fumigants, and avicides; most petroleum-derived products, including tar, asphalt, petroleum distillate herbicides, petroleum oils, aromatic petroleum solvents such as kerosene, gasoline, benzene, naphthalene, xylene, and toluene; highly toxic botanicals such as nicotine, tobacco dust, sassafras oil, and oil of citronella; naphthalene flake and mothballs; natural mineral poisons such as lead and arsenic; photo-degradable and biodegradable plastic mulches; and synthetic growth regulators. Special note: land covered with petroleum-derived pavement or road oils cannot be certified organic for three years after removal or application.

The preceding lists are not all-encompassing, so the prudent grower wishing to produce certified organic crops should check with the appropriate regulating agencies to determine acceptability of specific treatments, practices and products.

SAFETY IN THE USE OF PESTICIDES

Because pesticides are an important tool in pest management, they are used widely in horticultural production, as well as in other areas of agriculture. Licensing of pesticide applicators is required in many places. Pesticides vary greatly in their toxicity to humans. They may enter the human body through the skin, nose, eyes, mouth, or by other paths. Safety *always* must be considered regarding the mixing, application, disposal and other handling of pesticides. When used properly, pesticides can be quite safe, but improper usage can be dangerous. Growers must also consider time elapsed before worker reentry into fields and greenhouses and the time that must pass between pesticide application and harvest of plants. Different crops and situations often cause unique concerns; the application of pesticides to plants in a shopping mall (Figure 16-50) poses quite different problems from those encountered when chemicals are applied to an isolated field.

FIGURE 16-50. Indoor plantings in this shopping center require special attention to safety if pesticides must be used.

The following safety tips must be considered for any application of pesticides.

1. Only use pesticides that have been cleared for use on the crop or species to which they will be applied. That is the law.
2. Only use pesticides that are cleared for that specific locality. In the United States, the pesticides that may be used on a particular crop often vary from state to state.
3. Always read the label on the pesticide container carefully every time you use the chemical and do not deviate from these instructions. The only legal uses are those shown on the label.
4. Store the pesticide only in its original container. It will then be in its properly labeled container, which will greatly reduce the risk of misuse.
5. Follow the label directions carefully, and mix the pesticide according to the instructions.
6. Calibrate equipment properly to ensure that proper amounts are applied. Excess amounts are illegal and may cause **phytotoxicity**.
7. Always wear protective clothing. This clothing must be worn properly; tucking pants into boots, for example, makes it easier for a pesticide to get into the boots and be absorbed through the feet, so pants must be worn outside of boots.
8. Avoid contact of the chemical with the body or clothing.
9. Never smoke or eat during a pesticide application, in an area (e.g., greenhouse) in which a pesticide was recently applied, or until protective clothing has been removed and washed.
10. If spraying or dusting outdoors, remember that winds can cause chemicals to drift, possible causing damage to an adjoining area. Therefore, spray or dust only under calm conditions such as in the early morning.

11. When applying pesticides, be sure to cover or remove food and water containers for pets and livestock.

12. When selecting and applying pesticides, consider the activity of beneficial animals such as honey bees and their sensitivity to specific pesticides.

13. If any sickness or unusual feeling is experienced either during or after handling a pesticide (mixing, applying, disposing of, or cleaning up), *call a doctor* immediately. Always know how to contact the local Poison Control Center.

14. Whenever possible, work in pairs. If there is some type of accident, the partner can help.

15. Dispose of excess chemicals and containers properly.

16. Always wash thoroughly with soap and water after handling pesticides. Wash clothes separately from the clothes of other household members.

17. Always keep a written record of the chemicals applied, the rate of application, the plants or areas that were treated, and the date. This record will have several uses, including legal protection.

SUMMARY

Pests can damage plants and reduce the yield and quality of horticultural plants. To grow competitively priced plants successfully, pests must be controlled. Although control measures vary among pests and crops, an integrated approach is a sound method. An integrated approach includes proper genotype selection and use of appropriate growing conditions, sanitation, and horticultural practices, along with various control measures. Only with knowledge of these factors and experience in the techniques and art of growing can a horticulturist produce the high-quality plants required for success in the horticultural enterprise.

REFERENCES

Agrios, G. N. 1988. *Plant Pathology* (3rd ed.). Academic Press, San Diego, CA. 803 pp.

Agrios, G. N. 1997. *Plant Pathology* (4th ed.). Academic Press, San Diego, CA. 635 pp.

Anonymous 1981. *Weeds of the North Central States*. North Central Regional Research Publication No. 281. Urbana, IL.

Barbosa, P. 1989. *Introduction to Forest and Shade Tree Insects*. Academic Press, San Diego, CA. 639 pp.

Bernays, E. A. (Ed.). 1989. Insect-Plant Interactions. CRC Press, Boca Raton, FL. 2 vols.

Chaube, H. S., and U. S. Singh. 1991. *Plant Disease Management: Principles and Practice*. CRC Press, Boca Raton, FL. 319 pp.

Chet, I. (Ed.). 1987. *Innovative Approaches to Plant Disease Control*. Wiley, New York. 372 pp.

Dixon, G. R. 1984. *Plant Pathogens and Their Control in Horticulture*. Macmillan, London. 253 pp.

Elzinga, R. J. 1981. *Fundamentals of Entomology* (2nd ed.). Prentice-Hall. Englewood Cliffs, NJ. 422 pp

Gwynne, D. C., and R. B. Murray. 1985. *Weed Biology and Control in Agriculture and Horticulture*. Batsford, London. UK. 258 pp.

Hornby, D., and R.J. Cook (Eds.). 1990. *Biological Control of Soil-Borne Plant Pathogens*. CAB International, Wallingford, UK. 479 pp.

Klingman, G. C., and F. M. Ashton. 1975. *Weed Science: Principles and Practices*. Wiley, New York. 431 pp.

Miller, J. R., and T. A. Miller (Eds.). 1986. *Insect-Plant Interactions*. Springer-Verlag, New York. 342 pp.

Mukerji, K. G., and K. L. Garg. 1988. *Biocontrol of Plant Diseases*. CRC Press. Boca Raton, FL. 2 vols.

Pfadt, R. E. (Ed.). 1985. *Fundamentals of Applied Entomology* (4th ed.). Macmillan, New York. 742 pp.

Pirone, P. P. 1978. *Diseases and Pests of Ornamental Plants* (5th ed.). Wiley, New York. 566 pp.

Pyenson, L. 1981. *Plant Health Handbook*. AVI, Westport, CT. 241 pp.

Roberts, D. A., and C. W. Boothroyd. 1984. *Fundamentals of Plant Pathology* (2nd ed.). W.H. Freeman, New York. 432 pp.

Ross, M. A., and C. A. Lembi. 1985. *Applied Weed Science*. Burgess, MN. 340 pp.

Singh, D. P. 1986. *Breeding for Resistance to Diseases and Insect Pests*. Springer-Verlag, New York. 222 pp.

Smith, E. H., and D. Pimentel (Eds.). 1978. *Pest Control Strategies*. Academic Press, New York. 334 pp.

Trigiano, R. N., M. T. Windham, and A. S. Windham. 2004. *Plant Pathology*. CRC Press, Boca Raton, FL. 413 pp.

CROP

SCHEDULING

his textbook has focused on genetic and environmental factors that are used to produce horticultural crops. Examples were chosen from various commodity groups to emphasize that similar considerations apply to different crops. Nowhere in the textbook was a crop taken through all the steps from propagation through production, harvesting, and marketing. Recognizing that a horticultural crop is of little value until marketed or until it has reached its final use, this appendix addresses this important topic. The procedures for one crop from each horticultural commodity group are described in table form from propagation through marketing. The exception is tree fruits. Both apple and orange were chosen because cultural considerations are quite different for temperate and subtropical fruit crops.

Students should look for similarities and differences in growing the various crops. In all cases genotypes are chosen and the plants are given various cultural conditions to produce a high-quality product (desirable phenotype). All of the crops require fertilizer, and all must be propagated. Fertilizer requirements vary with the crop and soil conditions, and propagation techniques are specific to the given crop species.

The crops chosen here are commonly grown, but subtle differences in cultural considerations may vary with the cultivar. Consulting reference books and experienced growers will prove invaluable for growing successful crops.

With these schedules as a guide, students may schedule their own favorite crops. Although similarities exist among the tables for the various crops, the headings in each table differ. Students should modify their tables accordingly as they consult various references to determine all of the factors that must be considered to produce a successful horticultural crop.

SCHEDULE A   FIELD-GROWN PROCESSING TOMATO
(Lycopersicon esculentum)

PROPAGATION			
CLIMATE	**SITE SELECTION**	**SOIL**	**SOIL TESTING**
Tomato is a warm season crop that will not tolerate freezing temperatures. There must be at least three to four months from sowing seed to harvesting fruit. Photoperiod is not important, but tomatoes must be grown in areas with high sunlight for development of maximum flavor and sweetness.	A level, well-drained field with uniform soil conditions is best.	Tomatoes can be grown on many types of soil. A deep loam soil high in organic matter is best. There should be few soil pathogens. The tomato crop is therefore often rotated with other crops on a yearly basis. The pH can range from 5.5 to 7.5, ideally 6.0 to 6.5.	The soil should be tested to determine the pH and nutrient availability.

FERTILIZER AND pH ADJUSTMENT	**SOIL PREPARATION**	**PURCHASING SEED**	**PLANTING**
Fertilizer and possibly lime should be added based on the results of the soil test. Nitrogen is important for adequate vegetative growth. Amounts of other nutrients will depend on the soil test.	Fall and/or spring plowing is usually practiced. Soils are often plowed deeply. After plowing, the soil is disked or tilled to produce a fine seed bed.	Cultivars must be selected with consideration of fruit qualities for method of harvest (usually mechanical), uniform ripening, yield, fruit color, proper size, ability to withstand transportation, soluble solids content, acidity and pH, shape, pest resistance, and freedom from defects. Determinate cultivars that are not staked are used frequently for mechanically harvested processing tomatoes.	Although transplants are used sometimes, most processing tomatoes for mechanical harvest are directly seeded. Soil temperatures, weather, plant density, and spacing of rows must all be considered.

PRECISION PLANTERS

Various seeders are used to sow tomato seeds. Some planters apply a starter fertilizer high in phosphorus and containing a material to prevent soil crusting at the time of planting. Planters may plant clumps of four to seven seeds or plant single seeds.

GROWING THE CROP			
MULCHES	**IRRIGATION**	**WEED CONTROL**	**INSECT CONTROL**
Polyethylene mulches may be used for tomatoes to prevent weeds from growing, to reduce soil erosion, and conserve valuable soil moisture. Machines are used to lay the mulches.	Tomatoes may not need irrigation in areas with sufficient rainfall. Yields will be reduced and incidence of blossom-end rot will be high if tomatoes do not receive sufficient water. Over-head sprinklers can be used but may result in fruit rots when applied late in the season. Flooding irrigation is used frequently.	Although mulches may be used, weeds usually are controlled by mechanical cultivation and the use of herbicides.	Tomatoes are susceptible to several insect pests that can cause considerable crop losses. A variety of biological and chemical controls may be used, combined with crop scouting for density and location of the pests.

DISEASES		**HARVEST CONSIDERATIONS**	
Many pathogens attack tomatoes. Cultivars can be selected with resistance to pathogens and physiological disorders such as cracking. Monitoring pests and using an appropriate cultural and spray program are important for producing a successful crop.		Large weeds should be removed to allow for the mechanical harvesters. At the time of harvest, the crop should have a sufficient number of ripe fruit; otherwise a ripening agent such as ethephon must be applied to promote ripening.	

MARKETING			
HARVESTING	**FIELD SORTING**	**DRY SORTING**	**GRADING**
Fields are mechanically harvested if the soil is not too wet and the crop is at least 80 percent ripe.	The field-sorting crew removes defective and poorly colored fruit according to standards related to the intended use of the fruit.	Leaves, soil and defective fruit are removed within the processing plant.	Within the processing plant, tomatoes are sorted according to their size.
WASHING	**TRIMMING, SORTING, AND CORING**	**PEELING**	**PROCESSING**
Soil and other foreign substances are washed from the fruit.	Small green areas and rotten areas can be cut off the fruit. Fruit are sorted for different processing uses, including canned whole fruit, sauces, or paste. Cores can be removed by hand or machine.	Tomato fruit are peeled using steam, lye, or infrared radiation.	Tomatoes are then canned. Alternatively they may be crushed or chopped prior to extraction and canning of juice.

DISTRIBUTION
Canned tomatoes or tomato products are transported to warehouses where they will be held until sold in grocery stores throughout the year.

SCHEDULE B   GREENHOUSE-GROWN POINSETTIA
(*Euphorbia pulcherrima*)

	STOCK PLANTS		
ORDERING	**TRANSPLANT UPON RECEIPT**	**MEDIUM**	**FUNGICIDE DRENCH**
Plants in 5.5 cm ($2\frac{1}{4}$ in.) pots or rooted cuttings should be ordered to arrive in March, April, May, or June.	Transplant into pots or beds. Spacing or size varies with time of arrival. Those arriving earlier will grow larger and therefore require wider spacing. Bed spacing or pot size: 45×45 cm (18×18 in.), or 30 cm (12 in.) pots 38×38 cm (15×15 in.), or 25 cm (10 in.) pots 30×30 cm (12×12 in.), or 20 cm (8 in.) pots 20×20 cm (8×8 in.), or 15 cm (6 in.) pots Pots should be spaced at the same distances as plants in beds.	Must be pathogen-free. Excellent success has been achieved with light-weight, highly porous media.	Immediately after planting, drench with a fungicide or fungicide combination to control *both* water molds and non-water mold fungi.

FERTILITY	**LIGHT**	**HUMIDITY**	**TEMPERATURE**
Provide constant liquid feed (fertigation) to provide 250 ppm N, 44 ppm P_2O_5, 150 ppm K_2O), and 0.1 ppm Mo at each watering or use higher concentrations and an intermittent fertilization program.	During establishment, use a shade cloth or other shading to provide approximately 50 percent shade. After establishment, shading may be reduced somewhat. To avoid flower bud initiation, provide night interruption from (10:00 P.M. to 2.00 A.M. until May 15).	Relative humidity should be high during daylight hours during establishment to minimize plant stress.	During establishment provide 18°–21°C (65°–70°F) night temperatures. After establishment, provide 27°–29°C (80°–85°F) day temperatures.

PINCHING	**SURFACE DISINFESTATION**
Immediately after establishment, make a soft pinch to remove the apical growing point including one fully expanded leaf. As a result at least three buds will grow from below the pinch. When new growth is fully developed with good green color and at least four fully expanded leaves, pinch again. Pinch new shoots when they reach adequate size.	One or two days prior to taking cuttings, spray plants with fungicide combination plus wetting agent or dilute liquid chlorine bleach plus wetting agent.

ROOTING CUTTINGS

PURCHASING DIRECTLY	HARVESTING CUTTINGS	SANITATION	PLANT GROWTH REGULATORS
Rooted, callused, or nonrooted cuttings may be purchased from propagators. These generally are received in August and September.	These are generally harvested from June 15 until September 30. Select branches with no signs of diseases. Using a sterile, sharp knife, cut between the third and fourth fully expanded leaves from the tip. Do not remove leaves from cuttings. The optimum length is 6–7.5 cm (2½–3 in.). If another crop of cuttings is needed, leave two leaves on each stem from which a cutting is removed.	Workers must first wash their hands with soap and water, then rinse with disinfestant. Harvest cuttings with a sterile knife, and collect cuttings in a sterile container. Do not allow the sterile container to sit on the bench with the stock plants. Stick cuttings in disinfested propagation medium. The propagation bench should have been sterilized. Make holes or trenches with a clean tool in the medium prior to sticking cuttings. Other sanitary practices must be implemented (e.g., feet should *never* be placed on benches).	Some growers use none, but IBA in talc or liquid solution may be applied at the rate of 1500 to 2500 ppm. If plants show excessive growth in the propagation bench, they should be treated with a growth retardant.

HUMIDITY	ROOTING MEDIUM	TEMPERATURE	FERTILIZATION
Cuttings should be harvested from turgid stock plants in early morning or evening. Avoid water stress on collected cuttings. Use intermittent mist or fog in the propagation area. The mist frequency should be adjusted so the cuttings always have a film of water on their leaves. Shade the cuttings during the bright summer months to reduce evapotranspiration. Rooting should occur in 14 to 21 days. Misting intervals and frequency should be gradually reduced, then eliminated.	The rooting medium should be disinfested and well drained. Combinations are frequently used (e.g., perlite and vermiculite, perlite and moss peat, a combination with sand, or the final growing medium). The pH of the medium should be 5.5 to 6.0	The optimum air temperature for rooting is 21°C (70°F). The medium should be maintained at 21° to 27°C (70°–80°F). This is accomplished by providing bottom heat.	After one week a fertilizer drench may be applied. Some growers inject a mixture of ammonium nitrate and potassium nitrate in the mist.

(continued)

TRANSPLANTING

If plants are rooted directly into the pot where they will be finished, this step is not necessary. Well-rooted cuttings that have been hardened-off from the mist or fog may be transplanted. The pot size and type will vary with the cultivar, date of propagation, number of stems per pot, and market preference. Generally, the larger the pot, the more cuttings that are transplanted into that pot.

GROWING THE CROP

MEDIUM	WATERING	FERTILITY	LIGHT
Many different media have been used successfully. Media must be free from harmful microorganisms, insects, and weed seeds, and they should be well-drained, provide good aeration, hold adequate amounts of water and nutrients, and have sufficient weight for plant stability. The pH should be 5.5 to 6.5. Commonly peat-lite mixes or composted bark or leaf mixes are used.	Watering may be done by hand, by use of spaghetti microtubes or capillary mats, or by ebb and flow systems.	Constant liquid feed (fertigation) is used to provide 250 ppm N, 44 ppm P_2O_5, 150 ppm K_2O), and 0.1 ppm Mo at each watering, or use higher concentrations with an intermittent fertilization program, or top dress the medium with a dry fertilizer. Care must be taken not to "burn" the crop, or a slow-release fertilizer should be used.	Generally full sunlight should be used unless shading is necessary to control temperature. Depending on cultivar and latitude, it may be necessary to pull shade cloth for photoperiodic control of flowering. Poinsettia is a short day plant; shade cloth, when used, should be pulled for 14 hours each night.

TEMPERATURE	PINCHING	HEIGHT CONTROL	DISEASE CONTROL
The optimum daytime temperature is 27° to 29°C (80°–85°F). Minimum temperatures generally should not be below 15.6°C (60°F). Lower temperatures can affect crop scheduling, quality, and susceptibility to diseases.	Some cultivars are pinched (removal of apical bud) to promote branching. When pinching plants, leave the same number of nodes as the desired number of flowering branches. On free-branching cultivars, each node will generally produce a branch in reponse to pinching. Plants also can be pinched chemically.	Cultivars can be sprayed with chemical growth retardants to control height. Growth retardants may be effective as drenches or whole-plant sprays.	Complete sanitation is necessary to control diseases. Drench poinsettias regularly with fungicides to suppress organisms that cause stem and root rot. Air circulation is important to keep leaves and stems dry to prevent infection by *Botrytis* (gray mold). Fungicide sprays also are available to help control diseases of the above-ground portions.

INSECT CONTROL

A combination of biological and chemical control measures may be used. Prevention of introduction of the pests and control of weeds under benches that harbor insect pests is also helpful.

MARKETING

MATURITY	SHIPPING	DISTRIBUTION
Plants should not be sold before they are fully mature and have an intense bract color. Cyanthia (flowers) should still be attached.	Plants will show less deterioration when held at 10° to 12°C (50°–54°F) during shipping. Plants are often sleeved in paper or plastic before shipping. Epinasty (leaves curling under) may commonly occur during shipping from ethylene accumulation in boxes or trucks. Spacing the plants after arrival can correct this problem.	Plants are generally sold from the fourth week of November until the fourth week of December. Most growers wholesale the plants to retail outlets. Some growers sell directly to the customers at retail prices.

HOME CARE

Poinsettias should be lighted for at least 9 hours each day in the home. Fertilization is often not necessary because it will encourage unwanted new green growth. The plants should be thoroughly watered when the growing medium becomes dry and should be held at room temperature. The plants should hold their colored bracts for several months if cared for properly.

SCHEDULE C BALLED AND BURLAPPED SPREADING YEW
(*Taxus cuspidata*)

PROPAGATION	
CUTTINGS IN COLD FRAME	**CUTTINGS IN GREENHOUSE**
Relatively long (20–25 cm) cuttings with the lower needles stripped (for wounding) and with a heel or mallet of older wood at the base treated with 8000 ppm IBA in talc can be collected in the fall and placed in a cold frame. The cuttings will root during the following spring and summer.	Cuttings should not be taken from stock plants until one or several frosts have occurred. The bases of the cuttings should be wounded by stripping needles and should be treated with 8000 ppm IBA in talc plus a fungicide. Rooted cuttings may grow better if they are given a two-month cold period before lining out in the nursery. Rooting is quicker in the greenhouse than in the coldframe.

SITE SELECTION			
CLIMATE	**SITE SELECTION**	**SOIL**	**SOIL TESTING**
Taxus can grow in a variety of locations. A site with a long growing season, relatively mild weather conditions, and adequate rainfall is best.	A level, well-drained field with uniform soil conditions is preferred.	A soil that is high in organic matter that will not fall from the ball is best. A well-drained soil with good water-holding capacity that does not crust or heave is desirable.	The soil should be tested to determine pH and nutrient availability.

SITE PREPARATION		
SOIL PREPARATION	**PREPLANT WEED CONTROL**	**PLANTING DESIGN**
Nurseries typically are fall plowed. Organic matter can be incorporated at this time, or cover crops can be plowed under. The following spring, fertilizer and lime can be spread and disked in. The amounts will depend on the soil test.	Some nurseries use chemical sterilants or fumigants to kill soil-borne pathogens, nematodes, insects, and weed seeds. Preplanting herbicides often are applied to facilitate future weed control and help to produce high-quality plants.	The spacings between plants and rows depends on the size at which the plants will be sold. There is a variety of planting designs and some growers interplant spreading and upright plants in the field to make the most efficient use of space while allowing pathways for tractors to cultivate the field.

GROWING THE CROP			
PLANTING (LINING OUT)	**IRRIGATION**	**WEED CONTROL**	**FERTILIZATION**
Transplanting of rooted cuttings or seedlings (both called liners) into the field is called "lining out." Lining out can be done by hand or machines. It is important that rows be straight or follow a regular contour for cultivation.	Some nurseries rely solely on rainfall. However, many nurseries irrigate field-grown stock regularly to ensure high quality plants and rapid growth to shorten production time.	Generally, a combination of herbicides and cultivation by tractors is used. Some herbicides are applied with applicators that have guards to prevent contact with the crop. Hand weeding close to the plants is sometimes necessary.	Perennial nursery crops such as *Taxus* should be fertilized regularly, depending on soil tests. Typically fertilizer will be applied in the early spring, then again during the summer.

PEST CONTROL	PRUNING SHOOTS	PRUNING ROOTS	
If pests are a problem, a variety of biological and chemical controls may be used. These should be combined with nursery scouting for density and location of insects, diseases, or other problems.	Spreading *Taxus* plants typically are pruned or sheared in early spring before new growth and again during the summer as needed. The goal is to produce a well-shaped, vigorous, and full plant. A broad-spreading yew should be pruned so the height and spread are nearly equal at the time of sale.	Root pruning is often done by undercutting the plant using a *U*-shaped blade pulled by a tractor during the dormant season. A more compact and fibrous root system will result.	

	DIGGING (HARVESTING) THE CROP		
CONSIDERATIONS	**HAND DIGGING**	**MACHINE DIGGING**	**HANDLING**
To meet industry standards, the depth of the ball should be 60–75 percent of the diameter. Balls must be wrapped with burlap or other material and supported with twine or wire baskets to maintain a firm, rigid ball.	Digging is done in the autumn. Soil is excavated by pulling it away from the plant with a shovel. The ball is shaped, then wrapped with burlap or other material, pinned, and possibly tied.	Digging is done in the autumn. Machine diggers dig and power-lift the plants from the soil. The balls then are wrapped with burlap or other material and pinned by hand.	Balled and burlapped plants should be lifted by the ball and not by grabbing the branches. The root systems must be kept moist. The balls are often covered with chips, bark, or other material to prevent drying.

	TRANSPORTATION AND RETAIL SALE		
TRANSPORTATION	**UNLOADING**	**HANDLING**	**TRANSPLANTING**
Balled and burlapped *Taxus* are loaded onto trucks with care to avoid damaging the plants or allowing them to dry out.	Handling of balled and burlapped plants during unloading is critical. They should be lifted by the ball and gently set onto the ground. *The plants should not be thrown from the truck.*	The balls should not be allowed to dry out. The balls are often covered by sawdust, wood chips, or bark to prevent drying.	Holes should be dug approximately 50 percent larger than the ball. The burlap should be removed or loosened. The native soil should be used to refill the hole for best root growth into the surrounding hole. The yew should be planted at the same level as in the nursery, and soil can be mounded out in a circle to trap water. The plant should be watered thoroughly.

SCHEDULE D   APPLE, A TEMPERATE TREE FRUIT
(*Malus pumila*)

PROPAGATION			
SEEDLING ROOTSTOCKS			
SELECTION	**SEED EXTRACTION**	**SEED STORAGE**	**SEED TREATMENT**
Seedling rootstocks usually impart vigor to the tree. Common choices include French crabapple seedlings, domestic seedlings, and other crabapple seedlings.	Many apple seeds are found in the pomace that remains from pressing fruit for juice.	Apple seeds can be stored dry under refrigerated conditions for more than two years and remain viable.	Apple seeds must be stratified to overcome dormancy. Seeds can be fall-planted for winter chilling or stratified in a moist medium for up to 90 days at 2°–7°C (35°–45°F).
SOWING SEEDS	**ROOT PRUNING**	**CULLING**	
Seeds are sown in the nursery row and covered with soil. Soil crusting can impede emergence.	Seedlings often are undercut to sever the taproot and stimulate the formation of a fibrous root system.	If the size of some seedlings is unacceptable they are removed and discarded during the first year.	
CLONAL ROOTSTOCKS			
SELECTION	**CLONAL PROPAGATION**	**SOIL**	**PLANTING**
Clonal rootstocks impart specific characteristics to the tree, including size control, disease and insect resistance, early flowering, cold hardiness, anchorage, and propagation ease.	Mound layering is used most frequently, although rooting cuttings and micropropagation also can be used. We will focus on mound layering.	A soil that is high in organic matter, fertile, well-drained, and with good water-holding capacity is best.	Usually planting is in spring. Plants are spaced 30–45 cm apart and rows are 2–3 m apart.
PRUNING	**MOUNDING**	**HARVESTING LAYERS**	**HANDLING LAYERS**
The spring following planting, the plants are cut back to near ground level. This is done before bud break.	When new shoots are 7.5–30 cm tall, they should be mounded with soil using a plow, with sawdust, or with a soil-sawdust mixture to one-half the height of the shoots. Mounding is repeated in July, so that the mound is about 35 cm high, leaving approximately one-half the height of the shoots above the mound.	In the autumn the mounds are pulled down, and rooted shoots are cut close to the mother plant. The mounding begins again the following spring as new shoots form.	Layers frequently are lined-out into a nursery. They may be stored for spring planting or bench grafted during the winter.

GRAFTING AND BUDDING		
T-BUDDING		
DRAFTING AND BUDDING	**TIMING**	**CARE OF THE BUDDED PLANT**
Scions are grafted or budded onto seedling or clonal rootstocks using a variety of techniques. T-budding is one of the most common techniques used on apple.	Trees are usually budded during late summer or early fall. Bark must still be "slipping" on the understock because of vascular cambial activity.	The rootstock should be cut off just above the bud the following spring. Any shoots of rootstock origin must be removed. The new shoot should be staked.

GROWING THE CROP			
SOIL	**PLANTING SITE**	**PLANTING PLAN**	**SITE PREPARATION**
Apples can be grown on a variety of soil types, depending on the rootstock. Soil should be well drained and aerated, and the subsoil should not be impervious.	A gentle slope for air drainage is best. Low-lying sites where cold air will settle should be avoided.	The arrangement of the trees should be planned before planting. The topography of the land and selection of pollinizers are important considerations. Distances between trees must be considered and will vary with the rootstock, cultivar, and planned cultural management system.	Strips normally are tilled prior to transplanting.
TRANSPLANTING	**CARE AFTER TRANSPLANTING**	**PRUNING YOUNG TREES**	**TIME OF PRUNING**
Transplanting may be done in spring or autumn, but spring is most common. Tree roots should not dry out during the transplanting operation. Dead or injured roots should be pruned off. Normally, trees should be transplanted to the same depth as they were in the nursery.	If conditions are dry, trees must be irrigated to ensure survival. Trees may be fertilized following transplanting. Weeds must be controlled. Trees may be staked.	Trees may be pruned heavily at the time of transplanting to increase survival and stimulate branching.	Apple trees normally are pruned in late winter to early spring. Summer pruning controls excessive vegetative growth.

Pruning young trees	Pruning bearing trees	Fertility	Irrigation
The purpose is to establish the ultimate shape of the tree and to select strong, well-placed scaffold limbs with wide crotch angles. Branches that rub, grow back into the tree, and have narrow crotch angles are removed.	Older trees are pruned to control size and to allow for adequate light penetration to give a good yield of high-quality fruit.	Apple orchards are fertilized regularly depending on the age and vigor of the trees, soil and tissue nutrient analyses, soil type, cultivation or weed control practices, and presence or absence of a permanent sod.	Some apple orchards are irrigated, and some rely on rainfall. Frequency of irrigation depends on the soil type and evapotranspiration rate.

Weed control	Insect control	Diseases	Vertebrate pests
Weeds may be controlled by cultivation, herbicides, or a permanent sod in the orchard. Mulches have been used in some orchards.	Apples are attacked by a number of insect pests. Trees are sprayed regularly with insecticides. Biological controls also are becoming important.	Apples are susceptible to various bacterial and fungal pathogens. Removal of alternate hosts for pathogens and applications of sprays generally will result in higher quality fruit.	Mice and voles can attack and girdle roots and trunks. A metal or plastic guard around the trunk offers protection. Rabbits chew off the end branches. Various repellents are available for rabbits. Deer eat the ends of branches, and bucks rub their antlers on tree trunks. Exclusionary fences, chemical repellents, hanging bars of soap, or bags of human hair may be effective.

Biennial bearing	Pollination and fruit set	Fruit thinning	Preharvest drop
Some apple cultivars will bear a heavy crop one year and no crop the next. Properly thinning the fruit crop can reduce the tendency for biennial bearing.	If compatible pollinizers are present that bloom at the same time as the main cultivars, incompatibility should not be a problem. Honeybees pollinate apples, thus many growers rent hives during flowering. Cold winds will reduce bee activity. Pollen can be purchased and applied to flowers.	Apple fruit are thinned by hand or chemically to reduce biennial bearing and increase size and quality. Fruit are thinned within a few weeks following flowering.	A stopdrop must be applied to some cultivars where abscission of fruit occurs before maturity. Various auxins have been effective in controlling preharvest drop.

MARKETING			
MATURITY	**HARVESTING**	**PACKINGHOUSE**	**STORAGE**
Apples are harvested based on several indices of maturity, including brown seed color, skin color, soluble solids content, starch test, firmness, taste, and date. Harvesting at the proper time improves performance in storage and consumer satisfaction.	Apples are normally harvested by hand, although mechanical harvesters are available. Harvesters must be careful not to pull off spurs or cut or bruise the fruit.	Fruit are sized and graded, and culls are removed. Apples may be washed and waxed before being packed in baskets, boxes, or polyethylene bags.	Apples are usually cooled and kept refrigerated at a high relative humidity. Common cold (refrigerated) storage is typical, but storage time is limited. Controlled atmosphere (CA) storage where CO_2 is elevated and O_2 is reduced lengthens storage time considerably.

RETAIL SELLING

Apples are often sold out of refrigerated display racks in stores. Shelf life is shortened considerably if the fruit are not kept cool. Some cultivars store better than others and therefore are available year-round.

SCHEDULE E   ORANGE, A SUBTROPICAL TREE FRUIT
(*Citrus sinensis*)

PROPAGATION			
SEEDLING ROOTSTOCKS			
SELECTION	**SEED EXTRACTION**	**SEED TREATMENT**	**SEED STORAGE**
Many citrus species are used based on characteristics they impart to the tree, including vigor, disease and other pest resistance or tolerance, cold hardiness, tolerance to drought and salinity, effect on scion, fruit size, and quality. Freedom from seed-borne diseases including viruses is important.	Seeds generally are squeezed from the pulp and washed with water and sieved to remove the rind and pulp. Nonviable seeds are usually less dense than viable seeds and are floated off.	Seeds are immersed in agitated hot water (52°C or 125°F) for 10 minutes to eliminate the fungus *Phytophthora*. Seeds generally are treated with a fungicide to prevent fungal infection during storage.	Viability is usually highest in fresh citrus seeds. Seeds must be stored moist, or many will rapidly lose viability. They should not be stored in fruit juice. Seeds usually are stored in sealed polyethylene bags or moist sawdust or moss in unsealed containers. Seeds should be stored at 2° to 7°C (35°–45°F).
SEEDBED SOILS	**SEEDBED DISINFESTATION**	**SOWING SEEDS**	**SEEDBED IRRIGATION**
Deep, fertile sandy loam soils are best. The addition of organic matter will improve a less desirable soil type.	Many seedling diseases, insects, nematodes, and weeds can be avoided by disinfesting the seedbed soils chemically or with steam. Inoculation of the soil with an endomycorrhizal fungus or addition of phosphorus after disinfestation often will result in better seedling growth.	This can be done by hand or mechanically. Seeds should be spaced 3 to 5 cm apart. Seeds should be covered with 2 cm of clean river sand, sphagnum moss, or sphagnum peat.	Initially, irrigation should be frequent and light to keep the seeded area moist. Irrigation is less frequent once the seedlings are 7.5–10 cm tall.
SEEDBED FERTILITY	**SEEDLING PROTECTION**	**DIGGING SEEDLINGS**	**NURSERY SOILS**
Fertilizer usually is applied in low amounts on an "as needed" basis. Often complete fertilizers are applied.	Insects, mites, and weeds must be controlled. It is sometimes necessary to protect seedlings from frosts or freezes.	When seedlings are 20 to 60 cm tall, they are usually lifted with a spading fork and graded. The most healthy, straightest, and best seedlings are transplanted in the nursery. Nursery planting should be done as soon as possible after digging from the seedbed.	Deep, uniform nursery soils are best. If they have been planted to citrus previously, nursery soils must be fumigated prior to planting.

TRANSPLANTING IN THE NURSERY	NURSERY IRRIGATION	NURSERY FERTILITY	PEST CONTROL
Seedlings usually are transplanted 25 to 30 cm apart in straight rows that are spaced 1 to 1.5 meters. Alternatively, seedlings may be transplanted into containers.	Seedlings generally should receive 2.5 to 5 cm of water each week from rain and/or irrigation.	Seedling trees are usually fertilized with a fertilizer that is relatively high in nitrogen. Under some conditions, it is also necessary to apply certain trace elements such as zinc, manganese, copper, and iron.	Weeds, insects, mites, diseases, and nematodes must be controlled.

BUDDING

LOCATION ON UNDERSTOCK	BUDWOOD SELECTION	SANITATION	BUDDING
To avoid exposure of the scion to soil-borne fungi, budding is usually 15 to 20 cm above the soil line.	In addition to selection of the cultivar, the source trees must be free from viruses. The exocortis virus may be symptomless on the source tree, but may damage some scion–rootstock combinations, especially when trifoliate orange rootstocks are used.	Grafting tools must be disinfested regularly because they can transmit viruses and other causes of disease.	Budding should be done when the understocks reach sufficient size. The bark should be slipping because of vascular cambium activity. T-budding, hanging budding, and side budding are often used. Buds should be wrapped to prevent drying.

CARE OF BUDDED PLANT	DIGGING THE BUDDED PLANT		OTHER VEGETATIVE PROPAGATION METHODS
The new growth should be staked to provide support. Pruning is necessary if the development of lateral shoots in the nursery is desired.	Plants should be dug prior to a new flush of growth to maximize stored carbohydrates. The trees are dug bare root or are balled and burlapped. The root system must not be allowed to dry out.		Citrus trees are primarily budded but may be cloned by grafting, micrografting, rooting cuttings, layering, or micropropagation.

GROWING THE CROP

SOIL	CLIMATE	TRANSPLANTING	CARE AFTER TRANSPLANTING
Citrus trees grow best on deep, uniform fertile soils that are well-drained.	Temperatures at or below freezing will lead to tree damage or death and thus limit areas of production. Excessively high temperatures will damage fruit. High winds can lead to poor tree growth and loss of productivity.	The trees should be planted at the same depth as in the nursery to avoid soil contact with the scion and possible exposure to soil-borne disease organisms.	Irrigation must be adequate. Tree trunks are often wrapped to prevent sunburn or cold damage. Fertilization may not be necessary on fertile soils or may be applied lightly on other soils. Pests must be controlled.

(continued)

IRRIGATION	FERTILITY	FROST PROTECTION	PRUNING
Citrus is a salt-sensitive crop and can be damaged by irrigating with water high in salts. If only high salt water is available, irrigation must be frequent to prevent accumulations of salts in the root zone. Frequency and amount of watering depend on tree age and size, planting density, soil type, wind, relative humidity, and rainfall.	Fertilizer programs vary widely with the type, fertility, and cation exchange capacity of the soil and the size, age, and general appearance of the trees. Conditions such as rainfall and quality of irrigation water also influence fertilization programs. Leaf analysis for nutrients can serve as a guide for fertilization.	In areas of frosts and freezes, many approaches are used to protect trees from low temperature injury. These include planting on a slope for air drainage, various heating methods, wind machines, wind breaks, tree wraps, and overhead sprinkler irrigation.	Although some growers do not prune oranges, it can be beneficial. Pruning normally should be done in the early spring. It should be done on young trees to select strong scaffold branches that are well distributed around the tree. Pruning is also done to remove damaged branches and to allow better light penetration into the tree.

GIRDLING	DISEASES	INSECT CONTROL	NEMATODE CONTROL
A single knife cut around the base of the trunk that severs the bark and phloem can increase set of both parthenocarpic and seeded fruit. Girdling is usually done shortly after full bloom around the time of petal fall. Severe cutting into the xylem can result in tree death, so girdling should be done with caution.	Sanitation, planting of healthy trees, avoiding water or soils with high salts or parasites, avoiding crowding of trees, avoiding excessively wet soils, avoiding cold injury, and applying fungicides at the appropriate time will contribute to the production of healthy orange trees. Control of virus and phytoplasma-caused disease can be accomplished by quarantine, meristem culture, and resistance in scions and rootstocks.	Insects and mites can be controlled by quarantine of plants, eradication of introduced pests, pesticide sprays, and biological controls.	Nematodes can be controlled by soil fumigation, resistant rootstocks, nematicides, and biological methods.

VERTEBRATE PEST CONTROL

Birds and rodents can cause losses in the citrus industry. Rodents are a more serious problem than birds in the United States. Rodents can be controlled by eliminating weeds and grasses, using toxic baits and chemicals, and employing biological methods.

	MARKETING		
MATURITY	**HARVESTING**	**SHIPPING**	**PACKINGHOUSE HANDLING**
Oranges are harvested based on total soluble solids (sugar) : acid ratio. Oranges to be used for juice must have a minimum soluble solids content.	Mechanical harvesting of oranges is minimal. Most oranges are harvested by hand. Pickers often use hand clippers to remove citrus fruits from the trees; snap-picking can lead to decay.	Oranges are shipped in bins or on trucks to juice-processing plants or to packinghouses for fresh market sale.	Oranges may receive an ethylene degreening treatment, are soaked in a soap and fungicide solution, and then are washed and rinsed with fresh water. They are then waxed, dried, sorted, stamped, and sized. After being packed into cartons, they are cooled, stored, and then transported to market.

RETAIL SELLING

Oranges normally are sold out of refrigerated display racks in stores. Refrigeration increases shelf-life. In areas where citrus is grown, it is often not refrigerated at retail markets during the harvest season.

SCHEDULE F   KENTUCKY BLUEGRASS SOD
(Poa pratensis)

PROPAGATION			
SITE SELECTION	**SOIL**	**DRAINAGE**	**GRADING**
For sod production an even, slightly sloping site for drainage is best.	Organic or mineral soils are used. Muck or peat soils hold water well, and sod rolls are light in weight. Mineral soils result in heavy rolls, but the sod is better for high-wear areas and may be more similar to the soil where the sod is to be laid.	If there is poor subsurface drainage a tile drainage system must be installed.	The site may be graded to provide an even seedbed with a sufficient slope for surface drainage.
DEBRIS REMOVAL	**SOIL TESTING**	**FERTILIZER AND pH ADJUSTMENT**	**PREPLANTING WEED CONTROL**
No rocks, stumps, branches or other debris should remain. Even small stones can cause problems when harvesting sod.	The soil should be tested to determine the pH and nutrient availability.	Fertilizer and possibly lime should be added based on the results of the soil test.	Difficult-to-control weeds should be killed or controlled prior to sowing seed.
TILLING AND ROLLING	**PURCHASING SEEDS**	**STARTER FERTILIZER**	**SEEDING RATE**
The soil should be tilled within 24 hours of sowing seeds, then rolled to provide a firm seedbed.	Seeds may be a single cultivar or a blend. Seeds should be free from weed seeds as well as annual bluegrass and other grass seeds.	Either immediately prior to sowing seeds or at the time of sowing, a starter fertilizer should be applied at the rate of 1.5–2.25 kg of actual nitrogen per 100 m^2.	Normall Kentucky bluegrass seeds are sown at a rate of 1.5–2.25 kg per 100 m^2.
SEEDING PATTERN	**COVERING SEED**	**ROLLING**	**MULCHING**
To ensure an even coverage, one-half of the seed should be sown in one direction and one-half in the perpendicular direction.	Seed should be covered lightly (1–2 mm of soil) by raking lightly or dragging with a brush or mat.	To ensure better contact with the soil, the seedbed should be lightly rolled.	The newly seeded area may be mulched with straw or another material to reduce both erosion and soil drying.
IRRIGATION OF SEEDS AND SEEDLINGS			

The seedbed should be kept moist, and seeds and young seedlings should not be allowed to dry out. The surface 1 cm of soil should remain wet.

GROWING THE CROP			
GROWING	**FERTILIZATION**	**IRRIGATION**	**WEED CONTROL**
The grass should be mowed when it grows to a height one-third greater than the mowing height.	Kentucky bluegrass should be fertilized regularly using fertilizers and rates recommended for the geographic area where it is being grown.	To produce a healthy sod rapidly, the turf should be irrigated regularly. Young seedlings require lighter and more frequent watering than more established sods. Established turfgrasses require deeper irrigations. Amounts and frequency vary with the rainfall, soil type, and weather conditions.	A high-quality sod must be maintained in a weed-free condition. Pre- and postemergence herbicides are applied to control these weeds. Mowing at the proper height encourages a healthy sod that discourages weed growth.

DISEASE AND INSECT CONTROL

Turfgrasses are attacked by some disease-causing organisms and insects. To produce a high-quality sod, growers must control these pests by using sanitation and appropriate pesticides.

HARVESTING THE SOD			
MATURITY	**CUTTING THE SOD**	**HANDLING AND SHIPPING**	**TIMING**
Sod can be harvested when the stems and roots of the sod have knit together sufficiently so they can "hold a roll." This normally occurs within one year of sowing the seeds.	Sod-cutting machines will cut sod at various widths ranging from 30–60 cm. Strips are approximately 1–2 m long. Sod should not be cut too thick. Cutting to a depth of 1–2 cm is typical for Kentucky bluegrass.	The sod strips are folded or rolled, stacked on trucks, and transported to the site where they will be laid.	It is best if sod is harvested the day before it is used. Heating of the rolls and development of pathogens such as *Fusarium* can cause death of portions of rolls.

(*continued*)

LAYING SOD

SITE PREPARATION	LAYING THE SOD	ROLLING THE SOD	TOP DRESSING
The site should be prepared as outlined for seeding in this schedule. If the sod was grown on an organic soil and is to be transplanted onto a mineral soil, addition of organic matter to the mineral soil will aid root penetration. The soil should be moist prior to laying the sod.	The sod strips should be staggered and laid in a pattern similar to bricks in a building. Kneeling on a board placed on a previously laid piece of sod and progressively moving will prevent walking on the prepared soil. If sod is laid on a slope it may be pegged to prevent slipping.	To provide better sod-soil contact the laid sod should be rolled in a direction perpendicular to the strips. Tamping is also used.	Top dressing the newly laid sod with a good weed-free topsoil will fill in cracks between the strips.

CARING FOR THE SODDED LAWN

IRRIGATION	CULTURAL
Newly sodded lawns should be watered the same as a newly seeded lawn. The roots are in the upper portion of the soil. This upper 1–2 cm must remain wet. As the roots grow and the sod knits down, frequency of irrigation should be reduced.	Sodded lawns should receive the same care as seeded lawn regarding mowing, fertilizing, and pest control. Weeds should not be a problem if high-quality, weed-free sod was used.

ABSCISIC ACID (ABA) The most important hormonal inhibitor in plants. It is considered to be a "stress hormone," since various environmental stresses, such as drought, stimulate its biosynthesis. It functions, in part, to allow plants to adapt and survive environmental stresses.

ABSCISSION The shedding or dropping of plant organs, such as leaves, flowers, and fruits.

ACCLIMATIZATION The forcing of plants to adapt to conditions unlike those under which they were grown. Foliage plants are acclimatized for the low light conditions in homes, offices, and so on.

ACRE An area containing 43,560 square feet; 0.4 hectares.

ADULT PLANT A plant that is sufficiently mature to flower under normal inductive conditions. Contrast with juvenile plant.

ADVENTITIOUS A stem, leaf, or root that arises from an "unexpected" location. A shoot that grows from a leaf axil is not adventitious, but one that grows from a lateral root is.

AEROBIC That which occurs only in the presence of oxygen.

AIR LAYERING A form of plant propagation in which an attached stem is wounded and wrapped for the purpose of producing roots. The rooted shoot is subsequently removed and grown on, and the remainder of the plant continues to live and grow. Two or more separate plants result.

ALLELES Pairs or series of forms of a gene located at the same location in homologous (like) chromosomes.

ALLELOPATHY The phenomenon when plants give off substances that prevent or inhibit the growth of surrounding plants.

AMINO ACIDS Carbon-, oxygen-, nitrogen-, and hydrogen-containing compounds that are building blocks of peptides and proteins.

ANCYMIDOL A potent growth-retarding chemical.

ANGIOSPERM A taxonomic class of the plant kingdom, the flowering plants. The seeds of angiosperms are borne in an enclosed ovary.

ANNUAL Any plant that completes its life cycle within one growing season.

ANTHESIS The time when the anthers dehisce and shed their pollen. Generally used for the period of opening or expansion of a flower.

ANTITRANSPIRANT Substances that slow or stop the evaporative loss of water from plants.

APICAL DOMINANCE The phenomenon where the terminal bud exerts control over lateral buds by inhibiting their outgrowth.

APOMIXIS Seed (embryo) formation without sexual union of gametes. This occurs in Kentucky bluegrass and citrus.

ARBORICULTURE The study of the planting and care of trees. To distinguish from forestry, the focus is generally on landscape and urban trees.

ASEPTIC Sterile; an environment or area devoid of living organisms, especially microorganisms.

AUXIN A member of a class of plant growth substances that controls many growth and developmental processes in plants including cell elongation. Indole-3-acetic acid (IAA) is the major hormonal auxin.

AXILLARY BUDS Buds located where the leaf petiole is attached to the stem (the axil of the leaf).

BACTERIA Microscopic, single-celled organisms that do not contain chlorophyll. Some bacteria cause serious diseases in plants, including galls and soft rots; others are beneficial and fix atmospheric nitrogen for use by legumes.

BARE-ROOT A plant, usually woody or a herbaceous perennial, that is sold with little or no soil on its roots.

BEDDING PLANTS Herbaceous annual or perennial plants that are sold for use in flower or vegetable gardens.

BENZYLADENINE (BA) A synthetic cytokinin that stimulates cell division and adventitious shoot formation.

BIENNIAL Plants that require all or portions of two growing seasons to complete their life cycles.

BIOTECHNOLOGY This term is generally used to mean manipulating and transferring genes from one organism to another (*recombinant DNA technology*) and cloning of both plants and animals.

BLANCH The process of covering plants or plant parts to exclude light to prevent the formation of chlorophyll and to cause its loss in the covered tissue.

BOLTING The rapid elongation of a flower stalk from a rosette growth habit that occurs with many biennials and some annuals such as lettuce or spinach.

BOTANY The study of plant biology.

BOTTOM HEAT The heat applied to the root zone of plants. It is generally used to accelerate plant growth or rooting of cuttings.

BRACT A leaf that usually subtends a flower and which may be large and colorful, as in poinsettia.

BRASSINOLIDE Important hormonal brassinosteroid.

BRASSINOSTEROIDS A group of plant hormones that stimulates cell elongation and cell division, helps make plants more resistant to environment stresses, and stimulates seed yields.

BUD An undeveloped or compressed stem.

BUDDING The grafting of a bud onto an understock; a form of clonal (asexual) propagation.

BULB A cluster of fleshy leaves or leaf bases (scales) attached to a short, thick stem with basal roots. Bulbs surrounded by a papery covering (tunic) (e.g., onion, tulip) are called tunicate bulbs. Bulbs without tunics (e.g., lily) are called nontunicate or scaly bulbs.

BULBIL A small bulb that forms above the soil surface, frequently in leaf axils.

BULBLET A small bulb that forms below the surface of the soil.

CALIPER The diameter of a tree trunk measured 15 cm (6 in.) above the ground. If the caliper is larger than 10 cm (4 in.) the measurement is made 30 cm (12 in.) above the soil surface.

CALLUS An unorganized mass of cells that forms on plants or explants in response to wounding or application of some plant growth regulators, such as auxins and cytokinins.

CALLUS BRIDGE A connection that results from the intermingling of callus cells formed from the scion and understock as part of the formation of a graft union.

CALYX The collective name for the sepals of a flower.

CAMBIUM A secondary meristem. The vascular cambium divides, producing phloem cells on the outside and xylem cells on the inside, and is responsible for growth in girth of stems and roots. The cork cambium divides and gives rise to bark tissue.

CARBOHYDRATE Compounds that contain only carbon, hydrogen, and oxygen; their basic structural formula is $(CH_2O)_n$. This category includes sugars, starches, and cellulose.

CAROTENOIDS A group of pigments composed of carbon and hydrogen that reflect yellow to red light. The orange carotenoids are carotenes, the yellow pigments are xanthophylls.

CASTASTERONE Important hormonal brassinosteroid.

CATION EXCHANGE CAPACITY (CEC) The exchanging of positively charged ions (cations) on and off the negatively charged sites of surfaces of particles such as clay or organic matter. The CEC is expressed in milliequivalents per 100 g of dry soil.

CELLULOSE A carbohydrate polymer that is the main constituent of plant cell walls.

CHELATE A complex organic molecule that encloses certain trace elements such as iron and thus keeps them in solution.

CHEWING INSECTS Insects with chewing mouth parts that ingest plant parts.

CHILLING INJURY Low-temperature injury that occurs when the water that is present does not freeze.

CHIMERA A nongrafted organism composed of two or more genetically different tissues. One of the tissues is considered to be mutated.

CHLORMEQUAT A chemical growth retardant often used on floricultural crops.

CHLOROPHYLL A green pigment that is important for photosynthesis.

CHLOROPICRIN Tear gas that is used to disinfest soil chemically.

CHLOROSIS The yellowing of plant tissue.

CHROMOSOME Distinct granular bodies that are composed of DNA and proteins and bear hereditary factors.

CIRCADIAN RHYTHMS Daily biological rhythms that occur at 24-hour intervals, such as the "sleep" leaf movement of plants.

CLAY A small (less than 0.002 mm diameter), flat soil particle with a high cation exchange capacity.

CLOCHE A protective glass structure that can be placed over plants to protect them from the cold.

CLONE A population of plants derived asexually from one original individual.

COLD FRAME A low, box-like structure covered with polyethylene or glass without supplemental heating. Cold frames are often used to harden-off bedding plants.

COMPOSTING The process of converting plant and animal wastes into useful soil additives. This process involves the layering of herbaceous plant materials, soil, fertilizer, and lime and periodically turning the pile over for aeration.

CONIFER A cone-bearing gymnosperm tree or shrub. Pines, spruces, and firs are examples of conifers.

CORM The swollen base of a stem that is covered with dry papery leaves. Plant that have corms include *Gladiolus*, *Crocus*, and *Freesia*.

CORMEL A small daughter corm.

COROLLA The collective name for the petals of a flower.

CROWN The transition zone from root to shoot, generally located at the ground line. Foresters usually refer to the branching portion (top) of a tree as the crown.

CRYPTOCHROMES Blue light photoreceptors in plants. The cryptochromes are important in controlling stem elongation, the expansion of cotyledons and leaves, the daily rhythms of a plant (circadian rhythms), and flowering.

CULTIVAR A contraction of the two words *cultivated variety*. It is a group of plants within a species that is unique. Cultivars may be propagated either sexually or asexually.

CULTIVATION The act of loosening the soil by hand or with equipment.

CUTICLE The waxy covering on young stems, leaves, flowers, or fruits that prevents water loss from the plant.

CUTTING A portion of a stem, leaf, or root used for asexual propagation purposes.

CYTOKININS A group of plant growth substances that influence plant growth and development, including promoting cell division. Zeatin is the major hormonal cytokinin.

CYTOPLASM The contents of a cell, excluding the nucleus.

DAMINOZIDE A plant growth retardant used on floricultural crops.

DAMPING-OFF A seed or seedling disease in which decay occurs before emergence through the soil surface or after emergence, when the seedling stem collapses near the soil surface.

DECIDUOUS A plant that sheds all of its leaves at one time each year.

DEGREE DAYS See *growing degree days*.

DEHISCENCE The process of opening of a dry dehiscent fruit or an anther.

DEHISCENT FRUIT A fruit that splits open when ripe to release its seeds.

DEOXYRIBONUCLEIC ACID (DNA) A molecule found in living organisms composed of four repeating nucleotides — adenine, guanine, cytosine, and thymine. The order of the nucleotides forms the genetic code.

DIBBLE A pointed tool used to make holes in soil for sowing seeds or transplanting.

2,4-DICHLOROPHENOXYACETIC ACID (2,4-D) An auxin-type herbicide, frequently used for selective control of broadleaf weeds.

DICHOGAMY The condition when stamens shed pollen at a time that the stigma is not receptive.

DICOT An angiosperm that has two cotyledons.

DIF The difference between day and night temperatures; may be positive or negative. The more positive the DIF, the taller the plant. Used by greenhouse growers to control plant height.

DIKEGULAC A chemical pinching agent used on some floricultural crops.

DIOECIOUS A species that has separate male and female plants.

DIPLOID Vegetative cells that contain two matching chromosome sets.

DISBUDDING The removal of flower and/or vegetative buds from leaf axils so as to leave only one main stem.

DISEASE An interaction between an organism and its environment that results in an abnormal condition; can be either biotic or abiotic.

DISK FLOWERS Tubular flowers in the center of heads of most Asteraceae (Compositae), often surrounded by ray flowers.

DIVISION (1) A major taxonomic category of the plant kingdom, equivalent to a phylum of the animal kingdom. (2) A means of asexually propagating plants that grow in clumps. Division implies cutting the clump into pieces and then transplanting (contrast with separation).

DOMINANCE The genetic phenomenon when one allele manifests itself over another.

DORMANCY (REST) The condition where bud break or seed germination is inhibited by the plant's own physiology or anatomy. This condition is also known as endodormancy.

DOUBLE FLOWERS Flowers with more than the normal number of petals; full.

DOUBLE WORKING A grafting procedure where an interstem is grafted between an understock and scion usually to overcome incompatibility or to dwarf the plant.

DRIP LINE The circle that forms at the ends of the branches of a tree, where water would drip off the leaves onto the ground.

DRUPE Stone fruit, a fleshy one-seeded indehiscent fruit with the seed enclosed by a stony endocarp.

DWARFING ROOTSTOCKS Understocks onto which scions are grafted because they result in less stem elongation and thus a smaller tree.

ECODORMANCY When growth is inhibited because of unfavorable environmental conditions; growth will occur when these conditions improve. This condition is also known as quiescence.

EMBRYO A bipolar axis with root and shoot apical meristems that is present inside seeds.

ENDODERMIS An inner cylinder of cells in roots. The endodermis is located inside the cortex and surrounds the stele and generally has a waterproof Casparian strip.

ENDODORMANCY A dormant condition where growth is inhibited by physical characteristics of the seed or bud or by their hormonal balance, which typically happens in winter in temperate climates.

ENDOSPERM A starchy region of a seed that serves to store food. The endosperm forms from the fertilization of the two polar nuclei with a sperm nucleus and thus consists of $3n$ cells.

ENTOMOLOGY The study of insects.

ENZYME A protein that catalyzes biochemical reactions.

EPICOTYL The stem located above the cotyledons of a seedling.

EPIDERMIS The outer layer of cells covering young stems, leaves, flowers, roots, and other plant parts.

EPIGEOUS Seed germination in which the cotyledons emerge through the soil surface by hypocotyl elongation.

EPIGYNOUS The flower parts are borne on the ovary; the ovary is inferior.

EROSION The process of wearing away of a soil by the forces of water or wind.

ESPALIER A tree or shrub trained to grow in two dimensions, such as flat against a wall.

ETHEPHON A liquid ethylene-releasing agent that is widely used on many horticultural species for reasons including accelerating ripening, rooting cuttings of many crops, and initiating flowering in pineapple.

ETHYLENE A gaseous plant hormone that stimulates ripening, senescence, and abscission (C_2H_4).

ETIOLATION The development of a plant or plant part in the absence of light.

EVAPOTRANSPIRATION The evaporative loss of water from plants plus soil.

EVERGREEN A plant that retains at least some of its leaves year-round.

EXOCARP The outer layer of the ovary wall.

EXPLANT The portion that is removed from a stock plant for use in tissue culture.

FI HYBRID The generation that results from crossing two inbred parents with different genotypes. The first filial generation.

FAMILY A taxonomic grouping of genera based on their flower characteristics; the family name ends in -aceae

FASCICLES The dwarf shoots that contain a bundle of needles on pines.

FERTILIZATION (1) The application of fertilizer to a plant. (2) The union of male and female gametes.

FERTILIZER A nutritional spray or soil amendment that is applied to provide essential nutrient elements to plants. Fertilizers can be of organic or inorganic origin.

FIELD CAPACITY The maximum amount of water a soil can hold against the pull of gravity.

FILIAL Any generation following the parental. It is designated with the symbol F: The first filial generation is the F_1, the second is F_2, and so on.

FLORICULTURE The growing and study of flowers and foliage plants.

FLORIGEN A theoretical compound that stimulates the initiation of flowers.

FLURPRIMIDOL A chemical growth retardant that is active on turfgrasses and trees.

FOLIAGE PLANT A container-grown plant that is sold for its attractive leaves and growth habit for use indoors or on patios. Many house plants are foliage plants.

FOLIAR FEEDING The act of applying liquid forms of water-soluble fertilizers to leaves.

FROST POCKET A low area where cold air settles and is thus more prone to frosts than the surrounding hillsides.

FUMIGANT A volatile liquid or gas that is used to kill pests.

FUNGUS An organism that does not produce chlorophyll, vegetatively consists of strands called hyphae, and often reproduces by spores. Plural, *fungi*.

GAMETES Sex cells, such as eggs and sperm.

GENE A unit hereditary factor on a chromosome.

GENETICALLY MODIFIED ORGANISM (GMO) An organism that has been genetically engineered by having foreign genes inserted.

GENETICS The study of heredity.

GENOME One complete set of genes in an organism.

GENOMICS The study of the identification, function, and structure of many genes at the same time.

GENOTYPE The genetic makeup of an organism. The genotype interacts with the environment to give the phenotype.

GENUS A group of similar organisms representing a category within a family. A genus consists of one or more species.

GERMINATION A process that begins within a seed with the imbibition of water and which culminates with the emergence of the root radicle through the seed coat.

GIBBERELLIN A class of plant growth substances that is important for plant growth and development. Gibberellins induce many plant responses, including cell elongation and seed germination. Gibberellic acid (GA_3) is an example of a hormonal gibberellin.

GIRDLING Constricting or destroying the bark in a ring around the trunk or branch of a plant.

GLABROUS Smooth, without hairs, scales, or bristles.

GLAUCOUS Covered with a "bloom;" often composed of waxy substances, giving a whitish, bluish, or grayish appearance to plant parts.

GRAFTING The joining together of plants or plant parts so they grow as one. This is a common method of propagating plants asexually.

GREEN MANURE Fresh green plant material that is incorporated into the soil for the purpose of soil improvement.

GROUND COVER Low-growing plants that are used for control of erosion or for aesthetic reasons.

GROWING DEGREE DAYS Heat units that are used to calculate crop maturity and to estimate increases in the populations of various plant pests, including insects as part of integrated pest management.

GUARD CELL Specialized cell in the epidermis that is located to either side of a stomate. Each becomes turgid or flaccid as the stomate opens or closes.

GUTTATION The extrusion of water from plants in liquid form so that droplets appear at various locations, such as along leaf margins. The phenomenon can often be observed early in the morning.

GYMNOSPERM A taxonomic class of the plant kingdom. The seeds of gymnosperms are borne naked, with no ovary. Pines, spruces, and ginkgo are examples of gymnosperms.

GYNOECIOUS PLANT A plant that produces only pistillate (female) flowers.

HAPLOID The chromosome number present in the gametes. This is generally one-half the normal number of chromosomes and is represented by *n*.

HARDENING OFF The process of preparing plants for the harsh environmental conditions following transplanting. This generally involves reducing watering, fertilization, and temperatures.

HARDINESS The capacity of a plant to survive under harsh conditions, such as low temperatures.

HARDWOOD CUTTING A cutting taken from a woody plant in the late fall, winter, or early spring for the purpose of asexual propagation.

HEELING IN The act of covering the roots of plants with soil, sawdust, or mulch for the purpose of temporary storage.

HERBACEOUS PLANT Any nonwoody plant.

HERBICIDE A chemical used to kill plants, especially weeds.

HEREDITY The passing of genetic information from one generation to the next.

HERMAPHRODITIC FLOWER A flower with both male and female elements.

HETEROSIS The increase in size or vigor of a progeny over its parents; hybrid vigor.

HETEROZYGOUS The condition when homologous chromosomes within a nucleus contain unlike alleles.

HOMOLOGOUS Matching in structure or character, as in two homologous chromosomes.

HOMOZYGOUS The condition when homologous chromosomes within a nucleus contain like alleles.

HORMONE An endogenous plant growth substance that is an organic molecule, in low concentrations regulates plant physiological processes, is produced in one site in a plant and acts in another, and is not considered to be a nutrient or vitamin.

HOTBED A low boxlike structure covered with polyethylene or glass that has a source of supplemental heating.

HOT CAP Coverings that are placed individually over tender, recently transplanted plants to provide protection from frosts.

HYBRID A plant that results from crossing two parents that are different for one or more traits.

HYBRID VIGOR See heterosis.

HYDROCOOLING The process of rapidly cooling harvested produce in a water bath.

HYDROPONICS The cultivation of plants in water containing dissolved nutrients and oxygen. An inert support medium may be used.

HYDROSEEDING The process of sowing seeds with a mixture of water, fertilizer, and mulch. Hydroseeding is accomplished with specialized equipment.

HYPOCOTYL The stem tissue located between the seedling root and the cotyledons.

HYPOGEOUS Seed germination in which the cotyledons remain below the soil surface.

HYPOGYNOUS Floral parts borne at the base of the ovary; superior ovary.

IMBIBITION Absorption of water, accompanied by swelling, as in seeds.

INBREEDING The act of self-fertilization or breeding between closely related individuals.

INCOMPATIBILITY (1) A biochemical hindrance to sexual union of gametes between plants with a common sterility gene, among members of a clone, or within a single plant. (2) The failure to achieve a successful graft union because of physiological, growth, or biochemical differences between an understock and a scion. Graft incompatibility can appear quickly or several years after grafting.

INDEHISCENT Not splitting at maturity, as in fruits.

INDOLE-3-ACETIC ACID (IAA) An important hormonal auxin.

INDOLE-3-BUTYRIC ACID A synthetic auxin commonly used to root cuttings.

INFECTION The invasion of an organism by disease-causing microorganisms.

INFESTATION Pests on inanimate objects, on the outside of plants, or in soil.

INFERIOR Parts arising below other parts, as in an inferior ovary.

INFLORESCENCE A flower cluster or individual flower.

INHERITANCE The passing of genetic traits from one generation to the next.

INSECTICIDE A chemical used to kill insects.

INTEGRATED PEST MANAGEMENT (IPM) A system in which people utilize a range of strategies to control pests while considering the consequences of their actions.

INTERCALARY MERISTEM A growing point that is located between two nonmeristematic regions, such as the base of a leaf or stem. Grasses have intercalary meristems and grow from below, which allows mowing their tips without removing their growing points.

INTERNODE The sections of stems between nodes.

INTERSTEM (INTERSTOCK) A stem piece that is grafted between a scion and understock.

IN VITRO In glass; in horticulture it is frequently used synonymously with the term *plant tissue culture.*

ISOPENTENYLADENINE (2IP) A hormonal cytokinin.

JASMONATES Group of plant hormones that are important in plant resistance to insects and pathogens. They stimulate the coiling of tendrils, stimulate the formation of vegetative storage organs, and have some activities similar to ABA and ethylene.

JASMONIC ACID A hormonally active jasmonate.

JUGLONE A chemical produced by walnuts that some think is toxic to plants. Not all studies confirm this.

JUVENILE PLANT A plant that is growing vegetatively and cannot respond to flower induction stimuli that would normally lead to flowering in an adult plant. Contrast with adult plant.

KINETIN A synthetic cytokinin.

KINGDOM The major taxonomic class that separates living organisms on their general differences (e.g., the plant kingdom and the animal kingdom).

LATENT BUDS Preformed buds, usually located under the bark of trees, that grow into water sprouts in response to heavy pruning.

LATERAL BUDS See axillary buds.

LAYERAGE An asexual propagation technique in which shoots are induced to root while they remain attached to the mother plant.

LEACHING NUTRIENTS The downward movement in water of soluble nutrients through the soil column.

LEAF-BUD CUTTINGS Cuttings used for asexual propagation that consist of a leaf attached to a node and a short section of stem; single-node cuttings.

LEAF CUTTINGS Cuttings used for asexual propagation that consist of a leaf or portion of a leaf.

LEAFLET One segment of a compound leaf.

LEGGY PLANT A tall spindly plant that frequently results when grown under inadequate light.

LIFTING The process of digging a plant that is to be transplanted.

LIGHT Electromagnetic radiation, some of which can be perceived visually. Light is necessary for higher plants to survive and conduct photosynthesis.

LIGHT COMPENSATION POINT The photon flux at which the CO_2 fixed by photosynthesis equals the amount released by respiration. This is the minimum level of light for plant maintenance but is too low to support an increase in dry matter accumulation (growth).

LIME Agricultural lime usually consists of ground limestone, which is primarily composed of $CaCO_3$. Other liming materials are also available. Lime is used primarily to correct soil acidity (raise the pH of the soil) and is also a source of calcium.

LINERS Young plant material that is of adequate size to plant in a nursery.

LINING OUT The act of planting liners in a nursery.

LOAM SOIL A soil composed of sand, silt, and clay such that the characteristics of one do not predominate over the others. This is a good soil texture to support plant growth.

LOCUS The location of a gene on a chromosome (plural *loci*).

MACRONUTRIENTS Mineral nutrient elements that are required by plants in relatively large quantities. Nitrogen, phosphorus, and potassium are macronutrients.

MALEIC HYDRAZIDE (MH) A chemical growth retardant that nonselectively inhibits cell division. MH is used to prevent sprouting on potatoes and onions.

MANURE Plant and animal wastes that are used to fertilize or amend the soil.

MEFLUIDIDE A chemical turfgrass growth retardant.

MEIOSIS A form of nuclear division in which the 2*n* chromosome number is reduced to 1n; reduction division.

MEMBRANE A lipid bilayer with associated proteins that surrounds the cytoplasm and organelles in plant cells.

MEPIQUAT CHLORIDE A chemical growth retardant.

MERISTEM The plant tissues in which cell division occurs. Primary meristems result in elongation growth, and secondary meristems result in growth in girth.

MESOCARP The middle layer of the ovary wall.

MESOPHYLL The tissue in a leaf between the upper and lower epidermis, consisting of the palisade layer and the spongy mesophyll.

METHAM-SODIUM A chemical soil sterilant that is applied as a drench.

METHYL BROMIDE A fumigant used to disinfest soil. This is being phased-out worldwide because of its harmful effects on the earth's ozone layer.

1-METHYLCYCLOPROPENE (1-MCP) An ethylene action inhibitor that is used commercially to extend the storage life of many fruit crops and tomatoes.

METHYL JASMONATE A fragrant methyl ester of jasmonic acid that occurs as a gas at the temperatures at which plants grow. It gives flavor and scent to jasmine, flowers, and other herbs such as rosemary. It is a hormonally active jasmonate.

MICRONUTRIENTS Mineral nutrient elements that are required in very small quantities for normal plant growth. Iron, copper, zinc, boron, and molybdenum are examples of micronutrients.

MICROORGANISM A living thing that is too small to be seen with the naked human eye.

MICROSHOOT A stem with leaves that develops in tissue culture.

MIDRIB The central vein of a leaf.

MITICIDE A chemical that is used to control mites.

MITOCHONDRIA Organelles within a cell in which respiration occurs.

MITOSIS The process of cell division where the chromosomes replicate themselves and the nucleus divides, resulting in two daughter cells that have the same characteristics of heredity as each other and as the parent cell.

MOLLUSCICIDE A chemical used to control mollusks such as snails and slugs.

MONOCOT A subclass of the Angiospermae in which the plants have one cotyledon in their seeds, have parallel leaf venation, and have flower parts in threes or multiples thereof.

MONOECIOUS The situation in which the separate sexes are in different flowers but on the same plant.

MORPHACTINS Chemicals that are used to promote branching or retard plant growth.

MULCH A material spread on the soil surface to conserve soil moisture, influence soil temperature, and control weeds.

MULTIPLE FRUIT A fruit formed from the matured ovaries of two or more flowers, usually borne on a common receptacle (e.g., pineapple fruit and beet "seed").

MUTAGEN Something that causes a mutation.

MUTATION A sudden heritable change in a gene or in chromosome structure that can produce single or compound effects.

MYCELIUM A collective term for hyphae, the vegetative body of a fungus.

MYCORRHIZAE A symbiotic relationship between a fungus and a plant's root system that enables the plant to use soil nutrients more efficiently and offers a degree of protection against disease-causing organisms.

NAPHTHELENEACETIC ACID (NAA) A synthetic auxin used to root stem cuttings.

NARROW-LEAVED EVERGREENS Plants that retain a portion of their needle or scale-like leaves year-round. This category includes many conifers, such as the pines and junipers.

NECROSIS The death of plant tissues, frequently resulting in a brown color.

NECTAR A sweet secretion from any part of a flower.

NECTARY Organ that secretes nectar.

NEEDLES Narrow, slender, often pointed leaves, needlelike in appearance; common on most conifers.

NEMATICIDE A chemical used to control nematodes.

NITRIFICATION Transformation by bacteria (*Nitrosomonas*, *Nitrobacter*) of nitrogen compounds from the NH_4^+ to the NO_3^- form.

NODE The joint on a stem where a leaf is or was attached. Axillary buds are located at nodes.

NODULE A swelling on a root formed by the invasion of a root hair by a bacterium where nitrogen fixation occurs.

NOXIOUS WEED A weed that is defined by state law to be undesirable, troublesome, or difficult to control.

NUCLEOTIDE One of four repeating subunits of DNA or RNA that consists of a five-carbon sugar, phosphoric acid, and an organic base that contains nitrogen. The nucleotides are adenine, guanine, cytosine, and thymine for DNA; uracil substitutes for thymine in RNA.

NUCLEUS The organelle within a cell that contains the chromosomes.

NUT Hard, dry, indehiscent, one-seeded fruit somewhat like an achene.

NUTRIENT FILM TECHNIQUE A form of hydroponics in which a shallow stream of nutrient solution flows past the roots of a crop.

NYMPH The immature growth stage of certain insects.

OFFSHOOT A lateral shoot that forms near the base of the dominant stem of a plant that can be removed for the purpose of asexual propagation.

OLERICULTURE The study and production of vegetables.

ORGAN A plant part, such as a stem, leaf, or root, that is composed of different types of tissues.

ORGANIC MATTER Living and dead plants, animals, and microbes. Organic matter is frequently incorporated into the soil to improve soil structure.

ORGANOGENESIS A term frequently associated with tissue culture, meaning the initiation of adventitious organs.

OSMOSIS The diffusion of substances from high to low concentrations across a membrane.

PACLOBUTRAZOL A chemical growth retardant with activity on a wide range of species.

PANICLE Compound or branched raceme with each branch bearing a raceme.

PARADORMANCY Apical dominance.

PARASITE An organism that obtains its nourishment from another organism.

PARTHENOCARPIC A fruit formed without sexual union of gametes that often is seedless, such as navel oranges and bananas.

PARTHENOGENESIS Female reproduction without sexual fertilization by males. Parthenogenesis frequently occurs with aphids.

PATHOGENIC ORGANISM An infectious agent that causes a disease on a host.

PATHOLOGY The study of disease.

PEDICEL The stem of one flower in a cluster.

PEDUNCLE The stem of a flower-cluster or of a solitary flower.

PERENNIAL PLANT Any plant that can live for more than two growing seasons.

PERFECT FLOWER A flower that contains both pistillate (female) and staminate (male) parts.

PERIANTH The corolla and calyx, collectively.

PERICARP The wall of a fruit; matured ovary wall.

PERIGYNOUS Floral parts fused around, but not attached to, the ovary, as in some *Prunus* spp.

PERLITE A mineral mined from volcanic deposits that is expanded by heat treatment for use in a growing or propagating medium. Perlite is lightweight and well drained.

PESTICIDE A chemical or other substance used to control pests.

PETIOLE The stalk of a leaf.

PETIOLULE The stalk of a leaflet.

pH The negative logarithm of the hydrogen ion concentration; a measure of acidity or alkalinity. A pH of 7 is neutral, below 7 is acid, and above 7 is alkaline.

PHENOTYPE The outward appearance of an organism. The phenotype results from an interaction between the genotype and the environment.

PHENYLACETIC ACID (PAA) A hormonal auxin.

PHEROMONES The sexual attractants of insects that are sometimes used in pest management to lure insects.

PHLOEM The vascular tissue where the transport of food, other organic molecules, and recycled nutrients occurs.

PHOTOLYSIS The light-dependent splitting of water into oxygen and hydrogen that occurs during photosynthesis.

PHOTOPERIOD The number of hours of light each day.

PHOTOSYNTHESIS A metabolic process in which, in the presence of chlorophyll and light, water and carbon dioxide are converted into sugar and oxygen.

PHOTOTROPINS Blue light photoreceptors in plants. Phototropins are important light receptors for plants bending toward a light source.

PHOTOTROPISM The bending of plants or plant parts toward the source of light.

PHYLLOTAXY The arrangement of leaves or buds on a stem.

PHYSIOLOGY, PLANT The study of the internal mechanisms of plant function.

PHYTOCHROME A blue-green protein pigment that absorbs red and far red light and thus governs many aspects of plant growth and development, including flowering, seed germination, branching, and the onset of dormancy.

PHYTOPLASMA Microorganisms that lack cell walls and are surrounded by a membrane. They can cause plant diseases, including grape yellows, aster yellows, and coconut lethal yellowing.

PHYTOTOXIN A substance that is injurious to plants.

PISTIL The female portion of a flower that consists of the stigma, style, and ovary.

PISTILLATE Having pistils and no stamens; female.

PLANT GROWTH REGULATOR (PGR) A chemical that in low concentrations regulates plant physiological processes and which may or may not be produced by plants. Plant hormones are types of PGRs produced by plants.

PLANT GROWTH RETARDANT A chemical that stunts plant growth.

PLASMALEMMA The membrane that surrounds the protoplasm of a cell.

PLASMODESMATA The strands of cytoplasm that penetrate cell walls and connect adjacent cells.

PLUGS Seedling bedding plants that are grown in very small volumes of medium in special growing trays. Plugs offer the advantage of being easy to transplant, either mechanically or by hand.

POD A dehiscent dry pericarp; an unspecific term.

POLLEN Particles that contain the sperm nuclei and are produced by the male parts of a flower (anthers), often dustlike in size and appearance.

POLLINATION The transfer of pollen from one plant to another, such as bees or the wind.

POLLINATOR A vector that carries pollen from one plant to another, such as bees or the wind.

POLLINIZER A plant that serves as a source of pollen.

POLYPLOID Cells that contain more than two complete sets of chromosomes.

POMOLOGY The study and production of fruit and nut crops.

POT-BOUND The condition when the roots in a container circle the inside and become dense and matted. Pot-bound plants tend to dry out quickly and grow poorly.

PRICKING OUT The transplanting of seedlings from one container to another.

PRICKLE A small, spinelike body borne irregularly on the bark or epidermis, such as on roses or raspberries.

PROPAGATION The reproduction of plants by sexual or asexual means.

PROPAGULE A propagation unit, such as a seed or cutting.

PROTEOMICS The discovery and characterization of all proteins produced by different cell types and organisms.

PROTOPLASM The contents of a cell surrounded by and including the plasmalemma.

PROVENANCE The geographic origin of seeds.

PSEUDOBULB The bulblike formation on the base of the stem of some orchids; solid and borne above the ground.

PUBESCENT Covered with short, soft hairs; downy.

PUPA An immature, usually inactive insect; a stage of development that follows the larval stage (plural, pupae).

QUALITATIVE INHERITANCE Heredity of traits that show discontinuous variation.

QUANTITATIVE INHERITANCE Heredity of traits that fit a continuous array between extremes or have continuous variation such as yield or height.

QUARANTINE The isolation of plants for a period of time to allow officials to determine if they are free from certain pests.

QUIESCENCE The condition when a plant process, such as seed germination or bud break, is inhibited solely by unfavorable environmental conditions; also known as *ecodormancy.*

RACEME An indeterminate inflorescence with pedicelled or stalked flowers.

RACHIS Axis-bearing flowers or leaflets.

RADICLE The primary seedling root.

RAY FLOWERS Outer modified flowers of some composite family members. Ray flowers have an extended or straplike and usually colorful corolla.

RECEPTACLE The enlarged end of a stem on which some or all flower parts are borne.

RELATIVE HUMIDITY Expressed as a percentage of the amount of water vapor that the air can hold at a given temperature.

RESPIRATION A metabolic process in which oxygen is utilized, carbon dioxide and oxidation products are released, and energy is made available for cell metabolism.

RHIZOME A more or less subterranean, usually horizontal, stem that produces roots and shoots at the nodes.

RIBONUCLEIC ACID (RNA) A nucleic acid that is directly involved in the synthesis of proteins. RNA consists of four repeating nucleotides — adenine, guanine, cytosine, and uracil.

RIBOSOME An organelle that contains RNA and participates in the manufacture of proteins.

RINGING Removal of a thin ring of bark from around the bases of large branches or the trunk. This increases flower formation on vegetatively vigorous fruit trees and vines.

RODENTICIDE A chemical used to control rodents.

ROGUING The removal of unwanted plants because they are diseased or not true to type.

ROOT-BOUND See pot-bound.

ROOT-PRUNING The cutting of a portion of the root system to encourage new root growth or to check the growth of the top of the plant. Plants are often root-pruned to prepare them for transplanting.

ROOTSTOCK The lower portion of a graft onto which the scion or interstock is grafted. Rootstocks can impart various qualities to a plant, including pest resistance and size control.

ROSETTE A plant growth habit in which the leaves are arranged so that they radiate from the crown or center of a stem with short internodes. A rosette is usually at or close to the ground, as in spinach, lettuce, or dandelion.

RUNNER A specialized slender stolon that forms from axillary buds and which forms a new plant when its tip contacts the soil. Strawberries and spider plants have runners.

SAMARA An indehiscent winged fruit, as of maple and ash.

SAND Small rock grains (soil particles) between 0.05 and 2.0 mm in diameter.

SAPROPHYTE A plant or microorganism that obtains its nourishment from dead organic matter.

SAPWOOD The outer xylem of a tree or shrub that is light-colored and functions in the conduction of nutrients and water.

SCAFFOLD BRANCHES The large supporting branches from the main trunk of a fruit tree. Wide crotch angles enable scaffold branches to support heavy crops of fruit.

SCAPE A leafless peduncle arising from the ground; it may bear scales or bracts, but not foliage leaves.

SCARIFICATION A seed treatment to make a hard seed coat more permeable to gases, water, or the emerging seedling. Scarification can be accomplished mechanically or with acid soaks.

SCION The upper portion that is grafted onto an interstock or rootstock. The scion grows to become the top of the plant.

SEED COAT The outside covering of a seed that is produced from the integuments.

SEGREGATE A term in genetics meaning that the progeny do not resemble the parents and can be divided up into groups that are different for one or more traits.

SELF-INCOMPATIBLE A condition in which a plant is incapable of being fertilized by its own or its clone's pollen because of biochemical reasons that are controlled by certain sterility genes.

SEMIHARDWOOD CUTTING A stem cutting used for asexual propagation that is taken from a woody plant during the summer, just as the younger portions of the stem are becoming woody but are still somewhat succulent.

SENESCENCE The aging of a plant or plant part. Senescence in leaves is evident when they turn yellow and then brown.

SEPALS The modified leaflike structures that surround the petals of a flower. Sepals are often green.

SEPARATION An asexual propagation technique used on plants that grow in clumps that can be broken apart into several plants without cutting because they have a natural place for detachment (contrast with division).

SERRATE A margin that is saw-toothed with the teeth pointing forward. This term is frequently used to describe leaf margins.

SILT A textural class of soil with fine grains 0.002 to 0.05 mm in diameter.

SIMPLE (1) A leaf that is not compound. (2) An inflorescence that is not branched. (3) A fruit formed by the development of a single pistil or ovary.

SLIP A lay term for a cutting used for asexual propagation.

SLOW-RELEASE FERTILIZER A fertilizer that releases its nutrients gradually. This category includes many organic fertilizers and specially encapsulated forms.

SOFTWOOD CUTTING A stem cutting used for asexual propagation that is taken from the new, soft, succulent growth from a woody plant in the spring or early summer.

SOILLESS MEDIUM A medium for growing plants that does not contain any mineral soil. Many soilless media contain sphagnum peat or bark and an inert ingredient such as sand, perlite, or vermiculite.

SOIL STRUCTURE The arrangement or grouping together of primary soil particles (sand, silt, clay) into secondary units called peds.

SOIL TEXTURE The relative proportions of sand, silt, and clay in a soil. Loam is a soil textural class.

SOMATIC Refers to the vegetative (nonsexual) cells of an organism. This term is often used to describe a phenomenon (e.g., somatic embryogenesis is the formation of embryos from vegetative tissue without sexual union of gametes).

SOUTHWEST INJURY A type of winter injury that results in bark and cambial death on the south or west sides of the trunks of young, thin barked trees. This is caused by thawing and rapid refreezing when the sun sets or is obstructed (sometimes called sunscald).

SPADIX A thick spike that bears tiny flowers subtended by a spathe. This is characteristic of members of the Araceae, such as calla lily and *Dieffenbachia*.

SPATHE A bract subtending a flower cluster or spadix; sometimes colorful, and in *Anthurium* and calla lily.

SPECIES A taxonomic subdivision of genus represented by plants that resemble each other and interbreed freely.

SPECIFIC EPITHET In the binomial system for naming plants, the second name of the species. For example, *Pisum sativum* is the species name for the common garden pea; *sativum* is the specific epithet.

SPIKE A usually elongated, indeterminate inflorescence with stalkless or nearly stalkless flowers attached along its length.

SPINE A sharp, hard, modified leaf. Spines on barberry are modified entire leaves.

SPINDLY Leggy.

SPORT A spontaneous mutation that occurs in somatic (vegetative) tissue. Sports are called *bud sports* when they occur in an axillary bud and cause a branch or portion thereof to look different.

SPUR A stem with very short internodes on a woody plant that is often modified for flower and fruit production. Apple and ginkgo produce spurs.

STAMEN The male part of the flower, consisting of the anther and filament.

STAMINATE Having stamens and no pistils; male.

STARTER FERTILIZER A fertilizer that usually contains nitrogen, is high in phosphorus, and is used on new transplants or when sowing seeds.

STELE The central cylinder of a young root that is surrounded by the endodermis and includes the vascular tissue.

STIGMA The top of the pistil that receives the pollen.

STIPULE A leaflike basal appendage of a petiole.

STOLON A horizontal stem that is located above ground and may produce roots and shoots at the nodes.

STOMATE A tiny pore on a leaf that can open and close and is important for governing water loss from a plant and gaseous exchange (plural, *stomates* or *stomata*).

STRATIFICATION A moist chilling treatment used to overcome seed dormancy.

STRESS A condition that potentially can cause either reversible or irreversible injury.

STYLE The stalk of the pistil located between the stigma and ovary. Pollen tubes grow down through the style to the ovary, where fertilization occurs.

SUCKER An adventitious shoot that grows from a plant's roots.

SUCKING INSECTS Insects with piercing-sucking mouthparts that puncture and penetrate plant tissues to feed on sap.

SUNBURN (SUNSCALD) A condition on fruits in the summer when direct sunlight causes the fruit temperature to become so high that injury or death occurs on that side.

SYMBIOSIS A mutually beneficial relationship between two or more organisms.

SYRINGING Watering frequently and lightly to overcome heat stress or to establish new plantings.

TAP ROOT The primary root that develops from the original seedling radicle and penetrates to varying depths. The edible portion of the carrot is primarily tap root tissue.

TAXON The name applied to a group within the taxonomy (scientific classification) of organisms (plural, *taxa*).

T-BUDDING An asexual propagation technique in which a bud is grafted by making a T-shaped cut in the bark of the understock.

TENDRIL A stem or leaf modification that twists or twines around an adjacent support.

TERRARIUM A dish garden or other planting that is enclosed by plastic or glass to conserve moisture.

TESTA The seed coat.

TETRAPLOID A cell that contains four complete sets of chromosomes.

THATCH The accumulation of living and dead stems, leaves, and roots along the soil surface and beneath the topgrowth of turfgrasses.

THERMAL BLANKET A cloth or plastic covering that is used in a greenhouse for energy conservation. Thermal blankets are pulled over the crop at night during winter months and allow the grower to heat only the air immediately surrounding the crop.

THERMOPERIOD Alternating day and night temperatures.

THORN A sharp-pointed modified stem that arises at a branch tip or node.

TILLER (1) A branch that forms at the base of a monocot. (2) A machine used to cultivate soil.

TISSUE An organized group of cells in a plant that serves a particular function.

TISSUE CULTURE Maintaining, growing, or otherwise manipulating cells, tissues, or organs in vitro.

TONOPLAST The plasma membrane that surrounds a vacuole.

TOPIARY The pruning of plants into various shapes, such as animals.

TRACE ELEMENT See micronutrient.

TRANSCRIPTION The biosynthesis of RNA.

TRANSGENIC PLANTS Plants that have had foreign genes inserted. These genes often confer disease resistance, insect resistance, or herbicide tolerance.

TRANSLATION The biosynthesis of proteins.

TRANSLOCATION The movement of a substance from one location to another within a plant.

TRANSPIRATION The evaporative loss of water from a plant. Transpiration primarily occurs through open stomates.

TRANSPLANTING The act of planting a plant (a transplant) into a new location.

TRICHOME An outgrowth of an epidermal cell that commonly appears as a hair.

TRIFOLIATE Three-leaved, and in *Trillium*.

TRIFOLIOLATE A leaf with three leaflets, as in most clovers.

TRIPLOID A cell with three complete sets of chromosomes.

TUBER An enlarged or swollen underground stem, as in potato.

TUBEROUS ROOT An enlarged or swollen root that developed originally as a fibrous root, as in *Dahlia* and sweet potato.

TUNIC A loose membranous outer skin that is not the epidermis; the dried leaves that surround a corm or bulb.

TURGID Full of fluid, swollen. A term used to describe a cell or a plant part that is succulent and full.

UMBEL An indeterminate, often flat-topped inflorescence whose pedicels arise from a common point.

UNDERSTOCK See rootstock.

UNICONAZOLE A chemical growth retardant with activity on a large number of species at low concentrations.

VACUOLE A structure within a cell that contains the cell sap, which includes but is not limited to waste materials.

VARIEGATION A plant organ with two or more colors. The colors may appear as marks, patches, or streaks.

VARIETY A taxonomic subdivision of species. This is a botanical term and should not be confused or used interchangeably with *cultivar*.

VASCULAR SYSTEM The tissues in the plant that conduct water, mineral nutrients, and organic materials from one location to another. Xylem and phloem are vascular tissues.

VERMICULITE An expanded mica-like growing medium or additive that is lightweight, has a high cation exchange capacity, and provides some magnesium and potassium.

VERNALIZATION The effect of low temperature on the initiation of flowers. Many biennial plants require vernalization before they will flower.

VIABLE Alive; a term frequently used to describe seeds that have not germinated but are not dead.

VITICULTURE The growing and study of grapes.

WATERSPROUT A shoot that arises from stem tissue, often from latent buds located within branches or the trunk.

WEED A plant out of place. Weeds are troublesome because they compete with crop plants for water, minerals, and light.

WILTING The drooping of plant parts, especially leaves, generally because of a lack of water.

WINTER ANNUAL A biennial plant whose seeds germinate in the fall; it overwinters in the vegetative state and then flowers in the spring or early summer.

WITCHES' BROOM A growth that may occur in trees when many branches with short internodes arise from one point.

XENIA The direct effect of pollen on the embryo and endosperm of corn, so that a kernel will resemble its male parent.

XYLEM Part of the vascular system that is important for the conduction of water and nutrient elements. The woody portion of a tree or shrub is xylem.

ZEATIN A hormonal cytokinin.

ZYGOTE A cell that forms from the sexual union of gametes.

Y

yam, 33, 398
yellow rocket, 20
yew, 129, 358, 464–465
Yucca spp., 35

Z

Zea mays (*see also* corn), 20, 45, 62, 308
zeatin, 308
zinc (Zn), 254, 260, 271
Zinnia elegans, 56, 62, 129
Zoysia japonica (zoysiagrass), 102, 116, 129, 142, 375
zucchini, 82, 86, 328
Zygomycetes, 417, 418
zygote, 74